METHODS IN MOLECULAR BIOLOGY™

Series Editor
John M. Walker
School of Life Sciences
University of Hertfordshire
Hatfield, Hertfordshire, AL10 9AB, UK

For further volumes:
http://www.springer.com/series/7651

Xenopus Protocols

Post-Genomic Approaches

Second Edition

Edited by

Stefan Hoppler

Institute of Medical Sciences, University of Aberdeen, Foresterhill, Aberdeen, UK

Peter D. Vize

Department of Biological Science, University of Calgary, Calgary, Canada

Humana Press

Editors
Stefan Hoppler
Institute of Medical Sciences
University of Aberdeen
Foresterhill, Aberdeen, UK

Peter D. Vize
Department of Biological Science
University of Calgary
Calgary, Canada

ISSN 1064-3745 ISSN 1940-6029 (electronic)
ISBN 978-1-4939-5953-2 ISBN 978-1-61779-992-1 (eBook)
DOI 10.1007/978-1-61779-992-1
Springer New York Heidelberg Dordrecht London

Printed on acid-free paper

Humana Press is a brand of Springer
Springer is part of Springer Science+Business Media (www.springer.com)

Preface

The *Xenopus* system had a long and prominent history during the twentieth century as an accessible model system for studying vertebrate biology. Many discoveries in biochemistry, cell biology, and in particular in developmental biology have their roots in experimental investigation using *Xenopus.* Among the inherent advantages of the *Xenopus* model are the ease with which large numbers of eggs can be collected; the large size of those eggs and the embryos that develop from them after fertilization; a consistent fate map; the external development of these embryos and the speed with which the conserved vertebrate body plan is organized. The large cells in the early embryos were ideal for addressing cell biological questions but also allowed early embryonic patterning of the vertebrate embryo and the fate map to be unraveled. Many fundamental processes in vertebrate development were first discovered in *Xenopus* in areas such as axial development, mesoderm induction, neural patterning, and more recently organogenesis and regeneration, to name but a few. The absence of well-developed genetics has often been cited as the one disadvantage of the *Xenopus laevis* model system compared to mouse or zebrafish for instance, but as this volume shows very clearly this criticism has been thoroughly put to rest.

This volume of Methods in Molecular Biology on *Xenopus* comes at a crucial juncture for our model system. The recent publication of the draft genome for *Xenopus tropicalis* marks an important milestone. It reveals that the *Xenopus* genome is in itself an ideal model for vertebrate genomes since it manifests conserved vertebrate genomic organization and therefore reinforces the uniquely advantageous phylogenetic position of *Xenopus* as a general vertebrate model. The genome and the embrace of genomic approaches enabled by *X. tropicalis* as an experimental model have invigorated the entire *Xenopus* field, not just those working with *X. tropicalis.* This volume seeks to focus on these new approaches. *X. tropicalis* offers tractable genetics that will complement the traditional strengths of *Xenopus* as a model system, as do the powerful transgenics methods pioneered in *Xenopus.* The genome and new deep sequencing approaches open up a new chapter in the analysis of gene expression on a whole transcriptome level and of protein expression and interaction on a proteome level; here in particular the traditional advantages of the *Xenopus* model system synergize very directly with the new information technology available to enable completely novel approaches. However, this impact is also felt in new methodology to image these expressed transcripts and proteins in embryos and in cells, particularly with sophisticated fluorescence microscopy; and of course in new systems approaches and applications for stem cell technology and regeneration. As a consequence, the computer and online databases will become as much a tool for *Xenopus* researchers in the future as the micropipette, the injection needle, and the dissection scalpel have been for many years.

The new approaches inspired by the *X. tropicalis* genome combined with the unique experimental advantages of *X. laevis* and their distinctive phylogenetic position make them unique model systems for vertebrate embryonic development in the twenty-first century. The protocols in this volume written by the leading experts in each method provide a toolkit that will enable every laboratory to maximize the power of this extraordinary experimental system.

Aberdeen, UK *Stefan Hoppler*
Calgary, Canada *Peter D. Vize*

Contents

Contributors

ANITA ABU-DAYA • *National Institute for Medical Research, London, England, UK*
BONI A. AFOUDA • *Institute of Medical Sciences, University of Aberdeen, Scotland, UK; MRC Centre for Regenerative Medicine, University of Edinburgh, Scotland, UK*
ROBERT C. AKKERS • *Department of Molecular Biology, Nijmegen Center of Molecular Life Sciences, Radboud University Nijmegen, Nijmegen, The Netherlands*
BRYAN G. ALLEN • *Department of Biochemistry, The University of Iowa, Iowa City, IA, USA*
LEONARDO I. ALMONACID • *Molecular Bioinformatics Laboratory, Millennium Institute on Immunology and Immunotherapy, Pontificia Universidad Catolica de Chile, Santiago, Chile*
ENRIQUE AMAYA • *Healing Foundation Centre, Faculty of Life Sciences, University of Manchester, England, UK*
HÉLÈNE AUGER • *Faculty of Life Sciences, University of Manchester, Manchester, England, UK*
CAROLINE W. BECK • *Department of Zoology, University of Otago, Dunedin, New Zealand*
IRA L. BLITZ • *Department of Developmental and Cell Biology and the Developmental Biology Center, University of California, Irvine, CA, USA*
BOYAN BONEV • *Faculty of Life Sciences, University of Manchester, Manchester, England, UK*
JEFF BOWES • *Department of Biological Sciences, University of Calgary, Calgary, AB, Canada*
DANIEL R. BUCHHOLZ • *Department of Biological Sciences, University of Cincinnati, Cincinnati, OH, USA*
FRANK L. CONLON • *Department of Genetics, University of North Carolina, Chapel Hill, NC, USA*
ILEANA M. CRISTEA • *Department of Molecular Biology, Princeton University, Princeton, NJ, USA*
LANCE A. DAVIDSON • *Department of Bioengineering and Developmental Biology, University of Pittsburgh, Pittsburgh, PA, USA*
EAMON DUBAISSI • *Faculty of Life Sciences, University of Manchester, Manchester, England, UK*
FERNANDO FAUNES • *Center for Aging and Regeneration and Millennium Nucleus in Regenerative Biology, Pontificia Universidad Catolica de Chile, Santiago, Chile*
TIMOTHY J. GEACH • *National Institute for Medical Research, London, England, UK*
MICHAEL J. GILCHRIST • *Division of Systems Biology, MRC National Institute for Medical Research, London, UK*
ROBERT M. GRAINGER • *Department of Biology, University of Virginia, Charlottesville, VA, USA*
TODD M. GRECO • *Department of Molecular Biology, Princeton University, Princeton, NJ, USA*

MATTHEW GUILLE • *European Xenopus Resource Centre, School of Biological Sciences, University of Portsmouth, Portsmouth, England, UK*
SERGE HARDY • *UMR 6061 CNRS-Université de Rennes 1, Rennes, France*
RICHARD M. HARLAND • *Department of Molecular and Cell Biology, University of California, Berkeley, CA, USA*
ADAM E. HENDRY • *School of Biological Sciences, University of East Anglia, Norwich, England, UK*
DOUGLAS W. HOUSTON • *Department of Biology, The University of Iowa, Iowa City, IA, USA*
ALISSA M. HULSTRAND • *Department of Biology, The University of Iowa, Iowa City, IA, USA*
SHOKO ISHIBASHI • *The Healing Foundation Centre, The Faculty of Life Sciences, University of Manchester, England, UK*
ULRIKE G. JACOBI • *Department of Molecular Biology, Nijmegen Center of Molecular Life Sciences, Radboud University Nijmegen, Nijmegen, The Netherlands*
ALAN JAFKINS • *European Xenopus Resource Centre, School of Biological Sciences, University of Portsmouth, England, UK*
SAGAR D. JOSHI • *Department of Bioengineering, University of Pittsburgh, Pittsburgh, PA, USA*
ERIN KALTENBRUN • *Department of Genetics, University of North Carolina, Chapel Hill, NC, USA*
CLAIR M. KELLEY • *Department of Pathology, St. Jude Children's Research Hospital, Memphis, TN, USA*
MUSTAFA KHOKHA • *Department of Pediatrics and Genetics, Yale University School of Medicine, New Haven, CT, USA*
HYE YOUNG KIM • *Department of Bioengineering, University of Pittsburgh, Pittsburgh, PA, USA*
KRISTEN L. KROLL • *Department of Developmental Biology, Washington University School of Medicine, St Louis, MO, USA*
EMIN KULIYEV • *Department of Pathology, St. Jude Children's Research Hospital, Memphis, TN, USA*
JUAN LARRAIN • *Center for Aging and Regeneration and Millennium Nucleus in Regenerative Biology, Pontificia Universidad Catolica de Chile, Santiago, Chile*
ROBERT LEA • *The Faculty of Life Sciences, University of Manchester, Manchester, England, UK*
DASFNE LEE-LIU • *Center for Aging and Regeneration and Millennium Nucleus in Regenerative Biology, Pontificia Universidad Catolica de Chile, Santiago, Chile*
YOU E. LI • *Department of Biochemistry, The University of Iowa, Iowa City, IA, USA*
SARAH LOUIE • *Department of Biology, University of Virginia, Charlottesville, VA, USA*
NICK R. LOVE • *The Healing Foundation Centre, The Faculty of Life Sciences, University of Manchester, England, UK*
PAUL E. MEAD • *Department of Pathology, St. Jude Children's Research Hospital, Memphis, TN, USA*
FRANCISCO MELO • *Molecular Bioinformatics Laboratory, Millennium Institute on Immunology and Immunotherapy, Pontificia Universidad Catolica de Chile, Santiago, Chile*

AGNÈS MEREAU • *UMR 6061 CNRS-Université de Rennes 1, Rennes, France*
YANA MITEVA • *Department of Molecular Biology, Princeton University, Princeton, NJ, USA*
ANNA NOBLE • *European Xenopus Resource Centre, School of Biological Sciences, University of Portsmouth, Portsmouth, England, UK*
HARUKI OCHI • *Graduate School of Biological Sciences, Nara Institute of Science and Technology (NAIST), Takayama, Ikoma, Nara, Japan*
HAJIME OGINO • *Graduate School of Biological Sciences, Nara Institute of Science and Technology (NAIST), Takayama, Ikoma, Nara, Japan*
DAVID J. OLSON • *Department of Biology, The University of Iowa, Iowa City, IA, USA*
NIKI PANAGIOTAKI • *The Faculty of Life Sciences, University of Manchester, Manchester, England, UK*
NANCY PAPALOPULU • *The Faculty of Life Sciences, University of Manchester, Manchester, England, UK*
NICOLAS POLLET • *Institute of Systems and Synthetic Biology, Genopole, CNRS, Université d'Evry Val d'Essonne, Evry, France*
YASUSHI SAKA • *Institute of Medical Sciences, University of Aberdeen, Scotland, UK*
DEREK L. STEMPLE • *Wellcome Trust Sanger Genome Institute, Hinxton, England, UK*
RAPHAËL THURET • *University of Manchester, Manchester, England, UK*
MATTHEW L. TOMLINSON • *School of Biological Sciences, University of East Anglia, Norwich, England, UK*
CHIHIRO UCHIYAMA • *Graduate School of Biological Sciences, Nara Institute of Science and Technology (NAIST), Takayama, Ikoma, Nara, Japan*
GERT JAN C. VEENSTRA • *Department of Molecular Developmental Biology, Nijmegen Center of Molecular Life Sciences, Radboud University Nijmegen, Nijmegen, The Netherlands*
FLORENCIA DEL VISO • *Department of Pediatrics and Genetics, Yale University School of Medicine, New Haven, CT, USA*
PETER D. VIZE • *Department of Biological Science, University of Calgary, Calgary, Canada*
LAUREN WALDRON • *Department of Genetics, University of North Carolina, Chapel Hill, NC, USA*
DANIEL L. WEEKS • *Department of Biochemistry, The University of Iowa, Iowa City, IA, USA*
GRANT N. WHEELER • *School of Biological Sciences, University of East Anglia, Norwich, England, UK*
WARIF EL YAKOUBI • *University of Manchester, Manchester, England, UK*
DONALD A. YERGEAU • *Department of Pathology, St. Jude Children's Research Hospital, Memphis, TN, USA*
JOHN J. YOUNG • *Department of Molecular and Cell Biology, University of California, Berkeley, CA, USA*
YANAN ZHAO • *The Healing Foundation Centre, The Faculty of Life Sciences, University of Manchester, England, UK*
HAIQING ZHU • *Department of Pathology, St. Jude Children's Research Hospital, Memphis, TN, USA*
LYLE B. ZIMMERMAN • *National Institute for Medical Research, London, England, UK*

Part I

Xenopus Tropicalis

Chapter 1

Xenopus tropicalis as a Model Organism for Genetics and Genomics: Past, Present, and Future

Robert M. Grainger

Abstract

Xenopus tropicalis was introduced as a model system for genetic, and then genomic research, in the early 1990s, complementing work on the widely used model organism *Xenopus laevis*. Its shorter generation time and diploid genome has facilitated a number of experimental approaches. It has permitted multigenerational experiments (e.g., preparation of transgenic lines and generation of mutant lines) that have added powerful approaches for research by the *Xenopus* community. As a diploid animal, its simpler genome was sequenced before *X. laevis*, and has provided a highly valuable resource indispensable for all *Xenopus* researchers. As more sophisticated transgenic technologies for manipulating gene expression are developed, and mutations, particularly null mutations, are identified in widely studied genes involved in critical cellular and developmental processes, researchers will increasingly turn to *X. tropicalis* for definitive analysis of complex genetic pathways. This chapter describes the historical and conceptual development of *X. tropicalis* as a genetic and genomic model system for higher vertebrate development.

Key words: *Xenopus tropicalis*, Experimental model system, Genetics, Natural mutations, Mutagenesis screen, Genomics, Genome sequencing

1. Introduction

Amphibian species have held a prominent position in embryological research for well over a century because their large, abundant, externally developing embryos proved ideal for studying embryological phenomena. Although they share these features with fish embryos, amphibian embryos are more resilient to many surgical manipulations, and, as tetrapods, are more closely related to mammals. However, in other model systems, including fish, the powerful impact of genetic approaches for understanding complex developmental phenomena in the late twentieth century left a gap in the analytical approaches available to most developmental biologists using amphibians. While loss of gene function could be elicited

Stefan Hoppler and Peter D. Vize (eds.), *Xenopus Protocols: Post-Genomic Approaches, Second Edition*,
Methods in Molecular Biology, vol. 917, DOI 10.1007/978-1-61779-992-1_1, © Springer Science+Business Media, LLC 2012

by antisense reagents and dominant negative constructs, it became increasingly important to be able to develop genetics in an amphibian system to parse developmental phenomena in a way only feasible by studying mutations. While the amphibian model *Xenopus laevis* has had a very distinguished history in elucidating developmental phenomena (1), its long life cycle and allotetraploid genome made conventional genetic analyses more challenging. Although Zebrafish and other teleosts have been very successful genetic model systems, teleosts too have undergone a genome duplication that complicates both genetic studies and analysis of genome organization. Thus the introduction in the 1990s of *Xenopus tropicalis*, a species closely related to *X. laevis*, but which is diploid and requires a much shorter time to reach sexual maturity, offered the welcome prospect of a model that could provide the power of genetics to the *Xenopus* research community; especially so since it has a genome organization much like mammals. The development of the *X. tropicalis* system from that time to the present is described here.

2. Early Amphibian Genetic Research

It is important at the outset to highlight that important genetic research has been done in amphibian species before the emergence of *X. tropicalis*. In *X. laevis*, the *anucleolate* mutation leads to the loss of the nucleolus, and was instrumental for identifying the nucleolus as the site of the ribosomal RNA gene locus (2). In addition, screens of outbred *X. laevis* and animals generated from nuclear transplantation have revealed a number of interesting developmental mutations (3–5). The Mexican axolotl *Ambystoma mexicanum* has also been a valuable genetic model for which some molecular genetic resources have been generated, for example large collections of ESTs, enabling valuable genomic work (6), despite being an unlikely candidate organism for genome sequencing due to its extremely large genome, like many other amphibians.

3. The Search for an Amphibian Genetic Model System

Those of us who were thinking about a genetically tractable alternative to *X. laevis* in the early 1990s were also discussing potential model systems. For example, in my own laboratory I was exploring *Rana pipiens* as a system. It is possible to generate androgenetic haploid embryos in this species by mechanical removal of the egg nucleus from the very large eggs of this species (7) and from these embryos cell lines had been made and studied extensively for somatic cell genetics (8). Such cells could be a source of material for mutagenesis, for example by homologous recombination, and the nuclei from such cells used to reprogram development by transplantation back into the egg. The possibility of transplanting mutant haploid nuclei back into the egg had been discussed in a 1970 paper by Freed and Metzger-Freed (8), based on the previously published nuclear transplantation work of Briggs and King in *Rana* (9), work that was followed by the extensive studies on this subject by Gurdon and colleagues using *X. laevis* (10). The *Rana* system has the disadvantages, however, of not being so easily used to generate embryos year-round as *Xenopus*, nor does it have a large core of researchers working on cell and developmental problems. *Hymenochirus*, a genus of small rapidly developing pipid frogs, relatively close relatives of *X. tropicalis* (11), was also being discussed but is distant enough from *Xenopus* and studied by few researchers so that they were unlikely to be adopted by the *Xenopus* community.

4. Introduction of *Xenopus tropicalis* as a Genetic Model System

X. tropicalis was in some ways a very obvious choice for a new amphibian genetic model. This West African species is an evolutionarily close relative of the South African species *X. laevis*, having diverged approximately 50 Mya (12). This close relationship to *X. laevis* would mean that it was likely to share developmental features, and have conserved gene sequences and functions and thus be attractive to the large group of researchers working on *X. laevis* (14). Indeed many features are shared by the two systems (15, 16), including gene expression patterns. However, there is considerable remodeling of the post tetraploid transcriptome in *X. laevis* (17) that provides an advantage for *X. tropicalis* in comparative gene function and regulation studies, where the diploid gene structure and function is more likely to be conserved with mammalian species. *X. tropicalis* is, as well, the only known diploid species in the *Xenopus* genus.

Before the 1990s research on *X. tropicalis* was primarily restricted to studies on the evolutionary and cytogenetic relationships among the various *Xenopus* species (e.g., (11, 18)). While *X. tropicalis* were used and housed in several laboratories, and held in the Swiss repository for *Xenopus* lines at the University of Geneva (D. Runnger, personal communication), prior to 1990, they were not widely available nor used for genetic screens. In 1990 a small group of *X. tropicalis* originally collected from the Ivory Coast were brought from the Geneva Center to the United States by Marc Kirschner (University of California, San Francisco), and shortly thereafter, a much larger population was imported from Nigeria by the Kirschner lab (M. Kirschner and E. Amaya, personal communication). It was difficult at that time to import animals from Africa and, after exhausting conventional contacts with academic institutions in Nigeria, Marc Kirschner serendipitously tracked down and enlisted the aid of a sympathetic Nigerian tribal leader, formerly an academic in the United States, to import Nigerian *X. tropicalis*. The animals of both Ivory Coast and Nigerian strains were raised under the supervision of Enrique Amaya in facilities at UC San Francisco and UC Berkeley.

These two strains from California were transported to my laboratory shortly thereafter and we initiated inbreeding of the two *X. tropicalis* lines with the intent of developing lines more suitable for stable genetic screens, and for possibly mapping of genes by crossing of divergent lines. At the same time, we also imported other wild caught populations of *X. tropicalis* from Nigeria that we thought might provide useful genetic diversity in future experiments. The animals from the Kirschner/Amaya collection and my laboratory have been distributed around the world, and other lines were acquired as well, for example one line imported by the Asashima laboratory in Japan that is somewhat diverged from the Nigerian/Ivory Coast group (19).

5. Early Genetic Screens for Natural Mutations

The wild caught animals in our laboratory were the first to be used in a genetic screen (20), presuming that these outbred animals would be a natural source of genetic variation that would reveal mutations. This allowed us to begin to do screening immediately while we were still perfecting methods for inducing mutagenesis for later screens. The first screen took advantage of being able to generate gynogenetically derived offspring in *Xenopus* (and other amphibians) to greatly facilitate rapidly identifying mutations (21). This procedure entails generating haploid embryos carrying only maternal genetic material by fertilizing with UV irradiated sperm, which activate embryonic development but do not contribute genetically to the embryo. Such haploid embryos would develop abnormally, but they can be diploidized by preventing release of the second polar body by high pressure, or more simply by a cold shock protocol developed in my laboratory (modified after ref. (22); see Subheading 3.2 in Chapter 5). Such embryos develop as diploid animals but if eggs carry a developmental mutation, segregation of maternal chromatid pairs reveals recessive mutations in a fraction of the gynogenetic embryos. This method saves a great deal of time, and animal housing space, when compared to the classical three generation screening method of interbreeding offspring of potential heterozygote carrier animals used in many systems for identifying recessive mutations.

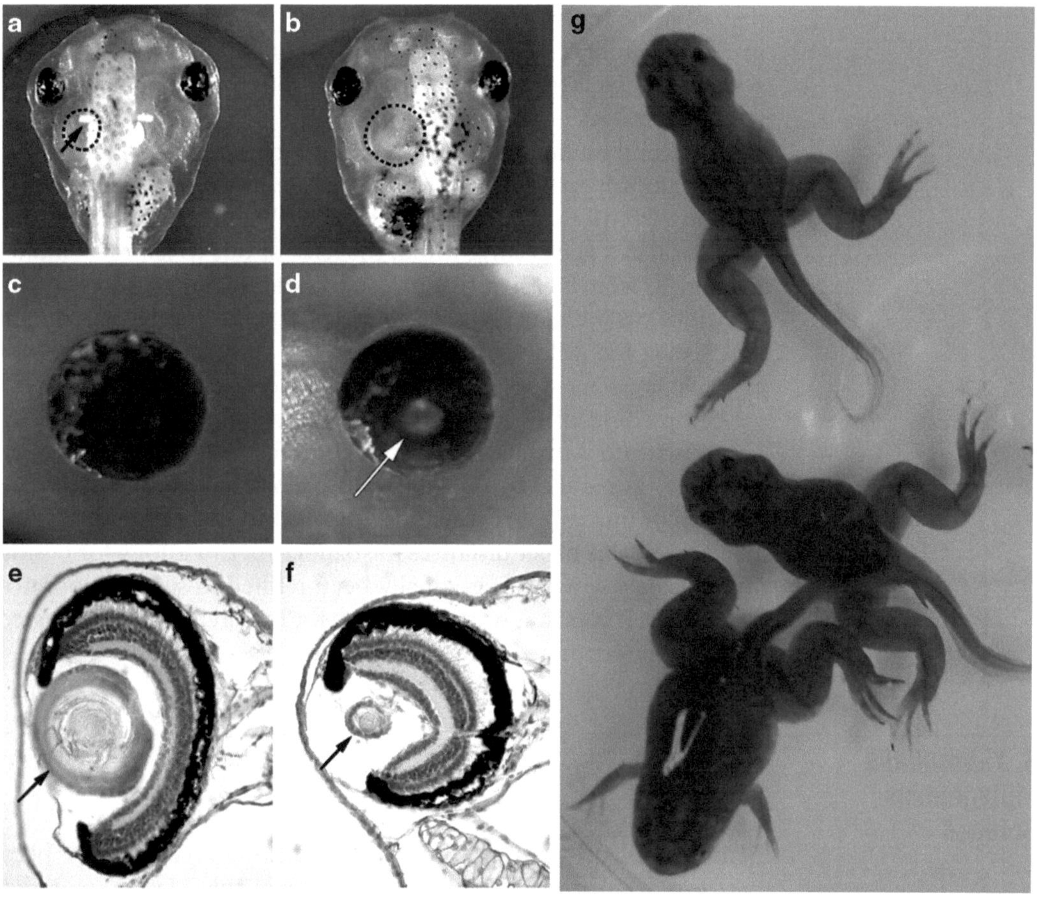

Fig. 1. Images of *Xenopus tropicalis* developmental mutants. Two *X. tropicalis* mutants identified during a gynogenetic screen of wild caught animals are shown in **a–f**. (**a**, **b**), images of wild type and mutant *little pitcher* embryos, respectively. Otoconia are missing in the mutant inner ear (demarcated by the *black arrow* in wildtype embryos) and the inner ear vesicle is enlarged (indicated by *dotted circle*). (**c**, **d**), images of wild type and mutant *cataract* embryos, respectively. The lens area appears very cloudy in the mutant embryos (*white arrow*). Sections of wild type and mutant embryos in (**e**) and (**f**), respectively, reveal a highly reduced lens size in *cataract* embryos (*black arrows*). (**g**) Metamorphosing *xenopus de milo* (*xdm*) froglets (two mutants; one wildtype). The striking absence of forelimbs illustrates the kind of mutation that is likely to be particularly insightful for clarifying tetrapod development. This mutation was found to interrupt the *nephronectin* gene, which produces a small integrin ligand, now clearly implicated in forelimb development. Image courtesy of L. Zimmerman (National Institute for Medical Research, Mill Hill, London).

In this initial screen a wide range of phenotypes was observed in mutations generated from several outbred lines. These include very broad defects, e.g., affecting left-right asymmetry, as well as alterations in specific tissues, e.g., in the inner ear. Figure 1 shows one of these mutations, *little pitcher*, in which the inner ear fails to develop any internal structure, including the otoconia characteristic of the early amphibian inner ear; embryos swim in circles due to failure to develop appropriate inner ear structures needed for balance. It is also valuable to note that, although *Xenopus* embryos are not transparent at pre-tadpole stages (as are zebrafish embryos), they become transparent as organogenesis proceeds. As can be seen in Fig. 1, embryos are indeed transparent at 5 days, when organ systems are well defined and subtle morphological defects can be recognized.

A second mutation identified during a gynogenetic screen of wild caught animals, *cataract*, shows an opacity in the lens area (Fig. 1) resulting from a greatly reduced lens size. Whether this defect is due to a lesion in a gene acting within the lens, or due to failure in inductive signals from adjacent tissues, can be determined because one can so readily make tissue chimeras in *Xenopus* (e.g., combining mutant lens tissue and wildtype inducers, or vice versa), as was done, for example, in other work on lens determination (23).

Inbreeding of *X. tropicalis* lines was being performed by Tim Grammer and Mustafa Khokha in the Harland laboratory (U.C. Berkeley) in parallel to our inbreeding efforts, and they described a group of mutations revealed during the inbreeding process (24). In addition this paper discusses husbandry methodology for raising *X. tropicalis*, which has gradually been refined in several laboratories so that it is now a routine process for researchers and discussed in detail in Chapters 2 and 3.

6. *X. tropicalis* Mutagenesis Screens

Mutagenesis using inbred lines as starting material has now been performed in several laboratories using the classic mutagen *N*-nitroso-*N*-ethylurea (ENU). In my group we have used spermatogonial mutagenesis (see Subheading 3.4.2 in Chapter 5), in which males are injected with ENU and F1 mutagenized animals are used to identify mutations by gynogenesis. Similar spermatogonial mutagenesis has been used in many model systems (25, 26). In the Zimmerman lab, mature sperm were mutagenized (see Subheading 3.4.1 in Chapter 5), in a manner similar to what has been done in zebrafish (27), a procedure that results in very high mutagenesis rates. However, first generation embryos generated in this manner are genetically mosaic since mutagenesis is DNA strand-specific when performed in this way and require sibling crosses in subsequent generations to generate conventional mutant lines. Using this protocol, the first large ENU screen was published (also using gynogenesis to identify mutant phenotypes), revealing 29 mutations in a wide diversity of organ systems (28). One mutant in this group, *xenopus de milo* (*xdm*) is mentioned here (and illustrated in Fig. 1) because of its striking effect on forelimb formation. This mutation illustrates an obvious advantage of *X. tropicalis* as a genetic system that can serve as a model for tetrapod-specific developmental features. In addition, limb regeneration in amphibians is one of the most carefully examined examples of regeneration in any organ system, and the genetic and transgenic approaches so well suited to *X. tropicalis* should be great assets that have broad relevance to the current surge of interest in regenerative medicine (29).

Mapping and cloning of *X. tropicalis* mutants has been facilitated by genomic resources, described below, particularly the availability of a reliable genetic map based on simple sequence length polymorphisms (SSLPs) (30). In addition, the initial steps of mapping genetic lesions in *X. tropicalis* is facilitated by gynogenetic screening, because the frequency of the appearance of a recessive mutation in a gynogenetically derived embryo population depends on its distance from the centromere (31) (*also* see Subheading 3.7.4 in Chapter 5). Gynogenetic mapping has also been used to develop methods to quickly identify the chromosome on which a particular mutation lies (32).

These methods have now permitted for the first time the positional cloning of a number of mutant genes in the past 2 years. This includes mutations in cardiac myosin *myh6* (resulting in a no heartbeat phenotype) (33), in the secreted integrin ligand nephronectin (resulting in missing forelimbs) (34), and in *unc45b* (resulting in paralysis) (35). Mutations identified in these screens appear to be commonly linked to human disease syndromes (e.g., (33)).

7. Development of the *X. tropicalis* Reseach Community and Resources for Genetic and Genomic Studies

Research on the genetic potential of *X. tropicalis* in the United States began in a few labs in the San Francisco area, as discussed above, in the early 1990s. While no formal meetings of the groups that studied this animal model occurred until several years later, there were informal gatherings, including the "First International *X. tropicalis* Meeting" in late 1992, an event that was held at one of the meetings of the Bay Area Research on Frogs (BARF) group that met during this period. Figure 2 shows the participants of this meeting, which merits highlighting here because the meeting included many of those who have played a key role in the development of *X. tropicalis* as a model system, and, more generally, contributed in very important ways to the development of *Xenopus* more broadly as a major system for studies of cell and developmental biology.

While work on *X. tropicalis* began in earnest during the 1990s in several laboratories, with transgenic lines created in this more rapidly developing *Xenopus* species, and enhancer/gene trap experiments underway (14), there was still only a small group working on this species, and no centralized efforts to generate community-wide resources nor for disseminating information about this new system, or more broadly about new technologies for *Xenopus* in general. The establishment of the Cold Spring Harbor *Xenopus*

Fig. 2. Participants in a Bay Area Regional Frog meeting in late 1992 in which the prospects for the *X. tropicalis* system were discussed. Participants included, from *left* to *right*, Marc Kirschner, Jeremy Green, Pascal Stein, Andre Brandli, Tabitha Doniach, William Smith, Kristen Kroll, Michael Sheets, Richard Harland, Enrique Amaya, John Gerhart, Robert Davis and Lyle Zimmerman.

Development Course by Hazel Sive and Richard Harland in 1994 was a productive step in that direction and resulted in the publication of a widely used manual of experimental technology for *Xenopus* (36).

An important turn of events occurred in 1999 when Harold Varmus, then Director of the United States National Institutes of Health (NIH), convened a meeting focused on Non-Mammalian Model Systems to establish priorities for centralized resources needed to develop these model organisms. Included in this group were representatives from the yeast, *Caenorhabditis elegans*, *Drosophila*, zebrafish, and *Xenopus* communities. Janet Heasman and I convened the *Xenopus* group and we were told by the organizers of this meeting that a primary reason *Xenopus* was included in the meeting because of the prospect of the development of genetic tools in this system, tools which were already a mainstay in the research of the other systems represented at the meeting. After surveying the research community, the recommendations of the *Xenopus* breakout committee (37) included genomic resources for producing ESTs, full-length cDNA sequencing, and microarray development. Regarding *X. tropicalis*, support was recommended for establishing its viability as a genetic system by support of several genetic strategies, development of large-insert DNA libraries, development of a genetic map, and sequencing of its genome. Regarding centralized resource centers, support of a database and an animal stock center and training center was recommended. The meeting helped to stabilize NIH infrastructure and support for *Xenopus* resources, which have grown significantly since this time (37, 38). Interestingly, and of great importance for the *Xenopus* research community, essentially all of the recommended resources from the 1999 meeting have now been realized.

Shortly following the 1999 meeting, the first large meeting of *Xenopus* researchers focusing on *X. tropicalis* research was held at the University of Virginia as a satellite meeting accompanying the Society for Developmental Biology meeting in Charlottesville. The meeting covered the history of genetics in *Xenopus*, strategies for performing genetic screens, transgenic strategies for studying gene regulation and for mutagenesis, review of new husbandry methods for raising *X. tropicalis* (39).

8. Sequencing of the *X. tropicalis* Genome

A key milestone in *X. tropicalis* research (and for the *Xenopus* community at large) was the sequencing of the *X. tropicalis* genome, undertaken by the Department of Energy's Joint Genome Institute (JGI) in 2001. *X. tropicalis* was an obvious choice for an amphibian genome sequencing project at the time, because of its diploid genome, contrasting the allotetraploidy of the *X. laevis* genome. By this time, the inbreeding of *X. tropicalis* lines was far enough along in our laboratory that a suitable animal was available for this project, and complementary work to produce BAC libraries important for both genome assembly and as a resource for projects examining gene organization. In addition to providing the first several generations of the *X. tropicalis* genome, resulting in publication of the work in 2010 (13), the JGI sponsored other genomically based efforts including identification of a very large number of ESTs. This work has provided the primary platform for research in genetics and genomics in *Xenopus*. The genome sequencing project confirmed that indeed *X. tropicalis* has a diploid genome, and showed that its genome has a remarkable degree of synteny with mammalian genomes, often in stretches of a 100 genes or more, far greater than that seen between fish and mammal. The genome organization is discussed further in Chapter 4. While the efforts of *X. laevis* researchers to have the genome of this widely used animal met with frustration early on, in large part because of the perceived difficulty in sequencing and assembling its allotetraploid genome, the *X. tropicalis* project has been extraordinarily useful for all *Xenopus* researchers, and important as a foundation for the new *X. laevis* genome project, now finally underway.

9. The Future of Research Utilizing *X. tropicalis*

X. tropicalis genetics is still a relatively new direction for the *Xenopus* community, though significant technological progress has been made, allowing forward genetic screens to be performed productively and identification of mutant genes by positional cloning to be accomplished effectively and relatively quickly. Of note is that many of the genes identified in screens in *X. tropicalis* bear on human diseases or syndromes (L. Zimmerman, pers. comm.). It is not clear whether there are the resources to undertake the large-scale screens performed in the zebrafish community in the 1990s. However, more focused screens, for example utilizing transgenic reporter lines to identify mutations in particular organ systems, offer a bright future for forward genetics even on a smaller scale. Novel insights from the *Xenopus* system are being made in many areas (proteomics, signal transduction, organogenesis, etc.) and provide unique areas where genetic analysis in the future will add important dimensions to what can be done in this, or any other system.

The challenges *Xenopus* researchers have had to face in setting up facilities suited to the genetic and transgenic manipulations that can be accomplished with *X. tropicalis* highlight the importance of a key community resource. The two large centers, The European *Xenopus* Resource Center in the United Kingdom and the National *Xenopus* Resource in the United States, provide sites for housing mutant lines, generating transgenic lines, and in general fostering new genetic technologies that will be vital for the development of genetics in *Xenopus*. They will provide efficient mechanisms for disseminating both animals and technologies to the community at large that would be difficult or impossible to establish in individual laboratories.

Many researchers look toward *X. tropicalis* as a system that will foster technologies to identify mutations in genes already known to be of high interest for studies in cell and developmental biology. As discussed earlier, there are at present no methods as yet developed in *Xenopus* for targeted gene inactivation as in the mouse. However, projects are underway to identify genes by screening for mutations in populations of mutagenized animals by high throughput sequencing (a process often referred to as TILLING (40), see *also* Subheading 3.6 in Chapter 5), a strategy now being pursued in a number of genetically tractable model organisms (41, 42). Mutations have now been discovered in *X. tropicalis* by this methodology (R. Grainger, S. Carruthers, D. Stemple, unpublished), which promises to become even more efficient and less expensive, as high throughput sequencing methods continue to improve. Other methods for gene inactivation in *X. tropicalis*, e.g., mutations induced by zinc-finger nucleases ((43) see *also* Chapter 7) and studies using RNA interference (44, 45), are showing promise as well.

X. tropicalis will also continue to contribute important biological insights because of the efforts to improve the sequence of its genome, and the utility of this genomic information. For example, analysis of the structure and function of enhancers, many of which can be a megabase or possibly further from the genes they regulate, is greatly aided by the long-range assembly available in *X. tropicalis*. This allows rapid identification of highly conserved elements (by comparative genomics), which can then be studying using the powerful transgenic methods available in *Xenopus* ((46), see *also* Chapter 15). Similarly, analysis of genome-wide epigenetic changes is ideally suited to *X. tropicalis* ((47), see *also* Chapter 17). These genomic approaches in *X. tropicalis*, combined with analysis of mutations in key genes in this species, will lead to novel insights about key gene regulatory networks that regulate complex cellular and developmental events.

Acknowledgements

The author gratefully acknowledges contributions to developing the *X. tropicalis* system from lab members Lyle Zimmerman, Nicolas Hirsch, Selina Noramly, Jei Chae, Hui Wang, Hong Jin, Hajime Ogino, Takuya Nakayama, Marilyn Fisher, Margaret Fish and Matthew Etzell. Research on *X. tropicalis* was supported by grants to R.M.G. from the National Institutes of Health RR013221, EY019000 and EY017400. Grants to R.M.G. from NIH also support a National *Xenopus* Resource (RR025867) and National TILLING Resource (HD065713).

References

1. Harland RM, Grainger RM (2011) Xenopus research: metamorphosed by genetics and genomics. Trends Genet 27(12):507–515
2. Brown DD, Gurdon JB (1964) Absence of ribosomal RNA synthesis in the anucleolate mutant of Xenopus laevis. Proc Natl Acad Sci U S A 51:139–146
3. Krotoski DM, Reinschmidt DC, Tompkins R (1985) Developmental mutants isolated from wild-caught Xenopus laevis by gynogenesis and inbreeding. J Exp Zool 233(3):443–449
4. Droin A (1992) The developmental mutants of Xenopus. Int J Dev Biol 36(4):455–464
5. Graf JD, Kobel HR (1991) Genetics of Xenopus laevis. Methods Cell Biol 36:19–34
6. Voss SR, Epperlein HH, Tanaka EM (2009) Ambystoma mexicanum, the axolotl: a versatile amphibian model for regeneration, development, and evolution studies. Cold Spring Harb Protoc 2009(8):pdb emo128
7. Porter KR (1939) Androgenetic development of the egg of Rana pipiens. Biol Bull 77:233–257
8. Freed JJ, Mezger-Freed L (1970) Stable haploid cultured cell lines from frog embryos. Proc Natl Acad Sci U S A 65(2):337–344
9. Briggs R, King TJ (1952) Transplantation of living nuclei from blastula cells into enucleated frogs' eggs. Proc Natl Acad Sci U S A 38(5): 455–463
10. Gurdon JB, Byrne JA (2003) The first half-century of nuclear transplantation. Proc Natl Acad Sci U S A 100(14):8048–8052
11. de Sa RO, Hillis DM (1990) Phylogenetic relationships of the pipid frogs Xenopus and

Silurana: an integration of ribosomal DNA and morphology. Mol Biol Evol 7(4):365–376

12. Hellsten U, Khokha MK, Grammer TC, Harland RM, Richardson P, Rokhsar DS (2007) Accelerated gene evolution and subfunctionalization in the pseudotetraploid frog Xenopus laevis. BMC Biol 5:31
13. Hellsten U, Harland RM, Gilchrist MJ, Hendrix D, Jurka J, Kapitonov V et al (2010) The genome of the Western clawed frog Xenopus tropicalis. Science 328(5978): 633–636
14. Amaya E, Offield MF, Grainger RM (1998) Frog genetics: Xenopus tropicalis jumps into the future. Trends Genet 14(7):253–255
15. Hirsch N, Zimmerman LB, Grainger RM (2002) Xenopus, the next generation: X. tropicalis genetics and genomics. Dev Dyn 225(4): 422–433
16. Khokha MK, Chung C, Bustamante EL, Gaw LW, Trott KA, Yeh J et al (2002) Techniques and probes for the study of Xenopus tropicalis development. Dev Dyn 225(4):499–510
17. Morin RD, Chang E, Petrescu A, Liao N, Griffith M, Chow W et al (2006) Sequencing and analysis of 10,967 full-length cDNA clones from Xenopus laevis and Xenopus tropicalis reveals post-tetraploidization transcriptome remodeling. Genome Res 16(6):796–803
18. Tymowska J (1973) Karyotype analysis of Xenopus tropicalis Gray, Pipidae. Cytogenet Cell Genet 12(5):297–304
19. Kashiwagi K, Kashiwagi A, Kurabayashi A, Hanada H, Nakajima K, Okada M et al (2010) Xenopus tropicalis: an ideal experimental animal in amphibia. Exp Anim 59(4):395–405
20. Noramly S, Zimmerman L, Cox A, Aloise R, Fisher M, Grainger RM (2005) A gynogenetic screen to isolate naturally occurring recessive mutations in Xenopus tropicalis. Mech Dev 122(3):273–287
21. Tompkins R (1978) Triploid and gynogenetic diploid Xenopus laevis. J Exp Zool 203:251–256
22. Kawahara H (1978) Production of triploid and gynogenetic diploid Xenopus by cold treatment. Dev Growth Differ 20(3):227–236
23. Henry JJ, Grainger RM (1990) Early tissue interactions leading to embryonic lens formation in Xenopus laevis. Dev Biol 141:149–163
24. Grammer TC, Khokha MK, Lane MA, Lam K, Harland RM (2005) Identification of mutants in inbred Xenopus tropicalis. Mech Dev 122(3): 263–272
25. Mullins MC, Hammerschmidt M, Haffter P, Nusslein-Volhard C (1994) Large-scale mutagenesis in the zebrafish: in search of genes controlling development in a vertebrate. Curr Biol 4(3):189–202
26. Justice MJ, Noveroske JK, Weber JS, Zheng B, Bradley A (1999) Mouse ENU mutagenesis. Hum Mol Genet 8(10):1955–1963
27. Riley BB, Grunwald DJ (1995) Efficient induction of point mutations allowing recovery of specific locus mutations in zebrafish. Proc Natl Acad Sci U S A 92(13):5997–6001
28. Goda T, Abu-Daya A, Carruthers S, Clark MD, Stemple DL, Zimmerman LB (2006) Genetic screens for mutations affecting development of Xenopus tropicalis. PLoS Genet 2(6):e91
29. Beck CW, Izpisua Belmonte JC, Christen B (2009) Beyond early development: Xenopus as an emerging model for the study of regenerative mechanisms. Dev Dyn 238(6): 1226–1248
30. Wells DE, Gutierrez L, Xu Z, Krylov V, Macha J, Blankenburg KP et al (2011) A genetic map of Xenopus tropicalis. Dev Biol 354(1):1–8
31. Reinschmidt D, Friedman J, Hauth J, Ratner E, Cohen M, Miller M et al (1985) Gene-centromere mapping in Xenopus laevis. J Hered 76(5):345–347
32. Khokha MK, Krylov V, Reilly MJ, Gall JG, Bhattacharya D, Cheung CY et al (2009) Rapid gynogenetic mapping of Xenopus tropicalis mutations to chromosomes. Dev Dyn 238(6): 1398–1446
33. Abu-Daya A, Sater AK, Wells DE, Mohun TJ, Zimmerman LB (2009) Absence of heartbeat in the Xenopus tropicalis mutation muzak is caused by a nonsense mutation in cardiac myosin myh6. Dev Biol 336(1):20–29
34. Abu-Daya A, Nishimoto S, Fairclough L, Mohun TJ, Logan MP, Zimmerman LB (2011) The secreted integrin ligand nephronectin is necessary for forelimb formation in Xenopus tropicalis. Dev Biol 349(2):204–212
35. Geach TJ, Zimmerman LB (2010) Paralysis and delayed Z-disc formation in the Xenopus tropicalis unc45b mutant dicky ticker. BMC Dev Biol 10:75
36. Sive HL, Grainger RM, Harland RM (2000) Early development of Xenopus laevis: a laboratory manual. Cold Spring Harbor Press, Cold Spring Harbor
37. Klein SL, Strausberg RL, Wagner L, Pontius J, Clifton SW, Richardson P (2002) Genetic and genomic tools for Xenopus research: The NIH Xenopus initiative. Dev Dyn 225(4): 384–391
38. Klein SL, Gerhard DS, Wagner L, Richardson P, Schriml LM, Sater AK et al (2006) Resources for genetic and genomic studies of Xenopus. Methods Mol Biol 322:1–16
39. Vogel G (1999) Frog is a prince of a new model organism. Science 285(5424):25

40. Stemple DL (2004) TILLING–a high-throughput harvest for functional genomics. Nat Rev Genet 5(2):145–150
41. Winkler S, Schwabedissen A, Backasch D, Bokel C, Seidel C, Bonisch S et al (2005) Target-selected mutant screen by TILLING in Drosophila. Genome Res 15(5): 718–723
42. Moens CB, Donn TM, Wolf-Saxon ER, Ma TP (2008) Reverse genetics in zebrafish by TILLING. Brief Funct Genomic Proteomic 7(6):454–459
43. Young JJ, Cherone JM, Doyon Y, Ankoudinova I, Faraji FM, Lee AH et al (2011) Efficient targeted gene disruption in the soma and germ line of the frog Xenopus tropicalis using engineered zinc-finger nucleases. Proc Natl Acad Sci U S A 108(17):7052–7057
44. Lund E, Sheets MD, Imboden SB, Dahlberg JE (2011) Limiting Ago protein restricts RNAi and microRNA biogenesis during early development in Xenopus laevis. Genes Dev 25(11):1121–1131
45. Chen CM, Chiu SL, Shen W, Cline HT (2009) Co-expression of Argonaute2 enhances short hairpin RNA-induced RNA interference in Xenopus CNS neurons in vivo. Front Neurosci 3:63
46. Ogino H, Fisher M, Grainger RM (2008) Convergence of a head-field selector Otx2 and Notch signaling: a mechanism for lens specification. Development 135(2):249–258
47. Akkers RC, van Heeringen SJ, Jacobi UG, Janssen-Megens EM, Francoijs KJ, Stunnenberg HG et al (2009) A hierarchy of H3K4me3 and H3K27me3 acquisition in spatial gene regulation in Xenopus embryos. Dev Cell 17(3):425–434

Chapter 2

Husbandry of *Xenopus tropicalis*

Alan Jafkins, Anita Abu-Daya, Anna Noble, Lyle B. Zimmerman, and Matthew Guille

Abstract

Xenopus tropicalis combine the advantages of *X. laevis*, for example using explants and targeted gain of function, with the ability to take classical genetics approaches to answering cell and developmental biology questions making it arguably the most versatile of the model organisms. Against this background, husbandry of *X. tropicalis* is less well developed than for its larger, more robust relative. Here we describe the methods used to keep and breed these frogs successfully.

Key words: *Xenopus*, *tropicalis*, Husbandry, In vitro fertilization, *Xenopus* health, Re-circulating aquaria

1. Introduction

Xenopus (*Silurana*) *tropicalis*, the Western clawed frog, is a small relative of the established laboratory model *X. laevis*. *X. tropicalis* is the only known diploid in the genus *Xenopus*, and at 1.5×10^9 bp, has one of the smallest genomes of any tetrapod. These species are sufficiently similar that viable hybrids can be produced, albeit only using *tropicalis* sperm on *laevis* eggs and not vice versa (1). Early studies of *X. tropicalis* mainly concerned phylogenetic analyses and suggested that *X. tropicalis* and the morphologically similar, tetraploid *X. epitropicalis* form an outgroup (sometimes denoted *Silurana*) with respect to other *Xenopus* species (2). It was not until the late 1990s that *X. tropicalis*' potential for adding genomics and genetics to *Xenopus*' traditional strengths in developmental and cell biology began to be exploited. Its small, sequenced genome (3), together with a genetic map (4), abundant embryo production, and relatively short generation time, are making genetic approaches increasingly practical (5–7). A number of mutations are now cloned (8–10). The simplicity of its genome has also encouraged genome-scale epigenetic

Stefan Hoppler and Peter D. Vize (eds.), *Xenopus Protocols: Post-Genomic Approaches, Second Edition*, Methods in Molecular Biology, vol. 917, DOI 10.1007/978-1-61779-992-1_2, © Springer Science+Business Media, LLC 2012

studies (11, 12) and morpholino oligonucleotide knockdown screens (13). The shorter generation time makes possible multi-generation transgenic strategies such as Gal4/UAS (14, 15), Cre/lox (16), and insertional mutagenesis (17).

The enormous toolkit of protocols developed in *X. laevis* for embryological, biochemical, and gain-of-function studies are also readily transferred to the smaller *X. tropicalis*, although not necessarily without modifications. The ability to combine the established strengths of *Xenopus* with *X. tropicalis*' robust genomics and emerging genetics makes it arguably the most versatile of all vertebrate model organisms.

Despite this versatility, *X. tropicalis* has not replaced laboratory use of the extraordinarily tough *X. laevis*, which has evolved to cope with a wide range of habitat extremes and whose rate of embryonic development can be conveniently regulated by temperature. Nevertheless, if maintained under appropriate conditions, *X. tropicalis* are also very robust frogs that can produce abundant embryos for more than a decade. Several things are likely to have contributed to problems in establishing *X. tropicalis* in some laboratories: (1) Cross-species contamination from *X. laevis* (microorganisms that are tolerated by one species are frequently pathogenic in another) or from wild-caught frogs. (2) Temperature stress, which can have effects on egg quality and disease susceptibility for months after an event such as heat- or cold-shock. (3) Diet during early life, as stocks that were undernourished or overcrowded, may never recover good fertility. Optimal husbandry regimes leading to good egg quality, rapid maturation, and low susceptibility to disease are particularly critical for the multi-generation genetic and transgenic strategies, which are less feasible in *X. laevis*. This chapter provides husbandry and breeding protocols which have been successful for large colonies over the course of >8 generations of *X. tropicalis*. Continuing to improve and share husbandry techniques for *X. tropicalis* should enable many more researchers to take advantage of the genomic resources and genetic applications available for this model.

2. Materials

2.1. *Xenopus tropicalis* Animals

X. tropicalis occur naturally in many countries of equatorial West Africa, but most laboratory stocks originate from Nigeria and Ivory Coast. Wild-caught animals are occasionally available, but extreme care must be taken to prevent disease introduction and taxonomic misidentification, as morphologically identical tetraploid species are known to occur in overlapping ranges. Most laboratories maintain outbred frogs; however, a few inbred strains of *X. tropicalis*

Table 1
Suppliers of *Xenopus tropicalis* adult frogs

Supplier	Webaddress	Distribution area
NASCO	http://www.enasco.com/c/xenopus/Xenopus+tropicalis	Worldwide
Xenopus express	http://www.xenopus.com/products.htm	Worldwide
US *Xenopus* resource center	http://www.mbl.edu/xenopus/index.html	North America etc.
European *Xenopus* resource centre	http://www.port.ac.uk/research/exrc/	Europe etc.

Current information about resource center stocks is provided by Xenbase (http://www.xenbase.org/other/obtain.do), but *X. tropicalis* are almost always readily available

exist (see Notes 1–3). *X. tropicalis* from different locations are kept under the same conditions and no significant differences in robustness or embryo quality have been reported (for some established suppliers, see Table 1).

2.2. Re-circulating Systems

There are two major suppliers that we have used successfully, Tecniplast and Aquatic Habitats. Systems from both suppliers are available either as large numbers of tanks linked to single water treatment units or as smaller stand alone systems. Which is best for a specific colony depends on the balance needed between the time taken to clean filters, a process which is much more efficient with a single treatment unit, and the risk of disease spreading between tanks.

The tanks in these systems can be chosen from a range of sizes all of which have the critical, secure lids and are of a plastic designed to be transparent to allow observation by care staff whilst minimizing the disturbance of frogs by movement around the tanks.

The re-circulated water is treated by a coarse filter, biofiltration, fine filter, activated charcoal, and ultraviolet light prior to being returned to the tanks. Around 10 % of water is changed automatically each day whilst conductivity, temperature, and pH are monitored and controlled automatically. Care staff can be called automatically on their mobile telephones if system parameters deviate from their norms.

2.3. Fill and Dump Systems

Most of these systems are home made and may be as simple as plastic baths with mesh lids or can be custom built. In the latter case, people have incorporated sloping floors to the tanks and special valves to aid drainage. Whatever the details of the tanks, it is vital that the lids are secure to prevent escape and that the water temperature is carefully regulated, particularly that the input water is free of chlorine and chloramine and that it is at the temperature of the dumped water.

Table 2
Marc's Modified Ringer's (MMR) solution

Chemical	1× Concentrations	20× Concentrations	For 20× stock solutions
KCl	2 mM	40 mM	40 mL of 1 M solution
$MgSO_4$	1 mM	20 mM	40 mL of 0.5 M solution
$CaCl_2$	2 mM	40 mM	40 mL of 1 M solution
NaCl	0.1 M	2 M	116.8 g
HEPES	5 mM	100 mM	23.832 g

Dissolve in 800 mL, adjust pH to 7.4 and take the volume to 1 L. Autoclave and store at room temperature

2.4. Food

1. Horizon XP 23 pellets (Skretting, Wincham, UK).
2. Reptomin (Tetra, Southampton, UK).
3. Sera Micron (Sera GmbH, Heisenberg, Germany).
4. Tropical flake (Aquarian, Marts Fishcare Europe, West Drayton, UK).

2.5. Sea Salt

Tropic Marine. Available from almost all Marine Aquarium specialist pet shops, we use Tropical Marine Centre (http://www.tmc-ltd.co.uk).

2.6. Bicarbonate

Sodium bicarbonate (Mistral Chemicals, Antrin, Northern Ireland).

2.7. Marc's Modified Ringer's (MMR) (18)

0.1 M NaCl, 2 mM KCl, 1 mM $MgSO_{4,}$ 2 mM $CaCl_2$, 5 mM HEPES (pH 7.8), 0.1 mM EDTA—most current formulations of MMR omit EDTA and are adjusted to pH 7.4 (see Table 2)

1× MMR—50 mL 20× MMR in 1,000 mL, pH 7.2–7.4, add 1 mL gentamycin (50 mg/mL). Gentamycin is added to a final concentration of 50 micrograms/mL to retard microbial growth. We keep a gentamycin 1,000× stock at 4 °C.

0.4× MMR—20 mL 20× MMR in 1,000 mL, pH 7.2–7.4, add 1 mL gentamycin (50 mg/mL).

0.05× MMR—2.5 mL 20× MMR in 1,000 mL, pH 7.2–7.4, add 1 mL gentamycin (50 mg/mL).

2.8. Human Chorionic Gonadotrophin

Chorulon (Intervet/Schering-Plough Animal Health, Milton Keynes, UK).

2.9. Software

Filemaker Pro (Filemaker, Santa Clara, California, USA).

3. Methods

3.1. Transporting and Delivering *X. tropicalis*

Moving *X. tropicalis* between labs is straightforward providing careful planning takes place. The necessary health checks and paperwork preparation must be arranged in advance, packaging obtained, and travel arrangements made with an experienced animal shipper. For short journeys (less than 10 h), we normally keep the frogs in water from their tanks in 25 L buckets ~1/3 full. These buckets have firmly clipped-on lids with 7 mm holes drilled both in the lid and in the sides of the bucket just below the lid to allow the frogs to breathe when the buckets are stacked. For longer journeys or those that include air travel, we pack the adults in wet moss in plastic containers (27 cm × 14 cm × 17 cm, max 10 frogs) with small breathing holes. It is critical not to recycle packaging between different species of *Xenopus*, as many microorganisms are benign in one species but infectious and pathogenic in the other. These plastic containers are themselves packed in a polystyrene box and cardboard box, both having breathing holes. To maintain the temperature, pre-warmed (or cooled) gel packs are added inside the polystyrene box. Packed like this, *X. tropicalis* can arrive in good condition even after journeys lasting 3 days in weather conditions that are not extreme. We check the predicted weather for the journey and occasionally postpone sending frogs if extremes of heat or cold are predicted at either transit or destination airports. Your shipper may have specific requirements for labeling of the boxes and this too should be checked well in advance. We have found that it is always advisable to arrange the shipment for the beginning of the week so that staff are on site to receive the frogs even if delays occur.

When frogs arrive it is important to know what feeding regime they had been kept under until that point. The supplier will normally be happy to provide a sample of the food that they are using and keeping to the same frequency of feeding as the frogs had previously experienced seems to prevent the occasional loss of embryo quality that can occur when frogs are moved. Changing to a new feeding regime should be done over a period of several weeks. Sperm and embryos may also be sent to avoid these problems (see Notes 4 and 5).

Since *X. tropicalis* may harbor asymptomatic infections which bloom after shipping stress, it is sensible to quarantine newly arrived stocks for a period of 3 months. A common approach is to segregate new arrivals from the main colony in either a separate room or at least rack of tanks with entirely separate nets, buckets, etc. to avoid transfer of contaminated water. People operate in this area only at the end of the working day so that there is no chance of pathogens being re-introduced to the main colony. Whilst in quarantine, the frogs need to be observed frequently for loss of weight, appetite, or skin condition. As more is discovered about the diseases to which *X. tropicalis* are prone and tests for them are developed, it may in time be possible to test the animals for specific diseases during this period. We currently do this routinely for *Batrachochytrium dendrobatidis* (below), but have not found it in *X. tropicalis*, only *X. laevis*. Frogs that show any sign of disease should be culled humanely or, if they are very valuable, then embryos should be produced by in vitro fertilization and the resulting adults added to the main colony, at least minimizing the chances of horizontal transmission of disease.

3.2. Keeping Adult Frogs in Re-circulating Systems

The systems we have are made either by Tecniplast or Marine Biotech. The parameters that we use for systems containing *X. tropicalis* are listed below:

- Temperature: 25.5 °C (range 24–26 °C). It is important that the temperature for adult frogs does not exceed 26 °C, since we and others have found that this causes a long-term decrease in egg/embryo quality without affecting the health of the adult frog. Temperatures of 21–24 °C are tolerated but are likely to result in depressed immune function. Larval growth may be more rapid at temperatures up to 30 °C. Sharp temperature fluctuations, e.g., when replenishing dump-and-fill systems with water more than 1 °C different from the tank content, can be extremely stressful to both adults and tadpoles and must be avoided.
- pH: 7.9 (range 7.5–9.0).
- Conductivity: 500–1,200 μS (conductivity as high as 1,500 μS is used by some laboratories), controlled by addition of aquarium sea salt.
- Water: Mains water (naturally hard in Portsmouth) or micro-filtered water.
- Tank volume: 2, 5, 10, 27, 40, or 400 L.
- Stocking Density: ~1–2 adults per liter (in the EU directive animals of snout to vent length <6 cm must have a minimum water depth of 6 cm with the first animal having >160 cm^2 of surface area and an additional >40 cm^2 for each other animal in the tank; for frogs of 6–9 cm in length the depth must be at least 8 cm, the first animal must have >300 cm^2 of surface area and each additional animal >75 cm^2).
- Flow rate: 1.33 L per minute.
- Lighting: fluorescent lights on a 13 h light/11 h dark cycle with night-light.
- 10 % of the water is exchanged every 24 h.
- Environmental enrichment: Dark-colored plastic guttering and downpipe tubes cut into approximately 20 cm lengths provide shelter that some frogs take advantage of. Artificial weed made from plastic gardening bags cut up and secured around a pebble

with a cable-tie also provide shelter and somewhere to sit partially out of the water, which again is behavior exhibited by some animals. The "weeds" are disposable and the pipes can readily be sterilized by autoclaving. Some plastics may leach endocrine disruptors, which are likely to affect gender ratios.

- Frogs in a re-circulating system are best fed small amounts of food frequently to minimize the uneaten food that the filters have to clear.

3.3. Keeping Frogs in Fill and Dump Systems

- Water temperature (as above) can be maintained by placing tanks in a heated room, by placing thermostatically controlled heat mats under each tank or by placing combined heater thermostats directly in the water. Place mesh around immersed heaters to prevent frogs from touching it directly and burning themselves.
- A central reservoir of water is needed to provide a continuous supply of water at the right temperature, pH, and conductivity from which chlorine and chloramine have been removed, either by standing or by using commercially available products.
- The frequency of water changes needs to be a compromise between minimizing the disturbance to the frogs and maximizing water quality. We test water quality weekly (nitrogen products and pH) using commercial aquarium kits in order to optimize the timing of water changes. The time between changes varies significantly depending on stocking density and feeding frequency.
- Frogs kept in fill and dump systems are best fed to satiety some little time prior to a water change.

3.4. Feeding

The regime below applies to animals in re-circulating systems. In fill and dump aquaria, the frogs are fed to satiety using the same food shortly before a water change. Frogs tend to ignore food that has been in the water for more than a few minutes.

- Adults: Horizon XP 23 pellets-fed 2–3 times per day, two pellets for each frog per feeding (5 days each week) or ~3 Reptomin sticks/frog (3 days each week), supplemented with whole fish flake 1×/week.
- Froglets: Horizon XP 23 pellets-fed 2–3 times each day, two pellets each and fish food (tropical flake; 7 days each week).
- All frogs may be given occasional supplements of lamb's heart or similar, cut to the appropriate size. A varied diet is always chosen for frogs with weight loss, or those that are observed to fail to compete for food. It has recently been reported that *X. tropicalis* tadpoles can fix on the food that they are fed after metamorphosis and so become difficult to feed thereafter. This suggests that a widely available food source should be chosen by frog users and suppliers to standardize on.

3.5. Health

One of the main health issues for *X. tropicalis* users is having frogs of very high quality so that the embryos they produce are of an excellent standard. This requires close monitoring of the recovery of female frogs between egg production cycles. Females can be ovulated up to six times each year, with a minimum of 6 weeks between ovulations depending on local regulations; males can be used more frequently. We ask ourselves the questions below when deciding if a frog is suitable for re-use:

- General appearance: with your experience do you consider it likely to lay well?
- Skin texture: is it smooth and free from abrasions and redness?
- Body shape: is this consistent with this frog's history (known to be a slim or plump frog?).
- Body weight: if it is possible to record body weight then this is a useful additional indicator of recovery; has this increased since the last use, has it returned to this animal's normal range?

3.6. X. tropicalis Diseases in Laboratory Colonies

Most health issues with lab-bred *tropicalis* stem from suboptimal husbandry regimes, with stress resulting in decreased resistance to normally benign microorganisms. In low-metabolism animals such as frogs, symptoms may appear months after a stress or infection has occurred, and secondary opportunistic infections are often present in dead or morbid specimens, further complicating diagnosis. It is extremely important to keep *tropicalis* and *laevis* colonies completely isolated, as many microorganisms that are benign in one species may be acutely pathogenic in the other.

Several species of Mycobacteria (which may be zoonotic and capable of producing skin lesions in human) have been identified in *tropicalis* colonies, often in healthy animals/water but in some instances resulting in catastrophic mortality, as no treatments are available. In one report, symptoms included a loss of diving reflex, bloating, and ulcerative skin lesions. The causative agent was initially characterized by Trott et al. (19) and named *Mycobacterium liflandii* (20); it has also been found in mainland Europe (21) in frogs imported from the US. Like many diseases of *Xenopus,* it is found in asymptomatic animals when they are housed in good conditions. A number of other mycobacterial species have been reported to infect *X. tropicalis* (e.g., (22)).

Much less dangerous to *X. tropicalis* is the chytrid fungus *B. dendrobatidis.* While this pathogen is fatal to many amphibian species, *X. tropicalis* can survive repeated high dose infections (23), although there are also cases where it has been pathogenic in colonies kept at high density (24). We have tested the prevalence of *B. dendrobatidis.* in *X. tropicalis* colonies from many European labs and cannot detect it using qPCR, although parallel testing of *X. laevis* colonies found *B. dendrobatidis.* in all but one lab

(Coxhead and Guille, unpublished). These data suggest that this fungus is not a significant risk to *X. tropicalis*, but it remains to be determined whether *tropicalis* can act as a reservoir for the pathogen. Since *B. dendrobatidis.* is thought to be one of the primary causative agents in the worldwide decline of many wild amphibian populations, transportation and sale of potential carrier species such as *Xenopus* and *Rana* are expected to come under increasing regulatory pressure.

3.7. Producing Embryos

Embryos can be produced either by natural mating or by in vitro fertilization; the latter is used when large numbers of 1 or 2-cell embryos are required synchronously, for example for transgenesis. Since injection of frogs and squeezing are procedures that require licenses in the UK and in many other countries, it is essential that prospective frog users check their local regulatory requirements. Our protocol is shown below:

- Females are primed by an injection of human chorionic gonadotrophin (HCG) (10 IU, 0.1 mL) into the dorsal lymph sac using a 30 g needle, 1–3 days before the eggs are required.
- A boosting dose of 100 IU of HCG (0.1 mL) is given into the dorsal lymph sac using a 30 g needle as early as possible on the day the eggs are required.
- Egg laying starts normally 3–5 h after the boosting dose and may last for several hours.
- Egg collection by squeezing (see below) should not be repeated until at least 45 min has passed.
- To provide testes for in vitro fertilisation males are sacrificed by terminal anesthesia using MS222.
- Testes are removed, blood vessels dissected away, and the testes stored in L15 supplemented with 10 % calf serum at 14 °C.
- Testes should be used as quickly as possible since there is a dramatic loss in fertility if the testes are stored for any length of time.
- Mash 1 testis in L15/10 % calf serum (0.2 mL) and add to squeezed eggs in a Petri dish using a pipette.
- Shake dish/or stir with pipette to ensure good mixing.
- After 5 min flood with 0.05× MMR.
- All solutions we use are pre-warmed to 25 °C.
- Some labs use solutions at 4 °C.

3.8. Squeezing Eggs from the Female Frog

This procedure should only be performed by trained individuals, since these small frogs are relatively fragile and should be treated gently. Squeezing seems to result in higher mortality than natural mating. If possible, only squeeze the amount of eggs required for the experiment, rather than emptying the frog.

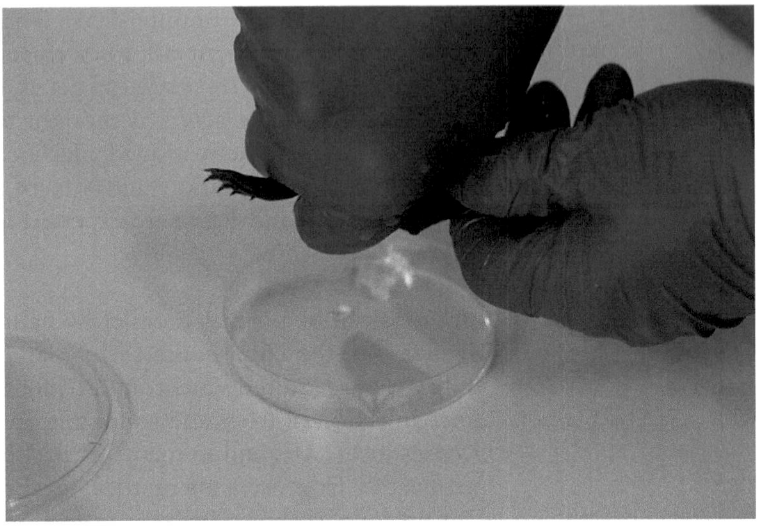

Fig. 1. The hold used to squeeze eggs from a female frog. The cloaca is placed above a Petri dish and the *left* back leg held forward gently using the *right* forefinger. Gentle pressure is applied to the abdomen with the *right* thumb and may be augmented with further gentle pressure with the *left* thumb and forefinger.

While the female frog is still under water, grasp in one hand, the head facing the operator, the index finger between the hind legs, but more to the left of the frog, (therefore, the right of right-handed operator) so that the cloaca is exposed.

After this initial manipulation, the frog is positioned so it is facing upwards, so that the cloaca is the lowest point of the frog's body. With gentle pressure on the abdomen, eggs will be extruded from the cloaca into a receptacle, usually a Petri dish. It is important that the dish remains dry or contains high salt buffer to ensure that eggs are not prematurely activated. The production of eggs can be increased by gentle pressure from the index finger and thumb of the opposite hand (see Fig. 1).

3.9. Natural Mating

Male *X. tropicalis* are stimulated to mate by priming and boosting injections identical to those used to induce ovulation in females, the same doses at the same times (see above).

- Males and females are kept apart until after the boosting doses.
- A pair is isolated in the main tank system (in a tank where the water flow is reduced or stopped altogether), or placed in individual glass or plastic tanks.
- After sufficient eggs are laid, the frogs are removed to a fresh tank and the eggs either left in the tank to grow on or removed to Petri dishes using nets for experimental work.

3.10. Raising Tadpoles

The care of young tadpoles is, in our hands, the most difficult period of the *X. tropicalis* life cycle and tadpole survival can easily be compromised if conditions are not ideal.

Eggs are initially squeezed into 90 mm plastic Petri dishes and fertilized as described above. Gastrulation of dejellied embryos is easily affected by overcrowding/sticking. Plastic dishes should be coated by rinsing with 0.1 % BSA to reduce sticking of dejellied embryos. Fertilized embryos should be kept at no more than 100/dish and spread to minimize touching each other. From hatching to a few days old, the tadpoles are kept in Petri dishes in an incubator in 0.05 × MMR containing Gentamycin (100 μg/mL) at 25 °C. As the embryos become larger, they are moved to glass Petri dishes of 200 mm diameter. We try not to overcrowd them and spread embryos from the same squeeze between a few Petri dishes (*c.* 100 embryos per dish). Fresh media at the same temperature should be exchanged daily and any dead tadpoles removed, or live tadpoles transferred to a fresh dish.

Feeding tadpoles (5 days post-fertilization) are kept in standing water (0.75 g/L sea salt (e.g., Tropic Marine for keeping marine fish) in distilled water) in 5 L tanks. It is important to remove tadpoles from antibiotic-containing media when they start feeding. ~50 % water is changed daily, taking care to equilibrate the temperature of the fresh water to that of the tank beforehand, to avoid sudden temperature shocks. For very young tadpoles, it is easier to start with a low water level and top up daily. After 2 weeks, an air bubbler is used to oxygenate the standing water gently.

Tadpoles are fed on Sera Micron (Sera; Heisenberg, Germany) powder, which is strongly recommended for standing water systems since it does not foul water quickly and can "bloom" in the presence of bright light, producing a healthy self-sustaining green plankton on which tadpoles can feed. During the first 2 weeks, very small amounts of Sera Micron are added several times a day, enough to maintain faintly green water. The amount of Sera Micron is increased for older tadpoles so the water maintains a deep green color. Frequent feeding (up to ten times each day) ensures that food is always available in the water column. Metamorphosing tadpoles and froglets are moved to flow-through systems and fed whole fish flake and crushed Reptomin sticks (pieces that are big enough for froglets to grasp with their forelimbs but small enough to swallow) daily. Metamorphs benefit from floating platforms in the tank on which they can rest out of the water. As the metamorphs become larger, they will eat the normal trout pellets/Reptomin.

3.11. Sexing *X. tropicalis*

X. tropicalis are more difficult to sex than *X. laevis* and even trained staff can make errors especially with immature frogs. There are subtle differences in the frogs, as follows:

(a) Mature male *X. tropicalis* usually have rough nuptial pads on the inner forelimb to aid the amplexus grip. This may appear as a darker (sometimes lighter) stripe, usually more distinct after HCG priming. The rough texture of the nuptial pad is the most reliable male feature.

Fig. 2. The smaller adult male (on the *right*) and female *Xenopus tropicalis*.

(b) Males tend to appear slimmer and smaller than females, which have a characteristic triangular shape if healthy and well-fed.

(c) The cloaca is more pronounced in females than in males, especially after hormone treatment.

(d) Males (Fig. 2, right) tend to be considerably more skittish, jumpy, and difficult to catch.

3.12. Identifying Animals and Record Keeping

The use of *X. tropicalis* for multi-generation genetic or transgenic studies requires careful recording of the genetic history of each animal or group of siblings; in addition, good husbandry and legal requirements in some countries need similar records. The method below has been used to keep records for a complex multi-generation colony.

The offspring of a specific mating or pair of parents are assigned a unique stock number as well as a genotype describing the stock (which may not be unique). Information about each stock is kept in a centralized database (e.g., Filemaker Pro) and includes the genotype of the stock (including the generation, e.g., F2, F3), date of birth/fertilization (DOB), maternal and paternal stock number and genotype, any local regulatory information including procedure number in the UK, and the number of animals. Some of this information (stock number, genotype, DOB, and license information) may be printed on adhesive waterproof tank labels to identify the tanks in which the frogs are housed. Labels have space for additional information, for example, the mating date of the frogs, so that the animals can be rested for an appropriate period prior to re-use.

Occasionally, frogs from a particular stock need to be identified and isolated as individuals, e.g., when a clutch of offspring contains both non-carriers and carriers of a particular mutation. This information can be added to the stock number (e.g., T300-C.1 = mutation-carrying individual 1 of the T300 stock, ~half of whom are non-carriers). Sometimes stocks with the same genotype, e.g., F2 and F3 transgenics from the same founder, can be combined to save space in the frog room, but that information is added to the database so the genetic history of stocks is not lost. Transgenic offspring carrying the same construct, but from different founders, should not be mixed.

One of the challenges of working with *X. tropicalis* is the similarity in patterning between individuals limiting the scope for identifying frogs visually using a photographic database as for *X. laevis*. We have collaborated extensively with a machine vision company to test whether the hue and patterns on *X. tropicalis* could be identified using state-of the-art technology but it proved impossible (P. Coxhead and M.J. Guille, unpublished data). As a result, we are investigating microchipping the animals. At least two labs (Harland and Mead) are using this technique routinely in *X. tropicalis* and have protocols on their lab websites (see *also* Chapter 6). Frogs can also be kept in smaller (2 L or 5 L) tanks as individuals or in pairs, but this may have negative effects on the welfare of the frogs, which seem to prefer living in larger groups.

4. Notes

1. The *N* strain derives from wild-caught frogs obtained by Marc Kirschner from Nigeria and subsequently inbred to the F12 generation in the Grainger lab and elsewhere (see Chapter 1). DNA from an F7 *N* frog was used to produce the draft genomic assembly (3), making it the strain of choice when targeting specific sequences, e.g., with morpholino oligonucleotides, since *N* frogs are likely to have fewer variations from the published genomic sequence. The *IC* line originates from frogs collected near Adiopodoume, Ivory Coast, and inbred in the Grainger and Zimmerman laboratories to F12.
2. *IC* frogs and embryos are usually less pigmented than *N*. Other stocks originating from Nigeria and Ivory Coast are not necessarily related to the *N* and *IC* strains. Other inbred lines include *TGB* and *Golden*.
3. Crosses between different strains are useful to generate map-cross frogs rich in interstrain polymorphisms, which facilitate genomic and meiotic mapping studies.

4. The complexity, cost, and stress to adult frogs involved in sending them long distances have resulted in the recent adoption by a number of labs of the practice of sending frozen sperm (25), testes, or even fertilized eggs of *X. tropicalis* lines from one lab to another.
5. Frozen sperm are sent on dry ice whilst fresh testes can be sent in 1× MBS in 50 mL screw capped tubes on frozen gel packs in polystyrene boxes; a similar approach but with 25 °C gel packs can be taken for fertilized eggs.

Acknowledgements

The European *Xenopus* Resource Centre is funded by the Wellcome Trust, BBSRC and NC3Rs. The authors are very grateful to Colin Sharpe for suggesting improvements to the manuscript.

References

1. Burki E (1985) The expression of creatine kinase isozymes in Xenopus tropicalis, Xenopus laevis laevis, and their viable hybrid. Biochem Genet 23(1–2):73–88
2. de Sa RO, Hillis DM (1990) Phylogenetic relationships of the pipid frogs Xenopus and Silurana: an integration of ribosomal DNA and morphology. Mol Biol Evol 7(4):365–376
3. Hellsten U et al (2010) The genome of the Western clawed frog Xenopus tropicalis. Science 328(5978):633–636
4. Wells DE et al (2011) A genetic map of Xenopus tropicalis. Dev Biol 354(1):1–8
5. Goda T et al (2006) Genetic screens for mutations affecting development of Xenopus tropicalis. PLoS Genet 2(6):e91
6. Grammer TC et al (2005) Identification of mutants in inbred Xenopus tropicalis. Mech Dev 122(3):263–272
7. Noramly S et al (2005) A gynogenetic screen to isolate naturally occurring recessive mutations in Xenopus tropicalis. Mech Dev 122(3): 273–287
8. Abu-Daya A et al (2009) Absence of heartbeat in the Xenopus tropicalis mutation muzak is caused by a nonsense mutation in cardiac myosin myh6. Dev Biol 336(1):20–29
9. Geach TJ, Zimmerman LB (2010) Paralysis and delayed Z-disc formation in the Xenopus tropicalis unc45b mutant dicky ticker. BMC Dev Biol 10:75
10. Khokha MK et al (2009) Rapid gynogenetic mapping of Xenopus tropicalis mutations to chromosomes. Dev Dyn 238(6):1398–1446
11. Akkers RC et al (2009) A hierarchy of H3K4me3 and H3K27me3 acquisition in spatial gene regulation in Xenopus embryos. Dev Cell 17(3):425–434
12. Akkers RC et al (2010) ChIP-chip designs to interrogate the genome of Xenopus embryos for transcription factor binding and epigenetic regulation. PLoS One 5(1):e8820
13. Kenwrick S, Amaya E, Papalopulu N (2004) Pilot morpholino screen in Xenopus tropicalis identifies a novel gene involved in head development. Dev Dyn 229(2):289–299
14. Chae J, Zimmerman LB, Grainger RM (2002) Inducible control of tissue-specific transgene expression in Xenopus tropicalis transgenic lines. Mech Dev 117(1–2):235–241
15. Hartley KO, Nutt SL, Amaya E (2002) Targeted gene expression in transgenic Xenopus using the binary Gal4-UAS system. Proc Natl Acad Sci U S A 99(3):1377–1382
16. Ryffel GU et al (2003) Tagging muscle cell lineages in development and tail regeneration using Cre recombinase in transgenic Xenopus. Nucleic Acids Res 31(8):e44
17. Abu-Daya A et al (2011) The secreted integrin ligand nephronectin is necessary for forelimb formation in Xenopus tropicalis. Dev Biol 349(2):204–212

18. Ubbels GA et al (1983) Evidence for a functional role of the cytoskeleton in determination of the dorsoventral axis in Xenopus laevis eggs. J Embryol Exp Morphol 77:15–37
19. Trott KA et al (2004) Characterization of a Mycobacterium ulcerans-like infection in a colony of African tropical clawed frogs (Xenopus tropicalis). Comp Med 54(3): 309–317
20. Mve-Obiang A et al (2005) A newly discovered mycobacterial pathogen isolated from laboratory colonies of Xenopus species with lethal infections produces a novel form of mycolactone, the Mycobacterium ulcerans macrolide toxin. Infect Immun 73(6):3307–3312
21. Suykerbuyk P et al (2007) Mycobacterium liflandii infection in European colony of Silurana tropicalis. Emerg Infect Dis 13(5): 743–746
22. Sanchez-Morgado JM, Gallagher A, Johnson LK (2009) Mycobacterium gordonae infection in a colony of African clawed frogs (Xenopus tropicalis). Lab Anim 43(3):300–303
23. Ribas L et al (2009) Expression profiling the temperature-dependent amphibian response to infection by Batrachochytrium dendrobatidis. PLoS One 4(12):e8408
24. Parker JM et al (2002) Clinical diagnosis and treatment of epidermal chytridiomycosis in African clawed frogs (Xenopus tropicalis). Comp Med 52(3):265–268
25. Sargent MG, Mohun TJ (2005) Cryopreservation of sperm of Xenopus laevis and Xenopus tropicalis. Genesis 41(1):41–46

Chapter 3

Generating Diploid Embryos from *Xenopus tropicalis*

Florencia del Viso and Mustafa Khokha

Abstract

A spectacular advantage of *Xenopus tropicalis* is the ease with which diploid embryos can be generated year round. By the simple administration of human chorionic gonadotropin, an investigator can generate many hundreds of synchronized embryos by *in vitro* fertilization or thousands of embryos from a mating pair. The ability to induce ovulations when desired facilitates many different experiments such as experimental embryology, molecular manipulation of gene products, and genetics.

Key words: *Xenopus tropicalis*, Egg, Sperm, Embryo, *In vitro* fertilization, hCG, Cysteine

1. Introduction

Since the 1950s, *Xenopus laevis* has been used as a model system to study developmental biology in vertebrates. *X. laevis* could be induced to ovulate and produce hundreds to thousands of embryos by the administration of human chorionic gonadotropin (hCG). No longer was experimental embryology seasonal, and the ability to generate embryos whenever needed facilitated the molecular transformation of experimental embryology (1). Induced ovulations and *in vitro* fertilization (IVF) facilitates many experiments such as microsurgery in early embryos (e.g. see Chapter 29), mRNA overexpression (e.g. see Chapter 20), gene knockdown studies using morpholino (MO) antisense oligonucleotides (e.g. see Chapter 8), transgenesis (see Chapters 12, 13, and 14), and more recently, high-throughput sequencing experiments (see Chapter 18).

Subsequently, *Xenopus tropicalis* has emerged as an amphibian genetic system. Its genome is relatively small and diploid (as opposed to *X. laevis*' allo-tetraploid genome) and a genome sequence is available (2). Additionally, many of the methods already optimized for *X. laevis* work well in *X. tropicalis* (3). For *X. tropicalis*, reasons

Stefan Hoppler and Peter D. Vize (eds.), *Xenopus Protocols: Post-Genomic Approaches, Second Edition*,
Methods in Molecular Biology, vol. 917, DOI 10.1007/978-1-61779-992-1_3, © Springer Science+Business Media, LLC 2012

to have large amounts of embryonic material are similar to *X. laevis*; but in addition, large ovulations equate to many meioses, which can greatly facilitate genetic mapping (see Chapter 5).

Here, we describe two ways of obtaining *X. tropicalis* embryos: by a natural mating and by IVF. In a natural mating, male and female frogs are stimulated with hCG injection and placed together for the male to fertilize the eggs laid by the female. In IVF, the male is sacrificed and a sperm suspension is applied to eggs collected from the female. Each has its advantages and disadvantages. Natural matings can yield the most embryos (as many as 5–6,000). Since the fertilization occurs over many hours, the resultant embryos are at a variety of stages. IVF yields fewer embryos (still hundreds to 1,000), but the embryos are synchronized which facilitates microinjection at a specific cleavage stage. Finally, IVF requires sacrificing a male, whereas with natural matings the colony is not depleted.

2. Materials

All solutions are prepared using ultrapure water (prepared by purifying deionized water to attain a conductivity of 18 MΩ cm at 25 °C) and analytical grade reagents. Follow all waste disposal regulations when disposing of waste materials.

2.1. Reagents and Buffers

1. 1/9× MR (Modified Ringer's solution): We keep a stock of 1/3× MR in distilled water and dilute threefold.

   1/3× MR: To make 20 L, add 39 g NaCl (to get 0.033 M), 66.7 mL of 100× MR salts, and 33 mL of 1 M HEPES pH 7.4.

   100× MR salts: 180 mM KCl, 200 mM $CaCl_2$, 100 mM $MgCl_2$.
2. 3% Ficoll in 1/9× MR: Dissolve 3 g of Ficoll in 100 mL of 1/9× MR and autoclave. Store at room temperature (see Note 1).
3. 3% Cysteine in 1/9× MR: Dissolve 3 g/100 mL of cysteine in 1/9× MR. Adjust pH to 7.8–7.9 with 10 M NaOH (see Note 2).
4. 1× MBS (Modified Barth's Saline): Prepare 1× MBS solution by mixing 100 mL of 10× MBS salts with 7 mL of 0.1 M $CaCl_2$, and adjust the volume to 1 L with distilled water. Store at room temperature for 6 months.

   10× MBS salts: 880 mM NaCl, 10 mM KCl, 10 mM $MgSO_4$, 50 mM HEPES (pH 7.8), 25 mM $NaHCO_3$, adjust pH to 7.8 with NaOH, autoclave and store at room temperature for 1 year.

   0.1 M $CaCl_2$: Dissolve 11.1 g of $CaCl_2$ in 1 L of distilled water, autoclave and store aliquots at −20 °C or 4 °C.

5. 1× MBS + 0.1% BSA: Prepare 1× MBS and add 0.1% (w/v) of BSA, filter sterilized and store at room temperature for 2 months (see Note 3).
6. 0.1× MBS: Dilute ten times 1× MBS and adjust pH to 7.8–7.9 with 0.1 M NaOH, within the same day of fertilization procedure (see Note 4).
7. hCG: Prepare stock solution of hCG (Chorulon) to final concentration of 1 U/μL in distilled ultrapure water.
8. 100× Benzocaine: Dissolve 5 g/100 mL in 100% ethanol and store at 4 °C.
9. 1,000× gentamycin: Dissolve 50 mg/mL in distilled water, filter sterilize, and store at 4 °C.

2.2. Supplies

Scissors and tissue forceps, Petri dishes, microcentrifuge tubes, microcentrifuge tube pestle, 30 gauge needles, 1 mL syringes.

3. Methods

All steps of the protocol should be carried out at room temperature.

3.1. Natural Mating

1. Prime 1 male and 1 female frog the day before the eggs are desired. To prime frogs, inject 10 U of hCG for both males and females. We recommend using 30 gauge needles for injections. Inject the hCG subcutaneously into the dorsal lymph sac (avoiding the lateral line sutures) (see Note 5). Place single pairs together in one tank with frog water at room temperature (22 °C) (see Note 6).
2. The day after priming, boost the males with 50 U and the females with 150 U of hCG, 3–5 h before eggs are needed. Alternatively, if you do not need early cleavage stage embryos, frogs can be boosted late in the afternoon and left to lay eggs overnight.
3. Prepare 3% cysteine solution (see Note 2).
4. If early cleavage stage embryos are needed and amplexus is occurring, collect the eggs from the tank using a transfer pipette shortly after the female starts laying eggs. Continue collecting new eggs at regular intervals. Alternatively, if later stage embryos are needed, allow the females to finish laying eggs and collect as many as desired at once. Eggs can be collected in Petri dishes or small Erlenmeyer flasks (see Note 7).
5. De-Jelly: To avoid dilution of the cysteine solution remove as much of the frog water as possible by gently decanting and replace with 3% cysteine solution, pH 7.8–7.9 (see Note 8). De-jellying can be easier if the embryos are placed in an

Erlenmeyer flask and gently swirled, especially if there are many embryos (see Note 9).

6. Periodically, inspect the embryos for removal of the jelly coat under the stereomicroscope (see Note 10).
7. For microinjections, remove the cysteine by washing three times with 1/9× MR and two times with 3% Ficoll in 1/9× MR. Alternatively, if embryos are not being used for microinjections wash three times in 1/9× MR.
8. After microinjections, or if uninjected embryos are used, incubate at the desired temperature (22–28 °C) in a final solution of 1/9× MR + 1× gentamicin to the desired stage of development.

3.2. In Vitro Fertilization

1. Prime 1–2 male and 2–3 female frogs the night before the eggs are desired (see Note 11). To prime frogs, inject 10 U of hCG for both males and females. We recommend using 30 gauge needles for injections. Inject the hCG subcutaneously into the dorsal lymph sac (avoiding the lateral line sutures) (see Note 5). We keep males separate from females overnight. Leave the frogs in frog water at room temperature (22 °C).
2. The day after priming, boost the males with 100 U and the females with 200 U of hCG, 3–4 h before eggs are needed (see Note 12).
3. Prepare 3% cysteine solution and 0.1× MBS (see Notes 2 and 4).
4. Confirm females are laying eggs.
5. Euthanize the male by placing it in 0.05% (1×) benzocaine solution for about 15 min. Once the frog can no longer move upon stimulation, remove and decapitate.
6. While waiting for the male, coat a few Petri dishes with 1× MBS + 0.1% BSA by adding the solution and then removing it immediately after. Also, collect and prepare dissection tools and have all solutions ready.
7. Dissect both testes and place into a 1.5 mL microcentrifuge tube filled with 0.5 mL of 1× MBS + 0.1% BSA (see Note 13).
8. Manually harvest eggs into 1× MBS + 0.1% BSA precoated dishes (diameter: 5 cm). Hold the female so the legs are pulled anteriorly with the cloaca at the vertex (Fig. 1). Gently squeeze the lower back and abdomen to expel the eggs. Usually, with a good female, simply holding her in this position will express hundreds of eggs. Examine the eggs to verify that they have a normal shape, a nicely pigmented animal pole and a pale vegetal pole (Fig. 2a) (see Note 14).
9. Release the sperm by mincing the testes using a microcentrifuge tube pestle until solution becomes cloudy.
10. Pour sperm on eggs and mix thoroughly with the pestle.

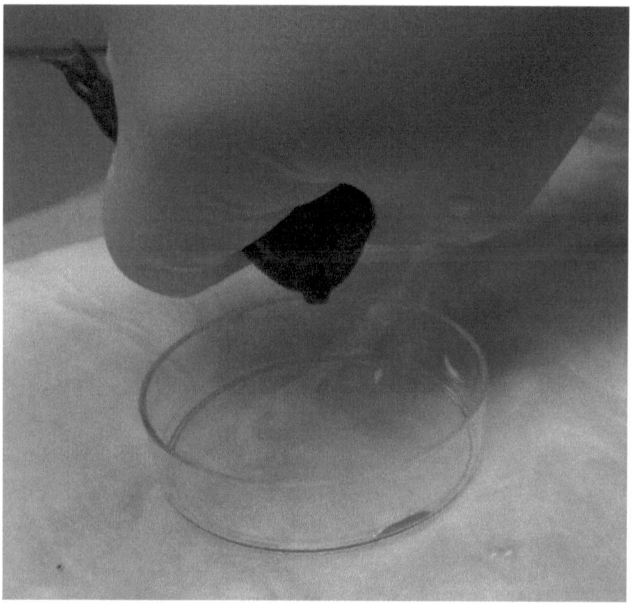

Fig. 1. Holding the female frog to squeeze out eggs. Hold the female so that the legs are pulled anteriorly with the cloaca at the vertex. Keep the cloaca low so that the eggs fall into the coated dish. To expel eggs apply gentle pressure on the lower back and abdomen.

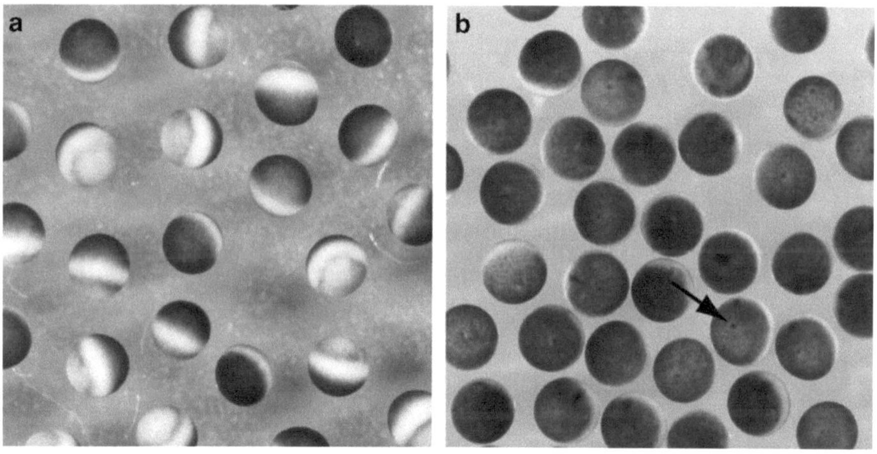

Fig. 2. Differences between unfertilized and fertilized eggs. Fertilized eggs (**b**) rotate with the pigmented animal hemisphere up, while unfertilized eggs are randomly oriented (**a**). The sperm entry point in the animal hemisphere is seen in some of the fertilized eggs as a black spot on the animal pole (*arrow*).

11. Incubate 3 min at room temperature, leaving the Petri dish at an angle to ensure all eggs are submerged in sperm solution.
12. Flood with 0.1× MBS pH = 7.8–7.9 (check pH of this solution THE SAME DAY, see Note 4), just enough to cover the eggs.
13. Incubate 10 min at room temperature (see Note 15).

14. Remove as much of the 0.1× MBS used to flood as possible by decanting and replace with 3% cysteine solution, pH 7.8–7.9 (see Note 8).
15. Swirl the embryos VERY gently until they become loose and separated. Fresh cysteine can be added or replaced to speed up the process (see Note 9).
16. To insure embryos are properly de-jellied (especially in using them for microinjections) visualize them under the stereomicroscope. They should pack tightly with each other without any space between them (see Note 10).
17. Remove the cysteine by washing three times with 0.1× MBS and two times with 3% Ficoll in 1/9× MR and leave in 3% Ficoll if using embryos for microinjection.
18. Alternatively, if embryos are not being used for microinjections they can be washed 3–4 times in 1/9× MR and left in this solution.
19. 1–2 h after microinjection or IVF, sort embryos to a new Petri dish with 1/9× MR+ 1× gentamicin.
20. Incubate at the desire temperature to raise embryos (22–28 °C).

4. Notes

1. Ficoll powder takes a long time (1 h or more) to dissolve in 1/9× MR.
2. Cysteine solution should be prepared fresh on the same day as the fertilization procedure. Otherwise, the cysteine precipitates.
3. We usually prepare not more than 100–200 mL of this solution to avoid long storage times.
4. 0.1× MBS solution should be prepared the same day, as the pH tends to fall during storage because of its weak buffer strength. It can be stored, but the pH should be carefully adjusted to 7.8–7.9 the same day of fertilization. In our experience, IVF rates are poor at a low pH.
5. To hold frogs for hCG injection, we recommend using the "iron claw" method. For this, grasp the frog so your index finger is between the legs, and your thumb and middle finger are flanking the legs. The head of the animal will be hidden in your palm. Use your thumb to draw the leg back toward the body so the frog cannot use it to kick. The frog should now be completely immobilized and is ready to be injected. We use small 30 gauge needles to minimize the injury to the frog's skin. Keep the bevel of the needle up and the needle parallel to the surface of the skin to avoid hitting the underlying muscle.

Sometimes when the skin is tough you may need to point the needle perpendicular to the skin to penetrate and then bring the needle parallel for injection. If the needle does not penetrate easily, it is not sharp and should be exchanged. Whenever handling *X. tropicalis*, we recommend wearing gloves (powder-free).

6. In our experience, natural matings work well in frog water, but salinity of the water could affect the fertilization, therefore a more defined salt solution (i.e., 1/9× MR) could be used. To get high fertilization rates, carefully choose males with obvious nuptial pads and females with prominent cloacas.
7. The eggs should not be left in the mating tank longer than 12 h because huge numbers of eggs do poorly in a relatively small container of frog media. In our experience, well-fed frogs usually do not eat the eggs after mating, but this would be another reason to remove eggs promptly.
8. *Xenopus* eggs are coated with thick, sticky protective jelly membranes, the jelly coat, that must be removed in order to easily handle or microinject individual embryos. Fortunately, a simple cysteine solution can gradually remove the jelly coat for many embryos at once and takes only a few minutes. It should take about 5–6 min to completely de-jelly and wash away cysteine. If it takes longer, the pH of the cysteine solution is likely to be too low. It can be sped up, but in our experience, *X. tropicalis* embryos do not tolerate even brief periods of pH > 8. If embryos are being used for microinjections, we recommend not hurrying through this step, as injecting improperly de-jellied embryos is quite inefficient and frustrating.
9. Embryos being de-jellied before the first cleavage may develop secondary axes if the embryos are swirled too vigorously. Therefore, if the de-jellying is being done at the one cell stage, very gentle or no swirling is recommended. For embryos past the first cleavage (e.g., embryos from a natural mating), we swirl the embryos gently in a flask, which greatly speeds up the process.
10. Depending on the purpose, not all the jelly coat needs to be removed in order to sort through the embryos and collect what is needed. If embryos are not being used for microinjection, you can partially de-jelly the embryos by treating for a shorter time in cysteine. In this case, the jelly coat should still be visible around the embryo under the stereomicroscope, but embryos should appear separate from each other in a monolayer. However, for microinjection, the entire jelly coat needs to be removed to allow for penetration of the microinjection needles. Completely de-jellied embryos should pack tightly with each other without any space between them. Completely de-jellied embryos can be sticky and stick to the bottom of the dish.

This can be avoided (although it is not essential) by adding 0.1% BSA or coating the dish with 0.1% BSA in 1/9× MR.

11. To get high fertilization rates, carefully choose males with obvious nuptial pads and females with prominent cloacas. We prime and boost two males, in case IVF with the first one does not work so we can repeat the procedure. We also prime and boost two females in case one does not lay eggs or lays abnormal ones. In *X. tropicalis*, one male is used for only one IVF round (one collection of eggs), whereas with *X. laevis*, one male is used for several IVF rounds, as testes are bigger and sperm can be stored. This can be expensive if you need to buy males. In our lab we raise our own colony of males to supply for IVF protocols.
12. We use a higher hCG dose to boost for IVF (compared to natural matings) to assure females lay in 3–4 h. With less hCG, it could take up to 6 h.
13. Once the male is completely anesthetized, remove the testis through an incision in the abdominal wall. Testes lie deep in the back of the abdomen flanking the aorta. Carefully put testes in the microcentrifuge tube, trying to remove surrounding tissue while avoiding crushing the testes. Alternatively, testis could be incubated with 1/9× MR + 0.1% BSA and plates for eggs could also be coated with this solution. Fertilization rates are comparable to the ones using 1× MBS + 0.1% BSA. We use a pair of testis to fertilize about 500–1,000 eggs to ensure a good fertilization rate. We use dissected testes immediately; however, we have not tried to store *X. tropicalis* testes for long periods of time.
14. The eggs are quite sticky and will run down your fingers and down the frog's legs so try to keep the cloaca at the lowest point so the eggs fall out and land in the dish. Observing the eggs quality is important in order not to waste sperm on poor quality eggs. After at least an hour of rest after the first squeeze, the females can be re-squeezed for a second fertilization if desired. In our experience, the first squeeze usually produces the highest quality eggs, although a second squeeze can be perfectly fine as well. If the first fertilization does not work we verify the pH of 1× MBS + 0.1% BSA and of 0.1× MBS buffers is within the 7.8–7.9 range. To repeat an IVF after re-squeezing a female, we sacrifice another male.
15. During these 10 min, fertilized eggs should contract their animal pole. This can be a useful indicator of the fertilization rate. After 30 min another measure of fertilization rate is turning: embryos tend to turn with the pigmented animal hemisphere up, while unfertilized eggs are impartial (Fig. 2b). One can also check the egg for a sperm entry point in the animal hemisphere. It will appear as a white/black spot on the animal

pole. While these are all useful measures of the fertilization rate, in our hands they are not always entirely predictive. Sometimes we see none of these indicators and still get fertilization.

Acknowledgements

Many thanks to the Grainger lab for their protocol and John Gerhart for help *with X. tropicalis* IVF at the very beginning. Also many thanks to the students and instructors of the CSH *Xenopus* course which inspired testing many different IVF protocols.

References

1. Fraser SE, Harland RM (2000) The molecular metamorphosis of experimental embryology. Cell 100(1):41–55
2. Hirsch N, Zimmerman LB, Grainger RM (2002) Xenopus, the next generation: *X. tropicalis* genetics and genomics. Dev Dyn 225: 422–433
3. Khokha MK et al (2002) Techniques and probes for the study of *Xenopus tropicalis* development. Dev Dyn 225:499–510

Chapter 4

Navigating the *Xenopus tropicalis* Genome

Ira L. Blitz

Abstract

The frog *Xenopus laevis* has for more than 60 years served as a model system for the study of vertebrate embryogenesis, molecular and cell biology, and physiology. Recently, there has been great interest in the related species *Xenopus tropicalis*, in part because it is diploid, unlike the allotetraploid *X. laevis*, and therefore amenable to forward genetics, adding to the strengths of the frog system. The genome sequence of *X. tropicalis* was published in 2010 and this resource is facilitating rapid progress in applying transcriptomic, genomic, proteomic, and systems biological approaches for which the *Xenopus* system is well suited. However, the availability of the primary nucleotide sequence is only the first step in using the genome. Accessing information embedded in the genome sequence requires well-annotated genes and knowledge of how to navigate the data. The current chapter provides a step-by-step guide for the novice to finding genes of interest in the current genome assembly, supplemented with detailed notes to improve understanding. Several publically available internet-based tools for examining the *X. tropicalis* genome are discussed, with special emphasis placed on the examination of synteny. Accurate determination of gene identity is enhanced by examination of orthology relationships with other organisms and thus synteny is a powerful tool. This chapter provides an access path into the *Xenopus* genome to enhance the researchers' ability to manipulate the organism.

Key words: Synteny, Genome browser, Gene model, Metazome, Ortholog, Xenbase

1. Introduction

The *Xenopus tropicalis* genome was sequenced to ~7.6-fold coverage by the Joint Genome Institute (JGI) and the most recent published assembly is known as version 4.1 (1). A more recent assembly, v7.1, has been released online (http://www.xenbase.org/fgb2/gbrowse/Xenopus_tropicalis_7_1/?source=Xenopus_tropicalis_7_1), but is currently being annotated, and therefore this chapter will focus on the better characterized v4.1 build. Assembly 4.1 consists of 19,759 DNA sequence fragments, or scaffolds, with scaffold numbers ascending according to decreasing sequence length. Half of the genome is located on the ~270 largest scaffolds. The genome

Stefan Hoppler and Peter D. Vize (eds.), *Xenopus Protocols: Post-Genomic Approaches, Second Edition*,
Methods in Molecular Biology, vol. 917, DOI 10.1007/978-1-61779-992-1_4, © Springer Science+Business Media, LLC 2012

has an estimated minimum of approximately 20,000 predicted protein-coding genes (1) with, at the time of writing, 4,711 genes containing some manual curation and an additional 11,133 solely having machine (automated) annotation (http://www.xenbase.org/other/statistics.do). Therefore, less than 25% of all genes have been manually confirmed by *Xenopus* community annotators and a minimum additional ~20% have not even been machine annotated. Furthermore, annotations applied in most cases are simply gene names and symbols to machine-generated predictions of transcript structures (gene models), which attempt to depict in the best cases all the components of a transcription unit (e.g., UTRs, exons, introns, translational start and stop signals). Therefore, it is currently still unreliable to search for genes by keyword alone. Consequently, it is important to understand how to find genes in the genome assembly and to establish gene identity with relative certainty before assessing in greater detail the accuracy of the models provided. This chapter introduces the application of several publically accessible internet-based tools that can be used to address these questions and examples illustrate some of the issues that arise. Methods to identify *cis*-acting elements in the genome are not discussed since there is an excellent review covering this problem by H. Ogino in Chapter 15 of this volume. Methods are also introduced to establish a gene's likely orthology and evolutionary history. Such methods utilize synteny relationships between *X. tropicalis* and other metazoan genomes, which is profoundly useful. Finally, changes to genomic resources are occurring continuously and new assemblies of the *X. tropicalis* genome will become available in the near future. It will be important to navigate new assemblies as they become available, including that of the soon-to-be-completed *Xenopus laevis* genome, and hence additional resources supporting present and future assemblies will also be introduced.

2. Resources

Resources to navigate the *X. tropicalis* genome are publically available from these internet sites:

1. Xenbase, the *Xenopus* community's online resource (www.xenbase.org).
2. JGI's *X. tropicalis* v4.1 genome browser (http://genome.jgi-psf.org/Xentr4/Xentr4.home.html).
3. NCBI BLAST (http://blast.ncbi.nlm.nih.gov/Blast.cgi).
4. NCBI Entrez Gene (http://www.ncbi.nlm.nih.gov/sites/entrez?cmd=search&db=gene).

5. Metazome, a synteny analysis and display tool (http://www.metazome.net/).
6. Genomicus (www.dyogen.ens.fr), an alternative synteny analysis and display tool.
7. UC Santa Cruz (UCSC) genome browser (http://genome.ucsc.edu/).

In the current chapter, greater emphasis has been placed on using JGI's genome browser and Metazome because many researchers may be unaware of the features of these resources. The UCSC genome browser together with Xenbase will likely become the main hubs for future *Xenopus* genome analyses.

3. Methods

3.1. Xenbase Gene Searches

Xenbase is a "one-stop shop" for *Xenopus* information (2, 3) and a detailed discussion of Xenbase features is presented by J.B. Bowes in Chapter 31 of this volume. Xenbase acts as a clearinghouse for gene annotations, providing these to NCBI, and currently houses individual web pages for nearly 16,000 annotated genes. Each gene has a dedicated Xenbase page containing links to data specific to that gene and there is a good chance that one's gene of interest may already have been identified.

In order to perform a keyword search of Xenbase follow this protocol:

1. Navigate an internet browser to www.xenbase.org.
2. Enter the gene symbol or name in the search window (see Note 1).
3. Run the search.
4. Examine the gene(s) found in detail to verify correct identification (see Note 2).
5. Access relevant information regarding the gene, its RNA or protein products, expression pattern, ESTs, publications, etc., by navigating the gene page links provided.
6. If a keyword search fails, select the "Blast Frogs" link in the menu bar. Paste sequence corresponding to the gene of interest into the window and adjust the alignment program and database pull-down menus for the type of BLAST search to be performed (see Subheading 3.2 for more detail on BLAST methods). Run the BLAST search. The resulting BLAST matches contain direct links to the corresponding Xenbase gene pages.

3.2. Gene Searches Using JGI's *Xenopus tropicalis* v4.1 Website

Because a large percentage of genes are not yet represented in Xenbase, alternative methods of gene finding are often needed. Two recommended approaches are to initiate keyword or BLAST-based searches within Metazome, detailed in Subheading 3.3, or using JGI's *X. tropicalis* website. JGI keyword and BLAST searches are outlined below.

3.2.1. Keyword Searching

To perform keyword searches using the JGI site follow this protocol:

1. Navigate an internet browser to JGI's *X. tropicalis* v4.1 website (http://genome.jgi-psf.org/Xentr4/Xentr4.home.html) and select the "Search" link in the menu bar.
2. Enter the gene symbol or gene name into the search box (Fig. 1 inset) and use either the default setting of "Gene Catalog" in the "Across" box, or the "All Models" setting within this pull-down menu (see Note 3).
3. Run the search.
4. Examine the output (Fig. 1) to determine which of the genes found corresponds to the gene of interest.

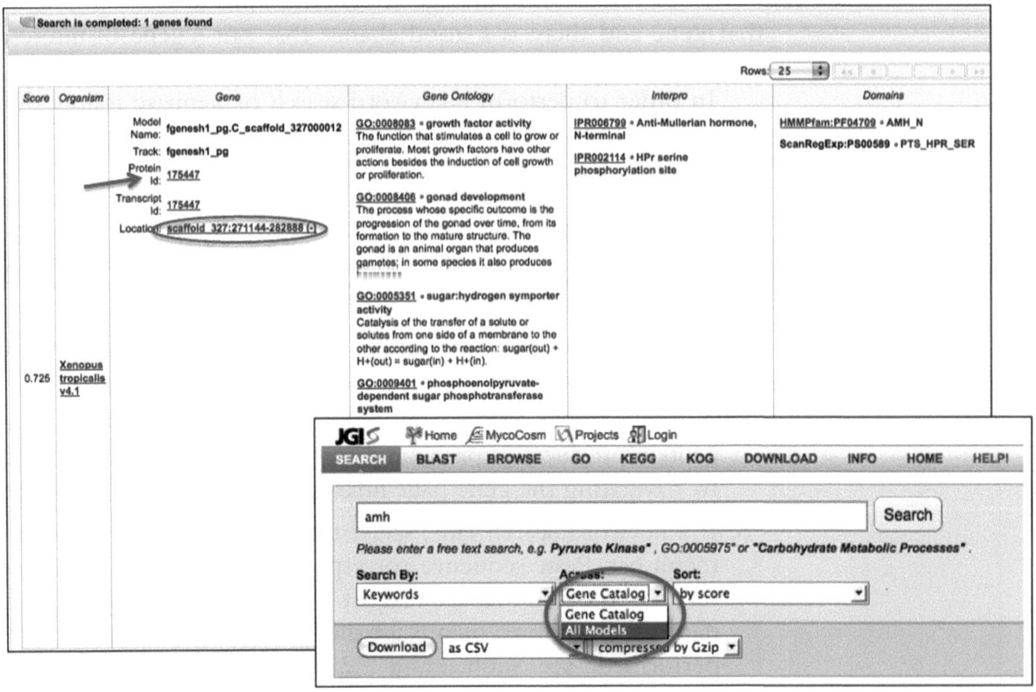

Fig. 1. Keyword searches using the JGI browser. The *inset* shows the search window containing "amh," the symbol for the gene *anti-müllerian* hormone, with pull-down menu to choose between searching the Gene Catalog or All Models. The single gene output from this search is displayed in the outer portion of this figure. Protein id links and scaffold information mentioned in the text are highlighted.

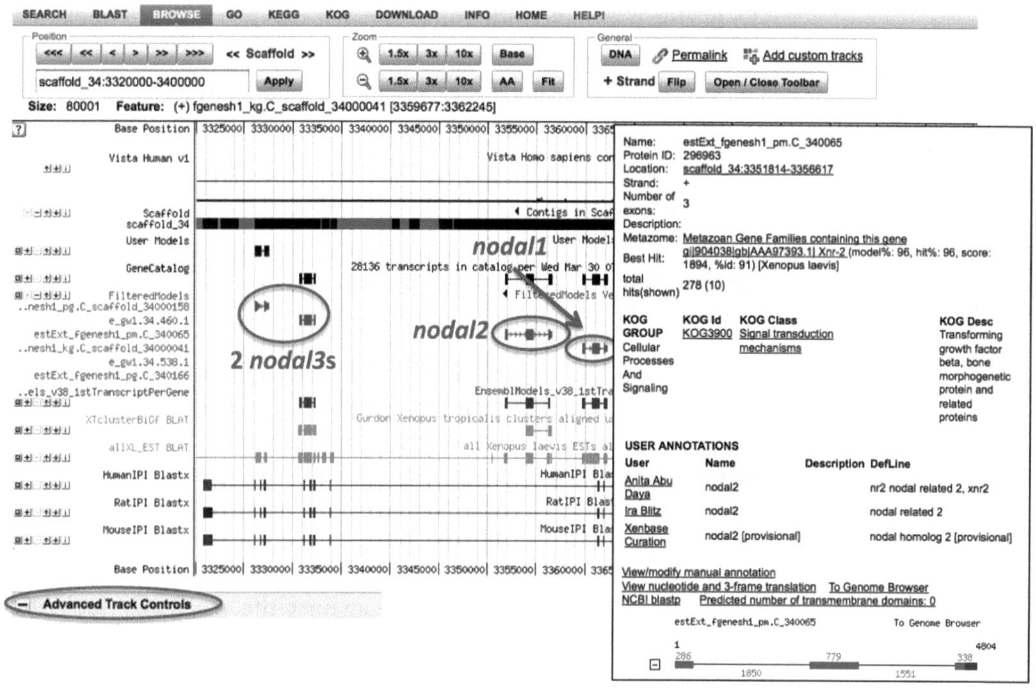

Fig. 2. Using the JGI browser. Selecting the link "JGI XT Genome 4.1" on the Xenbase gene page for *nodal2* (or the "protein id" link on the output page (e.g., Fig. 1) of a JGI keyword search for "nodal") opens the JGI gene page for *nodal2* shown in the *inset*. The outer figure contains the JGI browser view of the *nodal2*-containing region of scaffold 34 after zooming out 10×. Navigating to this browser view can be accomplished either from the Xenbase link "JGI browser" or the JGI gene page link "To Genome Browser" on the respective *nodal2* gene pages. Three other *nodal* genes can be found in the region and are circled. The Advanced Track Controls discussed in the text is also circled at the bottom.

5. Links to data on gene ontology, genomic (scaffold) location, the genome browser (Fig. 2) to view the genomic interval, and other useful information about the gene product are revealed in the search output (Fig. 1) (see Notes 4 and 5).

3.2.2. BLASTn Searching

Other means of searching are required if a gene remains missing following keyword searches, and directly BLAST (Basic Local Alignment Search Tool) searching, the genome is a good option for finding these genes. A number of different BLAST methods exist but BLASTn and tBLASTn are most useful in the current context (see Note 6). BLASTn directly compares specified nucleotide strings to nucleotide sequence databases, such as the genome sequence. In contrast, tBLASTn can be used to search the genome in an indirect manner: by comparing specified protein sequences to amino acid translations of nucleotide sequences. The most effective BLAST-based searches begin with prior knowledge *either* of some *Xenopus* (*tropicalis* or *laevis* will do) nucleotide sequence from the gene of interest, *or* protein sequence, perhaps from another organism but preferably closely related to *Xenopus*, to maximize likelihood of sequence similarity.

To perform a nucleotide query against the *X. tropicalis* nucleotide sequence (BLASTn)

1. Obtain nucleotide sequence for either *X. tropicalis* or *laevis* from a database (see Note 7).
2. Select the BLAST link at the top of any JGI *X. tropicalis* v4.1 page.
3. Enter the sequence (copy and paste) into the upper search window.
4. Run the BLAST search.
5. Evaluate the search output to determine how well the queried sequence matches a genomic region/gene model (see Note 8).

3.2.3. tBLASTn Searching

As knowledge of the primary nucleotide sequence from cDNAs or ESTs is not always available, a next best alternative is to query the genomic sequence using the protein sequence from an ortholog (see Note 9).

To perform a protein query against the *tropicalis* nucleotide sequence

1. Obtain protein sequence from a database such as NCBI (see Note 9).
2. Select the BLAST link at the top of any JGI *X. tropicalis* page.
3. Enter the sequence (copy and paste) into the upper search window.
4. Under the "Search Criteria" section below the BLAST window change the default BLASTn setting to tBLASTn using the pull-down menu.
5. Run the BLAST search.
6. Evaluate the search output to determine how well the queried sequence matches a genomic region/gene model (see Note 10).

3.3. Metazome Gene Searches and Synteny Relationships

Why does one need anything more than protein BLAST searches to demonstrate orthology? Why are synteny tools worth mastering? While the best hits from BLASTp searches most frequently *do* find orthologs, this is not always the case. Further evidence would be to perform the BLAST search in reciprocal fashion but basing corroboration of orthology solely on best BLAST matches has its perils (see Note 11). This is where synteny becomes especially useful. Metazome v3.0 (www.metazome.net) is an internet-based visualization tool for displaying microsynteny, which is defined here as colinearity relationships, or conservation of linkage, between small numbers of genes on a local level. This contrasts with macrosynteny, which is conserved linkage over large chromosomal intervals. Like the other resources, gene finding in Metazome can be performed by either keyword or BLASTp searches. One difference is that BLAST searches are run against *consensus* protein sequences produced from alignments between related ("clusters") genes, roughly corresponding to gene families (for detailed description, see http://www.metazome.net/Phytozome_info.php). The user also preselects organismal groupings to search. There are direct links to Metazome on both JGI and Xenbase gene pages. If a protein is a good match by reciprocal BLAST, and its gene also displays synteny, one can conclude that the gene is orthologous. But, importantly, due to evolutionary genome rearrangements, a lack of evidence for synteny on the micro (or macro) scale does not disprove orthology.

3.3.1. Keyword Searching

To perform a keyword search of Metazome v3.0 (see Fig. 3)

1. Select the node (branch point) corresponding to the phylogenetic grouping to be searched (see Note 12).
2. Select keyword search under "Choose a tool."

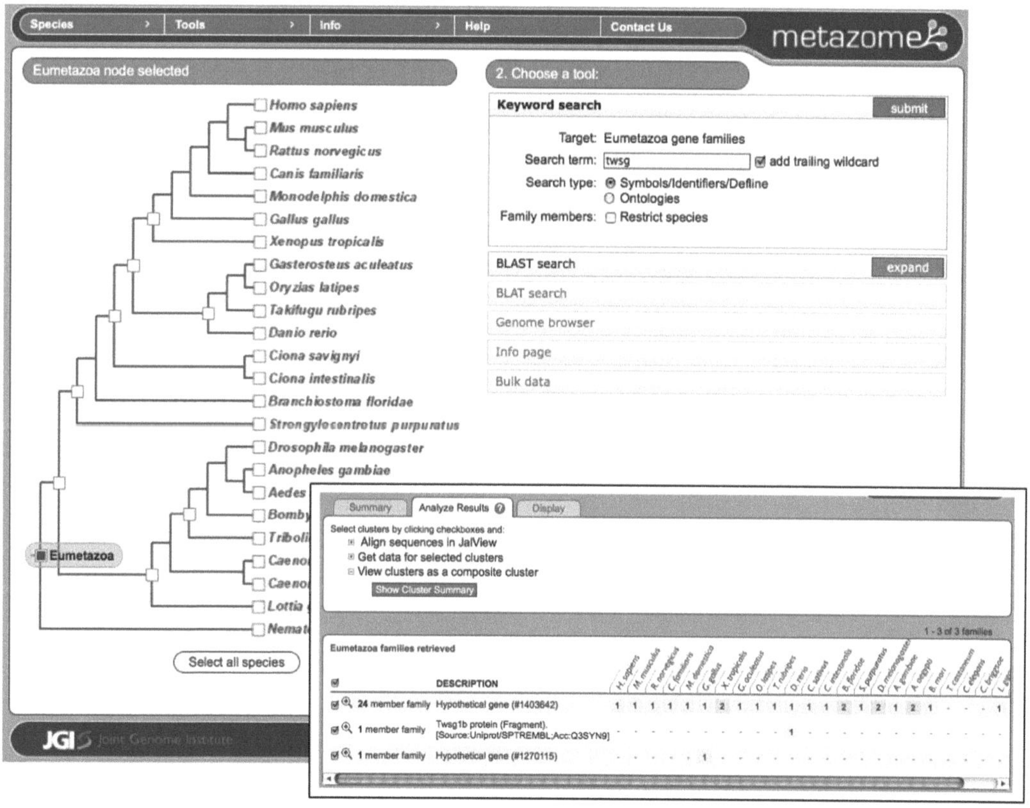

Fig. 3. The Metazome v3.0 homepage. Note the Eumetazoa node is highlighted when moused over and is selected on the left. The keyword search window is expanded in the upper right with "twsg," the gene symbol for *twisted gastrulation*, entered in preparation for submission. *Inset* shows the search output containing three "families" of genes across all the organisms represented by the Eumetazoa node. To display all of these as a single cluster, all the boxes to the left of families have been checked, the Analyze Results tab at the top has been opened, and View clusters as a composite cluster has been selected in preparation for selecting Show Cluster Summary.

3. Enter the keyword, the gene symbol, or name for the gene of interest (see Note 13).
4. Perform the search.
5. Examine the families retrieved (Fig. 3).
6. Decide which families to display as synteny relationships.
7. Examine synteny (Fig. 4), determine whether the gene of interest has been found, and whether there is a single gene or if it is a member of a multigene family (see Notes 14–16).

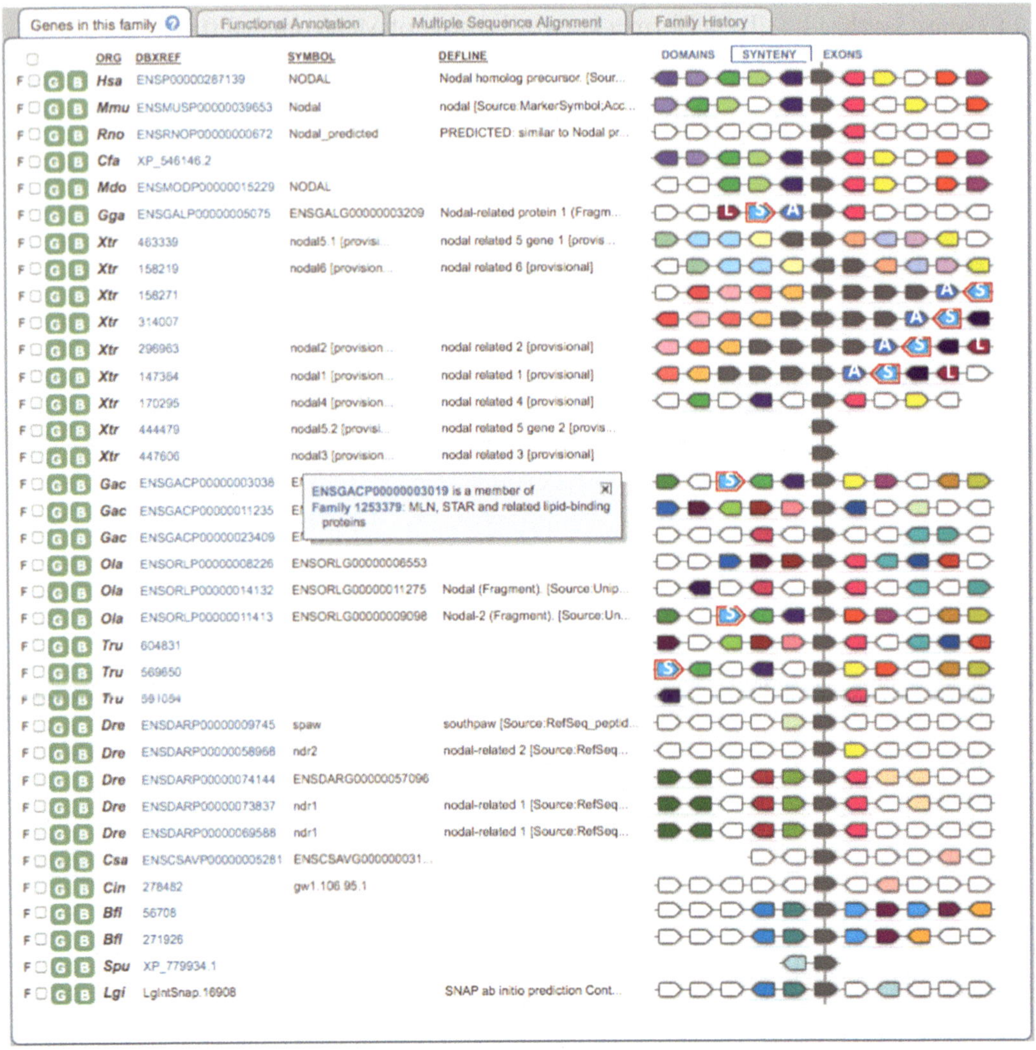

Fig. 4. A Metazome synteny display for *nodal* genes. This view was obtained following selecting the magnifying glass icon adjacent to the "Nodal (fragment)" family in the output to a "nodal" keyword search of Eumetazoa. The central genes in gray are the putative *nodal* orthologs. Colored genes to either side are shared with at least one other organism and indicate orthology. Mousing the cursor over the light blue gene to the right of *Xenopus tropicalis nodal2* highlights a box containing information about this gene, *star*, which is highlighted by an S in this figure. The *ash2l* and *lsm1* genes discussed in the text are highlighted here by As and Ls, respectively.

3.3.2. BLAST Searching

To perform a BLASTp search of Metazome

1. First select phylogenetic node.
2. Select the BLAST search option under "Choose a tool."
3. Enter the protein sequence to be queried into the window (see Note 17).
4. Perform the search.
5. Examine the search families retrieved.
6. Decide which families to display synteny relationships.
7. Examine synteny, determine whether the gene of interest has been found and whether there is a single gene or if it is a member of a multigene family (see Notes 18 and 19).

3.4. The UCSC Genome Browser

A resource that will be essential for *Xenopus* researchers is the University of California at Santa Cruz's (USCS) genome browser (http://genome.ucsc.edu/). This site has many useful features (see Note 20) and new genome builds will be supported by this browser, so it is recommended that researchers who plan to use the *X. tropicalis* genome become acquainted with this resource.

To begin using the UCSC browser:

1. After navigating to the website, select either the Genome link or the Genome Browser link.
2. Select Vertebrate from the clade pull-down menu and *X. tropicalis* from the pull-down menu under genome.
3. To find genes by keyword search, enter gene symbols into the *gene* search window (see Fig. 5) and select the "jump" button or return key.
4. Navigating to specific regions of the genome can also be performed by entering scaffold number and nucleotide positions into the "position or search term" window.
5. BLAST-like Alignment Tool (BLAT) searches (which are not as sensitive to sequence differences as BLAST) can be performed by selecting the BLAT link from the menu bar at the top of the page: DNA or protein sequence can be queried.

3.5. Conclusion

The current chapter introduces the novice to some of the many different paths to finding genes in the *X. tropicalis* genome assembly. The use of two genome browsers has been discussed. Many of the features for navigating the JGI browser are similar to the UCSC browser and once one has mastered the organization of either one of these, transitioning to using the other becomes significantly easier. Because the mission of the US Department of Energy, JGI's parent organization, is no longer considering animal genomes, it is likely that the UCSC and Xenbase browsers will be serving as the major genome hubs for navigation of future *Xenopus* genome

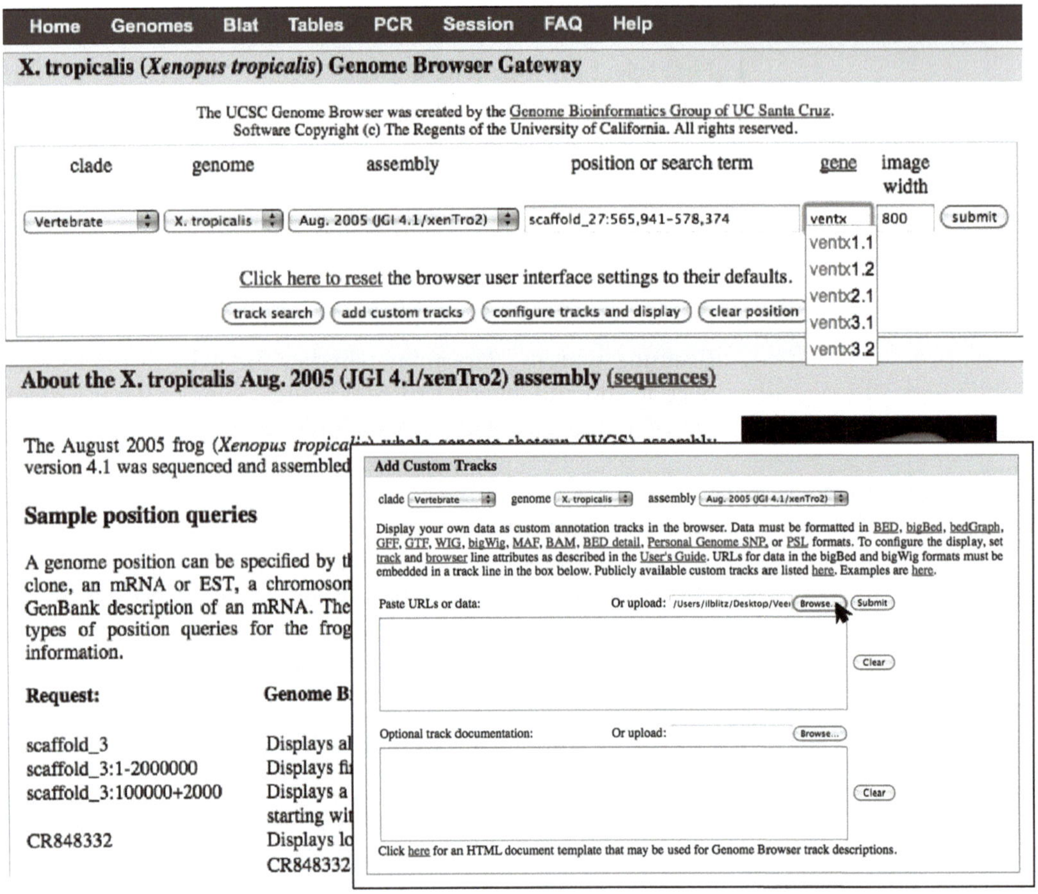

Fig. 5. The UCSC *X. tropicalis* genome browser. The outer figure shows the browser gateway that is obtained after one has selected the Vertebrate clade and *X. tropicalis* genome. "ventx" was entered into the "gene" search window for keywords searches and the six different ventx genes appeared in a pull-down menu below this window as options. To add new tracks, such as high-throughput sequencing results, selecting the button "add custom tracks" opens the window in the *inset*. After downloading wiggle, bed, or other compatible files to one's computer, using the "Browse" button to enter these data permits loading new tracks into the browser display, which is not shown.

builds. Considerable space was given to discussing Metazome because synteny relationships not only provide verification of gene identity but also give the researcher rich insights into the evolutionary origin of the gene of interest that may enhance thinking about research questions.

Within the next few years not only will more accurate and detailed *X. tropicalis* genome builds be produced (v7.1 is pending release at the time of writing) and the *X. laevis* genome sequence is anticipated to be completed. It is also clear that current *X. tropicalis* genomic information is in great need of improvement. Better gene annotations with much deeper wet biology evidence to support transcript structure (exons, introns, UTRs, etc.), along with the need to close sequence gaps and string scaffolds together into chromosome length pieces, will rapidly become available with high-throughput sequencing methods. The tools described here will be increasingly useful in navigating the frog genome(s), and mastering them will be valuable to researchers in many areas, including the application of genetics, studies on gene regulation, molecular and biochemical studies of gene function, and the evolution of genome organization, to name a few.

4. Notes

1. Nomenclature for *X. tropicalis* genes (http://www.xenbase.org/gene/static/geneNomenclature.jsp) was adopted by *Xenopus* community consensus and adheres to Human Gene Nomenclature Committee's (HGNC) gene names and symbols. Thus, common names for *X. laevis* genes in the literature are not always the official names in Xenbase. For example, a search for Vg1 (not shown), a TGFβ family ligand, produces a gene page for *growth differentiation factor 1* (*gdf1*), the official name for this gene. Synteny (Subheading 3.3) between frog and mammals makes it clear that the *Xenopus* Vg1 gene is orthologous to human *GDF1*. *Xenopus* exceptions to HGNC naming include genes not found in human (e.g., *vegt*) and gene multiplications where there is only a single human ortholog (e.g., *smad4.1* and *4.2*). It is also noteworthy that gene names can change. In addition to annotator-initiated corrections, HGCN, upon availability of new data or petitioning, sometimes modifies human genes names. Such changes result in concordant changes to frog gene names.
2. Regardless of the annotation found it is best to independently verify correct gene identification. Current gene names in the UCSC genome browser (Subheading 3.4) are especially inaccurate. Machine annotations can also be incorrect, particularly when genes are members of a large family. At a low frequency manually annotated genes may be misidentified. To begin verifying gene identity, BLASTn and/or BLASTp searches should be performed using NCBI's BLAST pages (http://blast.ncbi.nlm.nih.gov/Blast.cgi). Links to mRNA and protein sequences on Xenbase gene pages or JGI resources (discussed in Subheading 3.2) provide a starting point for these searches. BLAST will provide the first indication that a gene best matches orthologs from other species.

   When a NCBI BLAST search identifies a *X. tropicalis* sequence as the highest scoring match, it is not recommended that one relies on NCBI gene annotations for this sequence without further investigation. Since NCBI uses Xenbase-provided annotations, such apparent verification of a gene's identity would be the result of circular reasoning. Since *X. tropicalis* nomenclature is tightly linked to human, finding the highest scoring human match often provides an indication to a gene's identity. Once BLAST searches reveal that a gene *may* be correctly identified, further examination of synteny is recommended to more rigorously establish orthology (Subheading 3.3).

3. The JGI Gene Catalog is the collection of machine-generated gene models created by JGI, whereas All Models also included annotator suggestions for gene models that may be the product of other modeling efforts (e.g., Ensembl models or cDNA/EST sequences) or to other existing JGI models. Many JGI keyword searches of the Gene Catalog will yield a single gene output but in some cases there may be multiple outputs, particularly when a gene is a member of a multigene family. For example, a keyword search for "nodal" (not shown) finds 9 entries, and 8 of these are *nodal* genes.

   An advantage of searching All Models (Fig. 1 inset) instead of the default Gene Catalog search above is that the former may identify a gene(s) that is missing from JGI's Gene Catalog tracks, but is present and annotated in another track. One example is *anti-müllerian* hormone (amh), which is not found in Gene Catalog searches as of this writing. By keyword searching All Models for "amh," one finds a single entry (Fig. 1). Selecting the link on the search output page "Location link scaffold_327:271144-282888 (-)" opens a browser page (not shown) revealing that this region of the genome has no JGI models in the GeneCatalog or FilteredModels tracks. However, selecting either the "protein id" or "transcript id" link from the output page (Fig. 1) opens a gene page for *amh* because scaffold_327:271144-282888 (-) has been manually annotated as *amh*.

4. JGI gene pages, like Xenbase, provide links to useful information for verification of correct gene annotation. Selecting the "protein id" link (Fig. 1) opens a gene page specific for the gene found in the search. The inset in Fig. 2 shows an example, the gene page for *nodal2* that is opened after performing a keyword search for "nodal." A link is provided to perform NCBI BLASTp queries using the predicted amino acid translation from the represented gene model (which in this case is named estExt_fgenesh1_pm.C_340065, shown at the bottom of the inset). Putative mRNA sequence with three-frame amino acid translation, and exons, both with or without introns, can be found by selecting the link labeled "View nucleotide and 3-frame translation."

   Using the "To Genome Browser" link one can navigate to JGI's genome browser to compare the structures of various gene model predictions, ESTs, etc., in the context of the flanking genomic region (Fig. 2). Each gene has numerous modeling predictions, presented in separate color-coded tracks, with the top ranked JGI model being referred to as a "Filtered Model" (see Supplementary Note 4 of ref. (1) for description). The Filtered Model is the version found in the Gene Catalog, has been annotated, and is in Xenbase. The Ensembl model track and others may also be useful guides for possible gene structure. Similar examination of gene structure and alternative models can be performed within Xenbase, via the "Xenbase G-browse" link, including also models from v7.1. Discussion of Xenbase's G-browse by J. Bowes can be found in Chapter 31 of this volume.

5. In the current genome assembly approximately half of all gene models have truncated ORFs, which are incomplete at one or both ends (1). Reasons for incompleteness include (1) insufficient methodology in gene prediction produced a model lacking a complete ORF, (2) a poorly/incorrectly assembled genomic interval (due to sequence repeats and/or "in-scaffold gaps" within genes) can interfere with correct gene modeling, and (3) "between-scaffold gaps," where the gene is split between more than one scaffold and thus gene models cannot be completed. It is common that the missing portions of a gene, or more specifically, the protein-coding portion, are presvent nearby to the incomplete gene models. Sometimes missing portions of genes can be found by BLASTn queries using ESTs, cDNAs, or 5′/3′-RACE (*r*apid *a*mplification of *c*DNA *e*nds; ref. (4)) product sequences, if available. In some cases examination of the numerous model predictions for a gene may lead to a better understanding of the organization of a gene's transcription unit and the location of missing ORF information. The Filtered Model could be missing exons that might be found in other models (e.g., Ensembl). The presence of information in the human, rat, and mouse BLAST tracks in the JGI Genome Browser (Fig. 2) can provide clues to regions that might contain protein-coding similarities to *X. tropicalis* and tBLASTn using orthologous proteins from other organisms can help find the missing coding regions of an incomplete model. More details about navigating the *X. tropicalis* genome using these JGI resources can be found at http://genome.jgi-psf.org/help/overview.jsf. If desired, a predicted transcript structure might be experimentally validated by reverse-transcription (RT)-PCR using oligonucleotide primers derived from the various models or by employing 5′ and/or 3′ RACE.

   To visualize more gene models in JGI's browser page, one can take advantage of the "Advanced Track Controls" feature, accessible from the bottom of the browser page (Fig. 2). The default setting hides many models that JGI has created for a gene and these can be revealed by selecting "dense" or "full" buttons in the section titled "Model." Once selected, one must refresh the display using the "Refresh Browser" button within the Models section. Means of zooming the view in and out, and moving left and right, are apparent and one can also change the region viewed by manually modifying the scaffold number and/or nucleotide positions in the display window containing this information.

6. BLAST searches to compare sequences preferable over BLAT searches because BLAST is more sensitive to low sequence similarity. BLAT was primarily designed to permit large-scale sequence comparisons in short time intervals without extreme computational needs. For more details on the differences between BLAST and BLAT searches, see http://genome.ucsc.edu/FAQ/FAQblat.html. Another type of BLAST search discussed in the current chapter is BLASTp, which is a search for matches to a protein sequence database using a protein sequence string.

   BLAST searches of the *X. tropicalis* genome sequence can also be performed using NCBI's BLAST page (http://blast.ncbi.nlm.nih.gov/Blast.cgi). Beneath the heading "BLAST Assembled RefSeq Genomes" select the link "list all genomic BLAST databases," and then select "Other Vertebrates." Select the "B" button to the right of *Xenopus* (*Silurana*) *tropicalis* to open a BLAST page. The pull-down menu marked "Program" permits choosing BLASTn, tBLASTn, and other BLAST options.

   BLAST searches can be initiated in Xenbase using the "Blast Frogs" link in the navigation bar. Users can select JGI genome models as the preferred target, and these results are linked directly to Xenbase gene pages.

7. *Xenopus* nucleotide sequences can be obtained from a variety of databases, including NCBI, Xenbase (http://www.xenbase.org), NIBB (http://xenopus.nibb.ac.jp/), or M.J. Gilchrist's *Xenopus* RNA clusters database (http://genomics.nimr.mrc.ac.uk/online/xt-fl-db.html; see *also* Chapter 19 of this volume). By querying the genome using a cDNA or EST sequence, this approach ensures that the results are firmly "anchored" by the researchers prior nucleotide data—the found region/gene should be identical, or nearly so if using *laevis* sequence, to the gene of interest. This is in contrast to protein-initiated tBLASTn searches (see Subheading 3.2.3), where degeneracy of the genetic code may result in identification of closely related sequences that are not as highly matched as one might find using direct nucleotide searches.

   cDNA sequence data from *X. laevis* often performs well in searching the *X. tropicalis* genome because there is usually very high sequence identity, even in many 5′ and 3′ UTRs, and thus little ambiguity which *tropicalis* gene corresponds to which *laevis* cDNA/EST. One should be aware that sometimes *tropicalis* gene duplications result in more than one gene showing equally high sequence identity over significant stretches of cDNA.

8. BLAST search outputs show alignments of the queried sequence against potential matches and these are ordered from "best" to "worst" scores and expectation values (*E*-values). Evaluation of the output from a BLASTn search involves determination of whether the match is near 100%, as would be expected if one queries using a *X. tropicalis* sequence such as a cDNA/EST, or may be less than perfect if using a *X. laevis* DNA sequence query. For more information on how to evaluate BLAST searches, see http://www.ncbi.nlm.nih.gov/books/NBK21097/.

9. Because the *X. tropicalis* gene annotation is linked to human, BLAST queries using a protein sequence of human origin usually hastens finding the best match. To obtain a full-length protein sequence one can find the gene in NCBI Entrez Gene (http://www.ncbi.nlm.nih.gov/sites/entrez?cmd=search&db=gene), select the Protein link from the gene page and find the entry containing the longest protein sequence.

10. Once a tBLASTn output is obtained, one should determine whether the top hit is indeed the gene of interest or a closely related family member. The output of tBLASTn searches of the genome consists of alignments of the inputted amino acid sequence to amino acid sequence translated from the genome.

    When keyword searches fail to find a gene it is often because models are lacking, but these usually can be found in the genome assembly by tBLASTn search. For example, the genes encoding *anti-Müllerian* hormone (*amh*) and endothelin-3 (*edn3*) have no JGI gene models but are identifiable by tBLASTn search. More rarely, tBLASTn fails because the missing gene is not present in the genome assembly. One example is the *X. tropicalis* ortholog of *X. laevis soxd/sox15* (5). Reciprocal BLAST searches (see Subheading 3.3 and Note 11) of a sequence from *X. tropicalis* (NM_001126574) best match *X. laevis soxd/sox15*, but genomic BLASTn searches using this *tropicalis* cDNA fail to identify a corresponding genomic interval. It is important to note that while this gene is annotated as *sox15*, this is provisional. First, soxd's highest BLASTp match is to Sox15 proteins from other vertebrates and, second, no better evidence for another *sox15* gene can be found in the current *X. tropicalis* genome assembly. If, in future assemblies, *soxd/sox15* were found to be syntenic (Subheading 3.3) to *sox15* genes from other species then the annotation as *sox15* would be more firmly solidified.

11. Starting with a *tropicalis* protein sequence, the best BLASTp non-*Xenopus* hit in NCBI might be a protein from human. The reciprocal BLAST, using this best-match human protein to then tBLASTn search the entire *X. tropicalis* genome would find the best match in the *tropicalis* genome to the putative human ortholog. While reciprocal BLAST searches should be performed, this approach in isolation still suffers from a number of pitfalls. First, BLASTp searches often query using fragmentary protein sequence derived from *tropicalis* gene models that are currently incomplete, thereby best-matching different genes than the full-length sequence would match. Second, albeit rather infrequently, the best BLAST hit is not the true ortholog, which has diverged in sequence, but instead is a close family member, or a paralog (an ortholog that is the result of a gene duplication), that has diverged more slowly. An example is illustrated using a BLASTp search using *X. tropicalis* Vg1/gdf1 (not shown), which suggests that its closest matches are encoded by the *Xenopus* TGFβ Derriere (*der*) followed by human *GDF3*. Synteny, however, shows that Vg1 is the ortholog of *GDF1*, which is closely related to *GDF3*. The *der* gene lacks obvious synteny to any TGFβ family gene, hence it is unclear whether this is a *Xenopus*-specific TGFβ, or the ortholog of human *GDF3*, which is currently missing from the *tropicalis* assembly.
12. Metazome's home page (Fig. 3) displays a phylogenetic tree with branch points that are searchable nodes within the tree. Selection of a node specifies that the search will be performed against clusters of protein sequences that are roughly consensus protein family sequences for all the organisms on the tips of the branches emanating from that node, but not organisms diverging from points more basal to that node. For example, if one mouses over the box connecting *X. tropicalis* to the amniotes, the label "Tetrapod" appears, and selection of this box confines the search to those organisms displayed to the right that fall within the tetrapod grouping. The choice of which node one should search is a matter of personal interest. By limiting the search to tetrapods, one might miss interesting evolutionary origins of the gene that would only be revealed by a more inclusive search.

    When searching a broad grouping of organisms, one problem arises when the keyword search is too general. In this case, the keyword could return outputs that are too large (>300 genes) for synteny to be displayed by the Metazome site. In such cases, searches with more narrow organismal groupings, or with a different keyword, may be advisable to limit the output.
13. Entering a gene symbol into the Search term window allows for initiation of a search, but sometimes the HGCN symbol

fails, probably due to gene name changes. Choice of alternative keywords may require some guesswork, which can be facilitated by examining alternative names for the gene on NCBI's Entrez Gene page for the human ortholog. Alternatively, a keyword search in Xenbase may be successful as Xenbase stores all known synonyms for each gene, and direct links to Metazome are available on each gene page.

14. If one keyword searches for nodal using the Eumetazoa node, two families are retrieved, a 27-member family described as "protein PM5 homolog," and a 35-member family "Nodal (fragment)." Displaying the synteny relationships among the latter requires only selecting the magnifying glass icon to the left of the 35 member Nodal (fragment) family. Each gene is then displayed in its genomic context (Fig. 4), with *nodal* genes depicted as gray pentagons in the center of each line. All central Nodal genes are oriented with their transcription units oriented to the right. Five genes are displayed on either side as colored or white pentagons with their pointed ends indicating the direction of gene transcription relative to the central nodal gene. The colors are coded to permit easy comparison of orthology between all the organisms' chromosomes presented. The third, fourth, fifth, and sixth *X. tropicalis* (Xtr) entries in Fig. 4 all contain four gray pentagons because there are four nodal family genes in a cluster in this region of the genome. Scrolling down to the sixth entry labeled as *nodal1*, and selecting the light-blue colored gene located two genes to the right of *nodal1* highlights this nearby gene, *star* (marked in this figure with an "S"), in all organisms where *star* is located within this genomic window. Thus, one can see that a chick (Gga) gene located two genes to the left of the central *nodal* gene is also light blue, highlighted, and is the chick *star* ortholog. Similarly, *star* is highlighted to the left in the teleost fishes stickleback (Gac), medaka (Ola), and pufferfish (Tru). This gene is not seen nearby *nodal* genes in the mammals. Examining the genes to the right of frog *nodal1*, one can see dark blue (*ash2l*, marked with an "A") and magenta (*lsm1*, marked with a "L") colored genes that are also found to the left of chick *nodal*. None of the genes to the left of the frog *nodal* cluster show linkage to the nodals in the other organisms, at least within this 11-gene-wide window.

    X. tropicalis nodals have been named in accordance with their percent nucleotide sequence matches with *X. laevis* nodal orthologs to retain, as best as possible, continuity with the published literature. One exception is the presumptive *X. tropicalis* ortholog of *X. laevis nodal4*. Examination of the genes flanking both sides of *nodal4* shows it has synteny with the mammalian *nodals* (data not shown), and thus these are orthologous loci. Therefore *X. tropicalis nodal4* gene is now officially named *nodal* in Xenbase and NCBI, illustrating how *Xenopus* gene naming is linked with the human nomenclature. For more details on the likely evolutionary origins of the frog nodal gene clusters, see Supplementary Note 8 in Hellsten et al. (1).

To examine any of the genes displayed in more detail one can select the numbered link under the column DBXREF, which opens the JGI gene page corresponding to this gene. From there, NCBI BLASTp searches can be performed, or the transcript sequence can be copied from the "View nucleotide and 3-frame translation" link and pasted into an NCBI BLASTn search window.

15. If fewer than five genes are displayed on one or both sides of the gene of interest this is because the gene is near the end of a scaffold.

    Sometimes Metazome displays back-to-back gene multiplications where in fact there is only one gene. Metazome uses the gene models available to it and sometimes a *Xenopus* gene is fragmented into two or more separate genes modeled in tandem. Thus, when one finds more than one gene copy of interest in Metazome, in tandem, or even separated by another gene in the middle, a BLASTp search for the protein sequences from each model is recommended. If one model matches the N-terminal portion of the encoded protein and the other identifies the C-terminal portion, then it is likely these are not separate genes but instead back-to-back models that should be combined into one gene model.

16. Outputs often contain multiple families of related genes and one would like to combine these into a single synteny display. An example is shown in the inset to Fig. 3. A keyword search of Eumetazoa for the BMP regulator *twisted gastrulation* retrieves three families that all contain *twisted gastrulation* genes. In human, mouse, and other mammals there is a single gene, *TWSG1*, but two *twisted gastrulation* genes have been found in *X. laevis* ((6); Blitz and Cho, unpublished). To combine these three families into a single synteny display, first select the boxes to the left of each family to be included in the display. Then, under the "Analyze Results" tab, select "view clusters as a composite cluster" and "show cluster summary." The resultant synteny analysis (not shown) reveals that the single Twsg1 gene in mammals is syntenic with the *X. tropicalis twsg1* gene on scaffold 294 and that this gene is also syntenic to amphioxus (Bfl) *twisted gastrulation*. Interestingly, the second gene found in frogs, *twsg2*, is shown to be syntenic to a chick gene, and also with *twisted gastrulation* genes from other metazoa. Furthermore, this second gene shows synteny to *twisted gastrulation* genes in sea urchin (Spu) and the owl

limpet *Lottia* (Lgi). BLASTp of the *Lottia* protein against NCBI's nonredundant database shows its best matches are vertebrate twisted gastrulation proteins (not shown), verifying that it is likely a *twsg2* ortholog, and suggesting that the two frog twisted gastrulation genes are not the result of a recent, frog-specific duplication, but instead have an ancient origin, probably arising from a gene duplication event predating the deuterostome–protostome divergence. These findings illustrate the power of incorporating as many phylogenetic groups as possible into synteny analyses.

If one suspects that keyword searches are failing to find all the members of a gene family, a BLAST search to probe Metazome can assist in the search for these. For example, a search of the Eumetazoa node for the orthologs of *ventx* family genes using this keyword only finds two entries for human, *VENTX* and a pseudogene, and the six *X. tropicalis ventx* genes, but fails to identify known orthologs in other species, such as zebrafish or amphioxus. BLASTp searches of node consensus sequences solve this (see Notes 12 and 17).

17. The choice of protein sequence to query in a Metazome search may determine the outcome of the search results, just as in tBLASTn searching the JGI genome sequence in the previous section. Use of the human protein sequence usually suffices, but sometimes one can obtain hits to more distantly related organisms if one uses a different species' protein in the query. One approach is to query using the protein sequence of the human ortholog, obtain a synteny cluster as described in the keyword search section (using "View clusters as a composite cluster"), and then generate a consensus sequence from this cluster to perform a new search. To do this, upon obtaining a satisfactory cluster select the "BLAST for similar families" link found under the "Find related families" tab. The consensus sequence from the families selected should appear in the BLASTp window and it is just a matter of submitting the BLAST. (At the time of this writing this feature in v3.0 appears to be faulty and thus the cluster consensus can also be obtained by selecting all the genes in the synteny display, selecting "Get Sequences (basic)" under the "Get Data" tab, and then selecting the "Family consensus sequence" button. This consensus can then be manually copied/pasted into the search window.) A series of BLASTp hits with decreasing scores is returned and if one selects all those with high scores, one can view these in a single cluster. Often trial and error is required to determine how far down the list one wishes to venture, and sometimes orthologs can be found at or below an *E* value of 10^{-20}. Applying this approach to the *ventx* example, retrieves human, dog, frog, zebrafish, stickleback, medaka, and amphioxus (*Branchiostoma*) orthologs (not shown).

18. Sometimes Metazome searches fail. Metazome searches require the presence of gene model predictions derived from genome projects as starting material. If a gene model fails to be predicted (as discussed earlier) in some position in the genome, this would result in the absence of that gene in the Metazome synteny display. Thus, sometimes a *X. tropicalis* gene is found by directly searching the genome (as described in Subheading 3.2), but this gene lacks a JGI gene model in the Gene Catalog and will be missing in the Metazome synteny display.

19. A lack of demonstration of synteny is a negative result that fails to prove a gene is *not* syntenic. Metazome displays only five genes on either side of the central gene under study, and when synteny fails to be found one would like to view genes over a wider window. It is possible within Metazome to view genes further to the left or right by selecting a gene (pentagon) to one side of the central gene and then, within the description box for the gene family that appears (see example of a description box for the *star* gene in Fig. 4) select the "Family link" within this description box. This will now shift the synteny display such that the newly selected gene now occupies the central position. Thus, one can move along the scaffold/chromosome, but this is a laborious procedure for scanning both sides of the original 11-gene window.

    Genomicus is another tool for viewing synteny analysis is (http://www.dyogen.ens.fr) (7) with a wider window of genes for display. This site permits keyword searching by gene name only but permits examination of 22 genes (using 2× zoom out) on either side of the gene of interest, producing a 45-gene-wide window. Genomicus displays synteny in a manner similar to Metazome and also permits examination of many more species of vertebrates. A disadvantage is that Genomicus does not include species evolutionarily more distant than ascidians (*Ciona* sp.), with the lone exception of *Drosophila melanogaster*. Thus, Genomicus cannot reveal the synteny between frog, sea urchin, and limpet *twsg2* genes, and fails to find relationships for this gene across the species covered (not shown). However, examination of the Genomicus synteny display of the *ventx* genes (not shown) provides a more revealing view of orthologs of this gene in species that are not present in Metazome. *Ventx* genes by Metazome analysis do not seem to be present in rodents (rat and mouse) though they are found in dog and man. Genomicus extends the view of the amniote *ventx* family as orthologs are also shown to be present in reptiles and birds, nonhuman primates, panda, and elephant. Furthermore, *ventx* orthologs from several fishes share linked genes with the mammals, but not *Xenopus*.

Genomicus contains links to *X. tropicalis* Ensembl models and Pubmed, but lacks BLAST capabilities and does not have direct links to the JGI *X. tropicalis* genome. Due to space limitations the current chapter does not detail how to manipulate Genomicus displays but once one masters the Metazome manipulations above, Genomicus will not be too difficult to navigate. More details on using Metazome and Genomicus can be found at http://www.metazome.net/help.php and using the Help and Documentation link on the Genomicus webpage, respectively.

Finally, if one wishes to examine synteny over an even larger region of a chromosome, one could do this work the "old fashioned way"—manual pairwise comparisons of gene order between *X. tropicalis* and, for example, the suspected orthologous human chromosomal region. Such analyses might be best performed comparing genome displays using the UCSC genome browser (Subheading 3.4).

20. Vertebrate Multiz (pronounced Multi-Zee or Multi-Zed) Alignment & Conservation tracks show regions of DNA sequence similarity between *X. tropicalis* and some other vertebrates, which can be modified by selecting options under Comparative Genomics further down on the page. This is useful where gene models are lacking, and also for finding conserved regulatory elements in noncoding regions (see Chapter 15).

Regions within the window can be magnified by highlighting the area of interest: hold down the mouse and drag across the region of interest before releasing. If one magnifies enough, the DNA sequence will be displayed with aligned bases to the other organisms in the Conservation track highlighted.

Display of the tracks can be modified using the controls at the bottom of the browser once one is viewing a genomic interval. The order of display of the tracks can also be modified by dragging tracks up or down into the position desired.

The ability to display user-created tracks is a big bonus. There are a growing number of high-throughput sequencing datasets and more will become available in the near future. As researchers place more of these datasets into the public domain for download, they can be easily loaded into the UCSC browser to permit viewing of mapped sequence reads. ChIP-seq and RNA-seq reads or peaks can be displayed to permit examination of these data in any area of the genome. To create new tracks, select the "add custom tracks" button under the display window (Fig. 5). Datasets of the appropriate files types can then be loaded using the browse option (inset to Fig. 5) and the window can be refreshed. For example, G.J. Veenstra's mapping of transcription start sites and other datasets can be downloaded from his website (http://131.174.221.43/gert janveenstra/resources.htm) as wiggle (.wig) or BED (.bed) files and loaded into the *X. tropicalis* UCSC browser window. A number of other file types are also acceptable and outlined on the "add custom tracks" page (inset) (see *also* Chapter 17). More details on navigating using the UCSC genome browser can be found in ref. (8) and tutorials are provided at http://www.openhelix.com//cgi/tutorialInfo.cgi?id=27.

Note Added in Proof

X. tropicalis v7.1 gene models are available in G-browse on Xenbase gene pages. This genome assembly is available via ftp://ftp.xenbase.org/pub/Genomics/JGI/Xentr7.1/. To view the genome in a browser, the files xenopus_tropicalis_v7.1.tar.gz and Xentr7_2_Stable.gff3.gz can be loaded into the Broad Institute's Integrative Genomics Viewer, available at http://www.broadinstitute.org/igv/.

Acknowledgements

The author would like to thank William Chiu, Margaret Fish, Amanda Janesick, and Rebekah Le for critical comments and numerous discussions prior to publication. I would also like to thank Jeffrey Boore, Mike Gilchrist, David Goodstein, Uffe Hellsten, and Astrid Terry for valuable technical discussions on the frog genome, Jeff Bowes for help with Xenbase, and the editors Stefan Hoppler and Peter Vize for their valuable comments on the manuscript.

References

1. Hellsten U, Harland RM, Gilchrist MJ, Hendrix D, Jurka J, Kapitonov V, Ovcharenko I, Putnam NH, Shu S, Taher L, Blitz IL, Blumberg B, Dichmann DS, Dubchak I, Amaya E, Detter JC, Fletcher R, Gerhard DS, Goodstein D, Graves T, Grigoriev IV, Grimwood J, Kawashima T, Lindquist E, Lucas SM, Mead PE, Mitros T, Ogino H, Ohta Y, Poliakov AV, Pollet N, Robert J, Salamov A, Sater AK, Schmutz J, Terry A, Vize PD, Warren WC, Wells D, Wills A, Wilson RK, Zimmerman LB, Zorn AM, Grainger R, Grammer T, Khokha MK, Richardson PM, Rokhsar DS (2010) The genome of the Western clawed frog *Xenopus tropicalis*. Science 328:633–636
2. Bowes JB, Snyder KA, Segerdell E, Gibb R, Jarabek C, Noumen E, Pollet N, Vize PD (2008) Xenbase: a *Xenopus* biology and genomics resource. Nucleic Acids Res 36:D761–D767
3. Bowes JB, Snyder KA, Segerdell E, Jarabek CJ, Azam K, Zorn AM, Vize PD (2010) Xenbase: gene expression and improved integration. Nucleic Acids Res 38:D607–D612
4. Zhang Y, Frohman MA (1997) Using rapid amplification of cDNA ends (RACE) to obtain full-length cDNAs. Methods Mol Biol 69:61–87
5. Mizuseki K, Kishi M, Shiota K, Nakanishi S, Sasai Y (1998) SoxD: an essential mediator of induction of anterior neural tissues in *Xenopus* embryos. Neuron 21:77–85

6. Oelgeschläger M, Tran U, Grubisic K, De Robertis EM (2004) Identification of a second *Xenopus* twisted gastrulation gene. Int J Dev Biol 48:57–61
7. Muffato M, Louis A, Poisnel CE, Roest Crollius H (2010) Genomicus: a database and a browser to study gene synteny in modern and ancestral genomes. Bioinformatics 26:1119–1121
8. Zweig AS, Karolchik D, Kuhn RM, Haussler D, Kent WJ (2008) UCSC genome browser tutorial. Genomics 92:75–84

Part II

Genetics in *Xenopus*

Chapter 5

Genetic Analysis of *Xenopus tropicalis*

Timothy J. Geach, Derek L. Stemple, and Lyle B. Zimmerman

Abstract

The pipid frog *Xenopus tropicalis* has emerged as a powerful new model system for combining genetic and genomic analysis of tetrapod development with robust embryological, molecular, and biochemical assays. Its early development closely resembles that of its well-understood relative *X. laevis*, from which techniques and reagents can be readily transferred. In contrast to the tetraploid *X. laevis*, *X. tropicalis* has a compact diploid genome with strong synteny to those of amniotes. Recently, advances in high-throughput sequencing together with solution-hybridization whole-exome enrichment technology offer powerful strategies for cloning novel mutations as well as reverse genetic identification of sequence lesions in specific genes of interest. Further advantages include the wide range of functional and molecular assays available, the large number of embryos/meioses produced, and the ease of haploid genetics and gynogenesis. The addition of these genetic tools to *X. tropicalis* provides a uniquely flexible platform for analysis of gene function in vertebrate development.

Key words: *Xenopus*, *Silurana*, *tropicalis*, Genetics, Development, Organogenesis, Gynogenesis, Genetic screens, Genetic mapping, Mutagenesis

1. Introduction

Xenopus embryos have been remarkably productive models for developmental biologists for over 70 years (1, 2). The dominant laboratory species, *X. laevis*, continues to be an outstanding system for embryological manipulations and gain-of-function gene assays, but its tetraploid genome and long generation time hamper many genetic and genomic approaches. The related diploid species *Xenopus* (*Silurana*) *tropicalis* shares *X. laevis*' advantages for experimental embryology while also being ideally suited to genetics and genomics.

Stefan Hoppler and Peter D. Vize (eds.), *Xenopus Protocols: Post-Genomic Approaches, Second Edition*,
Methods in Molecular Biology, vol. 917, DOI 10.1007/978-1-61779-992-1_5, © Springer Science+Business Media, LLC 2012

1.1. Why Xenopus Genetics?

Forward genetic screens have the unique ability to identify novel gene functions without bias toward previously known DNA sequence. Our understanding of animal development is founded on genetic studies of invertebrate fly and worm models (3, 4) which identified the transcriptional control networks underpinning the basic animal body plan, but it is expected that many differences may exist in vertebrates. The development of gene targeting for reverse genetic studies in mouse has been especially powerful (5, 6), but forward screens for embryonic mutations in mammals are costly and difficult due to intrauterine development. A large number of informative mutations have been identified in teleosts, particularly zebrafish, where screens have benefited from rapid development of externally fertilized, transparent embryos, high fecundity, and short generation time (7–10).

Unlike the genome of teleost fish, derived from an ancient duplication (11) or the allopolyploid genomes of other *Xenopus*, derived from hybridization and genome retention of two separate species (12, 13), the genome of *X. tropicalis* is that of a canonical diploid vertebrate. At ~1.5×10^9 bp it is one of the smallest tetrapod genomes, about the same size as zebrafish, and shows robust synteny with those of amniotes (14), simplifying orthology assignment, functional analysis, and identification of noncoding regulatory elements. Pilot forward genetic screens have already recovered a number of heritable mutants (15–17), several of which have now been mapped to specific genes (18–20). Genetic studies in *X. tropicalis* are facilitated by the production of up to 9,000 embryos from a single mating, sufficient meiotic recombination events to map a mutation or conduct a variety of phenotypic analyses. Extensive genomic resources are available, including a high-quality chromosome-scale draft genome assembly and more than one million ESTs (see Chapter 4). Gain-of-function, molecular, and embryological assays are readily transferred from the well-characterized *X. laevis* system. Transgenic rescue of mutant backgrounds with floxed constructs (21) offers a method for obtaining conditional alleles to delete gene functions in specific tissues or points in development. This uniquely flexible in vivo system now combines the conventional strengths of *Xenopus* with loss-of-function genetic backgrounds and enhanced genomic tools, multiplying the reach of what has already been described as "perhaps the best vertebrate model organism for functional genomics" (22).

This chapter surveys methods for genetic analysis of *X. tropicalis* development, including genome manipulations (haploid genetics, gynogenesis, and androgenesis), uncovering naturally occurring mutations, mutagenesis, screening protocols, mapping strategies, sequence-based reverse genetic strategies, and analysis of mutant phenotypes.

2. Materials

2.1. General Embryology

1. 0.05× MMR+BSA: 1 mg/mL Bovine Serum Albumin (BSA) in 0.05× MMR; Adjust 0.05× MMR to pH 8.3 prior to adding BSA to prevent protein accumulation on pH probe. BSA is slightly acidic and will bring pH down to ~7.7–7.9.
2. 2.2% cysteine: Cysteine hydrochloride (Sigma) in 0.05× MMR, adjust pH to 7.7–7.9 with 10 N NaOH. Use within 2 h.
3. 0.4% MS-222: Add 4 g of Ethyl 3-aminobenzoate methanesulphonate (MS-222) to 1 L 0.05× MMR. Adjust pH to 7.7 with 1 N NaOH. Store at 4°C and reuse up to ten times.
4. L15/CS: Leibovitz-15 (L-15) media (GibcoBRL) supplemented with 10% calf serum (CS) (GibcoBRL). Store 10 mL aliquots at −20°C.

These materials are used in a variety of different protocols. Materials for specific procedures are indicated below. All chemicals are obtained from Sigma unless otherwise specified.

2.2. Husbandry and Obtaining Embryos

1. Human Chorionic Gonadotropin (HCG) (Chorulon, Intervet). Make stock of 1,000 U/mL in sterile H_2O and dilute accordingly. Store at 4°C. Sterilize seal with ethanol before and after each use.
2. Sera Micron Powder (Sera; Heisenberg, Germany) tadpole diet.
3. Reptomin Sticks (Tetra; Melle, Germany) adult frog diet.
4. Tropical Fish Flake (Sinclair Animal & Household Care, Gainsborough, UK) or equivalent; diet for metamorphosing tadpoles froglets and supplement for adults.
5. Sorting tools: manual pipette pump and glass Pasteur pipettes (*X. tropicalis* embryos tend to stick to plastic transfer pipettes). Notch glass Pasteur pipettes with a diamond pen, break off, and blunt edges with a Bunsen burner flame.

2.3. Karyotyping

1. 27G hypodermic needles.
2. Microscope slides (e.g., positively charged Superfrost Plus from Fisher) and large coverslips.
3. Paper towels.
4. Distilled H_2O.
5. 60% acetic acid in distilled H_2O.
6. Hoechst 33342 stain (Sigma; Poole, UK), working stock 0.1 mg/mL in distilled H_2O.
7. 70% glycerol in phosphate buffered saline (PBS).

2.4. Mutagenesis

1. *N*-nitroso-*N*-ethylurea (ENU). 1 g Isopac (Sigma; Poole, UK).
2. 2-(*N*-morpholino)ethanesulphonic acid (MES, see Note 1) (Sigma; Poole, UK): Make two 100 mM stocks in dH_2O. Adjust one to pH 6.0 and one to pH 6.2 with 1 N NaOH, store at 4°C.
3. Lab coat, plastic wrist guards, gloves, facemask.
4. Decontamination bath: 10% sodium thiosulphate, 1% sodium hydroxide in H_2O.
5. Nutator or roller.

2.5. Mapping

1. Embryo lysis buffer: 50 mM Tris–HCl pH 8.8, 1 mM EDTA, 0.5% Tween-20, store stock at RT; add 200 μg/mL Proteinase K (Roche) immediately prior to use.
2. PCR-compatible 96 well plates.
3. Standard PCR reagents and equipment.
4. Super Fine Resolution Agarose (SFR) (Amresco; Solon, USA).

2.6. Acrylamide Gels and Silver Staining

1. Benchtop sequencing gel apparatus (e.g., Thistle Scientific Model 2) with matching glass plates; 0.4 mm spacers & shark's tooth combs.
2. 1 photographic developing dish large enough to fit glass sequencing plate.
3. Denaturing DNA loading buffer: 50 mL stock = 49 mL formamide, 1 mL 0.5 M EDTA, 0.1 g bromophenol Blue, 0.1 g xylene cyanol.
4. 1× Tris Borate EDTA (TBE) Buffer.
5. 3-(Trimethoxysily)propyl methacrylate.
6. 100% and 70% ethanol.
7. Acrylease (Stratagene).
8. 1 L 10% ethanol.
9. 1 L 1% nitric acid.
10. 1 L 2 g/L silver nitrate.
11. 1 L Developing solution: 29.6 g sodium carbonate, 450 μL 37% formaldehyde. Prepare in advance and keep on ice.
12. 1 L 10% acetic acid.

2.7. DNA Prep for Genotyping from Tissue

1. 0.07% MS-222: Add 0.7 g of ethyl 3-aminobenzoate methanesulphonate (MS-222) to 1 L 0.05× MMR. Adjust pH to 7.7 with NaOH.
2. Scalpels/razor blades.

3. Lysis buffer: 100 mM Tris–HCl pH 8–8.5, 200 mM NaCl, 0.2% SDS, 5 mM EDTA; 100 μg/mL Proteinase K added just before use (Roche).
4. Isopropanol.
5. 70% ethanol.

2.8. Sperm Freezing

1. Cryoprotectant: Disperse one egg yolk (about 15 mL) in an equal volume of distilled water; dilute to 20%v/v in 0.4 M sucrose, 10 mM sodium bicarbonate, 2 mM pentoxyfylline solution. Centrifuge for 20 min at 13,000 rpm and use supernatant (store at −20°C for up to 1 month).
2. Styrofoam box small enough to fit into −80°C freezer.

3. Methods

3.1. Husbandry and Obtaining Embryos

Multigeneration genetic studies critically depend on minimizing generation time and maximizing egg quality and fertilization success; diet and husbandry are extremely important, particularly for tadpoles/froglets and egg-producing females. *X. tropicalis* will not thrive in same conditions as *X. laevis*, and the two species should *never* be housed in shared water systems, even with filtration, due to the risk of trans-species infection. See Chapters 2 and 3 for in-depth discussions of husbandry and breeding; short protocols for in vitro fertilization are provided in Subheading 3.2.1.

3.1.1. Husbandry

Briefly, frogs should be maintained in 24–26°C water, avoiding sudden changes during cleaning. Temperatures as high as 30°C are tolerated, but egg quality from females maintained above 27°C diminishes sharply. Temperatures below 23°C may depress disease resistance; below 20°C is usually lethal for *tropicalis*. *Xenopus* tadpoles are filter feeders, so optimal growth is obtained with suspended nutrients, and flow-through recirculation systems should be turned off during feeding. Sera Micron powder, fed frequently, is an excellent diet for small tadpoles and causes minimal fouling of standing water systems. Larger tadpoles and metamorphosing froglets can be supplemented with crushed fish flake daily. Adult frogs do well on a variety of protein-rich diets; we use Tetra Reptomin sticks (aquatic turtle diet) supplemented with fish flake. Reptomin is crushed into fragments for smaller froglets to young adults (see Note 2).

3.1.2. *Xenopus tropicalis* Strains

Strains that differ from each other at many sequence loci (polymorphic strains) are valuable for genetic mapping, whether by conventional means or next-generation sequencing approaches. Two strains (*IC* (originating in the Ivory Coast) and *N* (Nigeria)) have been inbred for >12 generations and successfully used for mapping mutations (18); a third strain, *ICB*, is also being inbred (23). An inbred *N* animal was the basis of the draft genome assembly (14); these animals may thus be more effectively targeted by sequence-based interventions such as morpholino oligonucleotides than the *IC* strain. Wild caught animals are occasionally available, but extreme care must be taken to prevent disease introduction and taxonomic misidentification, as morphologically identical nondiploid species (e.g., *X. epitropicalis*) are known to occur in overlapping range.

3.2. Genome Manipulations: Haploid Genetics, Gynogenesis, and Androgenesis

The ease with which the entire genome can be manipulated in amphibians is a great advantage for genetic and transgenic applications. Simple and efficient procedures exist for generating both haploid *X. tropicalis* embryos, which can undergo several days of development, and viable diploid gynogenetic embryos derived solely from the maternal genome (see Note 7). Haploids derived entirely from the paternal genome, or androgenetic haploids, can also be obtained, but in much smaller numbers (24); in this chapter, "haploid" indicates maternally derived unless otherwise specified. These manipulations are extremely useful for a wide range of applications including high-throughput forward genetic screens, identifying polymorphisms, rapid low-resolution genetic mapping, making transgenes homozygous, and generating completely isogenic strains.

Xenopus eggs, like those of many other lower vertebrates, are deposited prior to completion of second meiosis; extrusion of the second polar body normally begins a few minutes postactivation. If sperm suspensions are UV irradiated prior to use for in vitro fertilization, egg activation, polar body formation, and cleavage can occur normally, but the crosslinked paternal genome cannot contribute to the zygote, resulting in formation of haploid embryos (Fig. 1). Haploid embryos are not viable beyond feeding stages and show a high level of gastrulation defects and posterior abnormalities. However, they form anterior structures well enough for many phenotypes to be scored and are also very useful for characterizing parental polymorphisms. If the mother is a heterozygous carrier of a recessive mutation, scorable phenotypes are expected to be uncovered in 50% of the haploid progeny.

In gynogenesis, embryos fertilized with UV-irradiated sperm are rescued to diploidy using one of two basic methods. Polar body formation can be blocked with a simple cold shock shortly after fertilization (early cold shock, ECS) leading to the retention of both sister chromatid products of meiosis II. This method rescues haploids to viable diploidy with high efficiency and is extremely useful for uncovering recessive phenotypes in the progeny of carrier females. Critically, Mendelian phenotypic ratios are *not* expected in ECS embryos (gynogenotes) due to recombination during meiosis I and vary between a maximum of 50% phenotypically

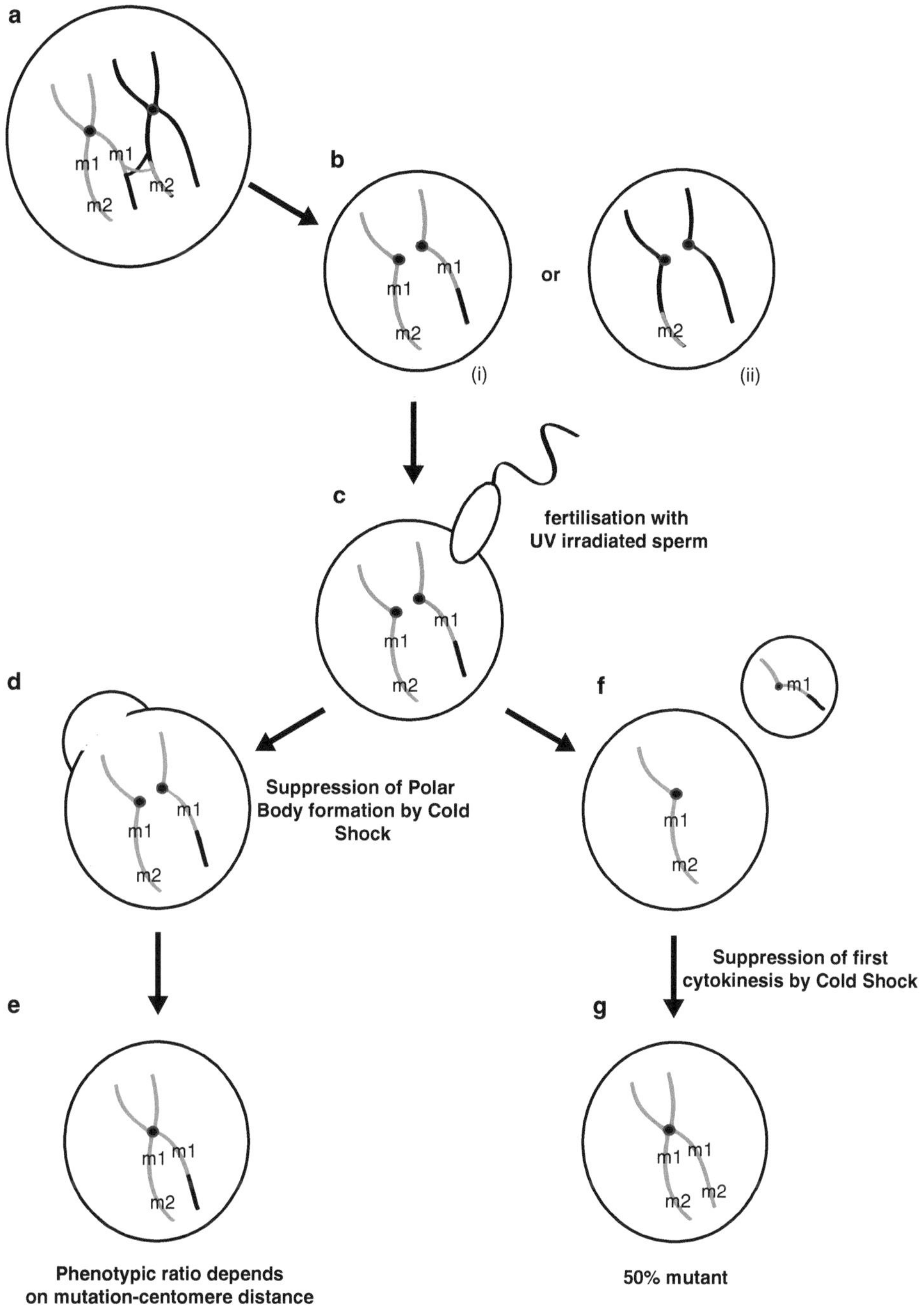

Fig. 1. Formation of haploid and gynogenetic embryos. (**a**) Diplotene oocyte in a female hybrid for mutagenized gray strain chromosomes and polymorphic black strain, showing a crossover event between mutant loci m1 and m2. (**b**) Unfertilized eggs showing segregation of sister chromatids after Meiosis I. Note that regions where centromeres hold sister chromatids together are homozygous. (**c**) UV-irradiated sperm activates development without paternal genetic contribution, forming haploid embryos (**f**) following polar body extrusion. (**d**) Early cold shock (ECS) suppresses formation of the second polar body, with the resulting gynogenote (**e**) rescued to diploidy and retaining both sets of sister chromatids. Recessive phenotypes at loci closer to centromeres (m1) are more likely to be uncovered than distal loci (m2), where recombination produces heterozygous wild type. (**g**) Late cold shock of haploid embryos following DNA replication prevents first cytokinesis, rescuing haploid to completely homozygous diploids.

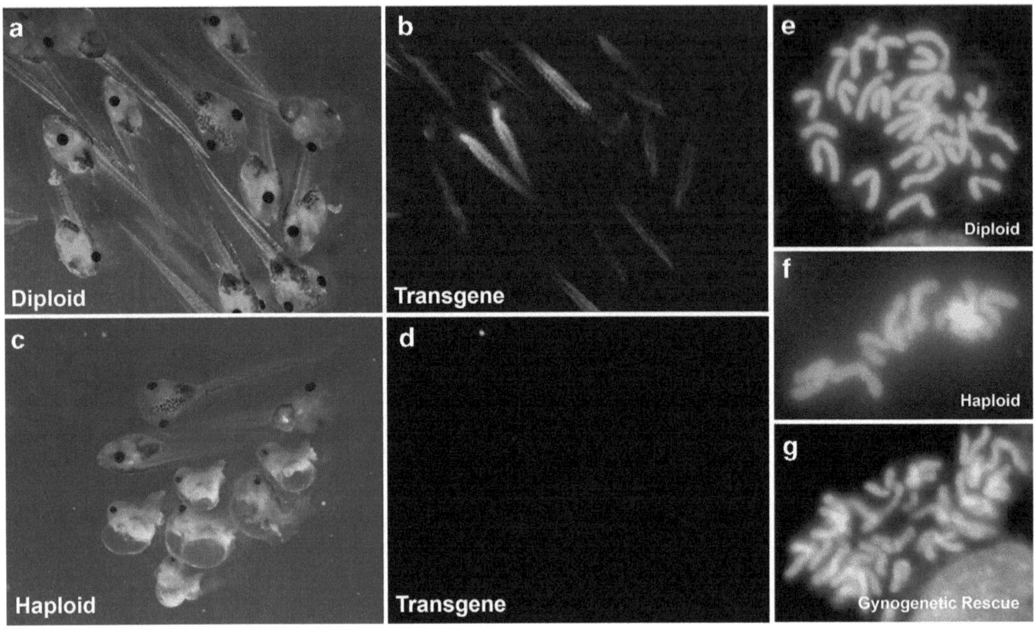

Fig. 2. Haploid and gynogenetic embryos. Wild-type eggs were fertilized using homozygous cardiac actin-RFP transgenic sperm, either untreated to form conventional diploid embryos (**a**) or UV irradiated to form haploids (**c**). Paternal transgene transmission is visible in the diploid clutch (**b**) but absent in the haploids (**d**). Haploids form anterior structures well, but posterior structures are truncated. Note that three spontaneously diploidized embryos without paternal transgene appear in the haploid clutch. (**e**–**g**) show the karyotype of outcrossed diploid (**e**), haploid (**f**), and ECS gynogenetic diploid embryos (**g**).

recessive mutant (for centromere-linked loci) and a minimum of ~5% for distal loci. The observed ratio can provide useful low-resolution map information (see Fig. 1 and Subheading 3.7.4). ECS, which relies on the retention of the maternal polar body, is not applicable to androgenetic haploids.

Alternatively, gynogenetic haploid embryos can be allowed to undergo the first round of DNA duplication, then rescued to diploidy by blocking the first cell division with late cold shock (LCS). This procedure is usually much less efficient than ECS, but can result in completely homozygous isogenic embryos and uncovers recessive phenotypes regardless of chromosomal location in 50% of the progeny of heterozygous carriers. LCS is also theoretically possible using androgenetic haploids, although the number of viable homozygous diploid embryos recovered is very small. These diploidization procedures were originally developed in *X. laevis* using pressure treatments (25, 26), but have been modified for the simpler cold shock technique in *X. tropicalis* by Rob Grainger's group (University of Virginia) (Fig. 2).

3.2.1. Production of Haploid Embryos

1. 12–72 h prior to procedure, prime two or more adult female *X. tropicalis* with 10 U HCG in 100 μL sterile water (see Notes 3 and 4).

2. On day of procedure, boost primed females with 100–200 U HCG in 100 μL sterile water (see Notes 3 and 4).
3. ~3 h after boosting females, kill two *X. tropicalis* males (see Note 5) and dissect testes into L15/CS.
4. Label two 10 cm culture dishes "haploid" or "control."
5. Place a few drops of 1× MMR into "haploid" dish for each female. High salt medium prevents premature activation of eggs.
6. Express eggs into media; avoid getting tank water from the frogs onto the eggs. Discard dead, lysing, or stringy eggs (see Note 4).
7. Transfer small number of eggs with pipette to "diploid" control dish.
8. Place testes in eppendorf tube containing 500 μL L15/CS, macerate with eppendorf pestle, add another 500 μL L15/CS and mix.
9. Allow testis fragments to settle, and place sperm suspension onto glass Petri dish; save large chunks of testis for diploid control.
10. UV-irradiate sperm suspension in Stratalinker (Stratagene) or equivalent with 50,000 μJ ("energy" setting 500).
11. Add irradiated sperm to "haploid" dish (see Note 5).
12. Add 500 μL L15/CS to the nonirradiated testis fragments in eppendorf tube, mix and use to fertilize "control" plate.
13. Gently shake dish to mix eggs and sperm, wait 5 min and flood with 0.05× MMR.
14. Dejelly (see Chapter 2) with 2.2% cysteine in 0.05× MMR pH 7.7–8.0, rinse, sort evenly cleaving embryos, and culture overnight at 25°C (see Note 6).

3.2.2. Early Coldshock (Gynogenesis by Suppression of Polar Body Formation)

As in haploid production (Subheading 3.2.1). In addition:

1. At least 1 h before in vitro fertilization, chill ~50 mL 0.05× MMR on ice per female being screened.

   For each female, label two 10 cm dishes "control," "haploid," plus one 6 cm dish "ECS" (*E*arly *C*old*s*hock).

As in haploid production (Subheading 3.2.1) steps 1–12.

13. Set timer for 5′, flood embryos with 0.05× MMR, start timer.
14. Transfer ~90% of embryos from flooded "Haploid" dish to ECS dish and remove media.
15. At 5′, add ice-cold 0.05× MMR to ECS dish and place in slushy ice bucket for 7′30″.
16. After 7′30″ remove ECS dishes from slush bucket and replace media with RT 0.05× MMR.

17. Wait >20 before dejellying and sorting. In ECS embryos cleavage will be delayed by 15–20′ relative to haploid and diploid controls (see Notes 5 and 6).

3.2.3. Late Cold Shock (Gynogenesis by Suppression of First Cleavage)

Suppressing cytokinesis after the first round of DNA replication in the fertilized embryo can also rescue *Xenopus* haploids to completely homozygous diploids. Time to first cleavage is critical and varies with egg batch as well as temperature. For optimal success, aliquots of each fertilization should be timed (see protocol below).

As in haploid production (Subheading 3.2.1); in addition

1. At least 1 h before squeezing females, chill ~50 mL 0.05× MMR per female in ice bucket as described in ECS section.
2. For each female, label four 10 cm dishes "diploid," "haploid," and "LCS" and "timer."

Procedure as in haploid protocol steps 1–9.

13. Set timer to count up and establish time to first cleavage by removing a small aliquot of sperm and eggs from each haploid dish to "timer" dish and flood with 0.05× MMR, start timer. Keep all dishes and media at the same temperature.
14. 3′ after flooding test fertilization, flood haploid and diploid control dishes.
15. Transfer ~90% of embryos from flooded "Haploid" dish to LCS dish.
16. After ~45′, remove media from LCS dish.
17. After 45′, observe "timer" dish every 1 or 2 min for onset of cleavage furrow. First cleavage can take between 45′ and 70′ at 23–25°C.
18. At first sign of cleavage furrow in "timer" dish, add ice cold 0.05× MMR to LCS dish and place in slushy ice for 5′.
19. After 5′ remove LCS dishes from ice bucket and replace media with RT 0.05× MMR.
20. Wait >20′ before dejellying and sorting (see Note 8).

3.2.4. Production of Androgenetic Haploids

1. 12–72 h prior to procedure, prime 4–6 adult female *X. tropicalis* with 10 U HCG in 100 μL sterile water (see Notes 3 and 4).
2. On day of procedure, boost primed females with 100–200 U HCG in 100 μL sterile water (see Notes 3 and 4).
3. ~3 h after boosting females, kill two *X. tropicalis* males and dissect testes into L15/CS.
4. Label 6 cm culture dishes "haploid" or "control."
5. Prepare sperm solution: place testis from 1 male in a 3 cm dish w/~400 μL L15 + 10% FBS and macerate w/forceps.

6. Perform a test fertilization to identify the females producing the best eggs:
   (a) Gently squeeze a very small number of eggs (~20–30) from each female into a dry dish; avoid getting tank water from the frogs onto the eggs. Discard batches with dead, lysing, or stringy eggs.
   (b) Add a drop of testis solution to remaining batches of eggs, flood with 1/20× MMR, wait ~10′ and inspect for activation (cortical contraction). Select female(s) showing the best fertilization.
7. Gently squeeze a very small number of eggs (~30–50) from selected female into a dry 6 cm dish; avoid getting tank water from the frogs onto the eggs. Proceed quickly through next steps to avoid drying out the eggs.
8. Using number 5 watchmaker's forceps, quickly transfer 10–30 good-looking eggs one by one to a 3 cm dish, animal pole up (for irradiation). Avoid eggs that lack a visible germinal vesicle.
9. Irradiate eggs in Stratalinker 1× 50,000 μJ ("energy" setting 500).
10. Cover eggs w/testis solution, then remove testis solution for re-use if more eggs are to be treated (step 13 below).
11. Add a small amount of testis solution to an aliquot of nonirradiated eggs as control.
12. Flood both irradiated and control eggs with 1/20× MMR.
13. Repeat with several more small squeezes, reusing the testis solution.
14. Fertilization rates in androgenetic haploids under optimal conditions can be nearly as good as with controls, but the total number of cleaving embryos produced is limited by manipulating individual eggs in jelly.

3.2.5. Karyotyping

Karyotying may be used to determine ploidy status of experimentally manipulated embryos as well as distinguish among *X. tropicalis* and similar nondiploid species. This protocol was developed by the Grainger lab (University of Virginia) and modified by M. Khokha (Yale University).

1. Place 10 Stage 24–34 tadpoles into a dish of deionized water.
2. With a scalpel or 27G needle, remove the yolky ventral portion of the tadpole and discard; allow remaining dorsal portions to stand for 20′.
3. Pipette the dorsal halves with as little water as possible into an eppendorf tube containing 0.2 mL of 60% acetic acid in water; let stand 5′.

4. Pipette all of the tissue (with minimal acetic acid) and place on a positively charged slide (e.g., Superfrost Plus from Fisher); blot away excess acetic acid.
5. Place a large coverslip on the slide. Fold a paper towel to the size of the coverslip and place it on top. Apply heavy pressure on top of the paper towel/coverslip for about 5 min using a lead brick or by pressing forcefully with a thumb, being careful not to move around (see Note 9).
6. After 5 min carefully remove the lead brick and paper towel.
7. Place slide on dry ice for 5 min, then remove from dry ice and use a razor blade to gently pry the coverslip from the still-frozen slide.
8. Place slide on a paper towel and stain chromosomes with Hoechst 33342 (1 μL Hoechst 33342 (stock: 0.1 mg/mL) in 1 mL distilled water) for 5′. Wear gloves when working with Hoechst.
9. Tip slide to allow stain to run off onto the paper towel.
10. Mount in a drop of 70% glycerol/PBS, add large coverslip and seal edges with clear nail polish.
11. Inspect slide under fluorescent illumination for stained metaphase spreads using a ×63 or higher power objective.

3.3. Mutagenesis Strategies

Although naturally occurring mutations may be recovered from untreated animals (16, 17), genetic screens normally begin with mutagenesis. The same mutagenized stocks can often be used for both phenotype-based forward and sequence-driven reverse genetic screening strategies. Methods for inducing mutations in *X. tropicalis* include X-ray deletions, insertion or mobilization of transgenes (19) (see Chapter 6), gene editing with engineered zinc finger nucleases (27) (see Chapter 7), and chemical mutagenesis (15). With the exception of gene editing, these methods produce randomly distributed genomic lesions to a first approximation.

Chemical mutagenesis is currently the most efficient method for inducing large numbers of simple sequence lesions to obtain a range of specific phenotypes. The combination of next-generation sequencing technology and improved genomic resources has greatly simplified cloning chemically induced point mutations. The size of deletions produced by X-irradiation is difficult to control, making it hard to associate phenotypes with specific gene functions. Gene disruption by insertional mutagenesis is an attractive strategy for rapid cloning of mutations using inverse PCR-related strategies to isolate endogenous sequences flanking known transgene constructs (28). Insertional mutagenesis by REMI nuclear transfer (see Chapter 11) has already identified a novel limb development gene function (19). Insertional mutagenesis via transposon-mediated transgenesis, which has the potential to greatly increase mutagenesis efficiency, is discussed in Chapter 6, and may eventually make gene trap and related strategies feasible (29). Since transgenesis can induce multiple lesions in the genome, it is important to recognize that identification of a transgene insertion does not supplant linkage analysis and/or independent functional correlation of the putative mutated gene with the phenotype.

3.3.1. Strategies Using Chemical Mutagens

Most extant *Xenopus* mutations have been induced using chemical mutagens. Positional cloning to identify affected genes (see Subheading 3.7) is greatly simplified by improved genomic resources including the chromosome-scale version 7.1 genome assembly (accessible via www.xenbase.org) and high density meiotic map (30). High-throughput sequencing of whole exomes, either using individual carriers or pooled mutant embryos, is poised to revolutionize identification of mutant lesions both in reverse genetic *T*argeting *I*nduced *L*ocal *L*esions *I*n *G*enomes (TILLING) strategies and underlying phenotypes identified in forward screens.

Chemical mutagens may be applied in vitro to mature sperm from dissociated testes or in vivo to target spermatogonia by injection into adult male frogs:

- Mutagenesis of dissociated testes followed by in vitro fertilization efficiently induces sequence lesions, and a useful rapid readout of mutagenesis efficiency is provided by the appearance of dominant phenotypes in the resulting embryos. The drawback is that the founding F1 generation is mosaic, because the chemical adducts (usually produced by alkylating agents such as ENU) on a single strand of the sperm DNA double helix are not repaired and permanently fixed on the complementary strand until the first somatic DNA replication or later. This mosaicism in the F1 generation precludes conventional 3-generation screens, since the likelihood of F2 sibling pairs carrying the same induced allele drops to 1/16 or lower, and the number of matings required to uncover recessive mutations becomes unworkable. Mosaic F1 animals are likewise unsuitable for direct analysis by high-throughput genomic sequencing, since a mutation present in a minority of reads at a given locus is likely to be invisible against the base call error rate. However, in vitro mutagenesis is compatible with haploid or gynogenetic screens using females of either mosaic F1 or nonmosaic F2 generations. The nonmosaic F2 generation is also suitable for tissue sampling for reverse genetic approaches.
- In vivo spermatogonial mutagenesis, in contrast, is compatible with either conventional 3-generation or haploid/gynogenetic screens. In practice, it lacks the immediate confirmation of mutagenesis provided by dominant phenotypes following in vitro mutagenesis. Mutagenesis rates can be scored by genomic sequencing in the next generation (see Subheading 3.6).

3.3.2. Insertional Mutagenesis

Insertional mutagenesis is an attractive strategy, since known transgene sequences facilitate identification of genomic integration sites and reduce reliance on positional cloning. The first insertional mutation in amphibians, producing a spectacular forelimbless phenotype, was recently described (19). As transgenesis protocols can introduce DNA damage elsewhere in the genome, linkage studies are still useful for confirming association of a phenotype with a specific insertion. A number of protocols have been described for mediating stable transgenesis in *Xenopus*, including transfer of sperm nuclei (10) (see Chapter 11), various transposable elements (28) (see Chapter 14), I-SceI meganuclease (31) (see Chapter 12), and phiC31 integrase (32) (see Chapter 13). However, the relative inefficiency of transgenesis in *X. tropicalis* has thus far precluded large-scale screens for insertional mutants. Genetic manipulation of transgenic lines is also potentially powerful. Many reporter lines have been established in *X. tropicalis* (33), which may be useful substrates for genetic screens focusing on specific tissues or processes, and binary and inducible systems are available for experimental manipulation of gene function (34, 35). Cre recombinase has also been shown to function in *Xenopus* stable transgenics (36). While targeted knock-in to engineer endogenous loci has not yet been demonstrated in *Xenopus*, null mutations are becoming increasingly available. Established methods may be used to introduce engineered transgenes and rescue the null background to viability, meanwhile creating extremely useful floxed conditional knockouts, tagged proteins, and other informative alleles.

3.4. Chemical Mutagenesis

ENU is highly carcinogenic and must be treated with extreme caution; all manipulations should take place in a fume hood, wearing lab coat, double gloves, and plastic wrist guards. All materials that come into contact with ENU solutions should go into decontamination bath for 24 h. ENU solutions are also highly labile, temperature- and pH-sensitive, and biologically effective dosage can be difficult to control. It is recommended that a titration series is performed for each batch prepared and aliquots frozen at –80°C.

3.4.1. ENU Mutagenesis

Preparation of ENU Stock Solution

1. Line a fume hood with absorbent benchcoat and place a decontamination bath and waste container/burn bin within.
2. Prepare 100 mL 5 mM MES solution from 100 mM pH 6.0 stock in dH_2O (see Note 1).
3. Remove ENU isopac bottle from protective canister (save can).
4. Using a 50 cc syringe with 18G needle, inject 85.4 mL 5 mm MES pH 6.0 into ENU isopac bottle, carefully withdrawing air from bottle into syringe while adding medium to avoid overpressurizing the bottle.

5. Return bottle to shipping canister (or cover with aluminum foil) and place on nutator or roller shaker in hood for several hours, occasionally monitoring.
6. When powder is all (or nearly) in solution, swirl, allow to settle, and freeze 1 mL aliquots at −80°C. Retain 20 μL for spectrophotometric determination of concentration if desired.

Determining ENU Concentration by Spectrophotometry (Optional)

1. 100 mM ENU = approx. 11.7 mg/mL (1 g/85.4 mL).
2. Dilute 20 μL ENU solution to 1 mL with 5 mm MES pH 6.0 (i.e., 1:50 dilution).
3. Using a disposable plastic cuvette, determine OD_{398}.
4. 1 mg/mL solution of ENU gives $OD_{398} = 0.72$.
   (a) Therefore, (ENU) mg/mL = $(OD_{398})(50)/0.72$
   (b) Or (ENU) mg/mL = $(OD_{398})(69.4)$

In Vitro ENU Mutagenesis of Mature Sperm

1. 12–72 h prior to procedure, prime 5 or more adult female *X. tropicalis* with 10 U HCG in 100 μL sterile water (see Notes 3 and 4).
2. On day of procedure, boost primed females with 100–200 U HCG in 100 μL sterile water (see Notes 3 and 4).
3. Prepare 10 mL of 3 mM MES pH 6.2 in L15 (without calf serum) (add 0.3 mL 100 mM MES pH 6.2 stock to 9.7 mL L15).
4. Thaw an aliquot of L15/10% CS.
5. Kill five males and dissect testes into L15/CS media.
6. Prepare four 15 mL tubes with L15/3 mM MES (see Note 1) (do not add ENU stock or sperm suspension until the last moment).

Sperm (mL)	ENU 100 mM stock	L15/3 mM MES6.2	f.c. (mM)
0.1	0	0.9 mL	0
1	0.1	0.9	5
1	0.15	0.85	7.5
1	0.2 mL	0.8	10

7. Thaw an aliquot of 100 mM ENU.
8. Macerate testis in 0.5 mL L15/3 mM MES pH 6.2 (no CS) using eppendorf and pestle, bring volume to 1.5 mL with L15/3 mM MES, swirl to mix.

9. Add 0.1 mL sperm solution to 0 mM ENU control and 1 mL to the 15 mL tubes with L15/3 mM MES corresponding to each of the ENU treatments.
10. Add ENU as indicated in step 6 and swirl to mix.
11. Place at 18°C for 1 h. Swirl to mix every 15′.
12. Add 10 mL L15 to each tube and spin down sperm for 5′ 1,000 rpm at RT in benchtop centrifuge.
13. While sperm solutions are spinning, squeeze eggs from females into a drop of 1× MMR; discard lysing/dead eggs, pool eggs from good females, mix, and split into three 15 cm dishes for the ENU doses and a smaller aliquot of eggs in a 10 cm dish for no-ENU control.
14. At conclusion of spin, carefully pipette as much of the supernatant as possible to decontamination bath without disturbing the sperm pellet, then resuspend sperm pellet in residual liquid by flicking. Repeat steps 12 and 14 2×, then gently resuspend in 1 mL L15/CS.
15. Remove remaining MMR from eggs, add treated sperm solution to eggs and mix by shaking briefly.
16. After 5′, flood with 0.05× MMR twice (removing first rinse to decontamination bath), dejelly after ~20′ postfertilization and transfer to fresh dishes. Eggs may now be treated as safe to handle normally.
17. *Important*: sort control and ENU-treated dishes at 4–8 cell stage, making sure to make comparable dishes of *regularly cleaving* embryos from all doses and controls. It is much simpler to identify regularly cleaving embryos at 4–8 cells than at later stages; dominant ENU-induced defects will only be scorable if equivalent regularly cleaving embryos are compared.
18. The following 3 days, sort viable embryos and score samples of control and mutagen-treated for dominant effects on gastrulation, death, and other abnormalities.

wt (%)	Gastrul. defect	Edema	Other	Dead
Control				
ENU 5 mM				
ENU 7.5 mM				
ENU 10 mM				

19. At feeding stage, select the dose(s) that result in a population of viable embryos, but also show clear dominant effects compared to controls. If desired, expand population of animals treated at this dose.

3.4.2. Spermatogonial Mutagenesis

Mitotic spermatogonia, rather than mature sperm, can be targeted for in vivo mutagenesis by injecting adult male frogs with ENU. Replication in the spermatogonial lineage then fixes mutations in the germline, avoiding mosaicism in the F1 generation. Animals usually need several months to recover after an injection series, during which time mutagenized mature sperm (which would contribute to an unwanted mosaic F1) will be cleared. Too much ENU can ablate the testes completely or result in repopulation from a small number of surviving spermatogonia, producing founder effects distorting mutagenesis rates.

1. Obtain five or more adult male frogs.
2. Weigh individual frogs (males are typically between 6 and 9 g).
3. Record weight and calculate dose of 100 mM ENU stock needed for injection (0.1 mg ENU/g frog). Dose per frog (in mL) = (frog weight) × (0.1 mg/g) × (1 mL/11.7 mg ENU), or 0.006333× frog weight.
4. Immobilize frogs by immersing for 2–5′ in a fresh stock of 0.07% MS222 pH 7.8 at RT until they visibly begin to slow down. Immobilizing frogs with anesthetic during injection reduces risk of serious accidents with ENU-contaminated needles.
5. Inject the volume calculated in step 3 to contain 0.1 mg ENU per gram of frog weight subcutaneously into dorsal lymph sac (see Notes 2 and 3).
6. Allow frogs to recover on wet paper towels in observation tank. They are usually awake and active in ~15–20 min.
7. Transfer frogs to observation tank with fresh water. Make sure to discard paper towels and liquid in ENU waste and treat appropriately as ENU waste material.
8. After several hours, discard frog water in ENU liquid waste and replace with fresh water.
9. The next day, discard frog water in ENU liquid waste and replace with fresh water. Do this throughout the day for two to three more water changes.
10. Return frogs to colony.
11. Re-inject once a week for a total of 3 doses.
12. Allow frogs to recover for >3 months before breeding.

3.5. Genetic Screens

Forward genetic screens

Conventional 3-generation breeding schemes to uncover recessive phenotypes are compatible with spermatogonial mutagenesis (Fig. 3). However, in vitro mutagenesis of mature sperm results in a mosaic F1 generation, making recovery of homozygotes by incrossing in subsequent generations prohibitively inefficient.

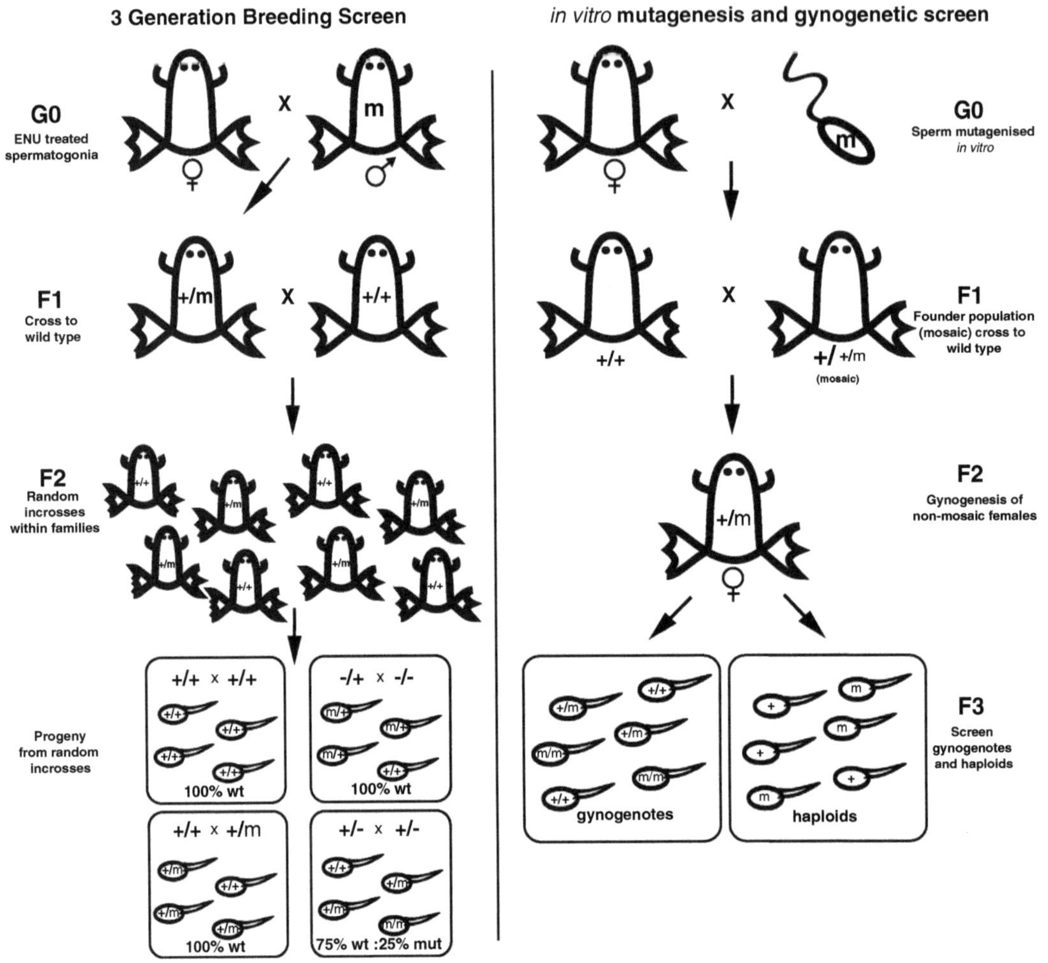

Fig. 3. Screening mutagenized populations. *Left*: 3-generation breeding screen. In vivo mutagenized males are crossed to wild-type females (G0), with the resulting F1 individuals outcrossed again to wild type to create F2 families. Progeny of random incrosses within families are then analyzed for mutant phenotypes. *Right*: In vitro mutagenesis and gynogenetic screen. Eggs are fertilized with mutagenized mature sperm, creating a mosaic F1 population, which is crossed again to create a population of nonmosaic F2 candidate carrier animals. F2 females are then screened by gynogenesis to uncover recessive mutations.

Populations derived from in vitro mutagenesis can still be effectively screened by gynogenesis. Mutations have been successfully recovered by screening the gynogenetic progeny of mosaic F1 females (15), but screens using nonmosaic F2 females are likely to be more efficient. Gynogenesis by polar body suppression (ECS) produces some bias towards recovery of centromere-linked alleles, as these loci will be uncovered in a higher proportion (up to 50% of gynogenetic progeny (gynogenotes)) than those produced by more distal loci (as low as ~10% phenotypically mutant for loci unlinked to centromeres, see Subheading 3.7.1). However, females in good condition can produce sufficient numbers of gynogenotes to recover mutations even in telomeric loci.

3.5.1. Early Coldshock Gynogenetic Screen for Recessive Phenotypes

Gynogenetic screens are primarily used to identify carrier females that are heterozygous for recessive mutations. Ideally, nonmosaic animals are screened, since a greater proportion of progeny will be homozygous for a given mutant locus and more readily detected. Adult females being screened may be housed individually or otherwise marked for the duration of the screen.

Early coldshock gynogenesis (see Subheading 3.2.2) should include haploid and diploid outcross (nonirradiated sperm) controls. The diploid outcross control serves to assess sperm and egg quality, reveals dominant effects, and can be raised as the next generation if desired. Haploid controls help evaluate efficiency of sperm irradiation (see Subheading 3.2.1) and are useful for identifying polymorphisms (Subheading 3.7.2). Some anterior phenotypes may also be scorable on the haploid background, where they would be observed in ~50% of the embryos.

The fraction of ECS gynogenotes in which a given recessive phenotype is observed is not Mendelian. Loci that are close to centromeres will be uncovered in up to 50% of the gynogenetic progeny of heterozygous females. More distal loci will tend to be observed with progressively lower frequency, with a plateau at ~10–15% due to presence of multiple crossovers.

ECS may not result in quantitative rescue of all haploid embryos to diploidy. Background abnormalities from remaining haploid and aneuploid embryos can make it challenging to identify preneurulation phenotypes. Postneurulation phenotypes can be screened efficiently by selecting morphologically perfect wild-type embryos from ECS and diploid control dishes on the morning after fertilization (St. 18–22) and monitoring these for subsequent appearance of abnormalities.

3.5.2. Morphological Screening Checklist

The following screen can be applied to embryos derived either from gynogenesis or from conventional matings; for simplicity, gynogenesis is described.

1. Compare sibling outcrossed diploid with ECS and haploid embryos for stage-specific developmental processes and to establish a baseline of egg-based nonheritable abnormalities and/or dominant phenotypes. Specific phenotypes that are uncovered in multiple embryos within a clutch are particularly convincing. If you see a phenotype in ECS or haploid dishes, separate those embryos and record specific defect(s) and number of phenotypically mutant and wild-type embryos. The phenotype might be lethal; isolating those embryos will make it easier to score the following day. Phenotypes that are scorable in haploids are expected at 50%. Single-gene phenotypes in a clutch of ECS embryos are expected at a maximum of 50% for centromere-linked loci, decreasing to ~10% for distal loci. Record all abnormalities on a scoresheet (example in Fig. 4). Collect both mutant and wild-type

Gyno Scoresheet Your Initials: MatingDate:

Mother Stock No: Genotype: Frog No.

A: Axis	B: dwarf	C:Circulation	D: Ear	E: Oedema	F: Eye
G: Gut	H: Heart	I: Head	J: pronephr	K: Cilia	L: lethal/dead
M: Motility	N: Neural	O: Other/notes	P: Pigment	Q: blastop.	R: pre-gast

ECS

Phenotype	**Day1**	**Day2**	**Day3**	**Day4**
Wild type				
Total				

Haploid

Wild type hap				
Diploid-like				
Total				

Diploid outcross

Wild type				
Total				

Notes:

Conclusion:

Fig. 4. Gynogenetic screening checklist. Sample form for scoring phenotypes during a gynogenetic screen.

ECS embryos in 96 well plates for use in low-resolution mapping and assignment of linkage group (see Subheadings 3.6.1 and 3.7). Also collect a small set (6–12) of haploid embryos for identifying polymorphic markers (Subheading 3.7.2).

Day 0: Gynogenesis and cleaving embryo sorting

1. Perform gynogenesis on potential mutant carriers as outlined in Subheading 3.2.2.
2. Sort regularly cleaving embryos from unfertilized embryos at 4–16 cells. Irregularly cleaving embryos will gastrulate poorly, increasing background noise and making it more difficult to detect specific phenotypes. Likewise, treat embryos with optimum care to minimize abnormalities caused by overcrowding or other mistreatment (see Chapter 2).

Day 1: Tailbud stage sort (~16–20 hpf; St. 18–24)

1. As early as possible, sort normal from dead/abnormal embryos in all dishes, again to obtain a low background of early defects upon which to recognize later-developing phenotypes. Record number of dead embryos and remove them. Remove abnormally developing embryos from ECS dish to fresh plate noting phenotype.
2. Check ECS dishes for any obvious axial or dorsoventral polarity defects. If these are seen in >50% of embryos, the defect is likely due to imperfect gynogenetic rescue or poor egg quality.
3. Sort haploid dishes and discard embryos that fail to develop reasonable heads, then score those with good heads for % of "diploid-looking haploids." True haploids typically display posterior truncations, failure of blastopore closure, and raised neural folds. Appearance of diploids can be due to either spontaneous polar body failure (in which case both diploid-appearing haploids and ECS can be scored for recessive phenotypes) or failure to inactivate sperm DNA, resulting in diploid embryos in the haploid control and triploid embryos in the ECS dish, reducing the proportion of embryos in which recessive phenotypes may be detected.

Day 2 (48 hpf; St. 35–40)

Check for:

1. *Axial defects*: Embryo shape, truncation/kinking, gross tissue defects.
2. *Mobility*: Swirl embryos to center of dish and gently poke with forcep tip. Wild-type embryos respond by twitching or swimming away.
3. Add a few drops of 1:1,000 MS-222 to dish, swirl, and repeat until embryos are immobilized.

   Score the embryos for the following defects:
4. *Cilia function*: do anesthetized embryos "glide" forwards due to coordinated beating of epidermal cilia?

5. *Heartbeat*: does heart beat at normal rate? Note that anesthesia can affect heart rate.
6. *Circulation*: Look at the tail above, below, and in between the somites for blood movement.
7. *Kidney*: Is pronephros forming/looping?
8. *Somites*: Are somites forming in comparable number, thickness, chevron-shape as wild type?
9. *Pigmentation*: Retinal pigment epithelium (RPE) defects? Have melanocytes formed and taken on the spreading star shape? Is there an increase in or strange patterns of pigmentation?
10. *Edema*: Check for edema in unusual or interesting places. Nonspecific edema often forms around the ventral abdomen or heart, but can also be associated with specific phenotypes, e.g. heart, vascular or kidney defects, and should not necessarily be disregarded.

    After screening transfer embryos back to fresh media without anesthetic.

Day 3 (72 hpf; St. 40–43)

Check for:

Repeat Day 2 checks. Embryos without heartbeat will probably display edema by now. If not, check for blood flow and note any accumulation of blood in the body cavity.

Gut defects: Check for occurence and direction of gut looping. Stage comparison is important.

Otolith/otic vesicle: Is the size and shape of otic vesicle correct? Are both otoconia present, with normal crystal morphology?

Day 4 (96 hpf; St. 43–46)

Repeat previous checks. Saccular and utricular otoliths will be much clearer on day 4.

Head morphology: By day 5 of development the head will have flattened and cleared. Compare jaw morphology and hindbrain segmentation with diploid controls.

3.6. Reverse Genetic Strategies in *X. tropicalis*

In contrast to forward genetics, where first mutant phenotypes are described and then the underlying sequence lesion is identified, in reverse genetics the phenotypic outcomes of mutations in previously known sequences are studied. In vertebrates, reverse genetics often refers to mouse knockouts via homologous recombination in ES cells. Equivalent procedures do not currently exist for *X. tropicalis*, although intriguingly, *Xenopus* oocytes and extracts efficiently perform extrachromosomal homologous recombination (37) and homology-flanked integration constructs have not been evaluated for targeting. More recently, chimeric zinc finger nucleases have been shown to be effective for targeted gene disruption in *X. tropicalis* (27) (and Chapter 7) and zebrafish (38).

Alternatively, reverse genetics can also refer to scanning populations of animals or plants to identify carriers of mutations in particular sequences. One such strategy, TILLING, shows promise in *X. tropicalis* (15, 39). Genomic DNA samples are obtained from a population derived from mutagenesis, from which coding regions of target genes are evaluated for induced mutations. This approach was initially developed with non-sequence-based methods such as Cel I digestion (refs) (see also Chapter 7) and denaturing HPLC to evaluate a discrete set of amplicons for specific genes from a large mutagenized population, and subsequently adapted for capillary sequencing. However, PCR amplification introduces allele bias and other errors, compromising high-throughput screens (40).

Strategies based on next-generation sequencing technologies can scale up this approach while limiting PCR-based allele bias. One attractive variation is based on directly sequencing the protein-coding space (exome) of genomic sequence. This exon capture involves solution hybridization of sheared genomic DNA with a set of synthetic baits representing the entire exon set (or a subset of specific genes of interest). Noncoding regions are discarded, increasing sequence density in genomic regions likely to give phenotypes and reducing dependence on PCR. Hybridized exomes are then subjected to next-generation sequencing and inspected bioinformatically for heterozygous lesions. Multiple indexed genomes can be run on a single lane of some available sequencing apparatus.

X. tropicalis is particularly suited for TILLING studies since mutagenized populations can be archived both as frozen sperm (see Subheading 3.10 (41)) and living stocks. These frogs are significantly longer lived than other vertebrate genetic models and can still breed when more than 10 years old. Several *X. tropicalis* TILLING resources are currently under construction.

In addition to reverse genetic strategies, direct identification of mutations underlying phenotypes recovered from forward screens can be accomplished by exon capture sequencing of tissue from carrier animals as described later. However, given the imperfect annotation of the *X. tropicalis* genome, not all mutations will be contained in the exon capture set. One very useful variation simultaneously obtains valuable map information. Sequencing exomes harvested from separate pools of ~50 mutant and wild-type embryos provides linked SNPs to define the mutation-containing genetic interval as well as possibly identify the mutation directly.

3.6.1. *Xenopus tropicalis* Exon Capture Sequencing

We have successfully used Agilent Sureselect technology to enrich for the whole *X. tropicalis* exome and sequenced the products on the Illumina HiSeq platform. It should also be possible to enrich for specific subsets of genes. Detailed protocols for high-throughput sequencing and solution hybridization kits are available from manufacturers. The following method describes our approach using the genomic resources available at the time.

1. Design an exome set comprising all *X. tropicalis* JGI v4.2 gene build Enseml exons, plus cDNAs and ESTs. 2× tile the exome with 120mer oligos in silico, and obtain biotinylated RNA baits corresponding to the 120mer sequences from Agilent. Pool all baits (84 Mb, covering the entire known *X. tropicalis* exome with some duplication due to cDNA/EST overlap with Ensembl exons) into a single tube.
2. Toe clip live nonmosaic F2 animals derived from in vitro ENU mutagenesis and prepare genomic DNA (see Subheading 3.9.1). F1 animals derived from spermatogonial mutagenesis are also suitable.
3. Shear DNA using a Covaris sonicator down to 150–300 bp.
4. Prepare a paired end Illumina sequencing library using standard Illumina reagents.
5. Hybridize the library for 24 h against the Agilent SureSelect baits using manufacturer's reagents and conditions.
6. Isolate hybridizing exome fragments by incubating with streptavidin-coated magnetic beads, wash away noncoding sequences, and digest RNA baits according to manufacturer's instructions.
7. Amplify enriched library using Illumina primers, quantify, load onto a single lane of the Illumina HiSeq flowcell, and sequence with paired-end runs (e.g., 76 bp).
8. Map Illumina HiSeq output back to exome; typically >60% of sequences are found to be on target.
9. Interrogate sequences that hit the exome with three SNP calling programs (e.g., Qcall, mpileup, and GATK) and merge the output from all three programs into a single .vcf file. Note: use of three different SNP callers is stringent and may exclude many legitimate SNPs.
10. Predict consequences of SNPs using the Ensembl genebuild and filter to identify nonsense and essential splice mutations.
11. Confirm high probability nonsense and essential splice mutations in genes of interest using capillary sequencing of amplicons or another independent SNP detection method (e.g., Kasp (KBioscience) or *d*erived *C*leaved *A*mplified *P*olymorphic *S*equence (dCAPS) (Subheading "Snip-SNP and Derived Cleaved Amplified Polymorphic Sequence Analysis")) to evaluate specific genomic DNA samples. Specific SNP detection methods are also valuable for identifying carriers in the next generation and genotyping embryos during phenotypic analysis.

12. Outcross animals carrying mutations of interest to obtain an F3 generation for phenotypic analysis. Alternatively, eggs from female F2 carriers can be subjected to gynogenesis to uncover mutations for preliminary studies.
13. Perform linkage analysis to associate candidate phenotypes with specific identified sequence lesions. Chemical mutagenesis can introduce multiple lesions per genome.

3.7. Mapping Mutations

Conventional positional cloning is based on associating a mutant phenotype with nearby naturally occurring differences (polymorphisms) between a mutation-carrying chromosome and a wild-type chromosome. To simplify identifying and refining mutation-containing genetic intervals, it is useful to obtain "mapcross" animals that are hybrids between the strain on which the mutation was originally induced and a polymorphic strain which differs at many sequence loci. Many of the mapping strategies developed in other genetic systems (42) can be applied directly to mapping in *X. tropicalis*. *X. tropicalis* has several unique advantages for positional cloning. The contiguity of the *X. tropicalis* genome sequence is steadily increasing (14), and a draft chromosome-scale assembly is accessible at www.xenbase.org. A meiotic map of ~2,900 Simple Sequence Length Polymorphisms (SSLPs) has been organized into ten linkage groups corresponding to the ten *tropicalis* chromosomes (30). Many phenotypes uncovered by gynogenesis can be rapidly assigned to one of the ten chromosomes using a small set of centromere markers (see Subheading 3.7.4 and ref. (7)). *Fm*, the fraction of phenotypic gynogenotes, also provides an estimate of the gene–centromere distance. Such low-resolution map information is useful for evaluating candidate genes. Higher resolution mapping is accelerated by the large numbers of embryos produced; upwards of 5,000 meioses can be scored routinely from a single cross. Figure 5 shows a flowchart with conventional mapping strategies. As the cost of sequencing continues to drop, mapping mutations by direct sequencing of all protein-coding genes using exon capture is also becoming increasingly feasible (see Subheading 3.6).

3.7.1. Embryo Genomic DNA Prep

Genomic DNA for mapping is prepared from whole embryos using a proteinase K based lysis buffer.

1. When embryos are at least 3 days old, sort phenotypic mutants into a separate dish using a flamed Pasteur pipette.
2. Place mutant embryos individually in wells of 96 well plates.
3. Collect ~12–24 wild-type embryos from the same breeding into clearly marked wells.
4. Remove excess media from each well and freeze at –80°C unless prepping genomic DNA immediately.

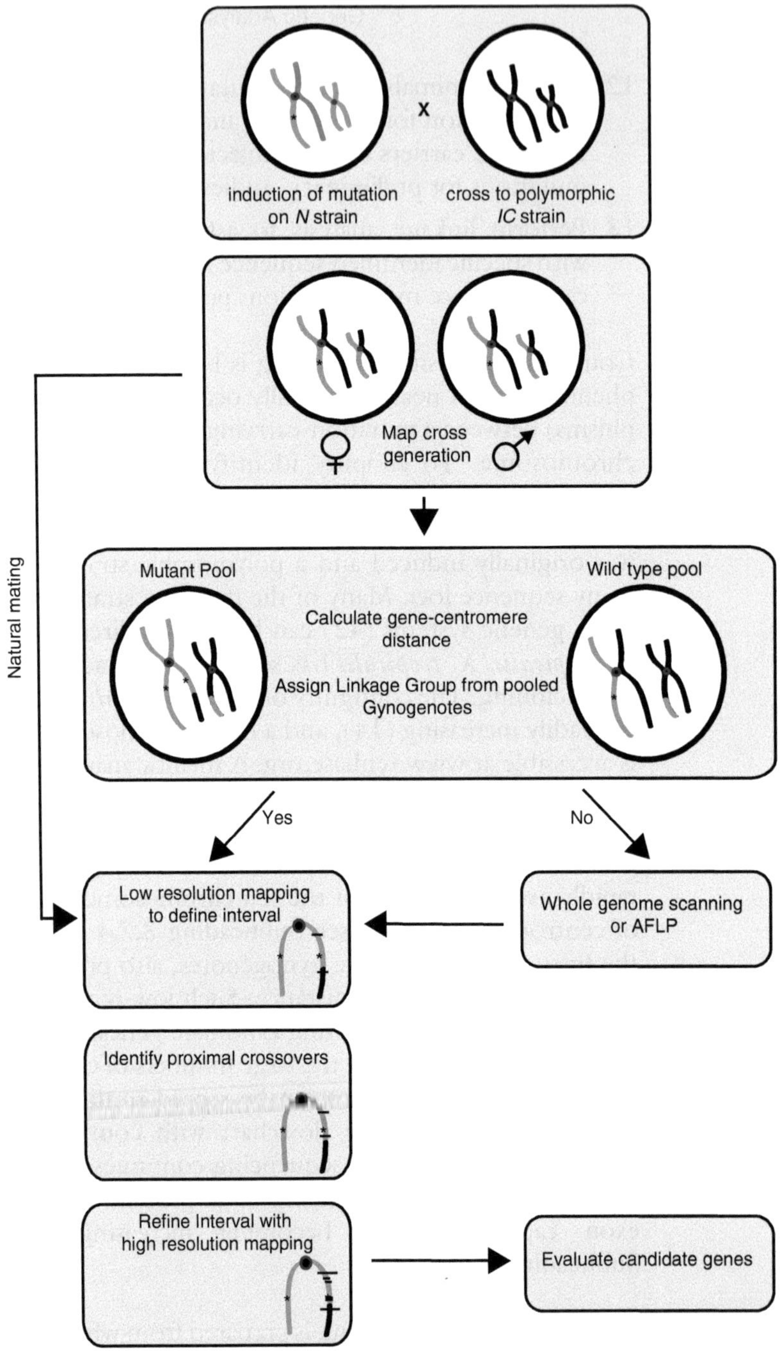

Fig. 5. Flowchart for genetic mapping in *Xenopus tropicalis*. A recessive mutation induced on one strain (*grey*) is crossed to a polymorphic mapping strain (*black*) to obtain hybrid map cross carrier animals. Gynogenetic embryos are obtained from map cross females to calculate gene–centromere distance and for bulk segregant analysis with centromere markers to identify linked chromosome. Conventional crosses between map cross carriers are performed for subsequent analysis. If chromosomal linkage cannot be assigned by bulk segregant analysis, whole genome scanning with polymorphic markers, or Amplified Fragment Length Polymorphism (AFLP) analysis can be used. Low resolution mapping with a small number of mutant embryos is used to identify markers ~3–10 cM apart flanking the mutation. These two flanking markers are then used to type large numbers (>500) of mutant embryos to identify those with crossover events between the flanking marker and the mutation. Small sets of recombinants can then be analyzed with further markers to refine the interval and number of genes contained within it. Candidate genes are then evaluated by changes in gene expression, spatial expression of transcripts, and cDNA sequence. Functional confirmation of any mutation found is accomplished by morpholino phenocopy and rescue with mRNA.

5. Add 50 μL of Lysis buffer w/proteinase K.
6. Incubate in PCR machine at 56°C for 4 h followed by 5 min at 95°C.
7. Use directly in PCR; no clean up required for most mapping applications (see Note 10).

3.7.2. Identifying Polymorphic Markers

Differences in the number of short di-, tri-, or tetranucleotide repeats, a.k.a. SSLPs are abundant between different strains of frogs (e.g., *N* and *IC*), so ideally mapping is conducted on the offspring of *N/IC* or other hybrid mapcross animals to locate markers linked to the phenotype. However, in all but the most inbred stocks, sufficient polymorphisms for low-resolution mapping are still likely to be present. It may be necessary to test several candidate SSLPs in a genomic region to identify those that are polymorphic in a given cross.

First prepare the following

1. Genomic DNA extracted from haploid embryos from a mapcross hybrid female (see Subheading 3.7.1).
2. PCR master mixes with range of potential polymorphic markers.

Procedure

1. Transfer 2 μL of DNA from six individual haploid embryos to fresh tubes.
2. Set up PCR master mix with primers for marker to test.
3. Add 8 μL of Master Mix to each individual haploid embryo DNA.
4. Run PCR under the following conditions 94°C—2 min, 35 cycles of (94°C—30 s 58°C—30 s 72°C—1 min), 72°C for 5 min, 4°C Hold.
5. Run 5 μL on 3% SFR agarose gel or polyacrylamide gel and silver stain (see Subheading 3.7.3).
6. If individual haploids produce different molecular weight PCR bands at ~1:1 ratio, the marker is polymorphic in the female parent and can be used for mapping.
7. Repeat for each candidate marker.

3.7.3. Polyacrylamide Gels and Silver Staining

While agarose gels are quick and convenient, resolution is limited and subtle polymorphisms may be missed. Single-base resolution can be obtained using standard sequencing-style denaturing 6% polyacrylamide/8% urea gels, visualizing DNA bands with silver nitrate. This protocol was adapted from ref. (43).

1. Thoroughly clean and dry both glass plates.
2. Coat small glass plate with mixture of 5 mL 100% EtOH, 75 μL 10% acetic acid, and 5 μL 3-(trimethoxysilyl)propyl methacrylate.

3. Wash with dH_2O followed by 70% EtOH, wipe and allow to dry.
4. Spray large glass plate with acrylease (Stratagene). Wait 5 min then wipe with clean wet tissue.
5. Pour a 6% acrylamide gel (containing 8 M urea) with shark tooth combs.
6. Prerun gel in 1× TBE for 30′ at 80 W.
7. Dilute PCR 1:2 with denaturing DNA loading buffer and heat to 95°C for 3 min in a thermocycler.
8. Load 5 μL of sample onto gel and run at 55 W. Run is complete when the buffer front passes through the bottom of the gel.
9. Split the glass plates apart with a razor blade.
10. Transfer glass plate containing gel to a large photographic developing dish. Gel side up.
11. Fix gel in 1 L 10% EtOH (this can be reused up to six times) for 10′.
12. Wash in 1 L 1% nitric acid for 3′ (this can be reused twice).
13. Rinse twice in dH_2O 3′ for each wash.
14. Stain for 20′ in 1 L silver nitrate.
15. Rinse twice in dH_2O 3′ for each wash.
16. Add 1 L developing solution and agitate gently until bands appear. This is usually within 5′ depending on temperature of the solution.
17. Stop the reaction in 10% acetic acid for 5′.
18. Wash gel in dH_2O for 10′.
19. Transfer to a light box to photograph with a standard digital camera.

3.7.4. Low-Resolution Mapping with Gynogenesis and Centromere Markers

Placing a mutation on a chromosome, combined with the rough gene–centromere distance provided by phenotype ratio in gynogenetic embryos, allows a genomic region to be inspected for candidate genes. The initial step in positional cloning usually entails defining the chromosome or genetic linkage group that contains the mutation. In many cases this can be accomplished rapidly by analyzing pools of mutant and wild-type gynogenetic embryos with polymorphic markers located near each of the ten *X. tropicalis* centromeres to identify one which segregates with the mutant phenotype (see ref. (23)). Examination of DNA from pools of mutant and wild type, know as bulk segregant analysis, simplifies rapid identification of markers linked to the mutant phenotype.

As outlined in Subheading 3.2.2, gynogenesis prevents second polar body extrusion allowing the postrecombination sister products of meiosis II to be retained. The genome of a gynogenote is therefore completely maternally derived, but not completely homozygous, analogous to half of a yeast tetrad (see Fig. 1 and ref. (44)). Polymorphic markers at the centromeres, where each pair of sister chromatids is held together during recombination, will be homozygous, with the different alleles segregating into different individual gynogenotes (see Fig. 1). Gynogenetic embryos that are phenotypically mutant for a recessive allele are also by definition homozygous at this mutant locus. If the mutation is located reasonably close to a centromere, a pool of mutant gynogenotes will also appear homozygous for the cognate centromeric marker derived from the mutagenized strain, while the wild-type pool will contain the alternative allele (see Fig. 6). For the chromosomes that do not contain the mutation, both centromere alleles will contribute equally to mutant and wild-type pools. In this fashion, linkage can be established using only the small set of markers corresponding to centromeres of the ten different chromosomes, and two pools of mutant and wild-type gynogenetic DNAs.

Conveniently, this apparent centromere-mutation linkage extends to more distal mutant loci. Consider a recessive mutation *m1* induced on the *N* background and crossed onto the polymorphic *IC* strain to create a heterozygous *N*/IC* carrier female (Fig. 1). The gynogenetic offspring of such a hybrid will thus each be homozygous *N/N* or *IC/IC* at all centromeres, and the mutant embryos will be N^{m1}/N^{m1} at the mutant locus. For a mutation *m2* further from its centromere, recombination events are more likely in the interval, resulting in gradual accumulation of wild-type N^{m1}/IC heterozygotes and decreasing the fraction of mutant gynogenotes. The wild-type pool will thus contain both *IC/IC* (from the original parental allele) and *N/N* centromeres (from single crossovers producing heterozygotes at the mutant locus). However, the reduced fraction of N^{m1}/N^{m1} mutant embryos is still likely to be homozygous *N/N* at the corresponding centromere. The exceptions derive from multiple crossover events; half of double crossovers will return linkage to the original centromere allele, while half may switch to the alternative nonparental allele (see ref. (7)). Only when the mutant locus is so distal that the majority of gynogenetic embryos contain multiple intervening crossovers will the mutation no longer appear linked to its centromere.

In addition, a rough estimate of the gene–centromere distance can be obtained from the proportion of mutant gynogenotes observed. If we assume that only single crossovers are present, then

$$\text{gene - centromere distance (cM)} = 50(1-(2\times Fm)$$

where *Fm* is mutant/total number of embryos.

In practice, this formula provides useful information for loci less than ~30 cM from centromeres (*Fm* > 0.2), where single crossovers predominate. If *Fm* < 0.2, the gene–centromere calculation only establishes that the locus is further than 30 cM from the centromere, as multiple crossovers are common in the longer chromosome arms. This rough map information can be used to refine candidate gene sets and to select markers for higher resolution linkage analysis.

Assigning Linkage Group by Bulk Segregant Analysis

First prepare the following

1. 2 pools of DNA from 10 to 20 phenotypically mutant gynogenotes (5 μL from each). Single pools and smaller numbers of embryos can be used, but risk of false positives increases.
2. 2 pools of DNA from 10 to 20 sibling wild-type gynogenotes (5 μL from each).
3. Primer stocks for SSLP markers closely linked to the centromere (see genetic locations of centromeres in ref. (7)).

Procedure

1. Identify polymorphic SSLP markers within 1.5 cM of *X. tropicalis* centromeres (see Subheading 3.7.2 and ref. (23)).
2. Aliquot 2 μL of DNA from the two mutant and wild-type pools into PCR reaction tubes for each polymorphic centromere marker being tested.
3. Add 8 μL of a standard PCR master mix for polymorphic centromere marker to each tube.
4. Run PCR under standard conditions for 35 cycles with an annealing temperature of 58°C (for all Wells et al. map SSLPs).
5. Run 5 μL on 3% SFR Agarose gel. Some polymorphisms are only scorable using higher resolution 6% polyacrylamide sequencing gels followed by a silver stain (see Subheading 3.7.3).
6. If a mutation is linked to a given centromere, one band will predominate in mutant lane; the corresponding wild-type lane will show either the other band (consistent with a tightly linked locus) or both species (consistent with a more distal locus). Unlinked centromeres will display identical mutant and wild-type bands (see Fig. 6).
7. Repeat until linkage is observed or polymorphisms at all ten centromeres have been tested (if no linkage is detected, see Subheadings 3.6 and 3.7.5).
8. Confirm by testing individual embryos with linked centromere marker, and determine linked chromosome arm by testing markers ~5 cM either side of centromere.

3.7.5. Alternate Strategies

Exon capture for direct sequencing to identify mutations is described in Subheading 3.6; in this section, we discuss inexpensive

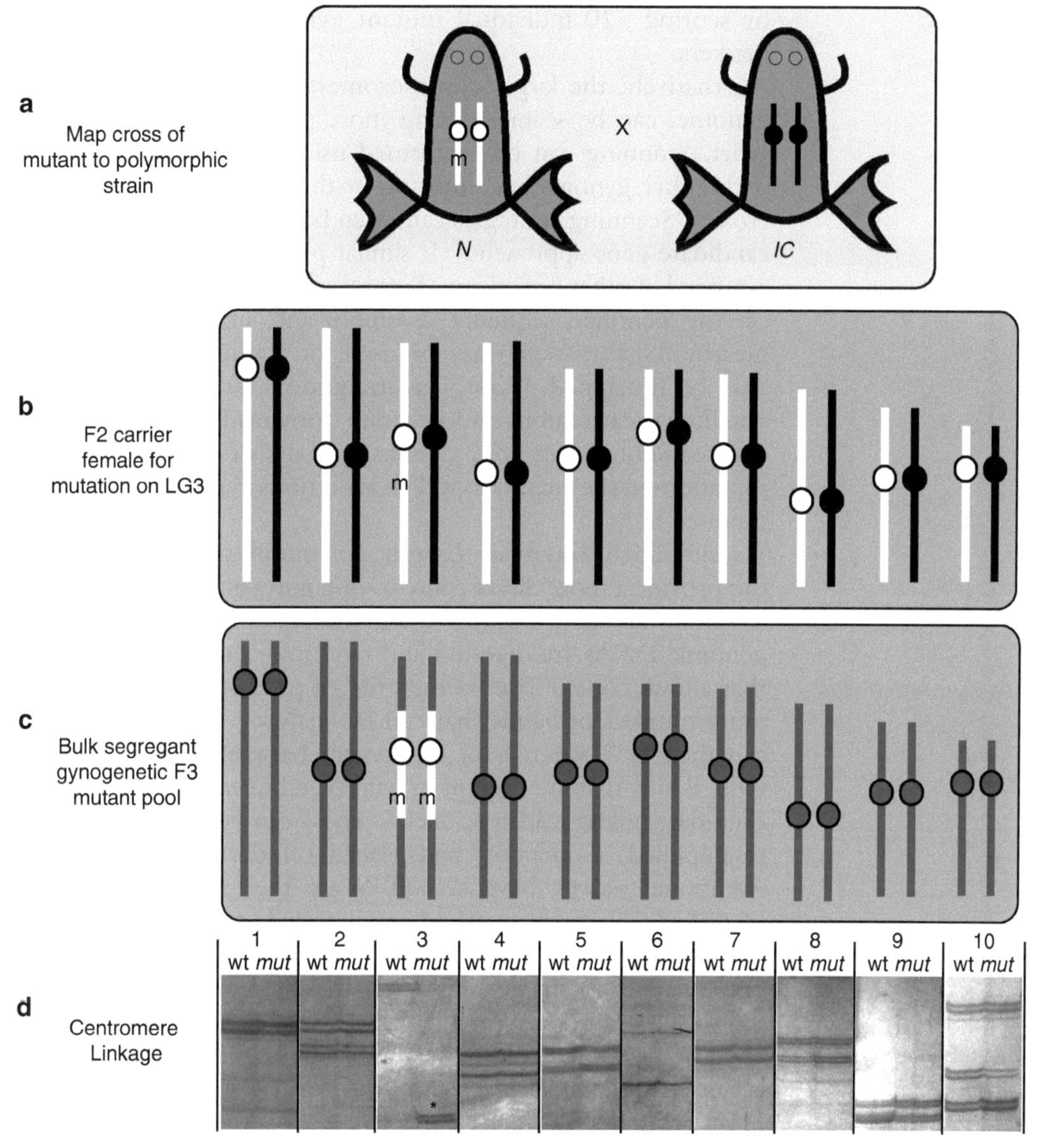

Fig. 6. Assigning chromosome linkage by bulk segregant analysis. (**a**) A frog carrying a recessive mutation m on the *N* strain (white chromosomes) is crossed to a polymorphic *IC* strain (black chromosomes). (**b**) "Mapcross" hybrid F2 carrier inherits one chromosome from each parent. (**c**) Pools of ~20 phenotypically mutant and wild-type gynogenetic embryos are collected (mutant pool represented). Unlinked chromosomes show equal contribution from white *N* and black *IC* alleles (gray chromosomes) in both mutant and wild-type pools. However, on the chromosome containing the mutation, the mutant pool is greatly enriched for the white *N* centromeric allele; the wild-type pool may contain either the *IC* allele or both *N* and *IC*. Centromere linkage can often be detected over large genetic distances in gynogenetic embryos. (**d**) Silver-stained gel showing pools of mutant and wild-type embryos scored with polymorphisms at the ten *X. tropicalis* chromosomes. Linkage is detected to chromosome 3.

alternative approaches for low-resolution mapping. Distal loci may appear unlinked to centromere markers in bulk segregant analysis, where a small minority of embryos with multiple crossovers can obscure linkage in pools. Weak linkage can sometimes be confirmed by scoring >20 individual mutant gynogenotes for centromere markers.

Alternatively, the larger chromosome arms, or indeed the entire genome, can be scanned using more polymorphisms 10–20 cM apart. Scanning can be performed using bulk segregant analysis with either gynogenetic embryos or the progeny of conventional crosses. Scanning strategies can often be efficiently combined with candidate gene approaches. If similar phenotypes have been characterized in other organisms, *X. tropicalis* orthologs may be located on the genomic sequence assembly and/or the meiotic map. Nearby SSLP markers may be used for scanning if the scaffolds have been mapped. If not, it is straightforward to identify microsatellite repeat regions and generate homemade "bespoke" markers (see Subheading 3.7.7). Linkage analysis can detect unrelated mutations in the vicinity as well as mutations in the candidate genes themselves.

*A*mplified *F*ragment *L*ength *P*olymorphism (AFLP) analysis can provide a more direct route to obtaining linked sequences, but is somewhat laborious. In brief, mutant and wild-type pooled genomic DNAs are digested and randomly amplified in a fashion that allows control over complexity of products (45). The maximum number of bands that can be analyzed on sequencing gels can then be inspected for differences between mutant and wild type. Bands that are present in wild type but not mutant lanes are candidate linked wild-type alleles. These can be cut out of the gel, re-amplified, sequenced, and placed on the genome assembly. Additional nearby bespoke SSLPs are then obtained from the identified sequence scaffold and tested to confirm linkage. Convenient AFLP kits are available, to facilitate AFLP analysis, for example e.g., Invitrogen AFLP Analysis System I.

3.7.6. Higher Resolution Mapping

Most of the considerations for subsequent steps in positional cloning are not specific to *X. tropicalis*. Gynogenetic embryos, which have fewer crossovers on the centromere side of the mutant locus, are less suitable for fine mapping than those derived from a conventional mating with useful crossover events on both sides. Conventional crosses can also provide larger numbers of embryos than gynogenesis; for successful positional cloning, at least 1,000 mutant embryos are often required.

After placing a mutation on a linkage group, the next step involves locating the mutation between two easily scorable flanking markers <10 cM apart. Initially, the linked chromosome can be scanned with markers spaced at ~10 cM intervals, using bulk segregant analysis of mutant and wild-type pools of ~20 embryos. Polymorphisms showing strong linkage are then evaluated using ~24–48 individual mutant embryos and 6–12 wild-type siblings, and other nearby markers from the meiotic map tested. Markers further from the mutation will yield more recombinants (heterozygotes) than closer markers. Importantly, markers on opposite flanks of the mutation give nonoverlapping sets of recombinants, and markers on the same side share recombinants. Flanking markers should be <~10 cM apart, and should be relatively easy to score, i.e. simple 2-allele systems, preferable distinguishable on agarose gels, as these will be used to genotype large numbers of embryos.

After flanking markers have been obtained, genotype the available mutant embryos with them to identify those that are recombinant. Mutant embryos that are heterozygous at a flanking marker have informative crossovers closer to the mutation. Once a set of >10–20 recombinants has been obtained, these can be typed with additional markers distributed evenly between the flanking markers to narrow the interval. The available resolution is determined by the number of meioses scored. Genomic regions corresponding to genetic intervals can be inspected for candidate genes easily using Ensembl BioMart (Subheading 3.8.1). Intervals can usually be subdivided rapidly with additional markers from the meiotic map, bespoke SSLP or SNP markers or SNPs (see Subheading 3.7.7).

In some regions, microsatellite repeat polymorphisms may be difficult to find. Single Nucleotide Polymorphisms (SNPs) are abundant in polymorphic crosses. These usually can be identified by sequencing random amplicons from intergenic regions from several mutant and wild-type individuals. SNP-containing regions from mutant embryos can be resequenced for linkage analysis, or identified SNPs can be converted into a variety of high-throughput assays. Many SNPs disrupt or create restriction sites, so alleles can be distinguished by digesting an amplification product (Cleaved Amplified Polymorphic Sequence, CAPS). A variation based on introducing a mismatched base in primers, "dCAPS" (46), can produce differentially cleavable alleles starting from any SNP sequence. Kasp and related genotyping techniques combine Fluorescent Resonance Energy Transfer (FRET) with allele-specific PCR to discriminate between SNPs using dye-linked allele-specific primers (www.kbioscience.co.uk). Many other techniques to discriminate between SNPs are continuously being developed.

3.7.7. Bespoke Markers for Mapping

The *X. tropicalis* meiotic map (30) currently arranges ~2,900 SSLPs in ten linkage groups corresponding to the ten *X. tropicalis* chromosomes. While this provides ample marker density for linkage assignment and rough mapping, higher resolution can require identification of bespoke (custom) markers. SSLP markers are convenient to identify and score, but high-resolution mapping may exhaust SSLP candidates in a region. Single Nucleotide Polymorphisms (SNPs) are abundant in most crosses and also useful for high-resolution mapping.

Obtaining Bespoke SSLPs

1. Download FASTA-formatted sequence corresponding to an *X. tropicalis* genomic region of interest from a genome browser (e.g. JGI, UCSD, Xenbase G-Browse, or Ensembl).
2. Go to Tandem Repeat finder website (http://tandem.bu.edu/trf/trf.html).
3. Click "Submit a Sequence for Analysis."
4. Click "Basic."
5. Copy and Paste or upload FASTA format sequence to website.
6. Click "Submit Sequence" button (may take a few minutes to load).
7. On following page click "Tandem Repeats Report."
8. Look through second column of table (Period Size) for 2, 3, or 4 (di, tri, or quad repeat).
9. Next look through third column (Copy Number) for 2, 3, or 4 period repeats and identify those with 10–30 copies
10. Click on link for these repeats in first column (Indices).
11. The following pages give the repeat and sequence flanking.
12. Search the original FASTA format .txt file for the repeat plus 30–50 bp of flanking sequence.
13. Go to Primer3 website (e.g., http://frodo.wi.mit.edu/).
14. Copy and paste the identified repeat plus flanking sequence into text box on Primer3 (leave parameters unchanged).
15. Click "Pick Primers" button.
16. Chose primers flanking the repeated sequence.
17. Repeat steps 12–16 for all period sizes of 2, 3, or 4 with a copy number of 10–30.

Bespoke SNPs

SNPs are abundant in intergenic noncoding regions and may be identified directly by sequencing PCR products (amplified with a high-fidelity polymerase) from mutant and wild-type embryos.

1. Design PCR primers to amplify ~400 bp from noncoding regions >5 kb away from exons.
2. Using a proofreading polymerase, amplify from three or more individual wild type and nonrecombinant mutant embryos, or >4 unsorted haploids.
3. Sequence all six fragments and compare, looking for consistent single nucleotide changes between wild type and mutant embryos.
4. Analyze by Snip-SNP, dCAPS, or by sequencing amplicons from individual mutant and wild-type embryos.

Snip-SNP and Derived Cleaved Amplified Polymorphic Sequence Analysis

Many SNPs identified by sequence result in RFLPs, or Snip-SNPs, that are simpler and cheaper to score than by sequencing multiple embryo DNAs. dCAPS allows virtually any identified SNP to be converted into a snip-SNP for high-throughput analysis (46) by generating PCR primer sequences in which a mismatch is introduced, converting one of the SNPs into a specific cleavable polymorphism. The PCR products can then be digested and compared. The most time-consuming part of this process can be generating mismatched primer sequences to create snip-SNPs; the authors of the dCAPS method have thoughtfully generated an online "dCAPS Finder" at http://helix.wustl.edu/dcaps/dcaps.html. This program also conveniently identifies enzyme choices if conventional snip-SNPs exist between two sequences. SNP-containing genomic regions from mutant and wild-type embryos can then be amplified by PCR, digested with the allele-specific restriction enzymes suggested by the dCAPS finder program, and distinguished on agarose gels.

3.8. Evaluation of Candidate Genes in Mapped Interval

Strong candidates for gene functions underlying mutant phenotypes can often be identified from larger gene lists based on previously known phenotypes or functions and expression data. Candidate genes are then evaluated in the mutant by expression analysis, cDNA and/or genomic sequence, and functional association (e.g., mRNA rescue or MO phenocopy).

3.8.1. Ensembl BioMart

Mutation-containing intervals can be rapidly inspected for candidate genes by downloading lists of gene models, including GO terms and protein domain information, using the Biomart tool in Ensembl (www.ensembl.org). At the time of writing, this resource references the *X. tropicalis* v4.1 assembly.

1. Go to www.ensembl.org/Xenopus_tropicalis/Info/Index.
2. Click on Biomart link (top right).
3. Choose database (a high number Ensembl or Vega).
4. Choose dataset "*X. tropicalis* genes."
5. Under "Filters" in the left hand menu, select "Region" and input scaffold information under 'Multiple Chromosomal Regions in the format 'Scaffold_Number:base-base (e.g. "scaffold_1:xxxxx-yyyyy") to obtain genes on the scaffold between polymorphisms flanking a mutation at base xx,xxx and yy,yyy. GO terms and external references can be added under the "Attributes" section on the left under the "GENE" subsection. Clicking "count" at top left gives the number of Ensembl genes in the set.
6. Generate a spreadsheet with all the genes and transcripts by clicking "Results" at the top left of the page.

3.8.2. Analysis of Candidate cDNAs

Compelling candidate genes in the mutation-containing interval can be evaluated in a number of ways. RT-PCR or in situ hybridization with 3′ probes may be used to detect changes in expression levels in mutant embryos. These are not necessarily the result of changes in transcription level; mutations which introduce stops are frequently degraded by nonsense-mediated decay (47). Likewise, immunostaining or western blot analysis is useful where antibodies are available. Sequencing specific cDNAs from the mutant is an informative and inexpensive option unless the gene is very large.

3.8.3. Confirmation of Candidate Genes

Many mutagenesis procedures will introduce multiple lesions per genome; induced base changes from chemical mutagenesis are detected as frequently as 1/50,000 bases. Even if a sequence lesion is identified in a coding region within the genetically defined interval, other mutations may also be present, and independent evidence is usually required to show that one gene is responsible for the mutant phenotype. Ideally, the phenotype can be rescued by a wild-type allele delivered by mRNA injection or as a transgene (see Chapters 11, 12, 13, and 14). Obtaining a specific phenocopy by morpholino oligonucleotide knockdown of the wild-type allele is also compelling. Microinjection techniques are similar to those used for *laevis*, with volumes and dosages adjusted for smaller *X. tropicalis* embryo. While all mRNA and morpholino oligonucleotides should be titrated, a starting point of 1/10th the dose used for *laevis* is useful, in an injection volume of up to 2 nL in one cell of a 2-cell embryo in filter-sterilized 3% Ficoll/0.05× MMR.

3.8.4. Phenotypic Analysis

Xenopus species are amenable to a broad range of strategies for analysis of gene function and expression. Techniques will vary with each phenotype, and a full review is beyond the scope of this chapter. In general, after effects on external morphology have been described, mutant embryos are processed for histology and fixed for wholemount in situ hybridization or staining with specific antibodies. Many published protocols for *X. laevis* are directly transferrable to *X. tropicalis.*

3.9. Genotyping Adult Frogs

Once a mutation has been identified or mapped to a narrow region, it may be simpler to identify carriers using a PCR or SNP-based approach using genomic DNA from an adult frog (rather than breeding a candidate carrier to known carriers).

3.9.1 Toe Clip for Genomic DNA

Large quantities of genomic DNA can be harvested nonlethally from adult *X. tropicalis* toes, which regenerate after about a month.

1. Anesthetize frog in 0.09% MS222 for 4–6 min. Times for this can vary so continually observe until frogs begin to slow their swimming motions, being careful not to over-anesthetize.
2. Remove frog immediately and briefly rinse in fresh water.

3. Place frog on damp paper towel and remove a single toe with a fresh scalpel/razor blade. Place toe in eppendorf tube containing 400 μL lysis buffer, or freeze tissue on dry ice and store at −80°C.
4. Rinse frog in fresh water and place in clean tank containing wet paper towels. Cover frog in a further damp paper towel. The frog should completely recover in ~1 h.
5. To extract DNA, incubate clipped toe in 400 μL lysis buffer at 55°C overnight with agitation.
6. Precipitate DNA by adding 300 μL isopropanol.
7. Spin at 4,000 rpm for 20′.
8. Wash pellet in 500 μL 70% ethanol.
9. Resuspend pellet in 100 μL nuclease-free water. Usually one toe clip yields ~10 μg genomic DNA.

3.9.2. Back Swab for Genomic DNA

Taking a small sample from the surface of the frog is an alternative noninvasive method for obtaining genomic DNA. However, DNA yield is only sufficient for a few reactions and usually requires amplification with nested primers.

1. Hold frog with gloved hand (to prevent DNA contamination).
2. Using a sterile bacterial inoculation stick or pipette tip, wipe across the skin on the frog's back approximately ten times.
3. Shake tip or bacterial stick in 500 μL of lysis buffer.
4. Incubate at 55°C for 2 h.
5. Add 1 mL 100% EtOH.
6. Spin at 13,000 rpm, 4°C for 10′.
7. Wash pellet in 70% EtOH and spin for a further 5′.
8. Resuspend in 14.3 μL nuclease-free water ready for PCR.
9. Add a master mix of primary PCR to this tube and cycle in a thermocycler under primer-specific conditions for 35 cycles.
10. Use 0.2 μL of this PCR as template for a second reaction with nested primers.

3.10. *X. tropicalis* Sperm Freezing

Storage of frozen sperm at −80°C facilitates archiving specific stocks and strains; shipping frozen sperm on dry ice is often simpler and more reliable than shipping adult animals. These protocols have been adapted from ref. (41).

3.10.1. Cryopreservation of Sperm

1. Prepare cryoprotectant.
2. Inject males with 100 U HCG 12–24 h before harvesting testes.
3. The next day, kill males and dissect out testes, rolling on clean paper towel to remove traces of blood.

4. Macerate both testes with eppendorf pestle in a single 1.5 mL eppendorf tube with 500 μL L15/CS.
5. Add 500 μL cryoprotectant.
6. Aliquot into 4–10 tubes.
7. Place tubes in small styrofoam box, wrap lid with aluminum foil.
8. Place the box in –80°C for at least 24 h, then transfer tubes to rack or box for long-term storage at –80°C or in liquid nitrogen (see Note 11).

3.10.2. In Vitro Fertilization Using Frozen Sperm

1. Prepare 25°C water bath and express eggs from females into a dry dish, discarding poor quality eggs.
2. Thaw frozen sperm in 25°C water bath. Remove immediately when thawed (<30″).
3. Dilute sperm with 2 mL distilled water and add to eggs.
4. Gently mix sperm and eggs with a pipette tip.
5. After 2 min, flood with distilled water.
6. Fertilization rates of 10–15% are typically obtained with a ½ testis frozen aliquot on ~1,000 eggs.

4. Notes

1. MES buffer (unlike many other common buffers such as Tris) does not contain amine groups that react with ENU.
2. A variety of diets work well for adults as long as they are relatively protein-rich and small enough for frogs of a given size to swallow easily, but large enough to grasp with their forelimbs. Live foods such as insect larvae or unprocessed offal carry some risk of transmitting parasites. Powdered fish flake is useful to supplement the diet of larger tadpoles in flow-through systems, but rapidly fouls standing water.
3. Injections are made into the dorsal lymph sac by inserting a 27G needle subcutaneously between the dorsal lateral line stripes. For optimal egg quality, females may be re-ovulated every 6 weeks to 6 months. Some groups re-use males as often as once a week, and females once a month, although it is likely that ovulating females more than 5–6 times per year may be stressful in the long term.
4. Always handle amphibians with wet hands to prevent damage to their skin. To squeeze eggs from frogs, grasp the female

dorsally, with her left leg between the index and middle finger of your right hand, and her left leg in your left hand. Gently massage her abdomen in an anterior-to-posterior direction with your thumb to express eggs. Eggs will activate prematurely if they contact low-salt solutions, so avoid dripping water from the frog while squeezing. Good quality eggs will be spherical, surrounded by clear jelly, with uniform pigmentation in the animal hemisphere except for a lighter polar patch overlying the germinal vesicle; poor quality eggs may be lysing or stringy. *X. tropicalis* in vitro fertilization is typically less efficient than with *laevis*; testes are smaller, contain less sperm, and appear to be more salt-sensitive. *X. tropicalis* embryos do not consistently undergo upward reorientation of the animal hemisphere as *laevis* do upon activation; cortical pigment contraction and germinal vesicle breakdown are more reliable indicators. When sacrificing frogs, use RT anesthetic solution; chilled solutions may immobilize animals without anesthetic effect.

5. If available, testes from males bearing fluorescent transgenes can be used to assess haploid formation; efficient UV irradiation will block paternal transgene transmission. Development of haploid embryos is strongly affected both by egg quality and genetic background. Haploid development typically includes deficits in axis elongation and posterior structures, incomplete blastopore closure, and other gastrulation defects, but anterior structures are often well formed (compare Fig. 2a, c). Diploid-appearing embryos are also observed in some batches of haploids. These may result from incomplete UV irradiation of sperm, but spontaneous diploidization of haploids has also been observed, probably due to failure of polar body formation (see Fig. 2a–d).
6. Wait at least 15′ after activation to allow completion of cortical rotation; dejellying during this early period can produce axial defects. BSA-coating of dishes prevents embryos sticking to the plates, which can affect gastrulation.
7. Depending on egg quality and genetic background, this procedure typically rescues 25% or more of the cleaving haploid embryos to viable diploids.
8. Haploid and outcrossed diploid controls should be reaching the 4-cell stage when LCS embryos undergo the first cell division; LCS embryos cleaving in synch with controls should be discarded. Rescue efficiency is strongly dependent on egg quality and genetic background. Use of transgenic sperm is recommended to control for efficiency of UV inactivation.

9. Heavy compression is important for getting good spreads, but the coverslip should not slide around. This technique can produce hundreds of stained nuclei, but finding those with complete countable spreads requires patience.
10. Providing the Proteinase K is adequately denatured (by the 5 min, 95°C step), this DNA can be immediately used in PCR reactions without the need for precipitation or cleanup. For mapping purposes, collect as many mutant embryos as are available. Assignment of mutant loci to specific chromosomes by centromere linkage can be accomplished with as few as 12 mutant and wild-type gynogenotes (see Subheading 3.7.4). Analysis of 1,000 mutant embryos from conventional crosses (~2,000 parental meioses) provides a theoretical resolution of 0.05 cM (see Subheading 3.7.5), which is usually sufficient to define the mutation-containing interval on a single sequence scaffold.
11. It is advisable to test-fertilize some eggs with aliquot of fresh sperm from each male; if fresh sperm is not capable of fertilization, frozen testis from that male should be discarded. This freezing method seems to work as well as more elaborate methods described in ref. (41).

5. Summary/ Discussion

We have tried to outline some useful strategies and protocols for developmental genetics using *X. tropicalis*. Amphibian embryos have been a valuable system for understanding vertebrate development using a broad range of embryological, molecular and, more recently, genomic tools. With its compact genome and short generation time, *X. tropicalis* offers the prospect of combining genomics and precision loss-of-function genetic approaches with the conventional *Xenopus* toolkit for a unique range of analysis of gene function in a single in vivo vertebrate model system.

Acknowledgments

Many colleagues contributed to these protocols. In addition to past and present members of the Zimmerman and Stemple laboratories, the authors would particularly like to thank Rob Grainger and Takuya Nakayama (University of Virginia), Richard Harland (UC Berkeley), and Mustafa Khokha (Yale). T.J.G. and L.B.Z. are funded by UK Medical Research Council U117560482; DLS by the Wellcome Trust WT 077047/Z/05/Z.

References

1. Gurdon JB, Hopwood N (2000) The introduction of Xenopus laevis into developmental biology: of empire, pregnancy testing and ribosomal genes. Int J Dev Biol 44:43–50
2. Harland RM, Grainger RM (2011) Xenopus research: metamorphosed by genetics and genomics. Trends Genet 27(12):507–515
3. Nusslein-Volhard C, Wieschaus E (1980) Mutations affecting segment number and polarity in Drosophila. Nature 287:795–801
4. Wang BB, Muller-Immergluck MM, Austin J, Robinson NT, Chisholm A, Kenyon C (1993) A homeotic gene cluster patterns the anteroposterior body axis of C. elegans. Cell 74:29–42
5. Doetschman T, Gregg RG, Maeda N, Hooper ML, Melton DW, Thompson S, Smithies O (1987) Targetted correction of a mutant HPRT gene in mouse embryonic stem cells. Nature 330:576–578
6. Thomas KR, Capecchi MR (1987) Site-directed mutagenesis by gene targeting in mouse embryo-derived stem cells. Cell 51:503–512
7. Streisinger G, Walker C, Dower N, Knauber D, Singer F (1981) Production of clones of homozygous diploid zebra fish (Brachydanio rerio). Nature 291:293–296
8. Kimmel CB (1989) Genetics and early development of zebrafish. Trends Genet 5:283–288
9. Driever W, Solnica-Krezel L, Schier AF, Neuhauss SC, Malicki J, Stemple DL, Stainier DY, Zwartkruis F, Abdelilah S, Rangini Z, Belak J, Boggs C (1996) A genetic screen for mutations affecting embryogenesis in zebrafish. Development 123:37–46
10. Haffter P, Granato M, Brand M, Mullins MC, Hammerschmidt M, Kane DA, Odenthal J, van Eeden FJ, Jiang YJ, Heisenberg CP, Kelsh RN, Furutani-Seiki M, Vogelsang E, Beuchle D, Schach U, Fabian C, Nusslein-Volhard C (1996) The identification of genes with unique and essential functions in the development of the zebrafish, Danio rerio. Development 123:1–36
11. Postlethwait JH, Woods IG, Ngo-Hazelett P, Yan YL, Kelly PD, Chu F, Huang H, Hill-Force A, Talbot WS (2000) Zebrafish comparative genomics and the origins of vertebrate chromosomes. Genome Res 10:1890–1902
12. Bewick AJ, Anderson DW, Evans BJ (2010) Evolution of the closely related, sex-related genes DM-W and DMRT1 in African clawed frogs (Xenopus). Evolution 65:698–712
13. Tinsley, RC, Kobel, HR (1996) The Biology of *Xenopus*, Oxford University Press, Oxford
14. Hellsten U, Harland RM, Gilchrist MJ, Hendrix D, Jurka J, Kapitonov V, Ovcharenko I, Putnam NH, Shu S, Taher L, Blitz IL, Blumberg B, Dichmann DS, Dubchak I, Amaya E, Detter JC, Fletcher R, Gerhard DS, Goodstein D, Graves T, Grigoriev IV, Grimwood J, Kawashima T, Lindquist E, Lucas SM, Mead PE, Mitros T, Ogino H, Ohta Y, Poliakov AV, Pollet N, Robert J, Salamov A, Sater AK, Schmutz J, Terry A, Vize PD, Warren WC, Wells D, Wills A, Wilson RK, Zimmerman LB, Zorn AM, Grainger R, Grammer T, Khokha MK, Richardson PM, Rokhsar DS (2010) The genome of the Western clawed frog Xenopus tropicalis. Science 328:633–636
15. Goda T, Abu-Daya A, Carruthers S, Clark MD, Stemple DL, Zimmerman LB (2006) Genetic screens for mutations affecting development of Xenopus tropicalis. PLoS Genet 2:e91
16. Grammer TC, Khokha MK, Lane MA, Lam K, Harland RM (2005) Identification of mutants in inbred Xenopus tropicalis. Mech Dev 122:263–272
17. Noramly S, Zimmerman L, Cox A, Aloise R, Fisher M, Grainger RM (2005) A gynogenetic screen to isolate naturally occurring recessive mutations in Xenopus tropicalis. Mech Dev 122:273–287
18. Abu-Daya A, Sater AK, Wells DE, Mohun TJ, Zimmerman LB (2009) Absence of heartbeat in the Xenopus tropicalis mutation muzak is caused by a nonsense mutation in cardiac myosin myh6. Dev Biol 336:20–29
19. Abu-Daya A, Nishimoto S, Fairclough L, Mohun TJ, Logan MP, Zimmerman LB (2010) The secreted integrin ligand nephronectin is necessary for forelimb formation in Xenopus tropicalis. Dev Biol 349:204–212
20. Geach TJ, Zimmerman LB (2010) Paralysis and delayed Z-disc formation in the Xenopus tropicalis unc45b mutant dicky ticker. BMC Dev Biol 10:75
21. Waldner C, Sakamaki K, Ueno N, Turan G, Ryffel GU (2006) Transgenic Xenopus laevis strain expressing cre recombinase in muscle cells. Dev Dyn 235:2220–2228
22. Amaya E (2005) Xenomics. Genome Res 15:1683–1691
23. Khokha MK, Krylov V, Reilly MJ, Gall JG, Bhattacharya D, Cheung CY, Kaufman S, Lam DK, Macha J, Ngo C, Prakash N, Schmidt P, Tlapakova T, Trivedi T, Tumova L, Abu-Daya A, Geach T, Vendrell E, Ironfield H, Sinzelle L, Sater AK, Wells DE, Harland RM,

Zimmerman LB (2009) Rapid gynogenetic mapping of Xenopus tropicalis mutations to chromosomes. Dev Dyn 238:1398–1446

24. Elsdale TR, Gurdon JB, Fischberg M (1960) A description of the technique for nuclear transplantation in Xenopus laevis. J Embryol Exp Morphol 8:437–444
25. Reinschmidt D, Friedman J, Hauth J, Ratner E, Cohen M, Miller M, Krotoski D, Tompkins R (1985) Gene-centromere mapping in Xenopus laevis. J Hered 76:345–347
26. Tompkins R, Reinschmidt D (1991) Experimentally induced homozygosity in Xenopus laevis. Methods Cell Biol 36:35–44
27. Young JJ, Cherone JM, Doyon Y, Ankoudinova I, Faraji FM, Lee AH, Ngo C, Guschin DY, Paschon DE, Miller JC, Zhang L, Rebar EJ, Gregory PD, Urnov FD, Harland RM, Zeitler B (2011) Efficient targeted gene disruption in the soma and germ line of the frog Xenopus tropicalis using engineered zinc-finger nucleases. Proc Natl Acad Sci U S A 108: 7052–7057
28. Yergeau DA, Kuliyev E, Mead PE (2007) Injection-mediated transposon transgenesis in Xenopus tropicalis and the identification of integration sites by modified extension primer tag selection (EPTS) linker-mediated PCR. Nat Protoc 2:2975–2986
29. Yergeau DA, Kelley CM, Kuliyev E, Zhu H, Sater AK, Wells DE, Mead PE (2010) Remobilization of Tol2 transposons in Xenopus tropicalis. BMC Dev Biol 10:11
30. Wells D, Gutierrez L, Xu Z, Krylov V, Macha J, Blankenburg K, Hitchens M, Bellot L, Spivey M, Kowis A, Ye Y, Pasternak S, Owen J, Tran T, Slavikova R, Tumova L, Tlapakova T, Seifertova E, Scherer S, Sater A (2011) A genetic map of Xenopus tropicalis. Dev Biol 354(1):1–8
31. Ogino H, McConnell WB, Grainger RM (2006) Highly efficient transgenesis in Xenopus tropicalis using I-SceI meganuclease. Mech Dev 123:103–113
32. Allen BG, Weeks DL (2005) Transgenic Xenopus laevis embryos can be generated using phiC31 integrase. Nat Methods 2:975–979
33. Hirsch N, Zimmerman LB, Gray J, Chae J, Curran KL, Fisher M, Ogino H, Grainger RM (2002) Xenopus tropicalis transgenic lines and their use in the study of embryonic induction. Dev Dyn 225:522–535
34. Chae J, Zimmerman LB, Grainger RM (2002) Inducible control of tissue-specific transgene expression in Xenopus tropicalis transgenic lines. Mech Dev 117:235–241
35. Hartley KO, Nutt SL, Amaya E (2002) Targeted gene expression in transgenic Xenopus using the binary Gal4-UAS system. Proc Natl Acad Sci U S A 99:1377–1382
36. Ryffel GU, Werdien D, Turan G, Gerhards A, Goosses S, Senkel S (2003) Tagging muscle cell lineages in development and tail regeneration using Cre recombinase in transgenic Xenopus. Nucleic Acids Res 31:e44
37. Lehman CW, Carroll D (1991) Homologous recombination catalyzed by a nuclear extract from Xenopus oocytes. Proc Natl Acad Sci U S A 88:10840–10844
38. Doyon Y, McCammon JM, Miller JC, Faraji F, Ngo C, Katibah GE, Amora R, Hocking TD, Zhang L, Rebar EJ, Gregory PD, Urnov FD, Amacher SL (2008) Heritable targeted gene disruption in zebrafish using designed zinc-finger nucleases. Nat Biotechnol 26:702–708
39. Carruthers S, Stemple DL (2006) Genetic and genomic prospects for Xenopus tropicalis research. Semin Cell Dev Biol 17:146–153
40. Showell C, Carruthers S, Hall A, Pardo-Manuel de Villena F, Stemple D, Conlon FL (2011) a comparative survey of the frequency and distribution of polymorphism in the genome of Xenopus tropicalis. PLoS One 6:e22392
41. Sargent MG, Mohun TJ (2005) Cryopreservation of sperm of Xenopus laevis and Xenopus tropicalis. Genesis 41:41–46
42. Bahary N, Davidson A, Ransom D, Shepard J, Stern H, Trede N, Zhou Y, Barut B, Zon LI (2004) The Zon laboratory guide to positional cloning in zebrafish. Methods Cell Biol 77:305–329
43. Bassam BJ, Caetano-Anolles G, Gresshoff PM (1991) Fast and sensitive silver staining of DNA in polyacrylamide gels. Anal Biochem 196:80–83
44. Streisinger G, Singer F, Walker C, Knauber D, Dower N (1986) Segregation analyses and gene-centromere distances in zebrafish. Genetics 112:311–319
45. Vos P, Hogers R, Bleeker M, Reijans M, van de Lee T, Hornes M, Frijters A, Pot J, Peleman J, Kuiper M et al (1995) AFLP: a new technique for DNA fingerprinting. Nucleic Acids Res 23:4407–4414
46. Neff MM, Neff JD, Chory J, Pepper AE (1998) dCAPS, a simple technique for the genetic analysis of single nucleotide polymorphisms: experimental applications in Arabidopsis thaliana genetics. Plant J 14: 387–392
47. Chang Y-F, Imam JS, Wilkinson MF (2007) The nonsense-mediated decay RNA surveillance pathway. Annu Rev Biochem 76:51–74

Chapter 6

Forward Genetic Screens in *Xenopus* Using Transposon-Mediated Insertional Mutagenesis

Donald A. Yergeau, Clair M. Kelley, Haiqing Zhu, Emin Kuliyev, and Paul E. Mead

Abstract

The class II DNA "cut-and-paste" transposons have been used to efficiently modify the *Xenopus* genome for transgenesis applications. Once integrated, the transposon is an effective substrate for excision and re-integration (remobilization) elsewhere in the genome by simply supplying the transposase enzyme *in trans*. We have used two methods to remobilize transposons resident in the frog genome: micro-injection of transposase mRNA at the one-cell stage and expression of the enzyme in the germline from a transgene. Double-transgenic frogs (hoppers) that harbor transgenes for both the substrate transposon and the transposase enzyme are outcrossed to wild-type animals and the progeny are scored for changes in reporter gene expression. Although both methods work effectively to remobilize transposons, the breeding-mediated strategy eliminates the time-consuming micro-injection step; novel integration events are produced by simply outcrossing the hopper frogs. As each outcross of *Xenopus tropicalis* typically produces 2,000, or more, progeny, this method can be used to perform large-scale insertional mutagenesis screens in this highly tractable developmental model system.

Key words: Transposon, *Sleeping Beauty*, *Tol2*, Insertional mutagenesis, Enhancer trapping, Gene trapping

1. Introduction

DNA cut-and-paste transposons are powerful tools for manipulating the vertebrate genome (1–14). Once integrated into the host genome, the transposon is a substrate for excision and re-integration (remobilization) following re-expression of the cognate transposase enzyme. This feature of transposon biology has been exploited for forward genetic screens in a variety of vertebrate model systems including mouse (13, 15–20), rat (21, 22), zebrafish (23–25), and *Xenopus* (26) and can be used to perform both somatic and germline screens. In mice, for example, the *SB* and *piggyBac* transposon

Stefan Hoppler and Peter D. Vize (eds.), *Xenopus Protocols: Post-Genomic Approaches, Second Edition*, Methods in Molecular Biology, vol. 917, DOI 10.1007/978-1-61779-992-1_6, © Springer Science+Business Media, LLC 2012

systems have been used to perform large-scale somatic screens to identify cancer-related genes (2, 27–31). Transgenic mice harboring a bidirectional gene trap transposon substrate were bred with animals harboring a ubiquitously expressed transposase enzyme. Random somatic remobilization of the transposon in the double-transgenic progeny resulted in tumor formation when the screens were performed on cancer-susceptible backgrounds. Cloning the transposon integration sites from tumor samples revealed genes that are involved in cancer formation and progression. Remobilization of transposons in the germline results in the production of novel transgenic lines. Germline remobilization in aquatic vertebrates can be achieved in two ways. First, one-cell embryos from transgenic parents can be injected with mRNA encoding the cognate transposase. The injected embryos are raised to adulthood and outcrossed and novel integration events are scored in the progeny. This method has been used successfully for *Tol2*-mediated enhancer trap screens in both fish and frogs (23, 25, 26, 32). Alternatively, to avoid the time-consuming micro-injection step, transgenic animals that express the transposase enzyme in germline cells can be outcrossed with transposon substrate animals to derive double-transgenic hopper lines. Expression of the transposase in the germline results in random remobilization events of the substrate transposon in the developing gametes. As different gametes will have different remobilization events, large numbers of potentially novel lines, each with a unique transposon integration event can be identified by simply outcrossing the double-transgenic hopper adults. This approach has been used to develop novel transgenics in mice (13, 19), rats (21, 22), zebrafish (23), and frog (26). The Kawakami laboratory used a heat-shock inducible *Tol2* transposase transgenic zebrafish model to achieve germline remobilization of a *Tol2* transposon harboring a gene trap construct (23). In *Xenopus*, we have used a ubiquitously expressed *Sleeping Beauty* (*SB*) enzyme transgenic frog to effect germline remobilization of a *SB* transposon to generate novel transgenic lines with unique integration events (manuscript submitted). The significant advantage of the breeding-mediated remobilization strategy is the removal of the time- and skill-intensive micro-injection step. Novel re-integration events are produced by simply outcrossing the double-transgenic hopper frogs. This strategy also exploits the remarkable fecundity of the frog, where clutch sizes routinely contain over 2,000 progeny, and can be in excess of 5,000 tadpoles (33). This means that, even when the remobilization frequency is low, large numbers of tadpoles with novel transposition events can be obtained with a relatively small pool of double-transgenic hopper frogs.

As transposon integration systems can harbor long DNA sequences (>10 kb), elaborate gene and enhancer trapping constructs can be engineered between the transposon arms. Genetic screens using enhancer traps can be useful for generating novel transgenic lines with unique reporter expression patterns.

The enhancer trap construct contains a minimal promoter element driving expression of a reporter gene (usually GFP or RFP). Integration of the enhancer trap near a gene regulatory element can confer the tissue-restricted specificity of the regulatory element onto the expression of the reporter (4, 11, 34). Gene trap constructs contain an RNA splice acceptor to generate a chimeric transcript when the transposon lands within a functional transcription unit. The expression of the fused reporter is dependent on the endogenous activity of the trapped gene, and thus the reporter pattern gives an indication of the normal expression of the trapped locus. Gene trap constructs are more likely to be mutagenic than standard enhancer traps as their expression is reliant on the interruption of an endogenous gene (35–40). A variation of the simple gene trap is the polyadenylation trap (pAT) (41). The pAT contains a promoter driving a reporter followed by a splice donor site and lacks a functional polyadenylation signal. The resulting transcript is unstable unless the pAT transposon integrates upstream of exon containing a functional polyadenylation signal. Splicing from the pAT trap splice donor to a splice acceptor upstream of the functional 3′-untranslated region (UTR) forms a functional transcription unit and expression of the reporter. The advantage of the pAT over the standard gene trap is that the expression of the reporter is independent of the expression of the endogenous gene as it is driven by the promoter element encoded in the pAT vector. If the endogenous gene is expressed at very low levels, or is expressed for a brief window of time in a small subset of cells in the body, then the reporter expression might be missed. Like the standard splice acceptor gene trap, the pAT construct is more likely to be mutagenic than an enhancer trap as its expression relies on integration with a transcriptional unit and co-option of a functional exon containing a polyadenylation signal (16, 21, 42). The pAT vector does not distinguish between active and inactive genes, and may trap pseudogenes, as it is reliant solely on integration upstream of an exon containing a functional polyadenylation signal.

Another advantage of the transposon remobilization strategy is that different DNA cut-and-paste transposons have different excision and re-integration activities (43). Different transposases have different site specificities and will thus target different regions of the genome. For example, *SB* requires a simple TA dinucleotide for integration (44) whereas *piggyBac* integrates at TTAA sites (45). In addition, different transposase enzymes have different propensities for local hopping; that is, re-integration near the donor site. *SB*, for example, has a strong tendency to re-integrate the substrate near the excision site. In frogs, we have shown that approximately 80% of *SB* remobilization events occur within 3 Mb of the donor site. The local hopping activity of *Tol2* is much less pronounced; only 20% of re-integration events occur near the transposon donor locus and 80% are randomly distributed throughout the genome (26).

In this chapter, we will outline the procedures for injection- and breeding-mediated germline remobilization strategies that form the basis of transposon insertional screens in the frog. As the methodologies for micro-injection were described in an accompanying Chapter 14, and methodology for setting up crosses is simple, this chapter will focus on the identification and molecular characterization of novel re-integration events, and the husbandry issues associated with performing forward genetic screen in *Xenopus tropicalis*. For simplicity, we will describe the injection-mediated approach using the *Tol2* system as an example and use the *SB* system to outline the breeding-mediated remobilization strategy. The methodologies are interchangeable for both transposon systems. As with any multigeneration screen, good husbandry practices are essential. Notes on caring for, and identifying, tadpoles, postmetamorphic and adult frogs are included.

2. Materials

Prepare all solutions with ultrapure nuclease-free water and analytical grade reagents. DNA and RNA stocks should be stored in small single-use aliquots to avoid repeated freeze–thaw cycles. All animal work should be done in accordance with local regulations and guidelines.

2.1. Injection-Mediated Transposon Remobilization Screens

The materials and methods for micro-injecting one-cell *X. tropicalis* embryos were described in an accompanying Chapter 14.

1. *Tol2* substrate transposon transgenic *X. tropicalis* adult frogs for setting up outcrosses to harvest embryos for micro-injection (see Note 1).
2. Capped *Tol2* transposase mRNA synthesized using mMessage mMachine in vitro transcription kits (Ambion, TX). Stored in single-use aliquots at −80 °C. Thawed samples are diluted to the appropriate concentration (usually 100 pg in 3 nL) and kept on ice until needed for injection.

2.2. Breeding-Mediated Transposon Remobilization Screens

1. Double-transgenic hopper frogs that harbor transgenes for both the substrate transposon and the cognate transposase expressed in the germline (see Note 2).

2.3. Components Needed for Outcross of Transgenic Frogs and Scoring Tadpoles

1. Breeding tanks; 8 qt/7.6 L polycarbonate tanks. Carlisle square food storage containers with lids (Carlisle, OK).
2. Human chorionic gonadotropin (HCG; 1,000 U/mL, Novarel). Store powder at room temperature, after reconstitution store at 4 °C for up to 1 month.

3. 1 mL syringes.
4. 27-gauge needles.
5. Frog system water (FSW). Fresh water drawn directly from the adult *X. tropicalis* aquarium. The water conditions are maintained at a salinity of 1,000 μS and pH 7.0 at 25 °C.
6. Tricaine methane sulphonate used at a final concentration of 0.015% (w/v). Stock solutions (~1% (w/v)) are prepared using sterile water and stored at 4 °C.
7. 2-Cell Mechanical Lab Counter (Carolina Biological Supply Company, NC).

2.4. Components Needed for Isolation of Genomic DNA from Tadpoles

1. Plastic disposable transfer pipettes.
2. Sterile 1.5 mL plastic Eppendorf tubes.
3. DNA extraction buffer; 20 mM Tris, pH 7.5, 100 mM NaCl, 10 mM EDTA, 1% (w/v) SDS. Stored for up to 1 month at room temperature.
4. Proteinase K (recombinant PCR grade, Roche Diagnostics GmbH, Mannheim, Germany). Stock solution ~20 mg/mL stored at 4 °C.
5. RNaseA stock solution (10 mg/mL) stored at 4 °C.
6. Phenol/chloroform/isoamyl alcohol (25:24:1 v/v/v).
7. Nutator mixing platform (Adams Nutator Mixer).
8. 10 M ammonium acetate. Stored at room temperature.
9. Isopropanol. Stored at room temperature.
10. 70% (v/v) ethanol. Stored at room temperature.
11. Sterile nuclease-free water. Stored at room temperature.

2.5. Components Needed for Implanting RFID Chips

1. Radio Frequency Identification (RFID) chips embedded in inert epoxy resin (~2 mm in diameter) such as mic3-TAG 64 bit RO (microSensys, Germany; www.microsensys.de).
2. RFID pen reader and handheld wireless computer to retrieve chip ID (microSensys, Germany).
3. Isoflurane (1-chloro-2,2,2-trifluoroethyl difluoromethyl ether; Forane, Baxter Healthcare Corporation, IL).
4. 1 L glass beaker with lid.
5. Sterile surgical scissors and flat forceps.
6. FSW.
7. 8 qt/7.6 L polycarbonate tanks. Carlisle square food storage containers with lids (Carlisle, OK).

2.6. Components Needed for Isolating DNA from Skin Swabs

1. DNA Buccal swabs (Isohelix; DDK-50 DNA preparation kit with 50 swabs with 2 mL tube and special release cap, individually wrapped, gamma treated).

2. 8 qt/7.6 L polycarbonate tanks. Carlisle square food storage containers with lids (Carlisle, OK).
3. FSW.

3. Methods

3.1. Injection-Mediated Transposon Remobilization Screens

Outcross of *Tol2* transposon transgenic adults is used to produce one-cell embryos for injection with *Tol2* transposase mRNA (100–500 pg; see *accompanying* Chapter 14 for protocols for micro-injection and tadpole husbandry). Injected embryos are raised to adulthood, outcrossed, and the progeny scored for novel reporter gene expression patterns. The general scheme for the injection-mediated transposon remobilization screen is outlined in Fig. 1.

3.2. Breeding-Mediated Transposon Remobilization Screens

The general strategy of the breeding-mediated transposon remobilization screen is depicted in Fig. 2. Double transgenic transposon hopper frogs are generated by incrossing transposon substrate

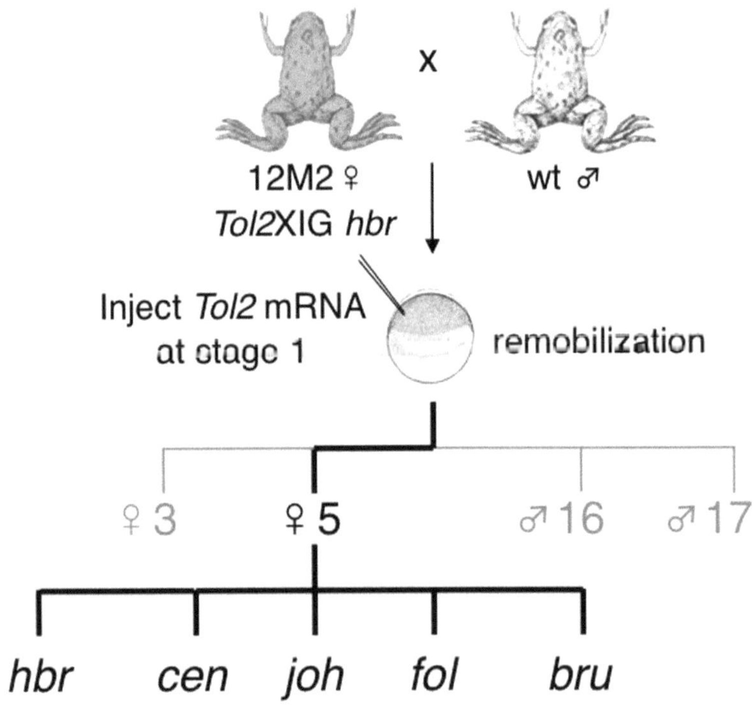

Fig. 1. Injection-mediated remobilization strategy. Transgenic embryos are produced by the outcross of a female frog harboring a single *Tol2* transposon integration, the *hbr* allele, with a wild-type male. Embryos are injected with *Tol2* transposase mRNA at the one-cell stage and the resulting animals are raised to adulthood and outcrossed. For example, the germline of remobilized frog ♀5 has four new independently segregating *Tol2* transposon alleles (*cen*, *bru*, *fol*, and *joh* (26)), in addition to the parental donor allele (*hbr*). Progeny of ♀5 inherit various combinations of the novel and donor alleles. The alleles are segregated into individual transgenic lines by serial outcrosses.

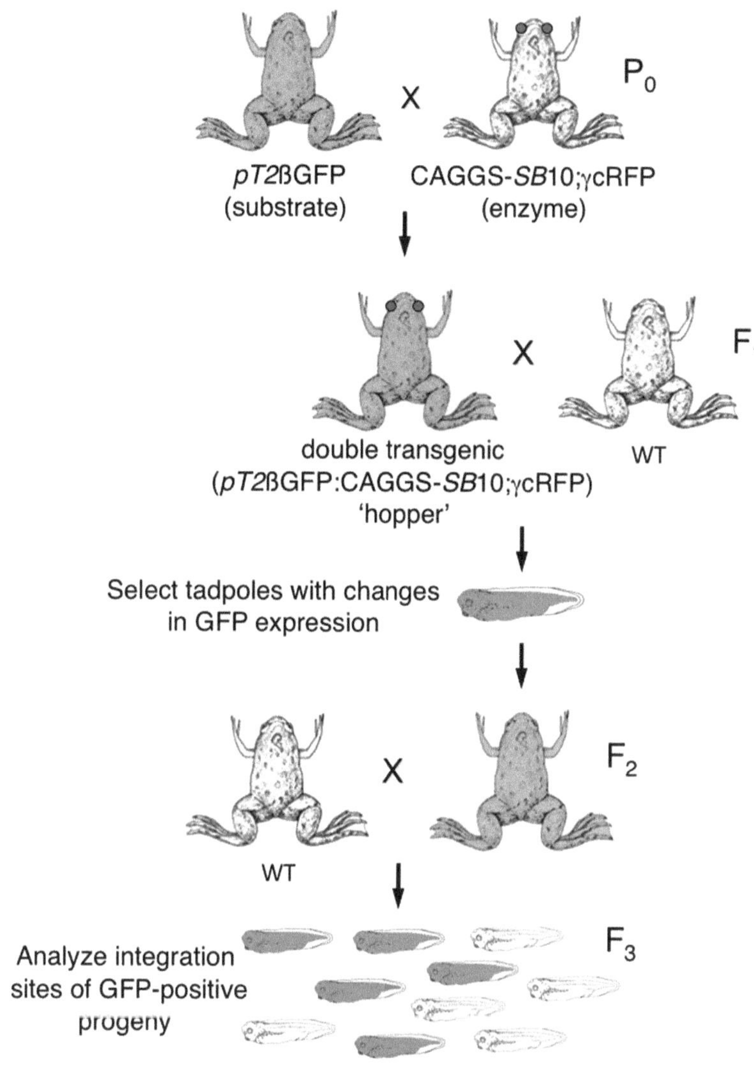

Fig. 2. Breeding-mediated remobilization strategy. Founder transgenic frogs (P_0) that harbor substrate transposon and transposase enzyme transgenes are incrossed to produce double-transgenic F_1 hopper frogs. The inheritance of the transposase transgene is followed by linking a surrogate reporter, gamma-crystallin RFP (γcRFP) that produces a red lens, to the enzyme mini-gene. Double-transgenic hopper frogs are outcrossed and the F_2 progeny are scored for changes in GFP expression indicative of modification of the endogenous parental transposon locus. Selected tadpoles are raised to adulthood and outcrossed, and the transposon integration events are characterized in the F_3 population.

transgenic frogs (males or females) with transposase transgenic animals (females or males, respectively). The double transgenic hoppers are outcrossed with wild-type frogs (of the opposite sex) and the progeny is screened for changes in reporter gene expression indicative of excision and re-integration of the substrate transposon (Fig. 3).

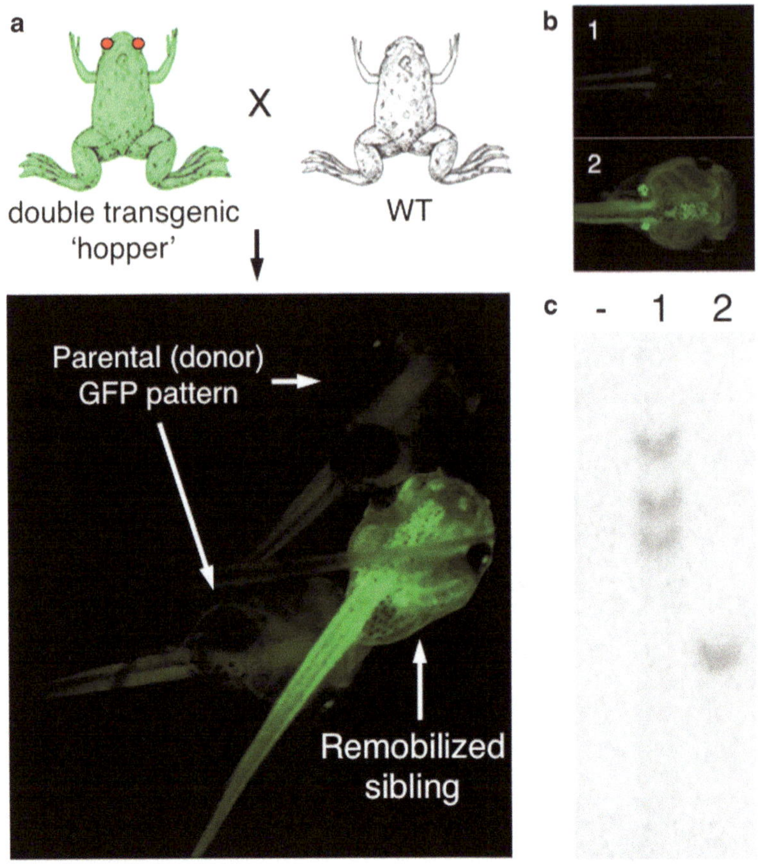

Fig. 3. Modification of the parental transposon locus by re-expression of the transposase alters the GFP expression profile in the outcross population. (**a**) A pool of tadpoles from the outcross of a hopper frog shows two tadpoles with the parental GFP expression pattern and a single tadpole with intense GFP expression, indicative of excision of the substrate transposon from the parental donor site and re-integration elsewhere in the genome. (**b**) GFP expression in tadpoles from the outcross of a double transgenic hopper. Tadpole 1 has the parental GFP expression pattern. The GFP expression in tadpole 2 is markedly higher than that in tadpole 1 suggesting that the parental (donor) transposon locus has been modified. Both images were taken using the same exposure time and aperture setting. (**c**) Southern blot analysis of a *SB* remobilization event. The parental locus has three copies of the *SB* transposon (pT2βGFP; lane 1 corresponds to tadpole 1 in (**b**)) and the remobilized sibling (tadpole 2) has a single transposon integration event. The lane labeled "–" contains DNA from a wild-type GFP-negative sibling tadpole.

3.3. Outcross of Xenopus tropicalis Transgenic Frogs and Scoring the Progeny

1. Egg laying is induced by injecting frogs in the dorsal lymph sac with HCG (200 U for females and 100 U for males). Transgenic animals are outcrossed with wild-type frogs. Pairs of primed frogs are placed in 8 qt polycarbonate breeding tanks containing ~4 L of FSW. A lid is placed on the tank and the frogs left undisturbed overnight at 25 °C. The following morning, the frogs are removed from the tank and the embryos are rinsed twice with ~8 L of FSW. The tanks are maintained at

25 °C until the tadpoles are freely swimming, and then moved to the nursery room (28 °C).

2. Excision of a transposon from one genomic locus and re-integration at a new locus frequently results in a change in the transposon reporter gene expression (see Fig. 3). Changes in GFP reporter expression can be readily observed using a dissecting fluorescent microscope. Free-swimming tadpoles (~stage 40–45; see Note 3) are immobilized with a low dose of tricaine methane sulphonate (0.015% (w/v)) and observed for reporter gene expression in small groups of 8–12 tadpoles per well in a 9-well depression slide (Pyrex, Spot Plate, 9 Depression, 85 × 100 mm). Batch scoring tadpoles allow large populations to be sorted quickly and tadpoles with subtle changes in GFP expression are more readily identified when compared side by side with those that have the "parental" unmodified GFP expression pattern. Wells containing tadpoles with distinct GFP expression profiles can be readily sorted to isolate individual animals. Tallies of GFP-positive and GFP-negative tadpoles in each batch are recorded using a 2-cell mechanical lab counter, and the final counts are used to determine the ratio of transgenic animals in the outcross population. Tadpoles with unique GFP expression profiles are raised in groups in static tanks in FSW at 28 °C.

3.4. Isolation of Genomic DNA from Tadpoles

Genomic DNA is harvested from anesthetized whole tadpoles, or tail clips (see Note 4), using standard overnight digestion with proteinase K followed by phenol/chloroform extraction and alcohol precipitation.

1. Tadpoles (stage 45–50) are anesthetized in FSW containing 0.015% (w/v) tricaine, photographed and transferred to individual 1.5 mL Eppendorf tubes. Excess liquid is removed from the tube with a transfer pipette, and 0.5 mL of DNA extraction buffer containing 0.1 mg/mL proteinase K is added to each tube.
2. The tubes are incubated overnight at 56 °C.
3. RNaseA is added to a final concentration of 0.02 μg/mL and the samples incubated at 37 °C for 1 h.
4. An equal volume (0.5 mL) of phenol/chloroform/isoamyl alcohol is added to each tube and incubated at room temperature for 5–10 min while shaking gently, laid horizontal on a nutator.
5. The organic and aqueous phases are separated by centrifugation at 16,110 × *g* at room temperature in a microfuge for 5 min.
6. The upper aqueous phase is transferred to a fresh tube and the DNA is precipitated by adding 0.2 volumes of ammonium

acetate and 1 volume of isopropanol. The samples are mixed by inverting several times and placed on ice for 30 min.

7. The precipitated DNA is recovered by centrifugation at 16,110 × *g* at room temperature in a microfuge for 5 min.
8. The liquid is aspirated from the tube and the pellet washed with 0.75 mL of 70% ethanol. After incubating at room temperature for 5–10 min, the tubes are centrifuged as above and the liquid is gently aspirated from the tube.
9. The DNA pellet is allowed to air dry for 10 min at room temperature and resuspended in 50–100 μL of nuclease-free water, depending on the size of the tadpole, and incubated overnight at 4 °C (see Note 5).
10. The DNA concentration is determined the following day by measuring the absorbance of duplicate diluted samples at 260 nm. Samples are stored at 4 °C.

3.5. Characterization of Transposon Integration Sites

Two methods are used to characterize the integration sites of remobilized transposons. First, Southern blotting is used to determine the number of integration events in transgenic lines produced by the transposon hopping strategies (Fig. 3, see Note 6). The injection-mediated remobilization approach, for example, can result in an increase in the number of alleles present in a given founder line. Remobilization of a single-copy *Tol2* transgenic by micro-injecting one-cell embryos with *Tol2* transposase mRNA resulted in progeny that contained multiple *Tol2* transposons in the germline. Remobilization events that occur in discrete blastomeres that give rise to germline tissue can increase the number of transposon alleles in the injected tadpole. In addition, excision and re-integration events that occur during the cell cycle after DNA synthesis but prior to cell division can increase the number of transposon alleles in an individual blastomere. With the *Tol2* systems, we have identified individual remobilized lines that inherit both the parental transposon allele and up to four discrete novel transposition events; injecting transposase mRNA into a substrate line that has a single transposon insertion resulted in progeny that have a total of five independently segregating alleles (26). Southern blot data is essential to verify the transposon content of each remobilized line.

Second, PCR-based strategies are used to clone the genomic DNA sequences flanking the transposon integration site. Several methods can be used to clone the transposon flanking sequences including inverse PCR, thermal annealing interlaced (TAIL) PCR, and extension primer tag selection (EPTS) linker-mediated (LM-) PCR. We have found that EPTS LM-PCR is the most robust method for cloning transposon integration sites (46). In EPTS LM-PCR, genomic DNA is digested with frequent cutter restriction endonucleases. A primer extension reaction using a biotinylated primer that hybridizes to the transposon sequence produces a single stranded copy of the transposon arm and the flanking genomic DNA. Excess biotinylated primers are removed from the completed reaction using size-exclusion spin columns and the primer extension products are purified on streptavidin magnetic beads. Double stranded linkers are ligated to the 3′ ends of primer extension products, and excess linkers are then removed by washing of the streptavidin beads. The ligated products, still attached to the beads, are used as substrates for standard PCR reactions using primers specific for the transposon arms and the synthetic linkers. When necessary, nested primer pairs can be used to re-amplify the PCR products to increase the yield for subsequent cloning steps. Several transposon insertion sites can be cloned from a single EPTS LM-PCR; we have cloned up to six integration events from a single reaction. A detailed method for EPTS LM-PCR can be found in Yergeau et al. (46) (see Note 7).

3.6. Tadpole and Juvenile Frog Husbandry

Tadpoles are raised in FSW in static tanks at 28 °C. For maximal growth rates, tadpoles are maintained at a density of no more than one tadpole per liter of water. Tadpoles are fed three times per day with a suspension of Sera Micron powder prepared freshly for each feeding. An airstone is added to the tank to provide gentle aeration. Each week, approximately one-half of the water is removed from the tank and replaced with fresh FSW. Tadpoles are raised in groups, and growing tadpoles in prolonged isolation should be avoided when possible. Juvenile frogs are kept in the static tanks for up to 1 month after metamorphosis. Moving young frogs to the adult aquarium too soon after metamorphosis can result in significant morbidity. Juvenile frogs are fed once a day with Jurassi-Diet Newt And Frog Food (Jurassipet).

3.7. Adult Frog Husbandry

Adult transgenic frogs are housed in a recirculating aquarium system at 25 °C. Water conditions are maintained at 1,000 μS and pH 7.0 by computer-controlled dosing pumps. Each day, one-tenth of the aquarium water is changed automatically using reverse-osmosis purified water (~800 L per day). Adult *X. tropicalis* frogs are fed three times a week with Nasco Frog Brittle for Post-Metamorphic *Xenopus* (Nasco, WI).

3.8. Identification of Individual Adult Frogs

Two methods are used to identify unique *X. tropicalis* transgenic animals. The simplest method is to house individual frogs in separate tanks. This method is an inefficient use of space and furthermore adult frogs do not thrive in prolonged periods of isolation. Communal housing can be achieved by adding wild-type frogs of the opposite gender to a tank containing a transgenic animal. Alternatively, we uniquely identify all transgenic animals in the system using implanted radio-frequency identification (RFID) chips (47, 48). The RFID chip carries a unique 16-bit alphanumeric code that can be used to identify the frog throughout its lifespan. The ID is read with a Bluetooth Pen Reader that displays the chip identity on a handheld computer device. We have developed a web-based database for our transgenic colony that links all data for an individual animal, including family tree, transposon integration site and Southern blot data, location in the aquarium and outcross history, to the unique 16-bit chip ID.

3.9. Implanting RFID Chips into Adult *Xenopus tropicalis*

1. Adult *X. tropicalis* frogs are anesthetized using isoflurane (see Note 8). This procedure is preformed in a chemical hood. A paper towel is prewetted with FSW and used to line the bottom of a 1 L beaker. Approximately 500 μL of isoflurane is pipetted onto the damp paper towel.
2. The frog to be chipped is placed into the beaker and a lid is applied to prevent escape during the brief excitation phase of anesthesia. The frog will be unresponsive to external stimuli (toe pinch) and lose the righting response in less than 1 min. The animal should be removed from the container immediately after anesthesia is induced to avoid over-exposing the animal to the anesthetic.
3. The frog is placed on a clean prewetted plastic sheet and a small incision (~3–5 mm) is made on the dorsal surface at the base of the neck.
4. A sterile epoxy-coated RFID microchip is inserted into the incision using a pair of flat tweezers. The chip is maneuvered into position, at least 5 mm anterior to the incision site, using the flat edge of the tweezers. Due to the small size of the wound, no stitches or skin-bonding agent is required and the incision is left to heal naturally (usually completely healed in 4–5 days). The chip ID is verified using the Pen reader and entered into the database.
5. The frog is placed in a labeled 8 qt polycarbonate tank containing ~2 cm of FSW. The tank is tilted slightly to ensure the nose of the frog is above the water line. The animal is observed continuously until it has recovered from the anesthesia and has begun to swim, usually in 5 min or less. Approximately 4 L of fresh FSW is added to the tank, and the frog is kept in the laboratory for up to 5 days to allow frequent observation during the healing period before being returned to the aquarium. The water is changed daily to prevent infection.

3.10. Genotype Adult *Xenopus tropicalis* Frogs Using Skin DNA Samples

It is sometimes necessary to genotype adult frogs to determine the number of transposon alleles present in an individual transgenic. This can be done by outcrossing the animal and scoring the progeny for transposon insertions. A quicker method is to obtain a DNA sample from the adult and perform PCR-based genotyping. Our preferred method of DNA isolation from adult frogs is to use a sterile buccal swab to collect skin cells from the surface of the frog. The swabbing procedure is noninvasive and can be done on frogs that are too young to outcross.

1. Hold the frog in left hand and gently swab the dorsal and ventral areas with an Isohelix SK-2 Buccal Swab. Swab the animal for 15–20 s to ensure that enough skin cells have adhered to the swab for DNA extraction.
2. If the transgenic frog has not been chipped, it can be kept in isolation in a lidded polycarbonate tank containing ~4 L of FSW until the genotyping is complete (1–2 days).
3. Follow the manufacturer's instructions for rapid DNA extraction following incubation of the swab tip in extraction buffer at 56 °C. All solutions required are provided with the kit (DSK-50; Isohelix).
4. Resuspend DNA in 50 μL sterile nuclease-free water and store at 4 °C.
5. Use 5 μL of DNA for PCR reactions with primers specific to all of the potential transposon integration sites in the frog (Fig. 4). Include PCR primers that will control for DNA recovery. We routinely use primers for GFP to verify the presence of the transposon and primers for an unrelated genomic locus to demonstrate adequate DNA recovery from each swab sample.

4. Notes

1. *Tol2* transgenic *X. tropicalis* adult frogs. Male or female transgenic animals can be outcrossed with wild-type frogs for microinjection experiments. We routinely use transgenic animals that are homozygous for the *Tol2* transposon allele so that all of the outcross progeny will be heterozygous for the substrate transposon.
2. Double-transgenic hopper frogs are produced by incrossing animals that are transgenic for a transposon substrate (GFP-positive) with those that are transgenic for the cognate transposase enzyme (see Fig. 2). In order to readily identify the transposase transgenic animals, we clone a surrogate reporter mini-gene, such as gamma-crystallin RFP or CFP, downstream of the transposase enzyme so that the transgenic animals can be scored for fluorescence in the lens of the eye.
3. We routinely score tadpoles after stage 40 as they large and easy to manipulate at this stage. Scoring can be done at earlier stages; however, the tadpoles are more sensitive to the anesthesia before stage 40, and some animals may be lost following prolonged exposure to tricaine.

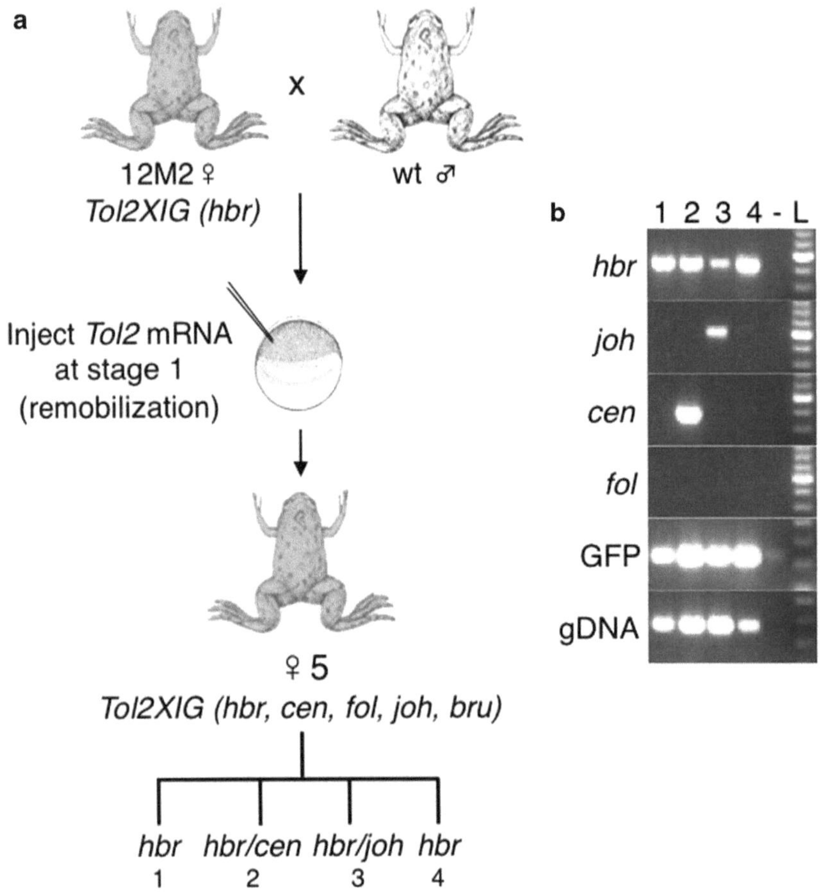

Fig. 4. Skin swabs can be used as a noninvasive method to harvest DNA from adult frogs for genotyping. (**a**) Remobilized *Tol2* transgenic frog ♀5 has five independently segregating *Tol2*XIG alleles (*hbr*, *cen*, *fol*, *bru*, and *joh*). The progeny from the outcross of ♀5 have different combination of these alleles. (**b**) PCR genotyping of adult progeny from ♀5. DNA samples from four adult frogs were harvested using skin swabs. Primer pairs specific to each allele were used for standard genomic PCR. All four individuals have inherited the parental *hbr* allele. Frog 2 has the *cen* and *hbr* alleles, frog 3 has the *joh* and *hbr* alleles, and frogs 1 and 4 have the *hbr* allele alone. Primers for GFP and a random genomic locus are used as positive controls.

4. Tail clip samples can be isolated from premetamorphic tadpoles by first inducing anesthesia and then clipping approximately 5 mm from the end of the tail with a razor blade or small surgical scissors, being careful to avoid the major vessels further up the tail. DNA is prepared as described earlier; however, the concentration will be low and may only be sufficient for PCR-based applications. Tadpoles are kept in individual crystallizing dishes containing ~100 mL of FSW with daily water changes.
5. It is important to allow DNA to fully rehydrate before any further manipulations are performed. This is best done by incubating the samples overnight at 4 °C.

6. Tips for Southern blotting. Use, at most, 3–5 μg of genomic DNA for restriction digests. Overloading lanes with DNA is the most likely cause for poor resolution of hybridizing bands. Measure the DNA concentration of each sample to accurately determine the amount of each sample needed for the restriction digest. Run 0.7% (w/v) agarose gels (11 × 13 cm) in 1× TAE (Tris/acetate/EDTA) buffer at 70 V for approximately 4 h.
7. Tips for EPTS LM-PCR. Digest up to 1 μg of genomic DNA with an appropriate frequent cutter restriction endonuclease for 2 h at the recommended temperature for each enzyme. In selecting an enzyme for fragmenting the sample, it is important to ensure that there are no sites for the enzyme within the transposon arm upstream of the primer extension oligonucleotide binding site. A frequent (4–5 base recognition sequence) cutter is used to produce DNA fragments that contain, on average, 250–1,000 base pairs of genomic sequence flanking the transposon arm. The short flanking sequence ensures that the PCR reaction will be completed efficiently during the extension phase of the PCR reaction, and that the PCR product will clone efficiently into the TOPO cloning vector. PCR products >1 kb clone less efficiently into the TOPO vector than smaller DNA fragments. For EPTS LM-PCR of *SB* and *Tol2* transposon integration sites, we commonly use, in order of preference, *Mae*III, *Nla*III, *Fnu*4HI or *Tse*I, and *Alu*I. For each genomic sample, we routinely perform EPTS LM-PCR with four different restriction enzymes to increase the likelihood that one enzyme will produce an appropriately sized PCR product. In general, ~50–200 bp of flanking genomic sequence is sufficient to uniquely identify the genomic integration site when the sequence is queried against the *X. tropicalis* genome sequence (49). For each integration site, we also generate PCR primers to amplify the integration site beyond the 5′ extent of the EPTS LM-PCR product. In addition, PCR primers are designed to amplify the flanking sequence at opposite end of the transposon, to further verify the precise locus of the integration event.
8. Our preferred method for anesthetizing adult frogs is to use isoflurane. Anesthesia is obtained in less than 1 min and the recovery time is usually around 5 min. By comparison, immersing frogs in solutions containing tricaine can take 10–15 min, or more, to induce sufficient anesthesia for minor surgical manipulations, such as implanting micro-chips. The recovery time with tricaine is likewise relatively long compared to isoflurane-induced anesthesia.

Acknowledgements

This work was supported by NIH grants HD042994 and MH079381 and by the American Lebanese and Syrian Associated Charities (ALSAC) to PEM.

References

1. Clark KJ, Geurts AM, Bell JB, Hackett PB (2004) Transposon vectors for gene-trap insertional mutagenesis in vertebrates. Genesis 39:225–233
2. Ding S, Wu X, Li G, Han M, Zhuang Y, Xu T (2005) Efficient transposition of the *piggyBac* (*PB*) transposon in mammalian cells and mice. Cell 122:473–483
3. Dupuy AJ, Clark K, Carlson CM, Fritz S, Davidson AE, Markley KM, Finley K, Fletcher CF, Ekker SC, Hackett PB, Horn S, Largaespada DA (2002) Mammalian germ-line transgenesis by transposition. Proc Natl Acad Sci U S A 99:4495–4499
4. Grabher C, Henrich T, Sasado T, Arenz A, Wittbrodt J, Furutani-Seiki M (2003) Transposon-mediated enhancer trapping in medaka. Gene 322:57–66
5. Horn C, Wimmer EA (2000) A versatile vector set for animal transgenesis. Dev Genes Evol 210:630–637
6. Ivics Z, Izsvak Z (2004) Transposable elements for transgenesis and insertional mutagenesis in vertebrates: a contemporary review of experimental strategies. Methods Mol Biol 260:255–276
7. Izsvak Z, Ivics Z, Plasterk RH (2000) *Sleeping Beauty*, a wide host-range transposon vector for genetic transformation in vertebrates. J Mol Biol 302:93–102
8. Johnson Hamlet MR, Mead PE (2003) *Sleeping Beauty* and *Xenopus*: transposons as genetic tools. Curr Genomics 4:687–697
9. Kawakami K (2007) *Tol2*: a versatile gene transfer vector in vertebrates. Genome Biol 8(suppl 1):S7
10. Koga A, Cheah FS, Hamaguchi S, Yeo GH, Chong SS (2008) Germline transgenesis of zebrafish using the medaka *Tol1* transposon system. Dev Dyn 237:2466–2474
11. Korzh V (2007) Transposons as tools for enhancer trap screens in vertebrates. Genome Biol 8(suppl 1):S8
12. Miskey C, Izsvak Z, Kawakami K, Ivics Z (2005) DNA transposons in vertebrate functional genomics. Cell Mol Life Sci 62:629–641
13. Takeda J, Keng VW, Horie K (2007) Germline mutagenesis mediated by *Sleeping Beauty* transposon system in mice. Genome Biol 8(suppl 1):S14
14. Yergeau DA, Mead PE (2007) Manipulating the *Xenopus* genome with transposable elements. Genome Biol 8(suppl 1):S11
15. Collier LS, Largaespada DA (2007) Transposons for cancer gene discovery: *Sleeping Beauty* and beyond. Genome Biol 8(suppl 1):S15
16. Dupuy AJ, Fritz S, Largaespada DA (2001) Transposition and gene disruption in the male germline of the mouse. Genesis 30:82–88
17. Largaespada DA (2009) Transposon mutagenesis in mice. Methods Mol Biol 530:379–390
18. Su Q, Prosser HM, Campos LS, Ortiz M, Nakamura T, Warren M, Dupuy AJ, Jenkins NA, Copeland NG, Bradley A, Liu P (2008) A DNA transposon-based approach to validate oncogenic mutations in the mouse. Proc Natl Acad Sci U S A 105:19904–19909
19. Keng VW, Ryan BJ, Wangensteen KJ, Baciunas D, Schmedt C, Ekker SC, Largaespada DA (2009) Efficient transposition of *Tol2* in the mouse germline. Genetics 183:1565–1573
20. Wang W, Bradley A, Huang Y (2009) A *piggyBac* transposon-based genome-wide library of insertionally mutated *Blm*-deficient murine ES cells. Genome Res 19:667–673
21. Kitada K, Keng VW, Takeda J, Horie K (2009) Generating mutant rats using the *Sleeping Beauty* transposon system. Methods 49(3):236–242
22. Lu B, Geurts AM, Poirier C, Petit DC, Harrison W, Overbeek PA, Bishop CE (2007) Generation of rat mutants using a coat color-tagged *Sleeping Beauty* transposon system. Mamm Genome 18:338–346
23. Urasaki A, Asakawa K, Kawakami K (2008) Efficient transposition of the *Tol2* transposable element from a single-copy donor in zebrafish. Proc Natl Acad Sci U S A 105:19827–19832
24. Choo BG, Kondrichin I, Parinov S, Emelyanov A, Go W, Toh WC, Korzh V (2006) Zebrafish transgenic enhancer TRAP line database (ZETRAP). BMC Dev Biol 6:5

25. Kondrychyn I, Garcia-Lecea M, Emelyanov A, Parinov S, Korzh V (2009) Genome-wide analysis of *Tol2* transposon reintegration in zebrafish. BMC Genomics 10:418
26. Yergeau DA, Kelley CM, Kuliyev E, Zhu H, Sater AK, Wells DE, Mead PE (2010) Remobilization of *Tol2* transposons in *Xenopus tropicalis*. BMC Dev Biol 10:11
27. Bestor TH (2005) Transposons reanimated in mice. Cell 122:322–325
28. Collier LS, Carlson CM, Ravimohan S, Dupuy AJ, Largaespada DA (2005) Cancer gene discovery in solid tumours using transposon-based somatic mutagenesis in the mouse. Nature 436:272–276
29. Dupuy AJ, Akagi K, Largaespada DA, Copeland NG, Jenkins NA (2005) Mammalian mutagenesis using a highly mobile somatic *Sleeping Beauty* transposon system. Nature 436: 221–226
30. Rad R, Rad L, Wang W, Cadinanos J, Vassiliou G, Rice S, Campos LS, Yusa K, Banerjee R, Li MA, de la Rosa J, Strong A, Lu D, Ellis P, Conte N, Yang FT, Liu P, Bradley A (2010) *PiggyBac* transposon mutagenesis: a tool for cancer gene discovery in mice. Science 330:1104–1107
31. Vassiliou G, Rad R, Bradley A (2010) The use of DNA transposons for cancer gene discovery in mice. Methods Enzymol 477:91–106
32. Parinov S, Kondrichin I, Korzh V, Emelyanov A (2004) *Tol2* transposon-mediated enhancer trap to identify developmentally regulated zebrafish genes in vivo. Dev Dyn 231:449–459
33. Hirsch N, Zimmerman LB, Grainger RM (2002) *Xenopus*, the next generation: *X. tropicalis* genetics and genomics. Dev Dyn 225:422–433
34. Balciunas D, Davidson AE, Sivasubbu S, Hermanson SB, Welle Z, Ekker SC (2004) Enhancer trapping in zebrafish using the *Sleeping Beauty* transposon. BMC Genomics 5:62
35. Balciunas D, Ekker SC (2005) Trapping fish genes with transposons. Zebrafish 1:335–341
36. Bronchain OJ, Hartley KO, Amaya E (1999) A gene trap approach in *Xenopus*. Curr Biol 9:1195–1198
37. Kawakami K (2005) Transposon tools and methods in zebrafish. Dev Dyn 234:244–254
38. Kotani T, Nagayoshi S, Urasaki A, Kawakami K (2006) Transposon-mediated gene trapping in zebrafish. Methods 39:199–206
39. Sivasubbu S, Balciunas D, Amsterdam A, Ekker SC (2007) Insertional mutagenesis strategies in zebrafish. Genome Biol 8(suppl 1):S9
40. Urasaki A, Kawakami K (2009) Analysis of genes and genome by the *tol2*-mediated gene and enhancer trap methods. Methods Mol Biol 546:85–102
41. Salminen M, Meyer BI, Gruss P (1998) Efficient poly A trap approach allows the capture of genes specifically active in differentiated embryonic stem cells and in mouse embryos. Dev Dyn 212:326–333
42. Clark KJ, Balciunas D, Pogoda HM, Ding Y, Westcot SE, Bedell VM, Greenwood TM, Urban MD, Skuster KJ, Petzold AM, Ni J, Nielsen AL, Patowary A, Scaria V, Sivasubbu S, Xu X, Hammerschmidt M, Ekker SC (2011) In vivo protein trapping produces a functional expression codex of the vertebrate proteome. Nat Methods 8:506–515
43. Craig NL, Craigie R, Gellert M, Lambowitz AM (2002) Mobile DNA II. ASM Press, Washington, DC
44. Liu G, Geurts AM, Yae K, Srinivasan AR, Fahrenkrug SC, Largaespada DA, Takeda J, Horie K, Olson WK, Hackett PB (2005) Target-site preferences of *Sleeping Beauty* transposons. J Mol Biol 346:161–173
45. Bauser CA, Elick TA, Fraser MJ (1999) Proteins from nuclear extracts of two lepidopteran cell lines recognize the ends of TTAA-specific transposons *piggyBac* and *tagalong*. Insect Mol Biol 8:223–230
46. Yergeau DA, Kuliyev E, Mead PE (2007) Injection-mediated transposon transgenesis in *Xenopus tropicalis* and the identification of integration sites by modified extension primer tag selection (EPTS) linker-mediated PCR. Nat Protoc 2:2975–2986
47. Waldner C, Roose M, Ryffel GU (2007) Marking transgenic *Xenopus* froglets with passive micro transponders. Transgenic Res 16:539–540
48. Yergeau DA, Kelley CM, Zhu H, Kuliyev E, Mead PE (2010) Transposon transgenesis in *Xenopus*. Methods 51:92–100
49. Hellsten U, Harland RM, Gilchrist MJ, Hendrix D, Jurka J, Kapitonov V, Ovcharenko I, Putnam NH, Shu S, Taher L, Blitz IL, Blumberg B, Dichmann DS, Dubchak I, Amaya E, Detter JC, Fletcher R, Gerhard DS, Goodstein D, Graves T, Grigoriev IV, Grimwood J, Kawashima T, Lindquist E, Lucas SM, Mead PE, Mitros T, Ogino H, Ohta Y, Poliakov AV, Pollet N, Robert J, Salamov A, Sater AK, Schmutz J, Terry A, Vize PD, Warren WC, Wells D, Wills A, Wilson RK, Zimmerman LB, Zorn AM, Grainger R, Grammer T, Khokha MK, Richardson PM, Rokhsar DS (2010) The genome of the Western clawed frog *Xenopus tropicalis*. Science 328:633–636

Chapter 7

Targeted Gene Disruption with Engineered Zinc-Finger Nucleases (ZFNs)

John J. Young and Richard M. Harland

Abstract

Reverse genetics in *Xenopus* has been limited to knockdown strategies using antisense morpholino oligonucleotides (MOs). Recently, engineered zinc-finger nucleases have been used to induce targeted mutations resulting in null alleles. Zinc-finger nuclease (ZFN) technology has been adapted to induce null mutations in many systems previously refractory to targeted gene inactivation. Here we provide a general protocol for inducing targeted mutations in *Xenopus tropicalis* using ZFNs, a method to detect resulting mutations, and the steps to generate homozygous mutant embryos.

Key words: *Xenopus*, Reverse genetics, Zinc-finger nucleases

1. Introduction

Reverse genetic manipulations have been available to the *Xenopus* community through the use of morpholino oligonucleotides (MOs). MOs have been invaluable in discerning gene function during early development through the beginning of organogenesis. However, this method is not ideal for probing gene function in late embryonic events as MO doses become diluted and less effective. Furthermore, incomplete knockdown of protein translation from intended target mRNA can confound experimental results. To overcome these limitations, the gene of interest needs to be inactivated at the genomic level.

Recent publications have demonstrated the effectiveness of designer zinc-finger nucleases (ZFNs) for inducing targeted mutations resulting in null alleles in fruit flies (1, 2), zebrafish (3–6), human T-cells (7), Chinese hamster ovary cells (8–11), rats (12, 13), and frogs (14). ZFNs are fusion proteins consisting of the nonspecific nuclease domain of Fok1 and zinc-finger proteins

Stefan Hoppler and Peter D. Vize (eds.), *Xenopus Protocols: Post-Genomic Approaches, Second Edition*,
Methods in Molecular Biology, vol. 917, DOI 10.1007/978-1-61779-992-1_7,

engineered to bind a specific genomic locus so as to induce a double strand break (DSB) (15, 16). DSBs are repaired via nonhomologous end joining (NHEJ), which is inherently error prone, resulting in small insertions and deletions (indels) (1). These indels often induce frameshift mutations and result in null alleles.

With the sequencing of *Xenopus tropicalis* and the development of reverse genetics via ZFNs, the frog will continue to be a powerful model for determining gene function. This protocol outlines delivery of commercially designed and validated ZFNs via mRNA injection and subsequent detection of mutations in *Xenopus* via the Cel-1 enzyme present in celery stalk extract. Additionally, we outline the steps necessary to generate homozygous mutant embryos in order to investigate gene function in amphibian development. The methods described herein will apply to mutagenesis via ZFNs as well as the recently developed transcription activator-like effector nucleases (TALENs).

2. Materials

2.1. Validated ZFNs or TALENs (See Note 1 and 2)

2.2. Subcloning ZFN Constructs

1. CS108 expression vector (Accession number: DQ649433).
2. EcoRI restriction enzyme and 10× EcoRI reaction buffer (NEB). Store at −20 °C.
3. XbaI restriction enzyme (NEB). Store at −20 °C.
4. QIAquick gel extraction kit (QIAGEN).
5. Ultrapure nuclease-free water (Ambion).
6. T4 DNA Ligase and 10× T4 DNA Ligase Buffer (NEB). Store at −20 °C.
7. Chemically competent XL1 Blue *Escherichia coli* strain or TOP 10 *E. coli* strain (Invitrogen). Store at −80 °C.
8. Sp6 sequencing primer: 5′-GATTTAGGTGACACTATAG-3′ and T7 sequencing primer: 5′-TAATACGACTCACTATAGGG-3′ diluted to 1 μM in nuclease-free water. Store at −20 °C.

2.3. RNA Synthesis

1. RNaseZap (Ambion).
2. AscI restriction enzyme and 10× NEB Buffer 4 (NEB). Store at −20 °C.
3. Phenol/Chloroform/Isoamyl Alcohol (25:24:1) pH 6.7 (Fisher Scientific).
4. 3 M sodium acetate pH 5.2.
5. Isopropyl alcohol.
6. 100 % Ethanol.

7. mMessage mMachine kit (Ambion). Store at −20 °C. Contents of this kit include the 2× NTP/CAP mix, 10× enzyme mix, 10× reaction buffer, 7.5 M LiCl, Turbo DNAse, and 5 M ammonium acetate.

2.4. Celery Extract

1. 500 g (approximately 1 lb) of fresh celery stalks (*Apium graveolens* var. *dulce*).
2. Omega® 8003/8005 masticating juicer. Chill to 4°C before use.
3. 1 M Tris Base (Sigma): 60.75 g in 500 mL water pH 7.7.
4. 0.1 M phenylmethanesulfonylfluoride (PMSF): 0.87 mg in 5 mL of isopropyl alcohol.
5. 5× Tris/KCl solution (0.5 M Tris Base/2.5 M KCl): 60.75 g Tris Base and 186.38 g KCl diluted in 1 L of water adjust to pH 7.7. Store at 4°C.
6. Ammonium sulfate.
7. Spectra/Por dialysis tubing 12–14,000 MWCO (Spectrum Medical Industries). Store at 4°C.

2.5. Obtaining Embryos

1. Inbred strains of *X. tropicalis.*
2. 4 L square plastic buckets with lids (Fisher Scientific).
3. Chorulon HCG (Intervet).
4. 1 mL luer lock syringes (BD Biosciences). 25 5/8 gauge and 30 1/2 gauge needles (BD Biosciences).
5. Cysteine hydrochloride (Sigma). Add 3 g to 100 mL water and adjust the pH to 7.8–8.0 with NaOH.
6. Plastic transfer pipettes (Fisher).
7. Glass Pasteur pipettes (Fisher).

2.6. Injection

1. 3.5 cm plastic dishes (BD Biosciences).
2. Synthetic fluorescent protein mRNA (eGFP or mCherry).
3. Ficoll 400 (GE Healthcare).
4. Glass thin walled capillary tubes with 1 mm filament (World Precision Instruments, Inc.).
5. Gentamicin (Sigma).
6. 1/9 MR: 11 mM NaCl, 0.2 mM KCl, 0.22 mM $CaCl_2$, 0.11 mM $MgCl_2$, 0.56 mM HEPES-NaOH (pH 7.6).

2.7. Detecting Mutations

1. pCR2.1 TOPO T/A cloning vector (Invitrogen). Store at −20 °C.
2. Fine scissors (Fine Science Tools catalog# 14058-11).
3. Proteinase K (Roche).

4. Lysis buffer: 50 mM Tris pH 7.5, 50 mM NaCl, 5 mM EDTA, 0.5 % SDS, 5 % Chelex 100 (Bio Rad), and 250 μg/mL proteinase K added immediately before use.
5. Celery extract.

3. Methods

3.1. Subcloning ZFN Constructs

1. Once ZFNs targeting the chosen site have been designed and tested in reporter assays (see Note 1), excise the ZFN using restriction enzymes used to cut the multiple cloning site of CS-108. Use EcoRI and XbaI for the 5′ and 3′ regions, respectively (see Note 3).
2. Resolve the cleavage products on a 1.6 % agarose gel and gel purify the linearized CS-108 and the ZFN template eluting in 30 μL of water (see Note 4).
3. Mix 2–3 μg of purified ZFN template with 600–700 ng of purified linear CS-108 and add 2 μL of 10× T4 ligase buffer and 1 μL of T4 ligase. Dilute the reaction to 20 μL with water and incubate at 16 °C overnight.
4. Transform 100 μL of chemically competent XL1-Blue or TOP10 cells with 5 μL of the cloning reaction.
5. Confirm successful subcloning by sequencing with Sp6 and T7 primers.

3.2. RNA Synthesis

1. Treat working area with RNaseZap.
2. Linearize 20 μL of plasmid DNA (>100 ng/μL) with an appropriate restriction enzyme in a 100 μL reaction. Incubate at 37 °C for 2 h (see Note 5).
3. Add 100 μL of nuclease-free water followed by 200 μL of phenol/chloroform isoamyl alcohol (25:24:1) and vortex.
4. Spin the solution in a tabletop centrifuge at 20,800 × *g* for 5 min.
5. Transfer the upper aqueous phase to a new tube and add 20 μL of 3 M sodium acetate (pH 5.2) and 200 μL isopropyl alcohol.
6. Shake the solution vigorously and spin in a tabletop centrifuge at 20,800 × *g* for 10 min.
7. Remove the supernatant and wash the pellet by adding 200 μL of 70 % ethanol and recentrifuge for 5 min at 20,800 × *g*.
8. Remove the supernatant and allow the pellet to air dry. Resuspend the pellet in 10 μL of nuclease-free water and ensure complete linearization by resolution on an agarose gel (see Note 6).

9. Use a mMessage mMachine kit (Ambion) for in vitro synthesis of RNA. Combine 10 μL 2× NTP mix, 2 μL 10× enzyme mix, 2 μL 10× reaction buffer, and 6 μL of linearized template DNA. Incubate at 37 °C for 2 h (see Note 7).
10. Remove the DNA template by adding 1 μL of Turbo DNAse (Ambion) to the reaction and incubate at 37 °C for 10 min.
11. Dilute the reaction to 50 μL with nuclease-free water and add 25 μL of 7.5 M LiCl. Incubate at −20 °C overnight.
12. Centrifuge at 20,800 ×*g* for 10 min to precipitate the RNA.
13. Remove supernatant and wash with 70 % ethanol as described in steps 6 and 7.
14. Resuspend the pellet in 100 μL of nuclease-free water.
15. Precipitate the RNA by adding 20 μL of 5 M ammonium acetate and 250 μL of 100 % ethanol and centrifuge at 20,800 ×*g* for 10 min (see Note 8).
16. Wash the pellet as described in steps 6 and 7.
17. Resuspend pellet in nuclease-free water to a concentration of 1 μg/μL.
18. Ensure that a proper transcript has been synthesized by running 1 μL of 1 μg/μL RNA aliquot on an 0.8 % agarose gel for 20 min. A successful synthesis will resolve a single discrete band.

3.3. Celery Extract

1. Cut 250 g (half pound) of fresh celery stalks and rinse thoroughly with distilled water. Perform all remaining steps in a 4 °C room. Juice the cut celery in a masticating juicer to produce 200 mL of celery juice. Use a mesh strainer to remove large particulate matter (see Note 9).
2. Chill a refrigerated centrifuge to 4 °C and spin celery juice at 2,600 ×*g* for 20 min to remove suspended solids. Transfer the supernatant to a fresh beaker and add 1 M tris (pH 7.7) and 0.1M PMSF to a final concentration of 0.1 M and 100 μM, respectively.
3. Slowly add 28.8 g of ammonium sulfate while gently stirring for 30 min. Centrifuge the solution in a rotor chilled to 4 °C at 13,000 ×*g* for 40 min. Transfer the supernatant to clean beaker. There should be approximately 200 mL of this solution.
4. Prepare 1× Tris/KCl/PMSF buffer by diluting 200 mL of 5× Tris/KCl/PMSF buffer and chill to 4 °C.
5. Gently stir the solution while slowly adding 78 g of ammonium sulfate for 30 min. Centrifuge the solution in a rotor chilled to 4 °C at 13,000 ×*g* for 90 min. Remove the supernatant and resuspend the pellet in 20 mL of 1× Tris/KCl/PMSF buffer.

6. Load celery extract into 15 cm of 12–14 KDa molecular weight cutoff dialysis tubing and dialyze against 2 L of 1× Tris/KCl/PMSF buffer for 4 h. Change the buffer hourly. Allow the final dialysis to go overnight (see Note 10).
7. Aliquot the dialyzed celery extract into 500 μL aliquots and store at −80 °C (see Note 11).

3.4. Obtaining Embryos

1. Inject female *X. tropicalis* with a priming dose of 10 U followed the next morning by a boosting dose of 200 U HCG (Chorulon) and 100 U HCG for males. Refer to Chapter 3 of this volume for injection protocol and additional embryo collection protocols.
2. Allow mating pairs to produce embryos for 4–7 h.
3. Collect freshly laid embryos using a plastic transfer pipette into a glass dish.
4. Dejelly the embryos by incubating in 3 % cysteine for 10 min at room temperature.
5. Rinse with several washes of 1/9 MR until all cysteine has been removed.

3.5. Injection

1. Prepare agarose coated injection dishes by pouring 1 % agarose melted in water into plastic Petri dishes and then quickly pouring it out. Inject all embryos in agar coated dishes to prevent them from sticking to the plastic (see Note 12).
2. Prepare a range of ZFN RNA dilutions for injection in RNAse free tubes diluted with synthetic fluorescent protein RNA (for use as a tracer) and nuclease-free water (see Notes 13 and 14).
3. Load ZFN RNA into pulled capillary tubes.
4. Carefully clip the end of the needle with a pair of watchmaker's forceps so that a >250 ms burst of air expels a 2 nL volume.
5. Pipette fertilized embryos into agar coated dishes and remove culture media slowly so that the embryos coalesce and orient with their animal poles facing up.
6. Inject each blastomere of two-cell embryos with 2 nL of diluted ZFN RNA mixture (see Note 15).
7. Following injection, and before embryos dry out, add 1/9 MR with 3 % Ficoll and incubate at room temperature for 1–2 h to allow embryos to recover.
8. Wash the embryos several times into 1/9 MR supplemented with 100 μg/mL gentamicin and culture injected embryos overnight at 28 °C.
9. Screen embryos for robust fluorescent protein expression and discard any uninjected embryos.

3.6. Detecting Somatic Mutations

1. Raise injected tadpoles to stage 40 and lyse them in 200 μL of lysis buffer, incubate at 55 °C overnight.
2. Extract genomic DNA by adding 200 μL of phenol/chloroform/isoamyl alcohol (24:24:1). Shake vigorously and centrifuge at 20,800×g for 5 min. Transfer the aqueous phase to a new tube and add 20 μL of 3 M sodium acetate and 200 μL of 100 % ethanol. Shake vigorously followed by centrifuging for 10 min at 20,800×*g*. Wash the pellet with 70 % ethanol and allow it to air dry (see Note 16).
3. Resuspend genomic DNA in 30 μL of 10 mM Tris–Cl (pH 8.5).
4. PCR amplify the targeted region using primers that will yield 300–500 bp amplicons with the target site offset from the center in a 25 μL reaction using 1 μL of genomic DNA (see Note 17).
5. Transfer 9 μL to a new tube and promote heteroduplex formation by incubating at 94°C for 5 min followed by cooling to 85°C at a rate of −2°C/s. Next, cool to 25°C at a rate of −0.1°C/s. Keep the reaction at 4 °C until digestion with celery extract or purified Cel-1 (see Note 18).
6. Digest heteroduplexes by adding 2 μL of celery extract to annealed amplicons and incubate at 42°C for 20 min (see Note 19).
7. Resolve digest products on a 1.6 % agarose gel and poststain with ethidium bromide (see Note 20). Alternatively, resolve the products on a 9 % acrylamide gel with a 6 % stacking layer (see Note 21).
8. Induced mutations will result in cleavage of heteroduplexes by Cel-1 and produce lower molecular weight bands on the gel.
9. Clone PCR amplicons from samples that produced Cel-1 bands into pCR2.1 TOPO T/A cloning vector (Invitrogen) by adding 4 μL purified PCR product to 1 μL of salt solution and 1 μL of the cloning vector. Incubate the reaction at room temperature for 5 min.
10. Transform the cloning reaction using XL1-Blue or Top10 cells and grow on agarose plates containing ampicillin.
11. Sequence the resulting plasmids using M13 forward and reverse primers (see Note 22).

3.7. Detecting Germline Mutations

1. Raise injected embryos to sexual maturity according to the guidelines outlined in Chapter 2 of this volume.
2. Outcross potential germline mutant carriers with uninjected siblings or frogs from the same inbred line as the injected individuals (see Note 23). Inject females with 200 U HCG and males with 100 U HCG and allow mating to occur in 2 L of

frog system water overnight. This will limit the possibility of polymorphisms that lead to spurious cleavage products in Cel-1 assays.

3. Grow offspring to stage 40 and lyse a subset for genomic DNA isolation as described earlier. Continue to raise siblings to adulthood to establish mutant line.
4. PCR amplify the targeted locus using the same primer set designed to detect somatic mutations.
5. Melt and re-anneal 9 μL of the PCR product and add 2 μL of celery extract as described earlier to digest any heteroduplexes. Resolve digestion products on a 1.6 % agarose gel. The presence of lower molecular weight bands indicates a heterozygous tadpole and therefore a mutation present in the germline of the ZFN injected parent.
6. Clone amplicons from heterozygous tadpoles into pCR2.1 TOPO and sequence miniprepped plasmids to confirm mutations (see Note 24).

3.8. Establishing Mutant Lines and Homozygous Mutants

1. Continue to raise the siblings of heterozygous tadpoles identified in Subheading 3.7, step 3 to stage 54 or later.
2. Use a pair of fine scissors to clip off 2 cm of the distal tail (see Note 25). Incubate the tail tip in 200 μL of lysis buffer at 55°C overnight.
3. Extract genomic DNA and identify heterozygous individuals as described earlier. Clone amplicons from identified heterozygous animals in pCR2.1 TOPO and sequence miniprepped plasmids (see Note 26).
4. Raise these animals to sexual maturity and intercross siblings to generate homozygous mutants for analysis of gene function (see Note 27).

4. Notes

1. ZFNs can be designed using a variety of methods. The ZFNs used to mutate the *noggin* locus in *Xenopus* (14) and the *no tail* and *golden* loci in zebrafish (3) were made by Sangamo Biosciences, and ZFNs based on this technology are available through Sigma-Aldrich. Another method involves using a bacterial one-hybrid system to select zinc-finger proteins that bind their target locus with high affinity (4). A variation of this method, oligomerized pool engineering (OPEN), uses a publicly available design protocol (17). A final method termed modular assembly combines precharacterized ZFPs to bind specified

9 bp sequences (18). Following design of ZFNs, it is useful to test their activity in a yeast-based reporter assay which provides good predictive value for the activity of the ZFN in other organisms (19). For each targeted locus, multiple ZFN pairs should be designed and tested as the activities can vary widely between them.

2. Recently, several groups have successfully targeted loci using TALENs. Originally discovered in the plant pathogen *Xanthomonas*, TALEs are used by the pathogen to induce transcription of host genes which aid in virulence. The DNA-binding domain of TALEs consists of tandem repeats that each bind one nucleotide (20) and can be altered to bind investigator specified sequences (21, 22). Fusing the Fok1 nuclease domain to engineered TALEs has proven effective at generating DSBs and inducing mutations in targeted loci (23, 24). The single basepair modularity of TALENs can offer an effective alternative to ZFN-induced mutagenesis.
3. Cloning the ZFN plasmid into CS108 so that Sp6 polymerase is used to make a sense message yields greater protein production than T7 transcribed RNA (25).
4. It is important to use unique restriction enzymes for the 5′ and 3′ portion of the insert. We have found that this cloning protocol works robustly without treatment of the vector with Calf Intestinal Phosphatase (CIP); however, one may treat the vector with CIP if cloning proves difficult.
5. The CS108 expression plasmid has a SV40 polyadenylation signal in the template. When linearized using AscI, sense strands transcribed with Sp6 polymerase will be adenylated following injection into embryos. This will result in more stable transcripts and higher expression of the resulting protein (25). However, it may be desirable to make nonadenylated transcripts that would result in an early burst of ZFN protein expression that is more transient than stabilized transcripts. In this case, linearizing using XhoI would be sufficient. In practice, we found that polyadenylated mRNA produces higher frequencies of somatic mutations (14).
6. A properly linearized plasmid will migrate at a higher molecular weight than an uncut control. Complete linearization is important to ensure efficient transcription of the desired product.
7. Including trace amounts of ^{32}P-labeled UTP in the in vitro transcription assay allows for very accurate measurements of RNA yield. To determine the RNA yield in a 20 μL transcription reaction, measure the total counts before precipitating the RNA and incorporated counts after the final precipitation and use the following formula: incorporated counts/total counts × 132 μg (incorporation of all available nucleotide in the described reaction would yield 132 μg of RNA).

8. It is very important to remove all traces of LiCl from the precipitation step. LiCl is a potent dorsalizing agent in embryos and any residual amounts can give unwanted and confounding phenotypes.
9. Masticating juicers are superior to centrifugal juicers because they do not generate heat in the juicing process, which could lead to decreased activity of the extract.
10. Be sure the dialysis tubing with the extract is fully submerged in the buffer. For this, we use a float at one end and a small stir bar at the other end both attached with small rubber bands, and dialyze in a 2 L beaker with the stir plate set to a moderate speed.
11. Working stocks can be kept at 4 °C and remain active for several weeks. Aliquots kept at −80 °C are active for longer than 1 year after isolation if thawing and refreezing is avoided.
12. *X. tropicalis* embryos will rapidly stick to plastic, making it difficult to transfer them without damaging the embryos.
13. Different ZFNs designed against the same target can have different activities and potential toxicities. Therefore, it will be necessary to test each pair at a range of concentrations and score the embryos for phenotypes associated with nonspecific toxicity such as gastrulation defects or kinked axes. We start with a low end at 10 pg total dose of ZFN RNA and a high dose at 100 pg.
14. The inclusion of fluorescent protein serves two purposes: to screen for embryos that were injected and to select embryos that had a similar injection dose within a cohort. We use an injection dose of 200 pg of mCherry RNA, as the protein is extremely bright and persists into tadpole stages.
15. Targeted injection of ZFNs is possible but unlikely to be useful for inducing tissue-specific mutations given the mosaic and variable nature of ZFN-induced mutagenesis.
16. It may be possible to amplify the targeted locus directly from the tadpole lysates and detect ZFN-induced mutations. We have found that phenol/chloroform isolation of genomic DNA greatly increases successful and repeatable detection of mutations.
17. Primers designed for Cel-1 assays are best when they amplify a 250–500 bp amplicon where the targeted site is offset from the middle of the amplicon. This will result in two lower molecular weight bands at predictable sizes following digestion with Cel-1 when a mutation is present.
18. Allowing the re-annealed amplicons to warm up before digestion can cause loss of heteroduplexes. It is best to keep the reaction at 4 °C and immediately continue to the digestion step.

19. We keep the re-annealed amplicons in a PCR machine block chilled to 4 °C and add the 2 μL of celery extract directly into the reaction.
20. Poststaining agarose gels with ethidium bromide provides for equal staining of all bands and allows for better detection of lower molecular weight bands.
21. Acrylamide gels will result in better separation of bands with small molecular weight differences; however, we have found 1.6 % agarose gels are sufficient for resolving Cel-1 cleavage products.
22. The induction of somatic mutations are often rare events, so analysis will require that several clones are sequenced from injected individuals that are positive for Cel-1 bands. A good starting point is to sequence 32 plasmids from one tadpole previously determined to have a mutation.
23. Outcrossing with unrelated frogs is possible if it has been determined that there are no SNPs or other polymorphisms in the region.
24. Given that the mutation must be heterozygous, it should only be necessary to sequence 5–6 plasmids from individual tadpoles.
25. Alternatively, the distal tip of toes of froglets and adults can be excised for genotyping using the protocol outlined earlier.
26. It is important to sequence the targeted allele in all identified heterozygotes as founders can be mosaic in their germline.
27. The first ovulation from female *X. tropicalis* often produces eggs that result in embryos with nonspecific anterior/posterior and dorsal/ventral defects (26) which will confound phenotypic analysis. Therefore, it is advantageous to allow a heterozygous female to ovulate once or twice before attempting to characterize any phenotypes associated with ZFN-induced mutations. If intercrossing is not possible due to an absence of fertile heterozygote males, gynogenetic diploid can be generated from heterozygous females as outlined in Chapter 5 of this volume.

Acknowledgements

We wish to thank the Conlon and Moens labs for the celery extract preparation protocols.

References

1. Bibikova M, Golic M, Golic KG, Carroll D (2002) Targeted chromosomal cleavage and mutagenesis in Drosophila using zinc-finger nucleases. Genetics 161(3):1169–1175
2. Beumer K, Bhattacharyya G, Bibikova M, Trautman JK, Carroll D (2006) Efficient gene targeting in Drosophila with zinc-finger nucleases. Genetics 172(4):2391–2403
3. Doyon Y, McCammon JM, Miller JC, Faraji F, Ngo C, Katibah GE, Amora R, Hocking TD, Zhang L, Rebar EJ, Gregory PD, Urnov FD, Amacher SL (2008) Heritable targeted gene disruption in zebrafish using designed zinc-finger nucleases. Nat Biotechnol 26(6):702–708
4. Meng X, Noyes MB, Zhu LJ, Lawson ND, Wolfe SA (2008) Targeted gene inactivation in zebrafish using engineered zinc-finger nucleases. Nat Biotechnol 26(6):695–701
5. Foley JE, Yeh JR, Maeder ML, Reyon D, Sander JD, Peterson RT, Joung JK (2009) Rapid mutation of endogenous zebrafish genes using zinc finger nucleases made by oligomerized pool engineering (OPEN). PLoS One 4(2):e4348
6. Siekmann AF, Standley C, Fogarty KE, Wolfe SA, Lawson ND (2009) Chemokine signaling guides regional patterning of the first embryonic artery. Genes Dev 23(19):2272–2277
7. Perez EE, Wang J, Miller JC, Jouvenot Y, Kim KA, Liu O, Wang N, Lee G, Bartsevich VV, Lee YL, Guschin DY, Rupniewski I, Waite AJ, Carpenito C, Carroll RG, Orange JS, Urnov FD, Rebar EJ, Ando D, Gregory PD, Riley JL, Holmes MC, June CH (2008) Establishment of HIV-1 resistance in CD4+ T cells by genome editing using zinc-finger nucleases. Nat Biotechnol 26(7):808–816
8. Santiago Y, Chan E, Liu PQ, Orlando S, Zhang L, Urnov FD, Holmes MC, Guschin D, Waite A, Miller JC, Rebar EJ, Gregory PD, Klug A, Collingwood TN (2008) Targeted gene knockout in mammalian cells by using engineered zinc-finger nucleases. Proc Natl Acad Sci U S A 105(15):5809–5814
9. Cost GJ, Freyvert Y, Vafiadis A, Santiago Y, Miller JC, Rebar E, Collingwood TN, Snowden A, Gregory PD (2009) BAK and BAX deletion using zinc-finger nucleases yields apoptosis-resistant CHO cells. Biotechnol Bioeng 105(2):330–340
10. Liu PQ, Chan EM, Cost GJ, Zhang L, Wang J, Miller JC, Guschin DY, Reik A, Holmes MC, Mott JE, Collingwood TN, Gregory PD (2010) Generation of a triple-gene knockout mammalian cell line using engineered zinc-finger nucleases. Biotechnol Bioeng 106(1):97–105
11. Malphettes L, Freyvert Y, Chang J, Liu PQ, Chan E, Miller JC, Zhou Z, Nguyen T, Tsai C, Snowden AW, Collingwood TN, Gregory PD, Cost GJ (2010) Highly efficient deletion of FUT8 in CHO cell lines using zinc-finger nucleases yields cells that produce completely nonfucosylated antibodies. Biotechnol Bioeng 106(5):774–783
12. Geurts AM, Cost GJ, Freyvert Y, Zeitler B, Miller JC, Choi VM, Jenkins SS, Wood A, Cui X, Meng X, Vincent A, Lam S, Michalkiewicz M, Schilling R, Foeckler J, Kalloway S, Weiler H, Menoret S, Anegon I, Davis GD, Zhang L, Rebar EJ, Gregory PD, Urnov FD, Jacob HJ, Buelow R (2009) Knockout rats via embryo microinjection of zinc-finger nucleases. Science 325(5939):433
13. Mashimo T, Takizawa A, Voigt B, Yoshimi K, Hiai H, Kuramoto T, Serikawa T (2010) Generation of knockout rats with X-linked severe combined immunodeficiency (X-SCID) using zinc-finger nucleases. PLoS One 5(1): e8870
14. Young JJ, Cherone JM, Doyon Y, Ankoudinova I, Faraji FM, Lee AH, Ngo C, Guschin DY, Paschon DE, Miller JC, Zhang L, Rebar EJ, Gregory PD, Urnov FD, Harland RM, Zeitler B (2011) Efficient targeted gene disruption in the soma and germ line of the frog Xenopus tropicalis using engineered zinc-finger nucleases. Proc Natl Acad Sci U S A 108(17): 7052–7057
15. Miller JC, Holmes MC, Wang J, Guschin DY, Lee YL, Rupniewski I, Beausejour CM, Waite AJ, Wang NS, Kim KA, Gregory PD, Pabo CO, Rebar EJ (2007) An improved zinc-finger nuclease architecture for highly specific genome editing. Nat Biotechnol 25(7):778–785
16. Urnov FD, Rebar EJ, Holmes MC, Zhang HS, Gregory PD (2010) Genome editing with engineered zinc finger nucleases. Nat Rev Genet 11(9):636–646
17. Maeder ML, Thibodeau-Beganny S, Sander JD, Voytas DF, Joung JK (2009) Oligomerized pool engineering (OPEN): an 'open-source' protocol for making customized zinc-finger arrays. Nat Protoc 4(10):1471–1501
18. Kim HJ, Lee HJ, Kim H, Cho SW, Kim JS (2009) Targeted genome editing in human cells with zinc finger nucleases constructed via modular assembly. Genome Res 19(7): 1279–1288
19. McCammon JM, Doyon Y, Amacher SL (2011) Inducing high rates of targeted mutagenesis in

zebrafish using zinc finger nucleases (ZFNs). Methods Mol Biol 770:505–527

20. Boch J, Scholze H, Schornack S, Landgraf A, Hahn S, Kay S, Lahaye T, Nickstadt A, Bonas U (2009) Breaking the code of DNA binding specificity of TAL-type III effectors. Science 326(5959):1509–1512
21. Moscou MJ, Bogdanove AJ (2009) A simple cipher governs DNA recognition by TAL effectors. Science 326(5959):1501
22. Hockemeyer D, Wang H, Kiani S, Lai CS, Gao Q, Cassady JP, Cost GJ, Zhang L, Santiago Y, Miller JC, Zeitler B, Cherone JM, Meng X, Hinkley SJ, Rebar EJ, Gregory PD, Urnov FD, Jaenisch R (2011) Genetic engineering of human pluripotent cells using TALE nucleases. Nat Biotechnol 29(8):731–734
23. Christian M, Cermak T, Doyle EL, Schmidt C, Zhang F, Hummel A, Bogdanove AJ, Voytas DF (2010) Targeting DNA double-strand breaks with TAL effector nucleases. Genetics 186(2):757–761
24. Miller JC, Tan S, Qiao G, Barlow KA, Wang J, Xia DF, Meng X, Paschon DE, Leung E, Hinkley SJ, Dulay GP, Hua KL, Ankoudinova I, Cost GJ, Urnov FD, Zhang HS, Holmes MC, Zhang L, Gregory PD, Rebar EJ (2011) A TALE nuclease architecture for efficient genome editing. Nat Biotechnol 29(2): 143–148
25. Turner DL, Weintraub H (1994) Expression of achaete-scute homolog 3 in Xenopus embryos converts ectodermal cells to a neural fate. Genes Dev 8(12):1434–1447
26. Grammer TC, Khokha MK, Lane MA, Lam K, Harland RM (2005) Identification of mutants in inbred Xenopus tropicalis. Mech Dev 122(3):263–272

Chapter 8

Reverse Genetic Studies Using Antisense Morpholino Oligonucleotides

Yanan Zhao, Shoko Ishibashi, and Enrique Amaya

Abstract

Here we present a protocol, which allows loss-of-function studies in *Xenopus* embryos using antisense morpholino oligonucleotides (MOs). Gene knockdown studies provide a critical method for assessing gene function in vitro and in vivo. Such studies are currently performed in *Xenopus* using primarily one of the two main methods: (1) overexpression of dominant negative constructs or (2) inhibition of gene function by using MOs targeting either the initiation of translation or mRNA splicing. While a dominant negative approach is very effective, it often suffers from specificity. Given that MOs target very specific nucleotide sequences in the target RNA, it suffers considerably less from issues of specificity. The most convenient method for introducing MOs into embryos is through microinjection, which is a simple procedure. Therefore, a reverse genetics approach in *Xenopus* using MOs is an extremely powerful tool to study gene function, particularly when taking advantage of available sequence data in the post-genomic era. Furthermore, given the well-established fate map in *Xenopus*, it is also very easy to generate mosaic knockdown embryos, where the gene of interest is affected in defined regions of the embryo. Finally it should be noted that MOs can also be used to block miRNA function and processing, so that it provides a convenient method to not only perform gene knockdown studies on protein coding genes, but also noncoding genes. The protocol we describe here is for both *Xenopus laevis* and *Xenopus tropicalis*.

Key words: *Xenopus*, Morpholino, MO, Loss-of-function experiment

1. Introduction

Antisense oligonucleotides are short synthetic nucleic acid polymers designed to inactivate selected messenger RNAs through a base-pairing mechanism. They were developed in the late 1970s for clinical therapeutic applications. Of those, MO oligonucleotides developed by Summerton and Weller (1) became particularly valuable due to their efficiency and specificity, resistance to nucleases, good aqueous solubility, extremely low toxicity, and low production costs. A MO has the riboside moiety of each nucleotide

Stefan Hoppler and Peter D. Vize (eds.), *Xenopus Protocols: Post-Genomic Approaches, Second Edition*,
Methods in Molecular Biology, vol. 917, DOI 10.1007/978-1-61779-992-1_8, © Springer Science+Business Media, LLC 2012

subunit converted to a morpholino moiety and uses phosphorodiamidate intersubunit linkages instead of phosphorodiester linkages. These features make MOs insensitive to nucleases, yet allow MOs to specifically and stably bind to its complementary nucleotide polymer, particularly RNA sequences, with high affinity to block access of cell components, resulting in inhibition of translation (2) and splicing (3) of mRNA or maturation of microRNA (miRNA) (4–6). They were first introduced for developmental biology studies in *Xenopus* by Janet Heasman and colleagues in 2000 (7) and since then have been used in variety of organisms, including sea urchin (8), *Ciona savignyi* (9), *Xenopus laevis* (10–12), *Xenopus tropicalis* (13), zebrafish (14–16), chick (17), and mouse (18). In *Xenopus*, the effectiveness of MOs in early tadpole stages has been well documented using transgenic animals expressing GFP (13).

2. Materials

1. Flaming/Brown Micropipet Puller (e.g., Model P-87, Sutter Instruments Co.).
2. Borosilicate glass capillaries (e.g., GC100TF-10, 1.0 mm OD × 0.78 mm ID, Harvard Apparatus), for pulling needles using a condition, $p = 50$, $v = 100$ and $t = 5$.
3. Microinjector (e.g., Picospritzer II, General Valve Corporation).
4. Dissection microscope (e.g., MZ6, Leica).
5. Micromanipulator (Märzhäuser Wetzlar, Germany, http://www.marzhauser.com/).
6. Fluorescence dissection microscope (e.g., MZFLIII, Leica) with GFP filters.
7. 60 mm Petri dish with mesh (840 or 1,000 μm Polypropylene Mesh for *X. laevis*, 200 μm Nylon Mesh for *X. tropicalis*, http://www.spectrumlab.com).
8. Dissection tools and Forceps (Dummont No. 5 or 55 can be obtained by any surgical tools supplier).
9. 10× MMR (Marc's Modified Ringer) stock solution (1 M NaCl, 20 mM KCl, 10 mM $MgCl_2$, 20 mM $CaCl_2$, 50 mM HEPES, pH 7.5).
10. 0.1× MMR.
11. 0.1× MMR/2% Ficoll 400 (Sigma, F-4375).
    - 2% (w/v) L-Cysteine (Sigma, C-7352) in 0.1× MMR, pH 8.0.

12. Pregnant mare serum gonadotropin (PMSG, 100 U/mL (e.g., P.G.600®, Intervet, Inc., 021825). Dissolve in water and stored at −20 °C).
13. Human chorionic gonadotropin (HCG, 500–1,000 U/mL (e.g., CHORULON®, Intervet, Inc., 057176). Dissolve in water and stored at 4 °C).
14. Morpholinos (Gene Tools, LLC) against the gene of interest and control, 1 mM.

3. Method

Both natural mating and artificial fertilization can be used to produce fertilized eggs for microinjection. We routinely collect fertilized embryos through natural mating for *X. tropicalis* and through artificial fertilization for *X. laevis*. Here we present a protocol for artificial fertilization of *X. laevis*. For protocols for natural mating and for artificial fertilization in *X. tropicalis*, see Chapters 2 and 3.

3.1. Experimental Strategies for Design of MOs

3.1.1. Translation-Blocking MOs

The most commonly used method of MOs knockdown targets the initiation of translation of mRNAs, since designing such MOs does not require genomic sequence information. Another advantage of targeting translation is that maternal transcripts are also susceptible to MOs aimed at blocking translation, while MOs targeting splicing (see below) only works on zygotic transcripts. However, one disadvantage of translation-blocking MOs is that assessing their efficiency requires the availability of good and reliable antibodies. Inhibition of translation is achieved through steric hindrance, which interferes with the ability of the translation initiation complex to scan through the 5′ UTR to the start codon. Therefore, in theory, MOs can be designed anywhere between the 5′ cap and the start codon of the gene of interest. However, it is nevertheless preferable to design MOs, which overlap the start codon. MOs may extend up to 25 base pairs downstream into the coding sequence, but it is essential that the start codon is covered, as once the large subunit of the ribosome binds at the start site and begins translation, MOs are no longer able to block the continuation of translation (2). This is why translation-blocking MOs are often also referred to as ATG MOs. Generally ATG MOs are designed to be around 25 bases in length with a 40–60% GC content. It is also important that they should not contain self-complementary sequences in order to prevent dimerization or intramolecular hybridization. One should also check available EST databases for possible single nucleotide polymorphisms (SNPs) around the target site, as this could affect the efficiency of binding (see Note 1). The efficiency of ATG MOs can be assessed using an antibody that specifically recognizes the endogenous gene product or with an epitope-tagged test construct (see Subheading 3.3.1).

3.1.2. Splice-Inhibiting MOs

Because MOs are small enough to freely diffuse into the nucleus, they can also target preRNA transcripts, prior to splicing (19). Draper and colleagues successfully used MOs targeted to exon–intron junctions to inhibit *fgf8* preRNA splicing in zebrafish (20). Like the ATG MOs, MOs targeting splice junctions also work by steric hindrance, but in this case they block access of the spliceosome to the splice junctions. Unlike ATG MOs, however, the efficiency of splice-inhibiting MOs on endogenous transcripts can easily be analyzed by RT-PCR (i.e., without requiring a target-specific antibody). The design of these MOs requires well-annotated genome sequence with information of variants and necessarily one or more introns in the target gene. A typical MO is 25 bases in length with a 40–60% GC content; however, they should cover an exon–intron or an intron–exon junction for splicing inhibition to be efficient.

3.1.3. miRNA-Blocking MOs

For the design and use of MOs targeting miRNA function or processing, see Chapter 25.

3.1.4. Control MOs

Uninjected embryos are not suitable as a negative control, since MO injected embryos often develop slower than uninjected siblings. Thus to analyze injected embryos at a particular stage, it is very important to have control embryos injected with a control MO. This will allow one to assess the specific MO phenotype relative to a control MO injected at a comparable developmental stage. We generally use the standard control MO from Gene Tools, which targets the human β-*globin* gene. However, a 4 or 5 base mismatch MO can also be designed and used as a negative control.

3.2. MO injection in *X. laevis*

1. Females are injected with 50 units of PMSG about 3–5 days prior to HCG injection.
2. Inject females with 500 units of HCG, 12–15 h before planned egg collection.
3. On the day of experiment, females are placed in 1× MMR high salt solution and allowed to lay eggs in the high salt solution.
4. Incubate 1 mM MO stock at 65 °C for 10 min (see Note 2).
5. Eggs are transferred onto an empty 90 mm Petri dish. It is important to get rid of most of the high salt solution on the dish. Alternatively eggs can be gently massaged from the ovulating female directly onto an empty 90 mm Petri dish. Macerate and rub the testes throughout the fresh eggs with forceps (see Note 3). Leave the sperm/egg mixture alone for 5 min, and then flood the dish with 0.1× MMR.

6. 10–30 Minutes after flooding the embryos with 0.1× MMR, dejelly the fertilized eggs with 2% Cysteine pH 8.0. This usually takes about 10 min.
7. While dejellying the eggs, backfill the needles with MO using a pipetman with a microloader tip, and place needle into a micromanipulator. Clip the tip of the needle to the size required to deliver the desired injection volume by adjusting the time of the injector.
8. Wash the dejellied eggs with 0.1× MMR three times and transfer them to a dish with mesh containing injection buffer (0.1× MMR/2% Ficoll 400).
9. Inject 1–8-cell stage embryos (see Note 4). We generally inject with 1 and 4 nL of 1 mM stock MO solution per embryo of *X. tropicalis* and *X. laevis*, respectively (see Note 5).
10. Transfer injected embryos to a fresh Petri dish containing 0.1× MMR, and incubate at 14–23 °C (*X. laevis*) and 20–25 °C (*X. tropicalis*) until they reach the desirable stage of development for molecular, cellular or phenotypic analysis.

3.3. Evaluation of Knockdown Efficiency of MOs

While performing knockdown studies with MOs, it is essential to evaluate the extent of knockdown efficiency, as this information will be valuable when interpreting the phenotypes that are uncovered. Here we present general protocols, with two specific examples, for how to evaluate the efficiency of MO knockdown.

3.3.1. Translation-Blocking MOs

If a good antibody is available that specifically recognizes the endogenous targeted gene product, immunostaining of embryos or immunoblotting from embryo extracts are the best ways to show the effect of injected MOs. The advantage of immunostaining is that it allows the assessment of the knockdown efficiency in embryos where the knockdown is performed mosaically, but it is not quantitative. For a quantitative measure of knockdown efficiency one should perform immunoblotting from embryo extracts on embryos where the MOs are present ubiquitously. Given that a good antibody is often not available or the level of endogenous protein is below the detection level of the antibody, Western blots on embryos overexpressing the gene of interest with or without MO can be used. If there is no antibody available, the gene product of interest should be tagged, for example, with HA or GFP, and an antibody against the tag can be used for immunohistochemistry or immunoblotting in combination with MO injection (see Box 1).

Box 1
Translation-Blocking MO Targeting Thymosin β4 Expression

Here we present an example from our recent work on *tmsb4x*/thymosin *β*4 (*T*β*4*). A full-length clone encoding *X. tropicalis Tβ4* (TGas019n14) was identified in the *X. tropicalis* full-length clone database (see Note 6). T*β*4 is a 43 aa peptide, and a molecular weight of 5 kD. Since a commercial antibody against T*β*4 did not work well in *Xenopus*, we generated a GFP-tagged T*β*4 (with the GFP tag on the C-terminus of T*β*4). *X. laevis* embryos were injected with 40 ng of either standard control MO or T*β*4 MO at the one-cell stage along with 500 pg *Tβ4-GFP* mRNA. Immunoblots on injected embryos were carried out with a commercially available GFP antibody (Fig. 1). T*β*4-GFP protein was detected when the RNA was coinjected with the control MO (Fig. 1, lane 2). However, microinjection of an ATG MO inhibited protein translation from *Tβ4-GFP* RNA (Fig. 1, lane 1).

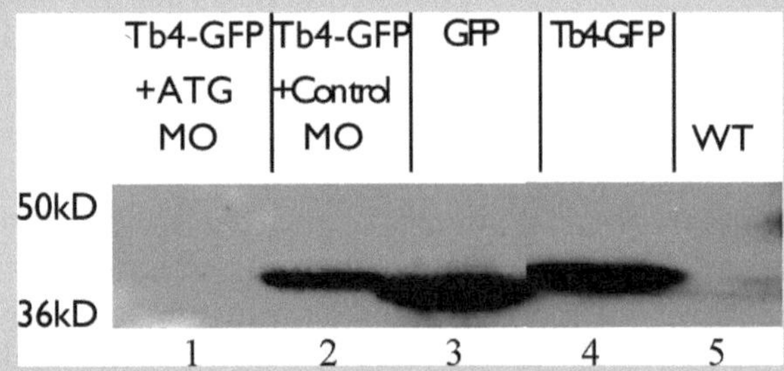

Fig. 1. Immunoblot on embryos injected with either control MO or Tβ4 MO along with Tβ4-GFP RNA. Protein extracts from injected embryos were separated by 12% SDS-PAGE gel and Tβ4-GFP was detected using an anti-GFP antibody. Note that Tβ4 MO abolished the expression of Tβ4-GFP. Also note that the Tb4-GFP fusion increases the molecular weight of GFP, relative to GFP alone (lane 2 and 4 versus lane 3).

3.3.2. Splice-Inhibiting MOs

The efficiency of splice-inhibiting MOs can be examined by RT-PCR using primers designed in the exons upstream and downstream of the target sequence. Furthermore, mis-spliced aberrant RNAs are often retained in the nucleus and can be detected by in situ hybridization. Since such RNAs might produce dominant negative or active forms of the gene product, it is highly advisable to perform RT-PCR, clone and sequence any aberrant bands to assess whether the products could produce active or dominant negative proteins (see Box 2).

Box 2
Splice-Inhibiting MOs Targeting *Notochord Homeobox (not)* Expression

To illustrate how splice-inhibiting MOs work, we present here an example using *not* (Gomperts and Amaya unpublished). The homeobox transcription factor *notochord homeobox (not)* is the earliest molecular marker of the notochord and pineal gland. In *Xenopus* there are two *not* genes, *nota* (previously known as *Xnot1)* and *notb* (previously known as *Xnot2)*, that are 90% identical in amino acid sequence and which represent paralogues genes as a result of the allotetraploid nature of the *X. laevis* genome. We cloned and sequenced both *not* paralogues and determined their exon–intron structure (Fig. 2a). The genes consist of three exons. MOs were designed that would target the exon 1–intron 1 boundary (e1i1), the intron 1–exon 2 boundary (i1e2), and the exon 2–intron 2 boundary (e2i2; Fig. 2b) as well as ATG MO. The *not* genes are divergent across the e1i1 and i1e2 junctions and so we made paralogue specific MOs for these sites. The e2i2 sequence is identical in all but one position and tests indicated that the *notb* e2i2 MO targeted both paralogues. To test the activity of the splice blocking MOs RT-PCR primers were designed to sequences in the first and third exons (Fig. 2b). Wild type fully spliced forms of *nota* and *notb* were predicted to have sizes of 678 and 665 bp, respectively.

The MOs (30 ng each) were injected into both cells at the two-cell stage and embryos were harvested for RNA at the gastrula stage. Following RT-PCR the products were cloned using TA cloning kit and sequenced to determine their identity. Uninjected embryos and those treated with ATG MO (*X2*) gave rise to wild type splice products (Fig. 2c, lanes 1 and 2). However, in embryos treated with the e1i1 MO we detected three splice forms (Fig. 2c, lane 3). The smallest band is the size of the fully spliced wild type forms of *nota* and *notb* and this band was present irrespective of MO dose. Since *notochord homeobox* mRNA is present in the egg these products may represent maternally deposited fully spliced *not* RNA, although we cannot exclude the possibility that some of the zygotic product is resistant to the MO. The two higher molecular weight bands derived from e1i1-injected embryos were 907 and 1,204 bp. These products are the size of *nota* and *notb* transcripts, respectively, in which intron 1 is retained. Embryos injected with the i1e2 MO gave rise to four PCR products (Fig. 2c, lane 4). As with the e1i1 MO, the two higher molecular weight bands represented *nota* and *notb* species in which intron 1 was retained. The fully spliced WT forms of *notochord homeobox* were also present. The smallest product consisted of transcripts from both *notochord homeobox* paralogs in which exon 1 was spliced directly to exon 3 (i.e., exon 2 was skipped). In this transcript, exon 1 remains unaltered but the aberrant splicing causes a shift in reading frame in exon 3 resulting in a STOP codon 37 amino acids downstream of the splice junction.

When we analyzed *notochord homeobox* splicing in e2i2 MO injected embryos we found four major products (Fig. 2c, lane 5). The highest molecular weight band consisted of *nota* and *notb* transcripts in which intron 2 was retained (980 and 901 bp, respectively). The wild type splice forms were present and we also recovered the small product with the frame shift observed in the case of the e1i1 MO. However, the major splice product recovered was one in which a deletion of 51 nucleotides (17 amino acids) had been introduced into exon 2 due to the uncovering of a cryptic

(continued)

Box 2
(continued)

splice site within exon 2. We reasoned that this transcript might retain some function as the most conserved domain of the homeobox remained intact. We therefore designed an additional MO (e2) that would block the cryptic splice event (Fig. 2b). On its own at 60 ng, e2 had no effect on

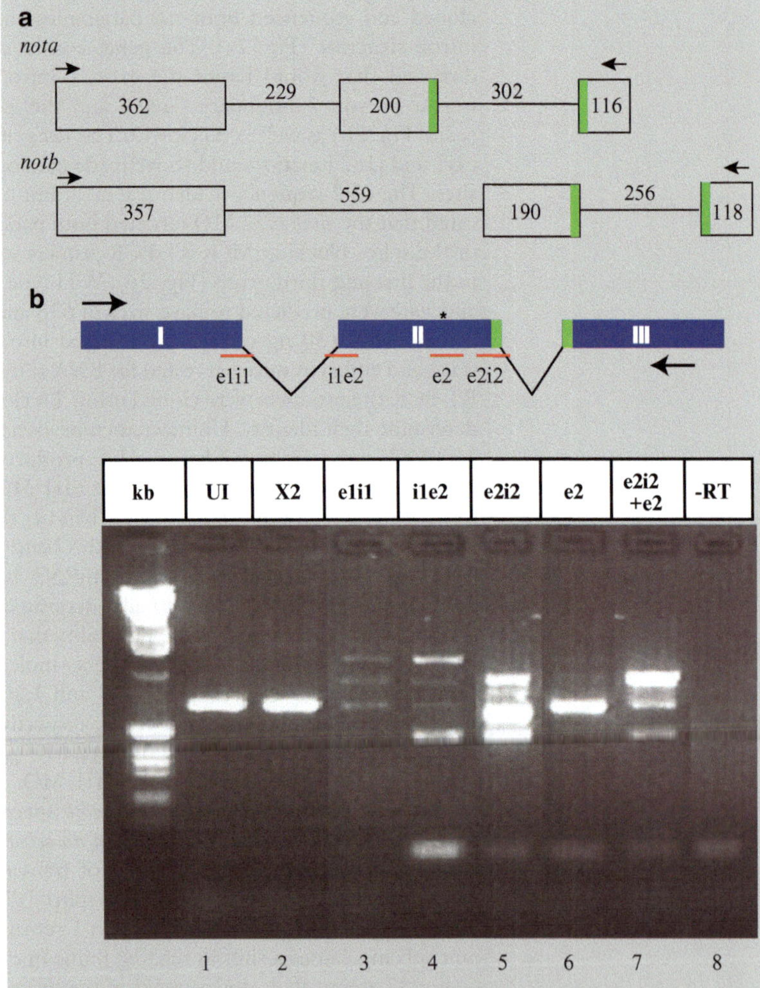

Fig. 2. MOs that bind exon–intron junctions cause aberrant splicing. (**a**) Schematic representation of genomic structure of *nota* and *notb*. The relative sizes of the exons and introns are indicated. The position of the homeobox is indicated in *green*. (**b**) The positions of MO sequences recognizing splice junctions are indicated in *red*. An *asterisk* indicates the position of the cryptic splice site revealed when splicing at the exon 2–intron 2 junction is blocked. *Black arrows* indicate the position of PCR primers. (**c**) MOs targeting splice junctions were injected into embryos at the two-cell stage to give a total of 60 ng MO per embryo. RNA was extracted at the gastrula stage. RT-PCR was performed using the primers indicated in **b**.

(continued)

Box 2 (continued)

notochord homeobox splicing (Fig. 2c, lane 6). When we injected e2i2 together with e2 into embryos at 30 ng per embryo each we found that the cryptic splice event had indeed been eliminated (Fig. 2c, lane 7). However, a compensating increase in the smallest product, which represents a complete loss of exon 2, was not evident. Instead, we observed an increase in abundance of the higher molecular weight products, which represent *notochord homeobox* transcripts in which intron 2 was retained. In summary this is an excellent illustration of the various effects on splicing one may encounter when using MOs targeting splicing sites. If one targets the first exon–intron junction, one may expect either intron 1 retention or uncovering of a cryptic splice site within exon 1. However, if targeting spice junctions downstream of the first exon–intron junction, then any one of the following may be encountered: 1. Intron retention, 2. Exon exclusion or skipping, or 3. Uncovering of a cryptic splice junction in an upstream exon. We encountered all of these in the above example. Thus, it is very useful to perform detailed RT-PCR analyses combined with cloning and sequencing of the resulting aberrant bands to determine the effect of the MOs on splicing and ascertain whether the resulting effect is likely to lead to a loss of function or aberrant function of the resulting protein.

3.4. Evaluation of Knockdown Specificity of MOs

Even though MOs are generally quite specific, possible non-specific effects have to be considered. The gold standard to show that an uncovered phenotype is specific to the intended target gene is to rescue the phenotype by resupplying the gene product of the gene targeted by the MO. This provides the most compelling evidence that the knockdown of the targeted gene is responsible for the phenotype.

For ATG MOs, synthetic RNA of your target gene can also be used to rescue the phenotype. However, if the target sequence of MO is present in the synthetic RNA, these methods cannot be used to rescue the phenotype, because the MO will bind to the synthetic RNA, as well as the endogenous target RNA and block translation of both. Therefore, in these cases, the target gene must be modified at its N terminus, in order to make the rescue construct resistant to the ATG MO. This can be done either by introducing an epitope tag in the N-terminus of the synthetic construct or by generating a mutation within the MO target site in the synthetic construct. In practice it is very difficult to get 100% rescue by injecting RNA or DNA, as it will be unlikely that the level of RNA injected or their localization and timing will mirror that of the endogenous RNA, resulting in overexpression or misexpression effects. In theory this can be addressed by rescuing the knockdowns by generating transgenic embryos that express the MO-resistant version of the gene under the control of its own promoter. However, this is not commonly done, as only few promoters have been characterized sufficiently to perform such rescues.

For splicing MOs, it is possible to simply inject synthetic RNA equivalent to the fully spliced mRNA of your target gene as a rescue construct, since the MOs only act on pre-spliced RNAs. However, as for the case discussed above, it is generally difficult to supply the synthetic mRNA exactly at the right level, location, and time as the endogenous gene, so a perfect rescue is rare. Therefore it is important to carefully titrate the injected RNA and attempt to localize the injections to the appropriate sites to mirror those of the endogenous gene. As discussed above, attempting to rescue using transgenic embryos expressing the target gene under its promoter would be another way to rescue the phenotype (21, 22). However, transgenic animals vary in expression level of the transgene, and plasmid DNA gives rise to a mosaic expression of the transgene. Therefore even these methods may not result in a perfect rescue.

An alternative and widely recognized method to show specificity is to use two or more independent MOs for the same gene target. The logic here is that if the different MOs produce the same phenotype, the probability is very high that the phenotype is specific.

4. Notes

1. In some cases it may be necessary to sequence the target site in the strain of animals that will be used for the study to ensure that the MO has full complementarity to the target sequence.
2. MOs are delivered with a data sheet containing sequence and mass. The MO arrives in a screwcap vial as a sterile lyophilized powder. We make a 1 mM stock in distilled water. The typical package size of a MO is 300 nanomoles. We dilute it with 300 μL volume of distilled water to the final stock concentration of 1.0 mM. Once diluted, MOs should be stored at RT or 4 °C. When stored cold, MOs can come out of solution. Always heat the solution 10 min at 65 °C and cool to room temperature before use. GeneTools suggest not freezing MOs, but to store them at 4 °C for long-term storage or at room temperature for frequent use. The solid MO powder can be frozen in a long term. However, we have found that evaporation can be an issue if MOs are stored for a long time at 4 °C; we have been freezing MO for long-term storage and found them perfectly active, provided they are heated 10 min at 65 °C and cooled to room temperature before use.
3. Testes can be isolated anytime and stored in a tube or vial filled with L-15 or 1× MMR solution at 4 °C. They can be kept

about 1 week or longer as an intact whole testis in a sterile condition.

4. If you wish to knockdown zygotic genes that are expressed at later stages in a particular organ, targeted injection at 8–32-cell stage referring to the fate map (23, 24) would produce a tissue-specific knockdown effectively using a small amount of MO. This might also minimize off-target effects of MO.
5. The MO solution is sticky and easily blocks the needle. Therefore you might need to check regularly for an injection bubble in the ficoll solution from your needle between embryo injections. It is good idea to co-inject with a tracer dye such as micro-ruby (Invitrogen, D7162), until you get reproducible results. Gene Tools also provide labeled MOs with fluorescein or lissamine (25).
6. The *X. tropicalis* full-length clone database can be accessed at: http://genomics.nimr.mrc.ac.uk/online/xt-fl-db.html.

References

1. Summerton J, Weller D (1997) Morpholino antisense oligomers: design, preparation, and properties. Antisense Nucleic Acid Drug Dev 7(3):187–95
2. Summerton J (1999) Morpholino antisense oligomers: the case for an RNase H-independent structural type. Biochim Biophys Acta 1489(1):141–58
3. Schmajuk G, Sierakowska H, Kole R (1999) Antisense oligonucleotides with different backbones. Modification of splicing pathways and efficacy of uptake. J Biol Chem 274(31):21783–9
4. Flynt AS et al (2007) Zebrafish miR-214 modulates Hedgehog signaling to specify muscle cell fate. Nat Genet 39(2):259–63
5. Kloosterman WP et al (2007) Targeted inhibition of miRNA maturation with morpholinos reveals a role for miR-375 in pancreatic islet development. PLoS Biol 5(8):e203
6. Bonev B, Pisco A, Papalopulu N (2011) MicroRNA-9 reveals regional diversity of neural progenitors along the anterior-posterior axis. Dev Cell 20(1):19–32
7. Heasman J, Kofron M, Wylie C (2000) Beta-catenin signaling activity dissected in the early Xenopus embryo: a novel antisense approach. Dev Biol 222(1):124–34
8. Howard EW et al (2001) SpKrl: a direct target of beta-catenin regulation required for endoderm differentiation in sea urchin embryos. Development 128(3):365–75
9. Satou Y, Imai KS, Satoh N (2001) Action of morpholinos in Ciona embryos. Genesis 30(3): 103–6
10. Audic Y et al (2001) Cyclin E morpholino delays embryogenesis in Xenopus. Genesis 30(3):107–9
11. Schweickert A et al (2001) Pitx1 and Pitx2c are required for ectopic cement gland formation in Xenopus laevis. Genesis 30(3):144–8
12. Sumanas S, Ekker SC (2001) Xenopus frizzled-7 morphant displays defects in dorsoventral patterning and convergent extension movements during gastrulation. Genesis 30(3):119–22
13. Nutt SL et al (2001) Comparison of morpholino based translational inhibition during the development of Xenopus laevis and Xenopus tropicalis. Genesis 30(3):110–3
14. Segawa H et al (2001) Functional repression of Islet-2 by disruption of complex with Ldb impairs peripheral axonal outgrowth in embryonic zebrafish. Neuron 30(2):423–36
15. Shepherd IT, Beattie CE, Raible DW (2001) Functional analysis of zebrafish GDNF. Dev Biol 231(2):420–35
16. Yang Z, Liu N, Lin S (2001) A zebrafish forebrain-specific zinc finger gene can induce ectopic dlx2 and dlx6 expression. Dev Biol 231(1):138–48
17. Kos R et al (2001) The winged-helix transcription factor FoxD3 is important for establishing the neural crest lineage and repressing

melanogenesis in avian embryos. Development 128(8):1467–79
18. Coonrod SA et al (2001) A morpholino phenocopy of the mouse mos mutation. Genesis 30(3):198–200
19. Partridge M et al (1996) A simple method for delivering morpholino antisense oligos into the cytoplasm of cells. Antisense Nucleic Acid Drug Dev 6(3):169–75
20. Draper BW, Morcos PA, Kimmel CB (2001) Inhibition of zebrafish fgf8 pre-mRNA splicing with morpholino oligos: a quantifiable method for gene knockdown. Genesis 30(3): 154–6
21. Tan X et al (2006) SmyD1, a histone methyltransferase, is required for myofibril organization and muscle contraction in zebrafish embryos. Proc Natl Acad Sci U S A 103(8):2713–8
22. Kunz M et al (2004) Autoregulation of canonical Wnt signaling controls midbrain development. Dev Biol 273(2):390–401
23. Moody SA (1987) Fates of the blastomeres of the 16-cell stage Xenopus embryo. Dev Biol 119(2):560–78
24. Moody SA (1987) Fates of the blastomeres of the 32-cell-stage Xenopus embryo. Dev Biol 122(2):300–19
25. Rana AA et al (2006) Defining synphenotype groups in Xenopus tropicalis by use of antisense morpholino oligonucleotides. PLoS Genet 2(11):e193

Chapter 9

Chemical Genetics and Drug Discovery in *Xenopus*

Matthew L. Tomlinson, Adam E. Hendry, and Grant N. Wheeler

Abstract

Chemical genetics uses small molecules to modulate protein function and has the potential to perturb any biochemical event in a complex cellular context. The application of chemical genetics to dissect biological processes has become an attractive alternative to mutagenesis screens due to its technical simplicity, inexpensive reagents, and low-startup costs. *Xenopus* embryos are particularly amenable to whole organism chemical genetic screens. Here we describe the basic protocols we have developed to screen small compound libraries on *Xenopus laevis* embryos. We score embryos either by observing phenotypic changes in the whole tadpole or by changes in gene expression pattern using automated wholemount in situ hybridization.

Key words: *Xenopus laevis*, Chemical genetics, Small molecule screens, Drug discovery, Embryo development, Tadpole, Wholemount in situ hybridization

1. Introduction

In recent years a great amount of effort has been focused upon the identification of novel molecular compounds with the capacity to modulate specific protein functions for drug development and new biological tools. The range of new drug targets needs to expand as it is getting harder to discover new novel drugs with conventional techniques. The recent discovery of a compound that can inhibit melanoma growth (1) is a good example of the power of chemical genetic approaches in *Xenopus* and zebrafish to discover new novel drug targets, unlikely to have been discovered with more conventional drug discovery approaches.

The advent of high-throughput chemical genetic screening has allowed researchers to quantitatively assess the phenotypic effects of hundreds of compounds in a short period of time. The benefits of screening in a whole organism such as *Xenopus* include being able to study cell to cell interactions, developmental processes,

Stefan Hoppler and Peter D. Vize (eds.), *Xenopus Protocols: Post-Genomic Approaches, Second Edition*,
Methods in Molecular Biology, vol. 917, DOI 10.1007/978-1-61779-992-1_9, © Springer Science+Business Media, LLC 2012

toxicology, penetrance, and drug metabolism. In particular, assays such as the frog embryo teratogenesis assay—*Xenopus* (FETAX) are well-established methods for developmental toxicity hazard assessment (2, 3).

Xenopus laevis and *Xenopus tropicalis* are highly amenable to medium–high-throughput screening (4–8). Forward chemical genetic approaches using *X. laevis* can be used to identify and investigate specific developmental abnormalities (7, 9). There are more specific benefits to using *Xenopus* such as its well-characterized fate map, ability to regenerate, and the possibility to do high-throughput in situ screening. Using *X. tropicalis* will also allow the linking of genetics and small molecule screening as is currently done with zebrafish. For instance, to perform enhancer/suppressor screens and assist in molecular target identification.

Xenopus was first established as a chemical genetic model in a proof of principle study looking at its utility with a set of compounds used in an earlier zebrafish chemical genetic screen (5). In the past few years a number of screens using *Xenopus* have been published looking at the vasculature, lymphangiogenesis, left/right asymmetry, neural crest, and pigment cell development (1, 5, 9–11). A study conducted by Tomlinson et al. (6) was designed to screen 2,990 compounds to observe the developmental abnormalities the compounds may cause. The initial screen identified 41 compounds displaying various phenotypes affecting general development, edema, eye development, melanophore migration, melanophore morphology, and overall pigmentation (6). Using the chemoinformatic Discoverygate algorithm upon the pigmentation altering compound NSC210627 (originally identified in *Xenopus*) demonstrated strong structural similarities to the DHODH inhibitor brequinar. Further chemoinformatic analysis and testing compounds cross species with zebrafish and human melanoma cells (Xenografts) ultimately led to the discovery of leflunomide, an arthritic drug now shown to have great promise in reducing melanoma growth in vitro and in vivo (1). These studies highlight the potential for forward chemical genetics in *Xenopus* to identify new therapeutic compounds using readily available compound libraries (see Note 1).

Here we will describe in detail the methods used in our lab for screening large libraries of small molecules. We shall describe two methods. One is the visual screen used in Tomlinson et al. (6). The other involves using high-throughput wholemount in situ hybridization to look for phenotypes associated with changes in gene expression patterns or cellular behavior with cell specific markers.

2. Materials

2.1. Components Needed for Generating Xenopus Embryos

1. *1.6% Ethyl 3-aminobenzoate methane sulfonate (to euthanize male)*: 0.5 g dissolved in 300 mL 0.1× MMR in a 500 mL beaker.
2. *2% Cysteine (250 mL)*: This needs to be made fresh on the day of use. Keep at room temperature and discard at the end of the day. 5 g l-cysteine dissolved in 250 mL 1× MMR. Adjusted to pH 8 with NaOH.
3. *10× MEM salts (1 L)*: Add 209.3 g MOPS, 76.08 g 20 mM EGTA, and 2.46 g 10 mM $MgSO_4$ to a 2 L beaker and make up to 1 L with deionized H_2O. pH to 7.4 with NaOH and sterilize via autoclave. Store at 4 °C protected from light/wrapped in foil.
4. *MEMFA (10 mL)*: 1 mL of 10× MEM salts and 1 mL of 37% Formaldehyde dissolved in 8 mL H_2O (see Note 2).
5. *10× MMR (1 L)*: Add 58.45 g 1 M NaCL, 1.5 g 20 mM KCl, 2.0 g 10 mM $MgCl_2$, 2.7 g $CaCl_2$, and 11.9 g 50 mM HEPES to a 2 L beaker. Make up to 1 L with deionized H_2O. pH to 7.5 using NaOH (see Note 3) and sterilize via autoclave.

2.2. Wholemount In Situ Hybridization Components

1. *0.1% PBST*: Add 5 mL 10% Tween detergent to 495 mL DEPC PBS and mix well (see Note 4).
2. *Proteinase K*: Add 1 μL stock proteinase K to 1 mL of PBST (see Note 5). Stock Proteinase K at 10 mg/mL is stored at −20 °C.
3. *Hybridization buffer (Hyb Buffer)*: 50% formamide, 5× SSC, 1 mg/mL Torula RNA, 100 μg/mL Heparin, 1× Denharts solution, 0.1% Tween 20, 0.1% CHAPS, 10 mM EDTA.
4. *20× SSC*: 175.3 g NaCl, 88.2 g Sodium citrate. pH adjusted to 7.0 and volume to 1 L with DEPC H_2O.
5. *1× MAB (Maleic acid buffer)*: 100 mM Maleic acid; 150 mM NaCl (pH 7.5).
6. *10% BMB (Boehringer Mannheim blocking agent)*: 10% (w/v) Boehringer Mannheim blocking agent in preheated (50 °C) 1× MAB, stirred until dissolved and then autoclaved, aliquoted, and stored at −20 °C.
7. *Blocking solution*: 2% BMB in 1× MAB.
8. *Antibody solution*: 2% BMB, 20% goat serum, anti-DIG Fab fragment (Roche, 1:2,000 dilution) in 1× MAB. This can be reused, if stored at 4°C.
9. *Alkaline phosphatase buffer*: 100 mM Tris (pH 9.5), 50 mM $MgCl_2$, 100 mM NaCl, 0.1% Tween 20.

10. *NBT (nitro blue tetrazolium)*: 75 mg/mL in 70% dimethylformamide (DMF).
11. *BCIP (5-Bromo-4chloro-3-indolyl-phosphate)*: 50 mg/mL in 100% DMF.

2.3. Small Compounds Used

1. Synthetic compound libraries can be sourced from many different companies (see Note 6). The synthetic compounds were obtained either prealiquoted in microtiter plates at 20 mM in dimethyl sulfoxide (DMSO) for screening or in powdered form and subsequently prepared as 10 mg/mL stock solutions in DMSO (see Note 7).

3. Methods

3.1. Determining Optimal Screening Concentrations

Carry out all procedures at room temperature unless otherwise specified.

1. Before beginning a chemical genetic screen the optimal screening concentration needs to be determined, to ensure a good discovery rate and low toxicity. This is achieved by randomly selecting a plate from the library, then assaying at serial dilutions, e.g., from 1 to 50 μM.
2. Each well is assayed and scored for the following effects on the five developing embryos: lethality, stunted growth and development, toxicity, and positive hits (effecting normal development) (5). The concentration which gives low toxicity but a good discovery rate should be used.

3.2. Screening Compounds

1. Embryos were grown in a Petri dish containing 10 mL of 0.1× MMR (supplemented with gentamycin sulfate to 50 mg/mL) at 18 °C until they reached stage 15 (see Notes 8 and 9).
2. Embryos were then arrayed by Pasteur pipette at 5 embryos per well (see Note 10) into 48 or 96 well plates, containing the compound in 1,000 or 200 μL of 0.1× MMR (supplemented with gentamycin sulfate to 100 mg/mL) respectively (see Note 11). Figure 1 shows a schematic plan of our screening strategy.
3. Arrayed embryos were grown at 18 °C (see Note 12) and examined visually with a dissecting microscope at 1, 2, and 3 days post fertilization (approximately stage 25, 32, and 38 respectively) (see Note 13 and Fig. 2, e.g., how the embryos might look).
4. Embryos were fixed at stage 38 (see Note 14) in their respective wells by removing as much of the dosing solution as possible using a Pasteur pipette and replacing with 1,000 or 200 μL MEMFA in 48 or 96 well plates respectively.

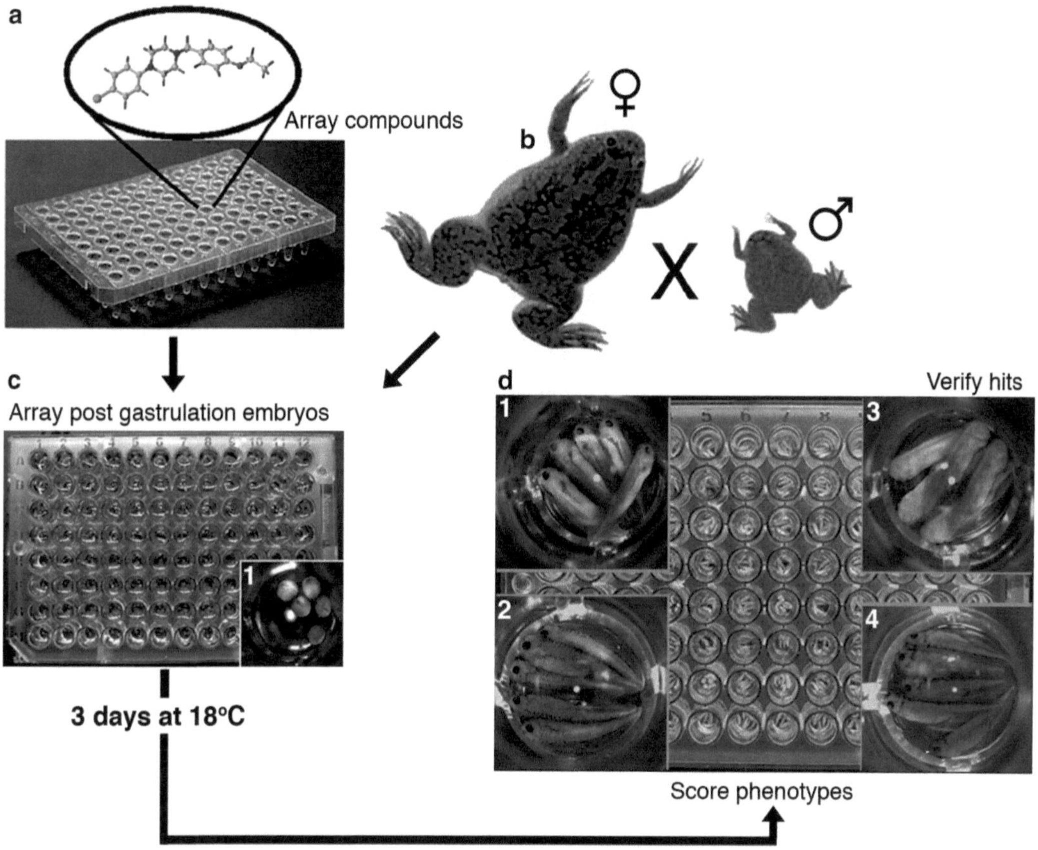

Fig. 1. Chemical genetic screening strategy with developing *Xenopus* embryos. Compound libraries are arrayed in 96 well format at 20 and 40 μM (**a**). Embryos are generated by in vitro fertilization (**b**) then arrayed post gastrulation at stage 15, 5 per well (**c** + **c1**). Plates are left to develop for 3 days at 18 °C. Once the embryos are post melanophore migration (stage 38) they are then visually scored (**d 1–4**) (Tomlinson et al. (6), Reproduced by permission of The Royal Society of Chemistry).

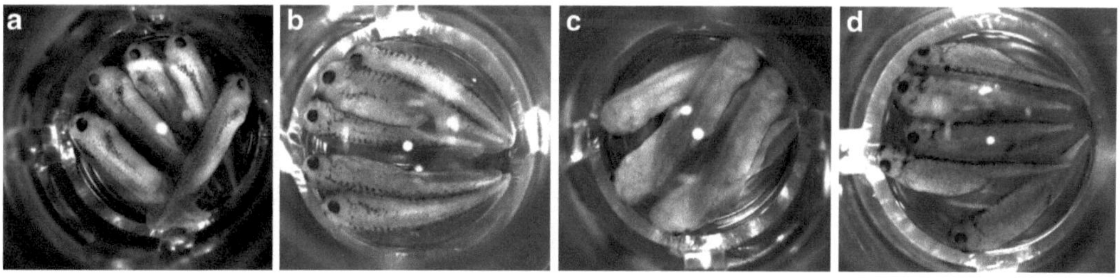

Fig. 2. Closer view of a range of pigmentation phenotypes. Embryos are shown in the screening plates, as they would be scored by the observer. (**a**) shows wild type embryos and typical pigmentation patterns. (**b**) shows a partial disruption (*stripes*) to the dorsal pigment stripe. (**c**) shows a complete loss of pigmentation. (**d**) shows a complete loss of pigment cell migration. All are at stage 38.

5. After fixation for a minimum of 2 h at room temperature the solution was replaced with PBST (see Note 15).
6. All positive hits need to be screened again in 5 cm Petri dishes with 20 embryos and 10 mL of 0.1× MMR with a serial dilution of the compound of interest. This will give statistical confidence and a dose response, revealing compound toxicity and efficacy.

3.3. Observing Embryos

1. Phenotypes were observed directly in the microtitre plate or by transferring the embryos identified in the primary screen (see Subheading 3.2, step 3) into a Petri dish containing a 2.5 mL 1% agar solution covered with PBS via Pasteur pipette (see Note 16).
2. Embryos were observed using a Q imaging 01-MP3.3-RTV-CLR-10 camera mounted on a Zeiss Stemi 5V6 microscope and processed with the Q capture software package (see Note 17).

3.4. Embryo Fixation

1. For long term storage, embryos in wells had as much PBST solution removed as possible using a Pasteur pipette and transferred to a 5 mL glass vial and covered with a PBST/methanol (50/50) solution, and left on a rocker for 5 min.
2. This wash step was repeated once again and twice using 100% methanol. Fixed embryos are labeled and stored at 4 °C (see Note 18).

3.5. High Throughput Combined In Situ Hybridization and Chemical Genetic Screen

The wholemount in situ hybridization technique (WISH) for localizing gene expression can be successfully coupled with chemical genetics to specifically identify compounds targeting individual cells or specific organs. An overview of the protocol is given in Fig. 3. We have developed a combined chemical genetic WISH screen to test small molecule compound libraries for their potential effects on embryonic macrophage migration with the aid of an automated in situ machine.

1. Developing embryos were applied to compounds at stage 15, as previously described and allowed to develop to the stage required for the WISH (in this case stage 25).
2. The embryos were then transferred to 96 well nylon net well plates, by Pasteur pipette, and assayed by in situ hybridization with the Intavis Biolane HT1 automated in situ machine, using a shortened WISH protocol.
3. The following steps were omitted from the standard WISH protocol used in the lab (see Note 19). The 0.1 M triethanolamine washing steps were removed. A 4-h incubation at room temperature with antibody solution may be sufficient to give a reproducible strong signal depending on the probe used.

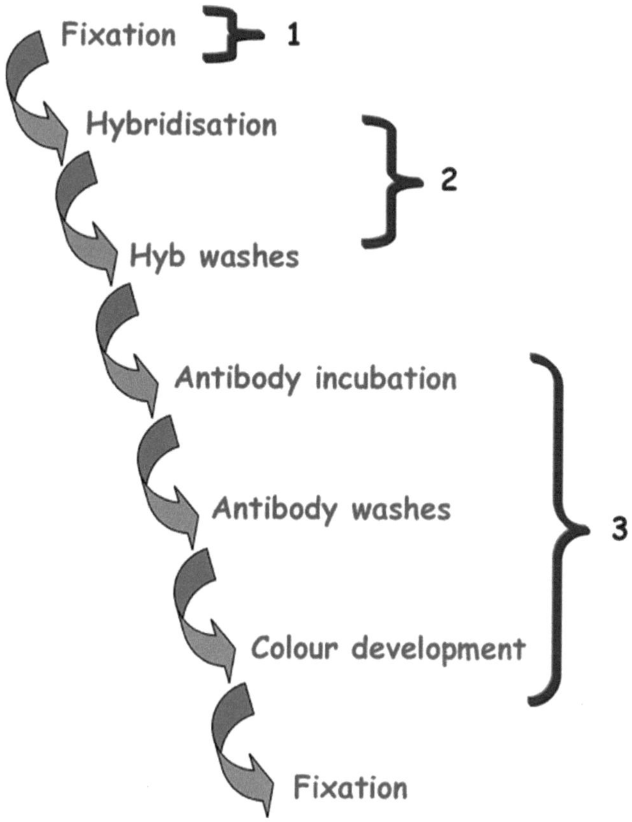

Fig. 3. Schematic overview of our automated in situ protocol. The different WISH stages are shown. *Numbers* indicate the days on which these stages are performed.

A last MAB wash (4 °C overnight) was omitted to increase the throughput of the screen. Typically four 96 well plates every 5 days were routinely assayed. A representative result is shown in Fig. 4 (12) and Tomlinson and Wheeler (unpublished results).

4. Fixation process—Day 1. Add MEMFA for 1 h with agitation at 20 °C. EtOH wash for 5 min with agitation at 20 °C, repeat three times. EtOH for 1 h with no agitation at 4 °C. Plates left at 4 °C to next day.
5. Rehydration and hybridization process—Day 2 (Start at mid-day) Initial Drain step. Wash with 100% MeOH for 5 min with agitation at 20 °C. The following steps are all for 5 min, 75% MeOH/25% H_2O with agitation at 20 °C. 50% MeOH/50% H_20 with agitation at 20 °C. 25% MeOH/75% PBST with agitation at 20 °C. PBST with agitation for 5 min at 20 °C, repeat again. Proteinase K/PBST for 10 min with agitation at 20 °C (see Note 20). PBST with agitation for 5 min at 20 °C, repeat twice more. 3.7% formaldehyde/PBST with agitation for 5 min at 20 °C. PBST with agitation for 5 min at 20 °C, repeat twice

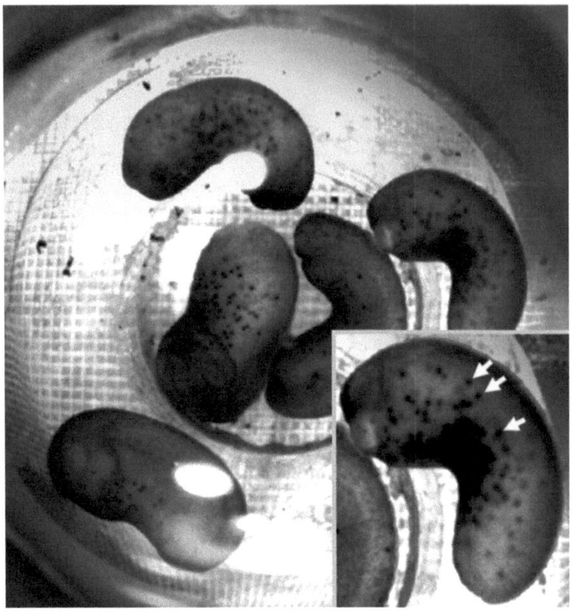

Fig. 4. A chemical genetic screen using the Biolane HTI in situ machine and the macrophage marker *mpo*. Five stage 25 embryos can be seen with stained macrophages migrating away from the ventral blood island. Insert shows a closer view of an embryo, *white arrows* highlight individual macrophages.

more. Hyb buffer with agitation for 5 min at 20 °C. Hyb/probe step at 60 °C with agitation for 2 h, then in hyb buffer for 17 h at 4 °C (will finish at 10 a.m. next day).

6. Washing process and antibody incubation stage—Day 3 initial drain step. Hyb Buffer for 10 min with agitation at 60 °C. 2× SSC for 20 min with agitation at 60 °C, repeat twice. 0.2× SSC for 30 min with agitation at 60 °C, repeat once. MAB for 5 min with agitation at 20 °C, repeat twice. Blocking with blocking solution was then done for 60 min with agitation at 20 °C. Add Antibody solution for 22 h with agitation at 4 °C or 4 h at room temperature.
7. Antibody washing process—Day 3 Initial drain step. Wash with MAB for 30 min with agitation at 20 °C, repeat three times more. Final MAB wash with agitation at 4 °C until ready for color development. Color development (with NBT and BCIP) should be done manually, in microtitre plates (nylon mesh can cause precipitation), and tracked carefully by the observer (see Note 21). Wash several times in PBST then visually score (see Note 22). Embryos can be fixed again and stored in ethanol as previously described

4. Notes

1. The two critical factors in performing a successful forward chemical genetic screen are selecting an appropriate compound library to source from and the biological question posed. This problem is illustrated by the vastness of chemical space (a representation of all the possible organic molecules below 500 Da that are believed to be biologically relevant to drug discovery), which is estimated to contain at least 10^{60} molecules, and has generated difficult questions as to which compounds should be chosen for experimentation (12). The comparison of drug properties and descriptors has led to the emergent field of chemoinformatics, and there are many different established methods in place that use a variety of molecular descriptors to screen whole compound libraries to assess which compounds have the highest potential for successful screening. The unification of chemoinformatics and high-throughput chemical genetic screening has provided a powerful tool for identifying and screening hundreds of drug like compounds in a vertebrate model. Conversely the NCI diversity set library, used in our study, has been screened successfully many times using different cells/organisms and appears to have not yet reached saturation (6, 13). This suggests that not only is chemical diversity important but also the biological question being addressed. With a complex organism like *Xenopus* undergoing a wide range of developmental process a rich resource is there to be mined.
2. MEMFA should be made fresh each day and stored at 4 °C with light excluded by aluminum foil.
3. 10× MMR has a high buffering capacity, the use of 10 M NaOH will help speed up the pH process.
4. The use of DEPC is only necessary to eliminate RNase contamination from a solution where this could be a concern such as with WISH, elsewhere it is not necessary.
5. Only make up proteinase K solution just before it is needed and keep at 4 °C until then. Each new batch of proteinase K has to be optimized for its activity by varying the amount of time embryos are treated in a control WISH. This becomes more critical when netwells are used.
6. As mentioned in Note 1 there are numerous and extensive chemical libraries to obtain compounds from and it is likely that only a select few will be appropriate for an individual experiment. They can contain collections chosen for their diversity or for known biological activity, such as patent expired

drug compounds. It is therefore important to take careful consideration as to which compound libraries are selected for each individual experiment. The author refers you to Bender et al. (14) for a detailed review on choosing appropriate compound libraries (13) and to Wheeler and Brandli which contains an extensive list of libraries used to date in zebrafish and *X. laevis* screens (8).

7. *Xenopus* has been shown to have a DMSO EC_{50} (malformation) of 1.92% (15). To improve solubility of some compounds either PBST or MeOH could be used instead of DMSO, these should be kept below 1% concentration when assaying embryos to avoid direct toxicity and use vehicle only controls as with DMSO.
8. Prior to stage 15 is the gastrulation phase of development. It is best to avoid screening during gastrulation as this stage of development is particularly sensitive to many compounds.
9. If by stage 15 more than 20% of embryos are dead or deformed than the screen should be abandoned as it is likely the embryos will not be robust enough to develop under the potential stress of a chemical genetic screen.
10. We find 5 embryos to be optimal in 96 well plates. In 48 well plates either 10 embryos can be cultured to stage 40 or 5 embryos to stage 44.
11. The Volume of 0.1% MMR transferred needs to be calculated and allowed for when determining compound concentration in the screen. Typically only 20 μL is transferred with the embryos.
12. *Xenopus* embryos will develop at different rates depending upon the incubation temperature. A *X. laevis* and *X. tropicalis* developmental timetable can be found in Xenbase (http://www.xenbase.org/other/methods.do). It is therefore particularly important to ensure that incubation temperatures remain consistent throughout the protocol.
13. Developmental effects can be noted at these stages which may be less obvious at later stages. Interesting effects can be reassayed and returned to later.
14. The stage at which to fix is dependent on what the phenotype is that is being screened for, i.e., stage 38 for pigment and stage 42 for vasculogenesis. A smaller developmental window in the screen may give more specific phenotypes. For example in looking at pigment cell migration, applying compounds at stage 24 to stage 38 will decrease the chance of nonspecific binding to other proteins or the same molecular

target at different developmental stages (e.g., the matrix metalloproteinases) (7).

15. If required embryos can be stored in PBST at 4 °C for up to 1 week with the addition of 0.5% sodium azide as an antimicrobial.
16. We have found that embryos are much easier to orientate under the microscope when using a hair loop. A glass pipette (with a rubber bulb), which has been flamed smooth may be preferable to a plastic Pasteur pipette for transferring fragile treated embryos as it has a larger bore and a smoother surface. For viewing the fixed embryos in dorsal or ventral views a shallow indent can be made into the agar surface.
17. Pictures can be further processed using software such as Adobe Photoshop.
18. A reverse of this procedure will rehydrate the embryos for further manipulation.
19. Optimization may be needed for each in situ probe used as to which steps in the protocol can be omitted or reduced. The Macrophage specific marker mpo/XPOX2 probe gives a high signal to background and can be reused eight times before significant loss of signal (16).
20. Optimization and care will be needed in the length and temperature of the proteinase K permeabilization stage as the nylon netwells can be abrasive to the embryos.
21. NBT/BCIP color development solution can be added to plastic trays slightly bigger than a netwell microtiter plate, these are then immersed. The whole tray and lid covered in aluminum foil to exclude light and rocked gently.
22. False positive hits were detected in our macrophage chemical genetic screen due to a number of compounds developmentally stunting the embryos. The screen was stopped and assayed when the control vehicle only embryos were at stage 25. Positive hits then have to be rescreened when the treated embryos have also reached stage 25.

Acknowledgements

The authors would like to thank Rob Field and Andrea Munsterberg for helpful discussions. AH is funded by a UEA/JIC joint studentship supported by AstraZeneca.

References

1. White RM, Cech J, Ratanasirintrawoot S, Lin CY, Rahl PB, Burke CJ, Langdon E, Tomlinson ML, Mosher J, Kaufman C, Chen F, Long HK, Kramer M, Datta S, Neuberg D, Granter S, Young RA, Morrison S, Wheeler GN, Zon LI (2011) DHODH modulates transcriptional elongation in the neural crest and melanoma. Nature 471:518–522
2. Song MO, Fort DJ, McLaughlin DL, Rogers RL, Thomas JH, Buzzard BO, Noll AM, Myers NK (2003) Evaluation of Xenopus tropicalis as an alternative test organism for frog embryo teratogenesis assay–Xenopus (FETAX). Drug Chem Toxicol 26:177–189
3. Longo M, Zanoncelli S, Della Torre P, Rosa F, Giusti A, Colombo P, Brughera M, Mazue G, Olliaro P (2008) Investigations of the effects of the antimalarial drug dihydroartemisinin (DHA) using the Frog Embryo Teratogenesis Assay-Xenopus (FETAX). Reprod Toxicol 25:433–441
4. Brandli AW (2004) Prospects for the Xenopus embryo model in therapeutics technologies. Chimia 58:694–702
5. Tomlinson ML, Field RA, Wheeler GN (2005) Xenopus as a model organism in developmental chemical genetic screens. Mol Biosyst 1:223–228
6. Tomlinson ML, Rejzek M, Fidock M, Field RA, Wheeler GN (2009) Chemical genomics identifies compounds affecting Xenopus laevis pigment cell development. Mol Biosyst 5:376–384
7. Tomlinson ML, Guan P, Morris RJ, Fidock MD, Rejzek M, Garcia-Morales C, Field RA, Wheeler GN (2009) A chemical genomic approach identifies matrix metalloproteinases as playing an essential and specific role in Xenopus melanophore migration. Chem Biol 16:93–104
8. Wheeler GN, Brandli AW (2009) Simple vertebrate models for chemical genetics and drug discovery screens: lessons from zebrafish and Xenopus. Dev Dyn 238:1287–1308
9. Kalin RE, Banziger-Tobler NE, Detmar M, Brandli AW (2009) An in vivo chemical library screen in Xenopus tadpoles reveals novel pathways involved in angiogenesis and lymphangiogenesis. Blood 114:1110–1122
10. Blackiston D, Adams DS, Lemire JM, Lobikin M, Levin M (2011) Transmembrane potential of GlyCl-expressing instructor cells induces a neoplastic-like conversion of melanocytes via a serotonergic pathway. Dis Model Mech 4:67–85
11. Dush MK, McIver AL, Parr MA, Young DD, Fisher J, Newman DR, Sannes PL, Hauck ML, Deiters A, Nascone-Yoder N (2011) Heterotaxin: a TGF-beta signaling inhibitor identified in a multi-phenotype profiling screen in Xenopus embryos. Chem Biol 18:252–263
12. Tomlinson ML, Garcia-Morales C, Abu-Elmagd M, Wheeler GN (2008) Three matrix metalloproteinases are required in vivo for macrophage migration during embryonic development. Mech Dev 125:1059–1070
13. Sierecki E, Sinko W, McCammon JA, Newton AC (2010) Discovery of small molecule inhibitors of the PH domain leucine-rich repeat protein phosphatase (PHLPP) by chemical and virtual screening. J Med Chem 53:6899–6911
14. Bender A (2010) How similar are those molecules after all? Use two descriptors and you will have three different answers. Expert Opin Drug Discov 5:1141–1151
15. Dresser TH, Rivera ER, Hoffmann FJ, Finch RA (1992) Teratogenic assessment of four solvents using the Frog Embryo Teratogenesis Assay–Xenopus (FETAX). J Appl Toxicol 1: 49–56.
16. Smith SJ, Kotecha S, Towers N, Latinkic BV, Mohun TJ (2002) XPOX2-peroxidase expression and the XLURP-1 promoter reveal the site of embryonic myeloid cell development in Xenopus. Mech Dev 117(1–2):173–86

Chapter 10

Maternal mRNA Knock-down Studies: Antisense Experiments Using the Host-Transfer Technique in *Xenopus laevis* and *Xenopus tropicalis*

David J. Olson, Alissa M. Hulstrand, and Douglas W. Houston

Abstract

The ability to inhibit the activity of maternally stored gene products in *Xenopus* has led to numerous insights into early developmental mechanisms. Oocytes can be cultured and manipulated in vitro and then implanted into the body cavity of a host female to make them competent for fertilization. Here, we summarize the methods for obtaining, culturing, and fertilizing *Xenopus* oocytes, with the goal of inhibiting maternal gene function through antisense oligonucleotide-mediated mRNA knock-down. We describe a simplified technique for implanting donor oocytes into host females using intraperitoneal injection. Also, we present optimized methods for performing the host-transfer procedure with *Xenopus tropicalis* oocytes.

Key words: *Xenopus laevis*, *Xenopus tropicalis*, Maternal genes, Antisense, Host-transfer, Oocyte, Oligonucleotides

1. Introduction

Maternal stores of mRNAs and proteins accumulated during oogenesis control many early developmental events in *Xenopus*. In some cases, these gene products are localized to different regions of the egg and can be differentially inherited during cleavage to influence early cell fate decisions. Zygotic transcription in *Xenopus* does not begin until the mid-blastula stage, about 8 h after fertilization (1, 2), and therefore maternal factors are particularly important for events occurring prior to this mid-blastula transition (MBT). These processes include dorsoventral axis specification, germ layer induction, and specification of germline fate (reviewed in ref. 3). Because these processes are initiated soon after fertilization, it can be problematic to interfere with the underlying

Stefan Hoppler and Peter D. Vize (eds.), *Xenopus Protocols: Post-Genomic Approaches, Second Edition*,
Methods in Molecular Biology, vol. 917, DOI 10.1007/978-1-61779-992-1_10, © Springer Science+Business Media, LLC 2012

molecular pathways by microinjecting reagents into fertilized eggs, as is commonly done in *Xenopus*. Maternal effect mutants have proven useful for studying early events in invertebrates, but these mutants are more difficult to generate in vertebrates, owing to longer life cycles and more complex genomes. Recently, several maternal effect mutants have been identified in zebrafish (4, 5), and screens are underway for maternal mutants in *Xenopus tropicalis*, a diploid relative of the allotetraploid *Xenopus laevis*. However, these methods are labor and cost intensive, precluding a routine assessment of maternal gene function in most vertebrates.

Fortunately, it is possible to readily study the activity of maternally stored gene products in *Xenopus* using the injection of antisense oligodeoxynucleotides (oligos) into oocytes to deplete maternally stored mRNAs (6–8). Antisense is a particularly successful strategy to assess maternal function in *Xenopus* because of several main advantages. First, oocytes can be easily obtained and manipulated in culture, something not readily done with oocytes from most species. Second, the biochemical makeup of oocytes favors the degradation of mRNAs in response to antisense oligo injection, as opposed to steric blockage of translation or stimulation of innate immune responses. Stage VI oocytes have endogenous RNase H-like activity (9, 10), therefore injection of DNA oligos results in cleavage and destruction of target mRNA strands of the resulting DNA:RNA hybrids. This leaves no doubt as to the mechanism of antisense action, which is not always clear in other antisense experiments. Third, and most important, the embryological outcome of maternal mRNA depletion can be assessed, since oocytes can be cultured through meiosis in vitro and then fertilized to initiate development (11–13). Last, the effects of maternal depletion persist at least until MBT, since zygotic mRNAs will not be resynthesized until then.

The main obstacle to using oligo-injected oocytes for studies of embryonic development is that ovarian oocytes are not directly competent for fertilization. Normally, ovulated eggs require proteolytic processing of the vitelline membrane and investment by jelly coats to attract and activate sperm and to facilitate sperm-egg binding. Methods exist to mimic the sperm activation process in vitro with jelly coat extracts (14, 15), and to bypass fertilization entirely using nuclear transfer or intracytoplasmic sperm injection (16, 17). Although these methods can successfully trigger cleavage and normal development from oocytes, their use in conjunction with antisense experiments has not been widely reported in the literature.

More commonly, oligo-injected oocytes are transferred into the body cavity of a host female to facilitate fertilization. In the coelomic cavity, the transferred eggs (along with normally ovulated host eggs) are moved by the action of peritoneal cilia into the oviducts (18), whereupon they acquire the necessary modifications for fertilization prior to egg laying. After the experimental eggs are recovered, they can be fertilized by normal methods to generate embryos lacking the mRNA of interest. The unusual robustness of *Xenopus* oocytes and adult animals to manipulation allows this procedure to be performed with a high rate of success. Thus, these methods allow maternal gene function in *Xenopus* to be analyzed without generating maternal effect mutations, something not easily done in other vertebrates.

Here we describe methods for obtaining, culturing, and injecting *Xenopus* oocytes with antisense DNA oligos to deplete maternal mRNAs, and for fertilizing these oocytes to obtain viable embryos. We detail a method for transferring in vitro cultured oocytes into host females by intraperitoneal (IP) injection. We also outline adaptations to the well-characterized host-transfer methods for *X. laevis* for use with *X. tropicalis* oocytes.

2. Materials

Prepare all solutions using deionized, ultrapure water and clean glassware, free from detergent residue. Also, use high quality and recently ordered reagents.

2.1. Surgery and Oocyte Collection

1. Buffered MS222 (ethyl 3-aminobenzoate methanesulfonate salt): 1.0 g/L MS222, 0.7 g/L sodium bicarbonate, adjust final pH to ~7.0.
2. Oocyte culture media (OCM): 70% Leibovitz L-15, 0.004% BSA, 1× Pen-Strep, 0.5 μg/mL (gentamicin): adjust pH to 7.6–7.8 with NaOH, make fresh, and store at 14–18 °C for up to 1 week.
3. Surgical and dissection instruments (Fine Science Tools). Scalpel handle (#3) and blade (#10 or 11), several pairs of Dumont forceps (#4 or #5, Biologie), Bonn iris scissors (curved or straight), Halsey or Olsen-Hegar microneedle holders.
4. 4–0 PDS II sutures: Violet monofilament with 17 mm ½ circle needle, RB-1 taper (Ethicon). 5–0 sutures can be used for *X. tropicalis* surgeries (Ethicon).
5. Frog water: tap water treated with Amquel (chloramine remover).
6. Povidone-iodine USP (10% solution; Betadine or equivalent): dilute 1:20 in distilled water before use.

2.2. Oocyte Transfer and Collection of Embryos

1. Progesterone: 10 mM stock in 100% ethanol, dilute to 1 mM in 100% EtOH to make a 500× working stock solution, store at −20 °C.

2. Human chorionic gonadotropin (hCG): 10,000 IU/vial, reconstitute with 10 mL sterile water, store at 4 °C up to 30 days.
3. Vital dyes: Stocks of blue, red, and brown are made up in 50 mL deionized water, incubated for 20 min with rocking and spun in a clinical centrifuge. Aliquots (~1 mL) are taken from the supernatants and stored at −20 °C.
   (a) 0.1% Nile Blue A: 0.05 g/50 mL.
   (b) 0.25% Neutral Red: 0.125 g/50 mL.
   (c) 1% Bismarck Brown: 1.0 g/50 mL.
4. 2 ml glass syringe (Tomopal Inc.), with metal luer lock adaptor.
5. 16 gauge, 1 in. sterile syringe needle.
6. 10× Marc's modified Ringers (MMR): 1 M NaCl, 18 mM KCl, 20 mM $CaCl_2$, 10 mM $MgCl_2$, 150 mM Hepes, adjust to pH 7.8 with 5 N NaOH, filter sterile, and store at 4 °C, readjust pH to 7.6–7.8 when 1 or 1.2× (high salt) solutions are made, dilutions of 0.3, 0.1, and 0.05× are made by diluting the 1× stock.
7. 2% cysteine in 0.1× MMR, pH to 7.8.
8. Teflon pestles for 1.5 mL microfuge tubes (Kimble-Kontes).

3. Methods

Carry out all procedures at room temperature unless otherwise specified. Be sure to follow all relevant animal care and use guidelines. See Fig. 1 for a graphical overview and general timeline of the procedure.

3.1. Surgical Isolation of Ovarian Tissue

1. Ovary is obtained surgically from anesthetized females. In most cases, survival surgery is performed so that ovary is removed from the same female multiple times. Survival surgeries should be done using aseptic technique (see Note 1).
2. Prepare fresh OCM media prior to obtaining ovary.
3. Prepare the surgical area and sterilize the instruments. Disinfect a dedicated surgical area using 70% EtOH. Place the tips of the instruments in a bead sterilizer at 250 °C for 20 s and then place on a sterile surface, such as a Petri dish, sterile gauze, or the inside of a suture package.
4. Select several females dedicated for ovary removal and place one in buffered MS222 anesthetic. A surgical plane of anesthesia will be reached in about 10 min. This can be checked by turning the frog onto her back or by pinching a toe. Anesthetized animals will be unresponsive. Frogs that are not fully anesthetized

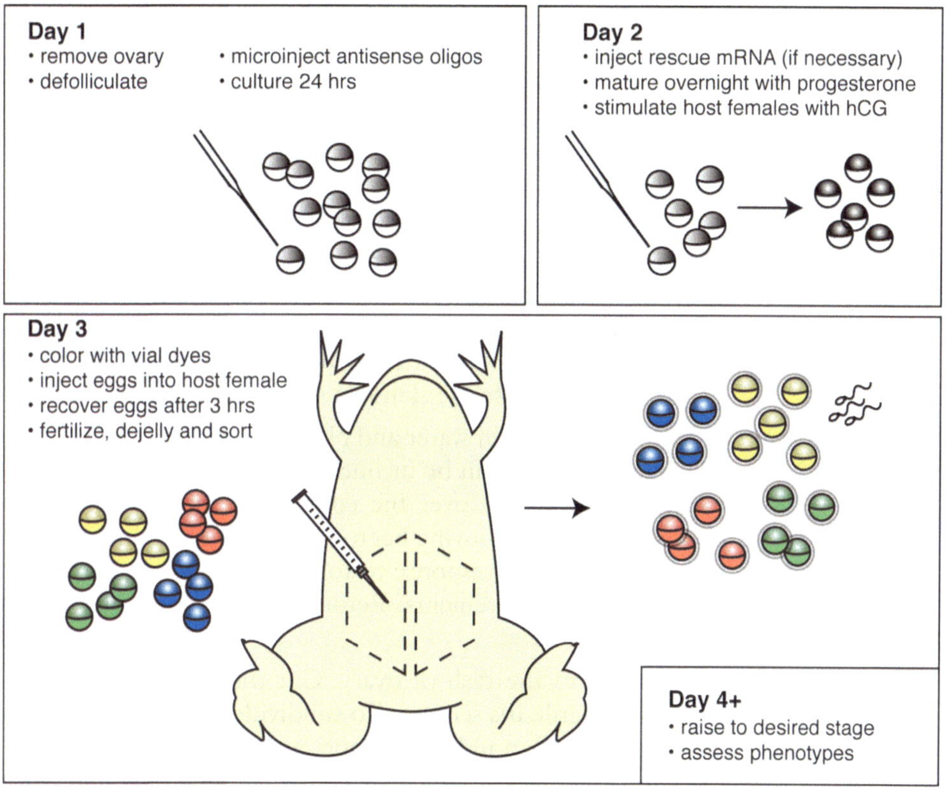

Fig. 1. Outline of the host-transfer procedure by intraperitoneal injection. On day one, healthy ovary is removed and oocytes are defolliculated. Oocytes are then injected with mRNAs or with antisense oligonucleotides targeting the endogenous mRNA of the candidate gene. On the second day, rescue mRNA is injected (if necessary), the oocytes are matured with progesterone and prospective host females are induced with hCG. On the third day, the oocytes are colored with vital dyes and injected into an egg-laying host female. Once the eggs are released, they are fertilized, sorted from host eggs and analyzed as desired. The dotted areas indicate optimal regions in the abdomen for inserting the experimental oocytes.

are placed back into MS222 and checked at 5-min intervals until anesthesia is complete. Females are operated one at a time until acceptable ovary is found.

5. Place the anesthetized female on her back on a damp wipe and cover the head and legs. Pour a small amount of diluted povidone-iodine (1:20 in distilled water) over the abdomen and gently blot with gauze (see Note 2).

6. Make a small (1 cm) incision in the skin in the lower part of the abdomen using small iris scissors. Incisions on the same frog should be performed on alternate sides, and can be made either parallel or perpendicular to the midline. The midline itself should be avoided, owing to the presence of large abdominal blood vessels. Use forceps to grasp the underlying muscle, pull upward and incise smartly to expose the body cavity. The ovary should be visible or may be found following careful exploration using blunt forceps.

7. The ovary has about 24 lobes, each containing hundreds of oocyte follicles (19). Pull several of the lobes through the incision and trim away at the level of the body wall. Place the tissue into a large (100 mm) Petri dish of OCM. Repeat until the desired amount is obtained. Six or eight large lobes are sufficient for a typical experiment (see Note 3).
8. Close the muscle and fascia layer first with simple interrupted suturing, using basic instrument ties to place surgeon's square knots (also called reef knots) several millimeters apart. After each suture, trim the material to leave a short tail before proceeding to the next suture. Once the body wall layer is secure, proceed to close the skin in a similar manner (see Note 4).
9. Rinse the frog in water and place in a shallow recovery bucket. The container can be inclined so the water does not cover the frog's nostrils. Cover the container and allow the animal to recover before moving her to a larger recovery bucket. Recovery is indicated by response to touch, eye bulging, and eventually purposeful movements. Return to the colony after the frog is fully recovered.
10. Return to the dish of ovary. Cut the ovary into small pieces using sterile iris scissors. To subdivide the ovary, first cut open one side of an individual lobe, and flatten out the tissue. The ovary is then cut into small pieces, about 2 cm^2, and pieces are transferred to a dish of clean OCM, keeping about six pieces per dish. Dividing the ovary in this way extends the life of the tissue in culture and makes defollicating a bit easier (see below). Culture up to 4–5 days in OCM at 18 °C.

3.2. Isolation, Culture, and Microinjection of Oocytes

1. Oocytes for host-transfer must be manually defolliculated since collagenased oocytes cannot be fertilized (20). Select only healthy, fully grown stage VI oocytes (1.2–1.3 mm diameter). These oocytes will have a uniformly pigmented animal hemisphere and may show a distinct equatorial band of lighter pigmentation. An ideal ovary will have >50 such oocytes per small ovary piece.
2. To begin manually defolliculating oocytes, select a piece of ovary and transfer to a separate dish of OCM. Grasp the connective tissue theca layer near a desired oocyte with forceps, using the non-dominant hand. Using a second pair of forceps, lightly grasp the ovary adjacent to the oocyte and tear apart the theca layer and continue pulling away from the ovary (see Note 5). The oocyte will be squeezed out between the follicle layers and can be teased gently away.
3. Successfully defolliculated oocytes will be somewhat flaccid and will lack blood vessels, which are part of the theca layer. Properly defolliculated oocytes can be easily microinjected, whereas oocytes remaining in their follicles are quite tough

(see Note 6). Experienced operators can easily defolliculate 200–400 oocytes per hour. Transfer groups of defolliculated oocytes to medium (60 mm) Petri dishes using a sterile Pasteur pipette. Culture groups of 75–200 oocytes at 18 °C (see Note 7).

4. Microinject oocytes with antisense oligos or RNAs as desired (see Subheading 3.7). The oocytes can be injected directly while in OCM. Typical injection volumes are 5–10 nL, delivering doses of 2–6 ng for DNA-based oligos, 5–100 ng for morpholinos or 20 pg–1 ng for mRNA. To affect localized molecules it may be necessary to target injections to the appropriate region of the oocyte.
5. Injected oocytes are transferred to dishes containing 8 mL of fresh OCM and cultured at 18 °C. Because of limitations on the number of colors that can be made using vital dyes (see below), only five-to-six groups (~75–200 oocytes each) can be transferred per female, so experiments must be planned with this in mind. Oocytes injected with DNA-based oligos should be cultured for at least 24 h to allow for degradation of the oligo before proceeding with the host-transfer. Culture can be extended to 72 h without affecting oocyte viability or developmental capacity of the resulting embryos.
6. If rescue experiments are being performed, inject the desired mRNA after the oligo has had time to degrade (~24 h). It is important for correctly interpreting the specificity of the oligo that the mRNA does not compete for binding to the oligo, but rather replace the function of the knocked-down protein. RNAs are typically injected the evening before the transfer is to be performed, irrespective of the length of oligo incubation.

3.3. Preparation of Donor Oocytes and Host Females for Transfer

1. On the evening before the transfer, usually the day after oligos are injected, oocytes are stimulated to undergo maturation by addition of progesterone to a final concentration of 2 μM (16 μL of 1 mM stock/8 mL OCM). The dishes are swirled briefly and returned to 18 °C overnight.
2. Around the same time, inject 3–5 females with hCG (1,000 U) to induce ovulation. Females are placed in 18 °C water and left at room temperature overnight. Host-transfer experiments work best in our hands if oocyte maturation and hCG injection are done about 10–12 h prior to oocyte implantation. Multiple females are injected to ensure that at least one will have suitable egg quality to serve as the host (see step 5 below).
3. The next morning, check that the oocytes have undergone maturation (look for germinal vesicle breakdown (GVBD)) and that the females have begun to lay eggs. Matured oocytes can be frozen on dry ice at this point for verification of mRNA or protein knock-down.

4. Vital dye stocks are thawed and spun for 2–3 min (see Note 8). Oocytes are stained with vital dyes by adding 80 μL of stock to each dish, followed by gentle swirling. When adding multiple dyes, add 80 μL of each and swirl to mix in the dish. Oocytes are incubated with rocking for 15 min and transferred to a large dish of OCM to wash until implantation.
5. While the oocytes are staining, choose a host female and place in buffered MS222 anesthetic. An ideal host female will have just started laying healthy eggs, and can be induced to release more eggs upon gentle squeezing. Females that are laying stringy eggs or that crush eggs upon squeezing should be avoided.
6. Place the anesthetized female on her back on a damp wipe. Cover the head and legs as above and gently swab with diluted povidone-iodine.

3.4. Oocyte Transfer by Intraperitoneal Injection

1. Oocytes can be transferred into the body cavity of the host by surgical implantation. Recently, our lab has devised a method to transfer oocytes by IP injection, using a large-bore hypodermic needle. Since the surgical method has been described previously (7, 21), the IP injection method will be described here (see Note 9). An overview of the method is shown in Fig. 2.
2. Attach a new 16-gauge, 1 in. disposable needle (16 G1) onto a 2 mL glass syringe fitted with a metal luer lock tip. Remove the plunger and set aside (it will not be used for injection). Rinse the inside of the syringe with OCM to coat the surfaces with BSA (in the OCM), reducing sticking of oocytes.
3. Holding the syringe at a 45° angle to the frog (vertically), with the beveled edge of the needle up, insert the tip into the lower abdomen of the host female, passing through both skin and muscle. Orient the needle toward the anterior. Placing a finger just above the injection site may be necessary to provide some support, allowing the needle to insert smoothly. The needle will have passed through both layers when resistance against the needle can no longer be felt. Do not penetrate too deeply into the body cavity.
4. Once the needle is inserted, hold the syringe steady at a 45° angle with one hand. Slowly introduce the colored oocytes into the upper part of the syringe using a Pasteur pipette. Add the oocytes against the side and let them drop to the bottom. They should collect in a ~1 mL volume of OCM, but should not enter the needle while the syringe is still inclined.
5. When all the oocytes have been added to the syringe, tilt the syringe upright, keeping the needle in the frog. Tap gently on the side of the syringe and the oocytes will begin draining into the body cavity. Adjust the needle depth up and down if oocytes do not flow right away. Flush with OCM if some oocytes

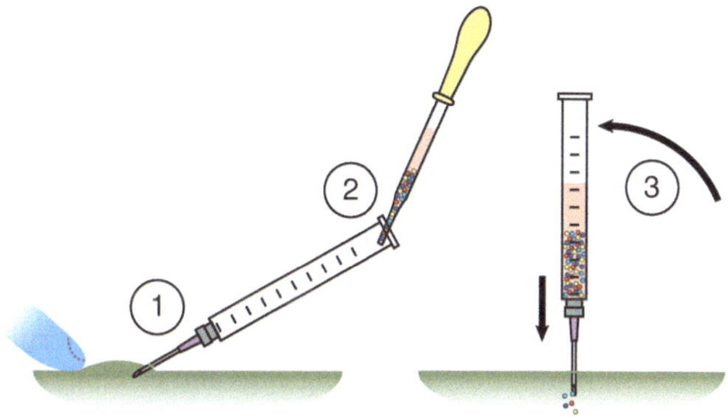

Fig. 2. Details of the intraperitoneal injection technique. (1) A 16-gauge needle is inserted into the body cavity at a 45° angle. The skin is braced by a finger. (2) The colored oocytes are back-loaded into the syringe, adding a minimal amount of excess OCM. (3) The oocytes will drain into the body cavity when the needle is brought vertical. Some gentle tapping or needle depth readjustment may be necessary to get the oocytes to flow. Figure shows a lateral view of the ventral surface of the abdomen, anterior is to the left.

become stuck around the edge of the syringe port. Excess OCM, up to 2 mL or so does not seem to be detrimental.

6. Return the syringe to a 45° angle and withdraw the needle. Oocytes should not spill out of the insertion site, although some OCM may drain out. Very little bleeding should occur. The female is rinsed in distilled water to wash off MS222 and placed immediately into a recovery bucket of water at 18 °C. Suturing or other means of closing the wound are not necessary.
7. Monitor the host's recovery from anesthesia as above. She should resume laying eggs soon after the procedure. Body cavity eggs and the implanted oocytes will be translocated by peritoneal cilia to the openings to the oviducts and the host should begin laying colored eggs 2–3 h after implantation. The female can be gently squeezed if colored eggs do not appear by 3–4 h. Rarely, the experiment can be lost if the host stops laying altogether, since it is problematic to recover the transferred eggs from the oviducts.

3.5. Recovery of Experimental Oocytes and In Vitro Fertilization

1. About 1 h after implantation, place the host female in a container of 1 L high salt MMR (1.2× MMR, pH 7.6). A 4 L Nalgene beaker is a convenient vessel for this purpose. Eggs released into high salt will remain competent for fertilization for many hours.
2. In the meantime (or beforehand) obtain testes from a male frog through non-survival surgery. Keep testes in OCM at room temperature throughout the day, and at 4 °C for longer-term storage.

3. Once colored eggs appear in the bucket, the female can be squeezed if desired. Eggs are squeezed into a dry Petri dish and the female is returned to the high salt MMR. Eggs are fertilized with a sperm suspension made in 1× MMR (or L15) for 4 min. The eggs are then flooded and rinsed with 0.1× MMR and left to develop. It is often convenient to dejelly and sort cleaving colored eggs at the 4-cell stage (see Note 10).
4. After the female has stopped laying (or after about 5 h), the eggs in high salt can be fertilized. Gently squeeze out any remaining eggs and carefully drain most of the high salt buffer. Rinse the eggs once in 0.3× MMR and carefully drain again. The eggs are transferred to a Petri dish and as much of the remaining buffer as possible is removed with a transfer pipette.
5. Fertilize the eggs by homogenizing half a testis in a microfuge tube using a Teflon pestle, in a volume of 500 μL 0.3× MMR. Add the homogenate to the eggs and swirl to mix thoroughly. After 10 min, flood and rinse with 0.1× MMR. Colored eggs are dejellied and sorted as above and left to develop to the desired stages for analysis (Fig. 3).

3.6. Modifications for Fertilization of Cultured *X. tropicalis* Oocytes

Oocyte transfers using *X. tropicalis* oocytes can be performed either by implanting oocytes back into *X. tropicalis* hosts or by heterotopically transferring into *X. laevis* females. The cross-species transfers are possible because *X. tropicalis* sperm can penetrate *X. laevis* jelly coats (22). This cross-species method is preferable since transfers are easier to perform on the larger *X. laevis* females. Additionally, *X. tropicalis* females lay eggs in larger bursts with many more host eggs, and often stop laying before many experimental eggs are recovered. The basic methods of obtaining ovarian tissue and isolating oocytes are similar for *X. tropicalis*, although the *tropicalis* oocytes are somewhat more difficult to defolliculate. We have optimized the following changes in the timing of the procedure to facilitate transfer of *X. tropicalis* oocytes into *X. laevis* females.

Isolate ovary from *X. tropicalis* females using the general methods described above. Manually defolliculate fully grown stage VI oocytes (these will be about one-third the size of *X. laevis* oocytes) and culture in OCM at 25 °C. Microinject as desired.

1. On the evening prior to transfer, 3–5 *X. laevis* females are injected with hCG (1,000 U) to induce ovulation. Females are placed in 18 °C water and left at room temperature overnight. *X. tropicalis* oocytes do not need to be treated with progesterone at this time; this will be done the following morning.
2. The next morning, progesterone is added to the *X. tropicalis* oocytes to a final concentration of 2 μM (16 μL of 1 mM stock/8 mL OCM). The dishes are swirled briefly and cultured at 25 °C for 2–4 h. *X. tropicalis* oocytes mature faster than

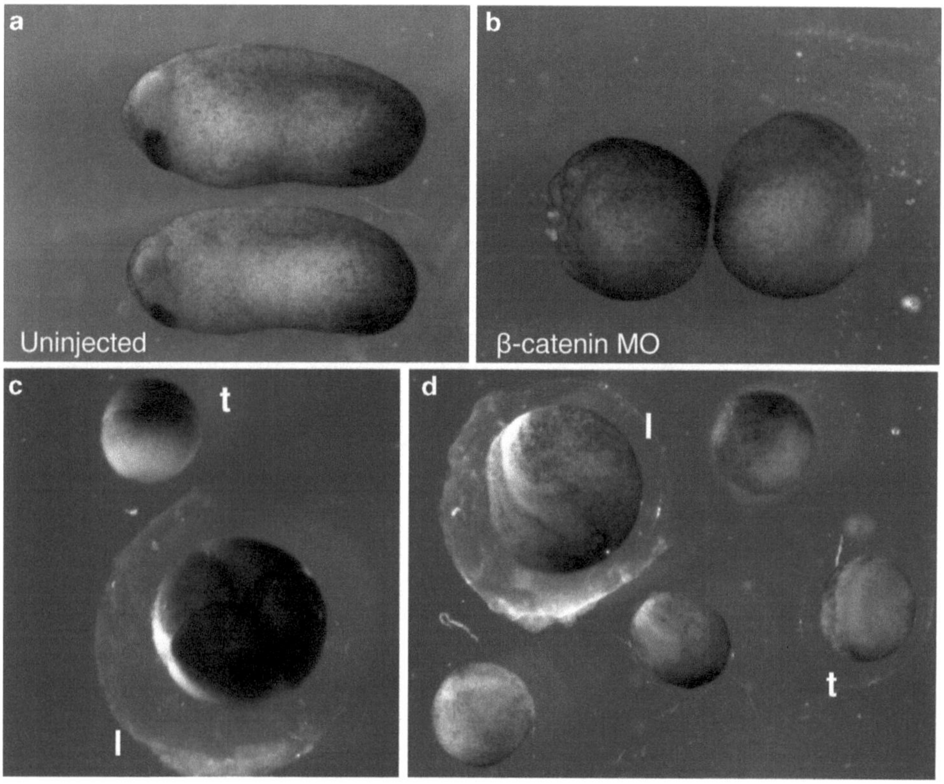

Fig. 3. Representative results of host transfers using *X. laevis* and *X. tropicalis* oocytes. (**a**, **b**) Maternal knock-down of β-catenin. Oocytes injected with 10 ng of β-catenin MO were matured and fertilized via the host-transfer method, as in ref. (23). (**a**) Uninjected and (**b**) MO-injected embryos at stage 25 demonstrate the knock-down phenotype. Uninjected embryos develop normally while knock-down embryos are ventralized (100%, $n=28$). (**c**, **d**) Representative results of *X. tropicalis* oocytes fertilized following host-transfer into *X. laevis* females. (**c**) *X. tropicalis* embryo at the 4-cell stage (t) shown next to a host *X. laevis* embryo (l). (**d**) *X. tropicalis* embryos at the neurula stage (t) shown next to a host *X. laevis* embryo (l).

X. laevis oocytes, therefore this treatment can be done the same day as the transfer.

3. Once the oocytes have matured, vital dye stain them and anesthetize a suitable *X. laevis* host female, as described above.
4. Rinse the donor oocytes in fresh OCM and transfer into the *X. laevis* female by IP injection as described above.
5. Squeeze or recover the oocytes in high salt MMR. The much smaller *X. tropicalis* eggs can be easily distinguished from the large *X. laevis* eggs. These are sorted into a separate Petri dish for fertilization using a wide-bore Pasteur pipette.
6. Rinse *X. tropicalis* eggs in 0.3× MMR, decant well and fertilize with a sperm suspension, made by homogenizing a *X. tropicalis* testis in L15. Incubate for 10 min and then flood with 0.05× MMR. Dejelly and sort dividing embryos and culture as desired (see Fig. 3 and Note 11). A variable number of fertilized host embryos, most likely interspecific hybrids, will also be generated by this method.

3.7. Use of Antisense Oligonucleotides in Xenopus Oocytes

Among the uses for this procedure, one of the most powerful is the depletion of maternally stored gene products in the oocyte followed by analysis of resulting embryos. This is often necessary because interference with maternal gene function after fertilization is frequently ineffective. This section describes the basic procedure for designing and testing for suitable antisense oligonucleotides.

1. Antisense oligos against a given mRNA are designed by trial and error. Paste the 5′ half of the gene of interest into primer or oligo design software (IDT, Primer3, MacVector; see Note 12). Set the parameters to identify 18–20 mer antisense oligos having a moderate GC content (40–60%) and a Tm of about 50–60 °C. Exclude oligos with GGGG motifs or multiple GG or CC dimers, since these can form tetraplex structures linked to drastic non-specific effects.
2. Order 3–4 DNA oligos modified with three phosphorothioate linkages on the 5′ and 3′ termini (chimeric oligos). The modified bonds on the ends will provide a suitable amount of nuclease resistance, while minimizing the potential side effects of a fully thioate-modified oligo (see Note 13). Obtain the smallest scale possible from your preferred oligo provider (50–100 nmol) and have the oligos HPLC purified.
3. To hedge against possible targeting of undesired mRNA, the oligo sequences are screened against *X. laevis* or *X. tropicalis* databases using BLAST (http://www.xenbase.org/genomes/blast.do). Sequences that show large stretches (>13 bases) of complementarity to other genes should be discarded and more oligos designed.
4. Oligos are dissolved in deionized water to a final concentration of 1 mM (~5–6 μg/μL), from which working stocks of 1 μg/μL are made. Aliquots of the concentrated stocks and working stocks are stored at −80 °C.
5. Oocytes for oligo screening are isolated as above. Test oligos are injected at several doses (e.g., 2, 4 and 6 ng) and the oocytes are cultured overnight (~16 h) at 18 °C. Remember to centrifuge oligos prior to injection, especially since oligo preps are prone to clogging the needle.
6. Levels of mRNA depletion are assessed by RT-PCR (preferred) or other method of RNA analysis. Oocytes are placed into microfuge tubes in minimal liquid, frozen on dry ice and either stored at −80 °C or processed immediately. Total RNA is isolated and single strand cDNA is prepared (23). Primers amplifying the target gene and a housekeeping gene are used and the degree of reduction in target gene expression relative to the control is determined. Realtime PCR is useful to obtain more quantitative data, but is not necessary to find usable oligos. Analysis of target protein levels by immunoblotting can also be done, if a suitable antibody is available (23).

7. Oligos that are effective (≥80% reduction in target mRNA levels) can then be used right away in pilot experiments and a larger synthesis is ordered. Usually 1–2 suitable oligos can be identified after a round or two of this exercise (see Note 14).

4. Notes

1. Guidelines for *Xenopus* survival surgery allow for five surgeries, followed by a sixth terminal procedure. Other surgical guidelines, such as the requirement for masks or maintenance of records, are regulated differently at different institutions, so be sure to check the guidelines established by your local animal regulatory committee.
2. Surgical scrubs containing detergents or ethanol are not appropriate for disinfecting the incision, as these will damage the sensitive skin of amphibians. Diluted iodine or chlorhexidine are acceptable.
3. Although the majority of females will have ovary suitable for these experiments, it is a good practice, however, to briefly inspect the ovary quality before proceeding. A small piece is removed initially and quickly examined under a dissecting microscope. Unhealthy ovary typically has many yellowish, speckled or dying/atretic oocytes, or has very few fully grown stage VI oocytes. The presence of many oocytes undergoing resorption, which are heavily covered in blood vessels, is another indication of poor ovary quality. If the ovary is deemed unsuitable, the female is sutured right away and another is tried.
4. The suturing technique is documented in Schneider et al. (21), and numerous internet-based videos are also available to demonstrating proper suturing. Additionally, institutional animal care offices should provide instruction in surgical methods if requested.
5. Manual defolliculating is labor intensive, but is a skill in which most students can become skilled with some practice. It is necessary to use polished and sharpened watchmakers' forceps. The tips should be properly shaped and must meet precisely at the tips. Forceps are first slightly blunted and sanded on a whetstone, followed by polishing with 800-grit sandpaper. When defolliculating, it is best to use extremely light pressure (just enough to close the tips) when tearing the theca; gripping too tightly generally results in pulling the oocyte away without actually removing the follicle layer.
6. Defolliculated oocytes should be free of marks and scratches, since even slightly damaged ones will not survive the host-transfer procedure. Also, perfectly manually defolliculated

oocytes retain a single cell inner epithelial layer of follicle cells, which are intimately associated with the vitelline envelope. Although not important for the host-transfer procedure, this fact should be kept in mind when interpreting RT-PCR or imaging results using manually defolliculated oocytes.

7. Since the maturation response rate varies with oocytes from different females, it is good practice to perform a test of the maturation rate prior to microinjection. Early in the day, treat 20 or so oocytes with 2 μM progesterone and assess the number that undergo GVBD, about 6 h later. If more than half the oocytes fail to mature, it may be wise to start again with new ovary.
8. The final vital dye concentrations are: Blue = 0.001% Nile Blue A, Red = 0.0025% Neutral Red, Brown = 0.01% Bismarck Brown. Five main colors are possible; each of the single colors plus Mauve (80 μL Blue + 80 μL Red) and Green (80 μL Blue + 80 μL Brown). A sixth color can also be made, Orange (80 μL Brown + 80 μL Red), but this can be very difficult to distinguish from the either Brown or Red alone and is only used as a last resort.
9. IP injection is less invasive than typical abdominal surgery and the female is under anesthesia for much less time. Also, IP injection is not usually considered a "survival surgery" for the purposes of animal use protocols and thus may be used without incurring additional regulatory burden. For conventional surgical implantation, females are incised as for ovary removal, oocytes are introduced using a Pasture pipette and the incision is sutured. In this method, it is essential that the operator maintain a grip on one side of the muscle layer at all times after introducing the oocytes, up to the completion of the first suture. Failure to do so will result in oocytes leaking out of the incision.
10. The vital dye colors are most easily distinguishable when viewed from the animal pole at the 2–8-cell stage. Afterwards it is necessary to roll the embryos over to see the vegetal pole. This can be laborious if the colors are faint or if there are numerous embryos.
11. A variable number of host *X. laevis* eggs will fertilize and develop normally, apparently as interspecific hybrids.
12. Oligos can also be chosen by eye, but it is useful to run these through primer design programs to screen out oligos with undesirable motifs, self-dimerization, hairpin formation, or unbalanced GC content.
13. It is also possible to test unmodified oligos, and then order modified version of successful ones. For this method, 10 or more oligos can be tested at 10 and 5 ng doses. Unmodified

oligos that deplete more than 50% of the target mRNA are reordered in modified and purified form. Oligos should be retested, since phosphorothioate modification can alter the binding of the oligo.

14. For the best interpretation of experimental results, it is desirable to identify two oligos targeting independent sequences in the target (non-overlapping sites). Finding more than one oligo that produces the same phenotype at similar levels of target mRNA depletion is one way to show specificity of the knock-down. The other, preferred method is to rescue the phenotype by injecting doses of mRNA that do not elicit overexpression effects.

Acknowledgments

This work was supported by NIH grant GM083999 (DWH) and The University of Iowa (DWH).

References

1. Bachvarova R, Davidson E (1966) Nuclear activation at the onset of amphibian gastrulation. J Exp Zool 163:285–295
2. Newport J, Kirschner M (1982) A major developmental transition in early Xenopus embryos: II. Control of the onset of transcription. Cell 30:687–696
3. Heasman J (2006) Patterning the early Xenopus embryo. Development 133:1205–1217
4. Dosch R, Wagner DS, Mintzer KA, Runke G, Wiemelt AP, Mullins MC (2004) Maternal control of vertebrate development before the midblastula transition: mutants from the zebrafish I. Dev Cell 6:771–780
5. Wagner DS, Dosch R, Mintzer KA, Wiemelt AP, Mullins MC (2004) Maternal control of development at the midblastula transition and beyond: mutants from the zebrafish II. Dev Cell 6:781–790
6. Hulstrand AM, Schneider PN, Houston DW (2010) The use of antisense oligonucleotides in Xenopus oocytes. Methods 51(1):75–81
7. Mir A, Heasman J (2008) How the mother can help: studying maternal Wnt signaling by antisense-mediated depletion of maternal mRNAs and the host transfer technique. Methods Mol Biol 469:417–429
8. Torpey N, Wylie CC, Heasman J (1992) Function of maternal cytokeratin in Xenopus development. Nature 357:413–415
9. Cazenave C, Chevrier M, Nguyen TT, Hélène C (1987) Rate of degradation of [alpha]- and [beta]-oligodeoxynucleotides in Xenopus oocytes. Implications for anti-messenger strategies. Nucleic Acids Res 15:10507–10521
10. Dash P, Lotan I, Knapp M, Kandel ER, Goelet P (1987) Selective elimination of mRNAs in vivo: complementary oligodeoxynucleotides promote RNA degradation by an RNase H-like activity. Proc Natl Acad Sci U S A 84: 7896–7900
11. Brun R (1975) Oocyte maturation *in vitro*: contribution of the oviduct to total maturation in Xenopus laevis. Experientia 31:1275–1276
12. Holwill S, Heasman J, Crawley C, Wylie CC (1987) Axis and germ line deficiencies caused by u.v irradiation of Xenopus oocytes cultured in vitro. Development 100:735–743
13. Smith LD, Ecker RE, Subtelny S (1968) In vitro induction of physiological maturation in Rana pipiens oocytes removed from their ovarian follicles. Dev Biol 17:627–643
14. Kloc M, Miller M, Carrasco AE, Eastman E, Etkin L (1989) The maternal store of the xlgv7 mRNA in full-grown oocytes is not required for normal development in Xenopus. Development 107:899–907
15. Elinson R (1973) Fertilization of frog body cavity eggs enhanced by treatments affecting the vitelline coat. J Exp Zool 183:291–302

16. Amaya E, Kroll KL (1999) A method for generating transgenic frog embryos. Methods Mol Biol 97:393–414
17. Schorderet-Slatkine S, Drury KC (1973) Progesterone induced maturation in oocytes of Xenopus laevis. Appearance of a 'maturation promoting factor' in enucleated oocytes. Cell Differ 2:247–254
18. Rugh R (1935) Ovulation in the frog. II. Follicular rupture to fertilization. J Exp Zool 71:163–194
19. Dumont J (1972) Oogenesis in Xenopus laevis (Daudin). I. Stages of oocyte development in laboratory maintained animals. J Morphol 136:153–179
20. Heasman J, Holwill S, Wylie CC (1991) Fertilization of cultured Xenopus oocytes and use in studies of maternally inherited molecules. Methods Cell Biol 36:213–230
21. Schneider, P.N., Hulstrand, A.M., and Houston, D.W. (2010). Fertilization of Xenopus oocytes using the host-transfer method. J Vis Exp, e1864.
22. Lindsay L, Peavy T, Lejano R, Hedrick J (2003) Cross-fertilization and structural comparison of egg extracellular matrix glycoproteins from Xenopus laevis and Xenopus tropicalis. Comp Biochem Physiol A Mol Integr Physiol 136:343–352
23. Heasman J, Kofron M, Wylie C (2000) Beta-catenin signaling activity dissected in the early Xenopus embryo: a novel antisense approach. Dev Biol 222:124–134

Part III

Xenopus Trangenesis

Chapter 11

Generating Transgenic Frog Embryos by Restriction Enzyme Mediated Integration (REMI)

Shoko Ishibashi, Kristen L. Kroll, and Enrique Amaya

Abstract

Here we present a protocol for generating transgenic embryos in *Xenopus laevis* and *Xenopus tropicalis*. The method includes three steps: (1) The preparation of high-speed egg extracts, which facilitates the replacement of protamines in sperm nuclei with nucleosomes and decondenses the chromatin of sperm nuclei; (2) The isolation of sperm nuclei; and (3) The mixing of sperm nuclei, restriction enzyme, and high-speed extract in vitro, following by nuclear transplantation into unfertilized eggs to generate the transgenic embryos. This procedure generates non-mosaic transgenic embryos at high frequency and efficiency.

Key words: *Xenopus*, Restriction enzyme mediated integration, Transgenesis

1. Introduction

The transgenesis protocol described here can be divided into three parts, (a) preparation of egg extracts, (b) sperm nuclei preparation, and (c) nuclear transplantation (Fig. 1). A crude egg extract is prepared using a low-speed centrifugation step. These extracts are driven into the interphase stage of the cell cycle by addition of calcium. A high-speed centrifugation is then performed to generate an interphase cytosolic fraction containing proteins required for the efficient decondensation of the sperm nuclei (Fig. 1a). In addition, sperm nuclei are prepared from isolated sperm by treatment with lysolecithin, which causes a gentle permeabilization of the sperm plasma membrane (Fig. 1b). The nuclear transplantation procedure involves (1) incubation of linearized plasmid DNA with sperm nuclei, (2) decondensation of sperm nuclei by addition of a high-speed egg extract containing a small amount of the restriction enzyme, and (3) the reaction mix is diluted and transplanted into

Stefan Hoppler and Peter D. Vize (eds.), *Xenopus Protocols: Post-Genomic Approaches, Second Edition*,
Methods in Molecular Biology, vol. 917, DOI 10.1007/978-1-61779-992-1_11, © Springer Science+Business Media, LLC 2012

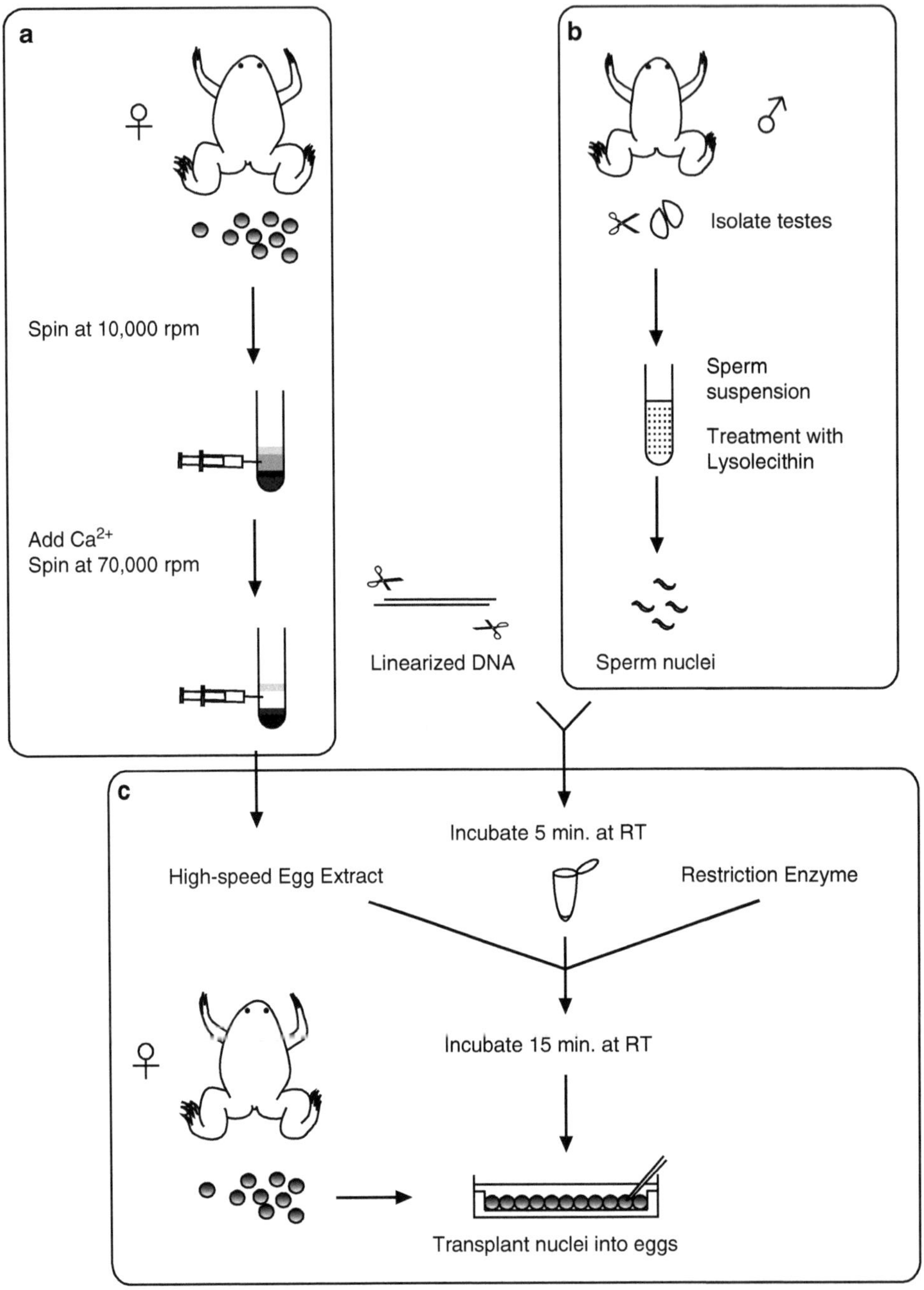

Fig. 1. Transgenesis procedure includes: (**a**) Preparation of egg extracts; (**b**) Sperm nuclei preparation; and (**c**) Nuclear transplantation. The egg extracts and sperm nuclei can be stored at −80°C. (**a**) Calcium is added to allow the crude egg extract (which are held in meiotic arrest) to progress to interphase, and a high-speed centrifugation is performed to obtain the cytosolic fraction. (**b**) Testes are macerated and filtered, and then the sperm suspension is treated with lysolecithin to disrupt the plasma membrane of the cells. (**c**) Sperm nuclei are incubated with linearized DNA for a brief period of time. High-speed egg extracts and a restriction enzyme are added. The egg extracts partially decondenses chromosomes and the restriction enzyme stimulates recombination by creating double-strand breaks, facilitating integration of DNA into the genome. Diluted nuclei are transplanted into unfertilized egg.

unfertilized eggs (Fig. 1c). The egg extract partially decondenses sperm chromatin and the restriction enzyme stimulates recombination by creating double-strand breaks, facilitating integration of DNA into the genome (1).

There are several advantages of the REMI method of transgenesis over other methods. First and foremost is that it allows the production of fully transgenic, non-mosaic embryos, without the need of propagating the transgene through the germline, since the transgene integrates into the genome prior or soon after fertilization. Second is its high efficiency. Within an afternoon, one can produce hundreds of fully transgenic embryos using this method, which can be analyzed and studied over the subsequent hours, days, or months. In particular, this method allows the rapid assessment of genomic regulatory regions (promoter and/or enhancer regions), the rapid labeling of tissues and/or subcellular structures in living embryos, and the rapid assessment of phenotypes resulting from misexpression experiments under tight temporal and spatial control. Thus, using this method, one can quickly and easily perform experiments on F0 transgenic animals. However, there are some disadvantages as well, particularly if the experiments rely solely on the analysis of F0 transgenic animals. Firstly, each F0 transgenic embryo is unique. Therefore analysis amongst F0 transgenic embryos will result in significant variability in transgene expression, due to differences in integration site and copy number. For this reason, F0 transgenic experiments should be limited to a quick assessment of expression patterns of promoters/enhancers or missexpresssion experiments. One should not perform these experiments at the exclusion of generating transgenic lines, which will provide results that are considerably more uniform and clean (2). In cases where misexpression experiments are lethal, the experiments will need to be designed using a binary system, such as the GAL4/UAS system (2). Another disadvantage to the REMI method of transgenesis is that the transgenes integrate as large concatemers (1). Thus this method should not be used to generate transgenic lines containing site specific or transposon elements, such as cre/lox, flp/frt, Tol2, etc. The close apposition of such elements in the concatemer insertion will cause havoc following the inducement of site-specific recombination and/or transposition in such lines.

2. Materials

2.1. High-Speed Egg Extract Preparation (Xenopus laevis Protocol, Also Used for Xenopus tropicalis Protocol)

1. Ultracentrifuge (Beckman Optima) and Rotor (SW 40Ti, Beckman) with Ultra clear tube, 14×95 mm (Beckman 344060).
2. Ultracentrifuge (Beckman optima TLX) and Rotor (TLA100.3) with Thickwall polycarbonate tube (Beckman 349622).

3. 1× Marc's Modified Ringer (MMR): 100 mM NaCl, 2 mM KCl, 1 mM $MgCl_2$, 2 mM $CaCl_2$, 5 mM HEPES, pH 7.5. Prepare a 10× stock, and adjust pH with NaOH to 7.5. Sterilize 10× solutions by autoclaving.
4. 20× Extract buffer (XB) salt stock: 2 M KCl, 20 mM $MgCl_2$, 2 mM $CaCl_2$, filter-sterilize and store at 4°C.
5. Extract buffer (XB): 1× XB salts (100 mM KCl, 0.1 mM $CaCl_2$, 1 mM $MgCl_2$; from 20× XB salts stock solution), 50 mM sucrose (1.5 M stock; filter-sterilize and store in aliquots at −20°C), 10 mM HEPES (1 M stock, titrated with KOH so that pH is 7.7 when diluted to 10 mM, should require about 5.5 mL of 10 N KOH for 100 mL, dilution drastically changes the pH of HEPES, so pH must be monitored after dilution; filter-sterilize and store in aliquots at −20°C). Prepare about 100 mL.
6. 2% (w/v) L-Cysteine hydrochloride 1-hydrate (Sigma, C7880): Made up in 1× XB salts before use and titrated to pH 7.8 with NaOH. Prepare about 300 mL.
7. CSF-XB: 1× XB salts (100 mM KCl, 0.1 mM $CaCl_2$, 1 mM $MgCl_2$), 1 mM $MgCl_2$ (in addition to $MgCl_2$ present in XB salts; final concentration 2 mM), 10 mM HEPES, pH 7.7, 50 mM sucrose, 5 mM EGTA, pH 7.7. Prepare 50 mL.
8. Protease inhibitors: Mixture of leupeptin (Roche, 1 017 101), chymostatin (Roche, 1004 638), and pepstatin (Roche 253 286), each dissolved to a final concentration of 10 mg/mL in dimethyl sulfoxide (DMSO). Store in small aliquots at −20°C.
9. 1 M $CaCl_2$ (filter-sterilize and store at 4°C).
10. Energy mix (store in aliquots at −20°C): 150 mM creatine phosphate (Roche, 621 714), 20 mM ATP (Roche, 519 979), 20 mM $MgCl_2$, store in 0.1 mL aliquots at −20°C.
11. Pregnant mare serum gonadotropin (PMSG): 100 U/mL PMSG (P.G.600®, Intervet, Inc., 021825). Dissolve in water and stored at −20°C.
12. Human chorionic gonadotropin (HCG): 1,000 U/mL HCG (CHORULON®, Intervet, Inc., 057176). Dissolve in water and stored at 4°C.

2.2. Sperm Nuclei Preparation (*Xenopus laevis* and *Xenopus tropicalis* Protocol)

1. Centrifuge and rotor (e.g., Sorvall HB-4 Swinging bucket rotor).
2. 14 mL tube (Falcon 2059).
3. Dissection tools; scissors and forceps.
4. Funnel.
5. Cheesecloth.
6. Fluorescence microscope.
7. Hemacytometer.

8. 1× MMR. Prepare as described in Subheading 2.1, item 3.
9. 0.1% Tricaine Methanesulfonate (MS222, aminobenzoic acid ethyl ester, Sigma A-5040), 0.1% sodium bicarbonate. Dissolve in water.
10. 2× Nuclear preparation butter (NPB): 500 mM sucrose (1.5 M stock; filter-sterilize and store aliquots at −20°C), 30 mM HEPES (1 M stock; titrate with KOH so that pH 7.7 is at 15 mM, filter-sterilize and store aliquots at −20°C), 1 mM spermidine trihydrochloride (Sigma S-2501; 10 mM stock; filter-sterilize and store aliquots at −20°C), 0.4 mM spermine tetrahydrochloride (Sigma S-1141; 10 mM stock; filter-sterilize and store aliquots at −20°C), 2 mM dithiothreitol (Sigma D-0632; 100 mM stock; filter-sterilize and store aliquots at −20°C), 2 mM EDTA (500 mM EDTA, pH 8.0 stock; autoclave and store at room temperature). On the day of the sperm nuclei preparation, make up 25 mL of 2× NPB for 1–2 males from the stock solutions.
11. 1× NPB: Make up 30 mL of 1× NPB by mixing 15 mL of 2× NPB with 15 mL of water and store on ice.
12. Lysolecithin: 100 μL of 10 mg/mL L-α-lyso-Lecithin, Egg Yolk (Calbiochem, 440154); dissolve at room temperature just before use. Store solid stock at −20°C. Discard the stock powder if it becomes sticky. Digitonin can be used instead of Lysolecitin. Digitonin is more specific for the plasma membrane leaving the nuclear membranes intact.
13. Bovine serum albumin (BSA): 10% (w/v) BSA (fraction V, Sigma A-7906). Make up 5 mL in water on the day of the sperm nuclei preparation.
14. 1× NPB, 3% BSA: Mix 5 mL 2× NPB, 3 mL 10% BSA, and 2 mL water and store on ice.
15. 1× NPB, 0.3% BSA: Mix 2.5 mL 2× NPB, 0.15 mL 10% BSA, and 2.35 mL water and store on ice.
16. 100% glycerol.
17. Sperm storage buffer: 1× NPB, 30% glycerol, 0.3% BSA. Make up by mixing the following solutions, 250 μL of 2× NPB, 15 μL of 10% BSA, 150 μL of 100% glycerol, and 85 μL water.
18. Sperm dilution buffer (SDB): 250 mM sucrose, 75 mM KCl, 0.5 mM spermidine trihydrochloride, 0.2 mM spermine tetrahydrochloride. Add about 80 μL of 0.1 N NaOH per 20 mL solution to titrate to pH 7.3–7.5 and store 0.5–1 mL aliquots at −20°C.
19. Hoechst No. 33342 (Sigma B-2261): 10 mg/mL stock in dH_2O, store in a lighttight vessel at −20°C.

20. PMSG: 100 U/mL PMSG (P.G.600®, Intervet, Inc., 021825). Dissolve in water and stored at –20°C.
21. HCG: 500 U/mL (*X. tropicalis*) or 1,000 U/mL (*X. laevis*) HCG (CHORULON®, Intervet, Inc., 057176). Dissolve in water and stored at 4°C.

2.3 Nuclear Transplantation Reagents and Equipment (X. laevis Protocol)

1. 2% L-Cysteine hydrochloride 1-hydrate in 1× MMR (titrate to pH 8.0 with NaOH). Make up freshly.
2. 1× MMR as described in Subheading 2.1, item 3.
3. 100 mM $MgCl_2$.
4. SDB. Prepare as described in Subheading 2.2, item 18.
5. 0.4× MMR, 6% (w/v) Ficoll (Sigma, F-4375), 10 μg/mL gentamycin (a 10 mg/mL stock solution is purchased from Gibco-BRL 15710-015). Sterilize by filtration.
6. 0.1× MMR, 6% (w/v) Ficoll, 10 μg/mL gentamycin. Sterilize by filtration.
7. 0.1× MMR, 10 μg/mL gentamycin.
8. Linearized plasmid (100 ng/μL in water): Any enzyme can be used for linearization of plasmid. Digest DNA using standard conditions, and purify by phenol/chloroform extraction and ethanol precipitation. There is no need to gel-purify the plasmid.
9. Restriction enzyme: Dilute in water before adding to the transplantation reaction. Although any enzyme can be used, do not use an enzyme that digests within regulatory or protein coding regions of the construct (see Note 1).
10. Agarose-coated injection dishes: 1.0% agarose in 0.1× MMR is poured into 60-mm Petri dishes. Before the agarose solidifies, a template is laid onto it. After the agarose has solidified, the templates are removed and the dishes are wrapped in parafilm and stored at 4°C until use. As a template we routinely use a 35-mm × 35-mm square, small weighing boat, which holds about 400 *X. laevis* eggs.
11. Transplantation needles: 30-μL Drummond MICROCAPS® (Cat. No. 1-000-0300) are pulled to produce large needles with long, gently sloping tips (Fig. 2). We use a Flaming/Brown Micropipet Puller Model P-87 (Sutter Instruments Co.) for pulling needles using a condition, $p=50$, $v=100$, and $t=5$. Needles are clipped with a forceps to produce a beveled tip of 80–100 μm diameter, using the ocular micrometer of a dissecting microscope for measurement (Fig. 2).
12. Transplantation apparatus: Most commercial injection apparatus used for RNA and DNA injections which are based on air pressure are not suitable for nuclear transplantation, due to the difference in needle tip size. Therefore we use an oil-filled

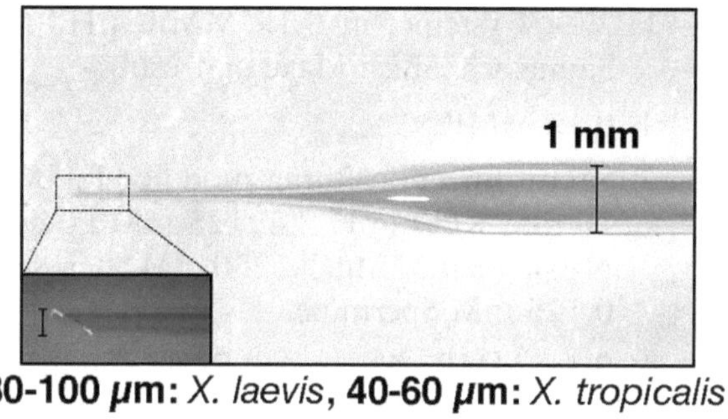

Fig. 2. Transplantation needle has a gently sloping tip and is clipped with forceps to produce a beveled, 80–100 μm wide tip for *Xenopus laevis* and a 40–60 μm wide tip for *Xenopus tropicalis*.

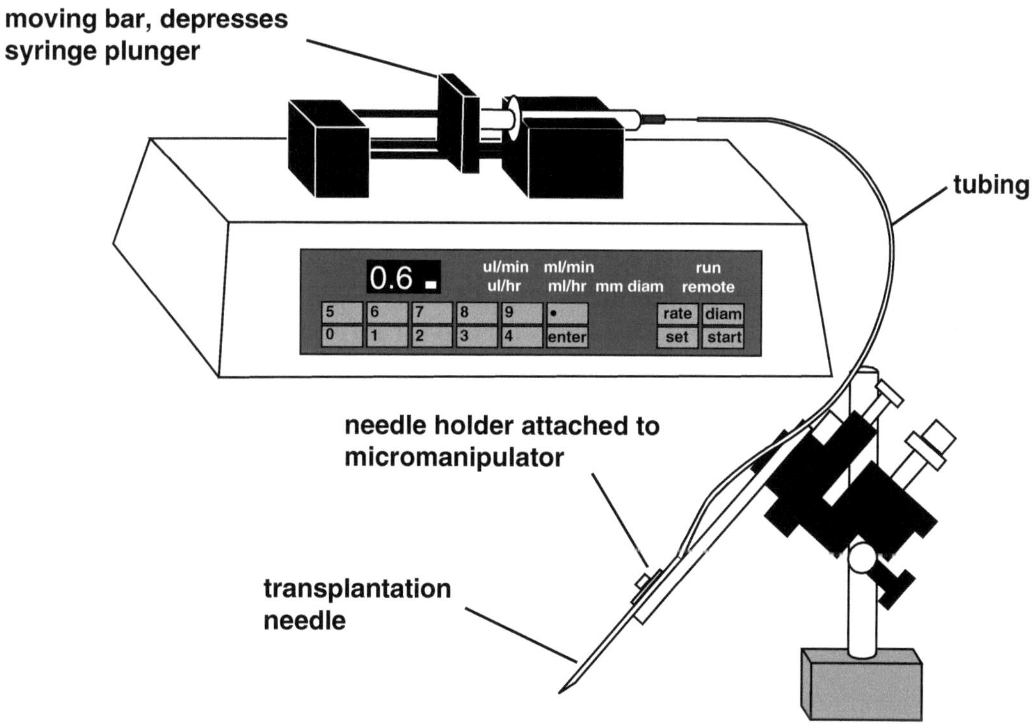

Fig. 3. We use an oil-filled injection system for the nuclear transplantations. The syringe and tubing are filled with mineral oil and the infusion pump depresses the syringe plunger, resulting in a constant, desirable flow rate (0.6 μL/min for *X. laevis*; 0.2 μL/min for *X. tropicalis*).

injection system, Harvard apparatus 22 syringe pump (NP 55-2222) with two 2.5 mL Hamilton Gas Tight Syringes and plastic tubing (ID = 0.7 mm, OD = 2.4 mm, Tygon® tubing, R3603) (Fig. 3). Two people can transplant nuclei at the same time. We use Mineral Oil (Sigma M-8410) in the system. The infusion pump allows us to control flow rate easily. We set the speed of flow at 0.6 μL/min (0.2 μL/min for *X. tropicalis*, see below).

2.4. Nuclear Transplantation Reagents and Equipment (X. tropicalis Protocol)

1. 2% L-Cysteine, in 0.1× MMR, pH.8.0 (titrate with NaOH, Sigma, C-7352). Make up freshly.
2. 0.1× MMR.
3. Sperm Injection Buffer modified for *X. tropicalis* (MOH) (3): 10 mM KPO_4, pH 7.2, 125 mM Potassium gluconate, 5 mM NaCl, 0.5 mM $MgCl_2$, 250 mM Sucrose, 0.25 mM Spermidine, 0.125 mM Spermine.
4. 0.1× MMR, 6% (w/v) Ficoll (Sigma, F-4375). Sterilize by filtration.
5. 0.01× MMR, 6% (w/v) Ficoll (Sigma, F-4375). Sterilize by filtration.
6. 0.01× MMR.
7. Linearized plasmid (100 ng/μL in water) as described in Subheading 2.3, item 8.
8. Agarose-coated injection dishes as described in Subheading 2.3, item 10.
9. Transplantation needles as described in Subheading 2.3, item 11. Needles are clipped with a forceps to produce a beveled tip of 40–60 μm for *X. tropicalis*, using the ocular micrometer of a dissecting microscope for measurement (Fig. 2).
10. Transplantation apparatus: Harvard apparatus 22 syringe pump (NP 55-2222) with two 0.1 mL Hamilton Gas Tight Syringes and Tygon tubing. The speed of flow is set at 0.2 μL/min for *X. tropicalis*.
11. Agarose-coated 24 well dishes: 1.0% agarose in 0.1× MMR is poured into 24 well dishes.
12. Agarose-coated 90 mm Petri dish. We use agarose-coated dishes to culture *X. tropicalis* embryos until they hatch at around stage 28, because their vitellin membrane is quite sticky.

3. Methods

3.1. Transgenesis Method for Xenopus laevis

3.1.1. High-Speed Extract Preparation

This protocol is an adaptation of Murray (4). Briefly, a crude cytostatic factor (CSF) arrested egg extract (cytoplasm arrested in meiotic metaphase) is prepared.

Calcium is then added to allow the extract to progress into interphase, and a highspeed spin is performed to obtain a purer cytoplasmic fraction. Cytochalasin is omitted from the protocol, since carryover of cytochalasin into the final extract used for sperm incubations interferes with normal development of transplant embryos. Use of high-speed rather than crude cytoplasmic extracts is advantageous, because high-speed extracts promote swelling of added sperm nuclei (and some chromatin decondensation), but do not promote DNA replication. Replication of sperm DNA incubated in these extracts occurs after transplantation of the nucleus into the egg rather than in the extract. High-speed extract can be stored frozen in small aliquots (at –80°C) and thawed before use.

1. Prime 8–12 female adult *X. laevis* about 3–5 days prior to HCG injection by injecting 50 U of PMSG into the dorsal lymph sac. The evening before the extract preparation begins, inject each frog with 500 U HCG and place two frogs/container into 2 L 1× MMR. Since one frog with lysing or activating eggs can compromise the whole extract preparation, we prefer to separate the frogs into pairs for the ovulation. The frogs are then placed at 15–18°C overnight (12–14 h). On the next morning, the egg quality from each container is screened before mixing all the eggs and starting the extract preparation. All the eggs released in a container with mottled, lysing, or dying eggs are left out of the extract preparation.
2. All solutions should be prepared before beginning the extract preparation, since the procedure should be carried through all steps promptly once it is initiated; optimally, the high-speed spin should begin within 45–60 min of dejellying the eggs. Gently, manually expel eggs from each frog into large beakers containing 1× MMR, and collect unbroken eggs with even pigmentation. Good eggs can also be collected from the 1× MMR in the frog buckets. Total volume of eggs should be 100 mL or greater before dejellying.
3. Remove as much MMR as possible from the eggs. Dejelly eggs in 2% cysteine in XB salts (no HEPES/sucrose). Add a small amount at a time, swirl eggs, and partially replace with fresh cysteine several times during dejellying. Remove broken eggs with a pipet during dejellying. Dejellying should be initiated separately for different batches of eggs, and batches that show breakage or egg activation are discarded. The rest of the eggs can then be combined.
4. Wash eggs in XB (with HEPES/sucrose). We use about 35 mL for each wash, and do four washes.
5. Wash eggs in CSF-XB with protease inhibitors. We do two 25 mL washes.
6. Using a wide-bore Pasteur pipet, transfer eggs into Beckman ultraclear tubes. For these volumes, we typically use 14×95 mm tubes (Beckman, 344060). If multiple tubes will be used, try to transfer an equal volume of eggs per tube. Allow the eggs to settle and remove as much CSF-XB as possible.
7. Spin for about 60 s at 1,000 rpm (150 g) in a Beckman SW 40 Ti Rotor (or similar rotor) in an ultracentrifuge. Remove the excess CSF-XB and then balance the tubes.

8. Spin the tubes for 10 min at 16,000× (10,000 rpm) at 2°C in a Beckman SW 40 Ti Rotor (or similar rotor) in an ultracentrifuge to crush the eggs. The eggs should be separated into three layers: lipid (top), cytoplasm (center), and yolk (bottom). Collect the cytoplasmic layer from each tube with an 18-gage needle by inserting the needle at the base of the cytoplasmic layer and withdrawing slowly. Transfer cytoplasm to a fresh Beckman tube on ice. If large volumes of darkly pigmented eggs are used, the cytoplasmic layer may be grayish rather than golden at this step. After a second spin to clarify this extract, it should be golden.
9. Add protease inhibitors to the isolated cytoplasm (do not add cytochalasin); recentrifuge the cytoplasm in Beckman tubes for an additional 10 min at 16,000× to clarify, again using a swinging bucket rotor. Collect the clarified cytoplasm as before. Expect to obtain 0.75–1 mL cytoplasm/batch of eggs collected from one frog.
10. Add 1/20 vol of the ATP-regenerating system (energy mix). Transfer the clarified cytoplasm into TL100.3 thick-wall polycarbonate tubes (Beckman, 349622). Tubes hold about 3 mL each and should be at least half full.
11. Add $CaCl_2$ to each tube to a final concentration of 0.4 mM; this inactivates CSF and pushes the extract into interphase. Incubate at room temperature for 15 min and then balance for the high-speed spin.
12. Spin tubes in a Beckman tabletop ultracentrifuge in a TL100.3 rotor (gold top; fixed angle) at 70,000 rpm for 1.5 h at 4°C.
13. The cytoplasm will fractionate into four layers, top to bottom: lipid, cytosol, membranes/mitochondria, and glycogen/ribosomes. Remove the cytosolic layer from each tube (about 1 mL if 2–3 mL were loaded into the tube) by inserting a syringe into the top of the tube through the lipid layer. Transfer this fraction to fresh TL-100 tubes, and spin again at 70,000 rpm for 20 min at 4°C.
14. Aliquot the high-speed cytosol supernatant into 10 μL aliquots in 0.5-mL Eppendorf tubes. Quick-freeze aliquots in liquid nitrogen, and store at −80°C until use. We typically obtain 1–2 mL of high-speed cytosol from preparations of this scale. Sperm nuclei should be incubated in an aliquot of extract and stained with Hoechst as described below in Subheading 3.1.2, step 16 to determine whether extract is effective. If active, interphase extract should cause nuclei to swell visibly (thicken and lengthen) within 10 min of addition at room temperature.

3.1.2. Sperm Nuclei Preparation (Xenopus laevis)

We follow the standard protocol of Murray (4), but omit the protease inhibitors leupeptin and phenylmethylsulfonyl fluoride from all steps to avoid transfer into the final mixture, which is diluted for egg injections. We always obtain better sperm nuclei when males are injected with hormones.

1. Prime one or two males about 3–5 days prior to HCG injection with 50 U of PMSG.
2. Inject the male or males 12–15 h before the preparation with 500 U of HCG.
3. Dissect and isolate the testes from the male:
   (a) Anaesthetize a male by immersion in 0.1% Tricaine Methanesulfonate/0.1% sodium bicarbonate for at least 20 min (immersion of the animal in ice water for 20 min may also be used), and pith it.
   (b) Cut through the ventral body wall and musculature, and lift the yellow fat bodies to isolate the two testes, which are attached to the base of the fat bodies, one on each side of the midline.
   (c) Remove the testes with dissecting scissors, and roll them on a dry paper towel to remove the blood, blood vessels and fat body.
   (d) Wash the testes briefly in a 60-mm Petri dish containing cold 1× MMR, removing any attached pieces of fat body or debris with a forceps. Take care not to puncture the testes, as this releases the sperm.
   (e) Rinse the testes in cold 1× NPB.
4. Move the cleaned testes to a dry 60-mm Petri dish, and macerate the testes well (until clumps are no longer visible to the naked eye) with a pair of clean forceps.
5. Add 2 mL cold 1× NPB and mix well by pipetting the solution up and down with a 10 mL plastic pipet.
6. Squirt the sperm suspension through four thicknesses of cheesecloth placed into a funnel, and collect the solution in a 14-mL tube (Falcon 2059).
7. Rinse the dish with an additional 3 mL of cold 1× NPB, and force this through the cheesecloth into the 14-mL tube.
8. Add 5 mL of cold 1× NPB and squeeze the cheesecloth by hand, wearing gloves, to get any remaining liquid through the funnel into the 14-mL tube.
9. Pellet the sperm by centrifugation at 3,000 rpm for 10 min at 4°C (we use a Sorvall HB-4 or similar swinging bucket rotor with the appropriate adapters). During spin, allow 1 mL of 1× NPB to equilibrate to room temperature. Note that the pellet of sperm should be primarily white, but it may contain a central

core of redness, due to pelleted erythrocytes. The erythrocytes are heavier than the sperm, so they pellet first.

10. Decant the supernatant and resuspend the sperm in 9 mL 1× NPB using a 10-mL plastic pipet and repellet by centrifugation at 3,000 rpm for 10 min at 4°C. During this spin, dissolve 1 mg of Lysolecithin in 100 μL of water (10 mg/mL) at room temperature.
11. Decant the supernatant and resuspend the pellet with a 1 mL blue tip in the 1 mL 1× NPB that has equilibrated at room temperature.
12. Add 50 μL of 10 mg/mL lysolecithin. Mix gently and incubate for 5 min at room temperature.
13. Add 10 mL of cold 3% BSA/1× NPB to the suspension to stop the reaction, and centrifuge at 3,000 rpm for 10 min at 4°C. Note that after this spin, the pellet should be softer, and more spread out, and it should not contain any redness, as the plasma membrane of the erythrocytes should have been disrupted, thus releasing the hemoglobin into the supernant.
14. Decant the supernatant and resuspend the pellet in 5 mL cold 0.3% BSA/1× NPB. Mix well by pipetting with 5-mL plastic pipet and centrifuge at 3,000 rpm for 10 min at 4°C.
15. Take supernatant carefully and resuspend the pellet in 500 μL of sperm storage buffer, and transfer suspension into a 1.5-mL Eppendorf tube.
16. Count the number of sperm nuclei using a hemacytometer: Cut off the end of a yellow tip with a razor blade and mix the sperm nuclei well by pipetting. Dilute 1 μL of the sperm nuclei with 100 μL of SDB, and add 1 μL of 1:100 diluted Hoechst stock to visualize the sperm nuclei under a fluorescence microscope. For a 1:100 dilution of our sperm stock, we typically obtain counts of 100–200 ($\times 10^4$ nuclei/mL) in 1-mm × 1-mm × 0.1-mm square of an improved Neubauer hemacytometer. At this concentration, the undiluted stock contains $1–2 \times 10^5$ nuclei/μL. If your sperm stock is substantially less concentrated (i.e., a count of <50 for a 1:100 dilution), let the sperm settle for a few hours or overnight, and remove some of supernatant. We leave fresh nuclei overnight at 4°C to allow the penetration of glycerol for best cryopreservation of the sperm. Then next day the sperm is aliquoted (20 μL/aliquot) and fast frozen in liquid nitrogen. The frozen aliquots are then stored at −80°C.

3.1.3. Transgenesis by Sperm Nuclear Transplantation into Unfertilized Eggs from Xenopus laevis

1. Prime two females 3–5 days before HCG injection with 50 U of PMSG.
2. Inject the females with 500 U of HCG, 12–15 h before egg collection.

3. Allow an aliquot of SDB to equilibrate to room temperature, and make up 2% Cysteine in 1× MMR, pH 8.0.
4. Turn on the switch of the Harvard Apparatus and start the infusion pump to get the flow to stabilize before injection.
5. Set up a reaction using a clipped yellow tip: mix 4 μL sperm stock (~4–8 × 10^5 nuclei) and 1–2 μL linearized plasmid (100 ng/μL), and incubate for 5 min.
6. Dilute 0.5 μL of a restriction enzyme in 4.5 μL of water, and mix 1 μL of the diluted enzyme with 18 μL of SDB, 2 μL of 100 mM $MgCl_2$, and 2 μL of high-speed egg extract.
7. Add the mixture to the sperm/DNA and mix well by gentle pipetting (using a clipped yellow tip). Incubate for 15 min at room temperature.
8. During the reaction, collect eggs in a beaker by squeezing frogs and dejelly them in 2% cysteine/1× MMR, pH 8.0. This usually takes about 10 min, so by the time the eggs are ready, the reaction is nearly complete. Squeeze eggs directly into a dry beaker and add cysteine immediately to keep egg quality. One may need to dejelly eggs from different females separately, as dejellying times vary or poor quality eggs make the solution dirty (see Note 2).
9. Wash the dejellied eggs with 1× MMR at least three times, and transfer the eggs to injection dishes containing 0.4% MMR/6% Ficoll using a wide-bore Pasteur pipet. We generally fill the square space with eggs so that no gap is left between the eggs. After about 5 min in 0.4× MMR, 6% Ficoll, the eggs will pierce easily. Transplantation should be performed at around 16°C. To achieve this we place the injection dish on the plastic box half-filled with ice and place the box and eggs under the injection microscope.
10. After the incubation with extracts, mix the sperm nuclei gently by pipetting with a clipped yellow tip. Then transfer 5 μL of the reaction into 150 μL of SDB that has equilibrated at room temperature.
11. Mix well but avoid making bubbles, using a clipped yellow tip with a piece of plastic tube attached (Fig. 4a). Fill the clipped yellow tip with the diluted sperm suspension, carefully detach the clipped yellow tip, keeping the tip horizontal and backfill a transplantation needle by attaching it to the tube (Fig. 4b). You can keep the yellow tip with the remaining nuclei, by placing it horizontally, in case you need to load another needle. Keep decondensed sperm nuclei at room temperature and transplant them within an hour, but preferably within 30 min.
12. Attach the needle to the tube filled with mineral oil that is connected to the syringe on the Harvard Apparatus.

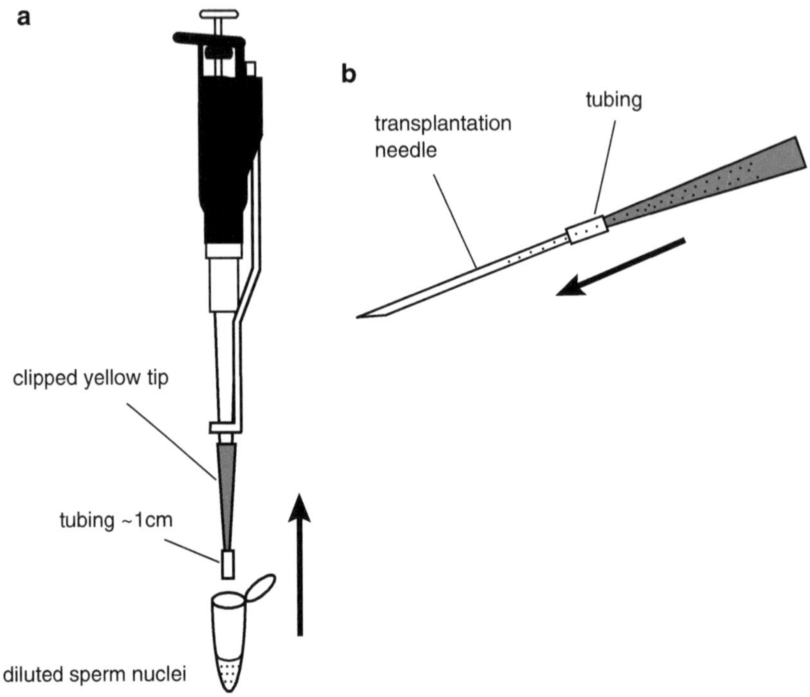

Fig. 4. Method for backfilling nuclei into a transplantation needle. (**a**) The diluted reaction mix is drawn into a clipped yellow tip containing about 1 cm of Tygon tubing. (**b**) The yellow tip containing the Tygon tubing and dilute sperm nuclei is carefully detached from the pipetman and connected to the back of the needle using the tubing. The needle is gently loaded with the dilute sperm nuclear reaction by gravity. This is done by slowly increasing the angle of the yellow tip/tubing/needle so that the mixture flows gently into the needle. Once the needle is completely filled, the needle is detached from the yellow tip and the needle is ready to connect to the infusion pump. The remaining sperm mixture can be set aside horizontally and used to reload another needle, if two people are injecting simultaneously or if a needle is accidentally damaged or blocked.

13. Check the flow and start injecting. Keep the needle inside each egg for approximately 0.5 s, and move the needle fairy rapidly from egg to egg piercing the plasma membrane of each egg with single, sharp motion. We usually transplant for about 15–20 min. If needle is blocked by debris during transplantation, change needle or try to fix by pinching the tube or cutting the tip of needle using forceps.
14. After injection, incubate embryos at 16°C.
15. When the embryos reach the 4-cell stage (about 3–4 h after injection at 16°C), gently transfer normally dividing embryos to 10-cm Petri dish containing 0.1× MMR, 6%Ficoll, 10 μg/mL gentamycin using a wide-bore Pasteur pipet (see Note 3).
16. The next day, when embryos are around stage 12, transfer healthy embryos to a new 90-mm Petri dish containing 0.1× MMR, 10 μg/mL gentamycin without Ficoll. Because of the large needle tip used for transplantations, embryos often develop large blebs at the site of injection. These blebs occur

when cells are forced out of the hole left in the vitelline membrane at the injection site, but they generally do not affect development. The blebs usually fall off at the neurula or tailbud stages.

17. Incubate embryos until they reach the stage that you need to analyze at 14–22°C. Approximately 10–30% of embryos proceeded to post-gastrula show stable expression of transgenes (see Note 4).

3.2. Transgenesis Method for Xenopus tropicalis

X. tropicalis has a diploid genome and shorter generation time than *X. laevis*, but retains many of the advantages of *X. laevis* (3, 5). The basic protocol for transgenesis in *X. tropicalis* is very similar to the procedure used for *X. laevis.*

3.2.1. High-Speed Extract Preparation

We use a high-speed egg extract from *X. laevis* (see Subheading 3.1.1) in the protocol for *X. tropicalis.*

3.2.2. Sperm Nuclei Preparation (X. tropicalis)

1. Prime two males the day before the sperm nuclei preparation with 10 U of PMSG.
2. Inject the males 3 h before the nuclei preparation with 50 U of HCG.
3. Dissect and isolate the testes from the male as described in Subheading 3.1.2. Take care not to puncture the testes, as this releases the sperm. This is particularly important for *X. tropicalis.*
4. Move the cleaned testes to a dry 60-mm Petri dish, and macerate the testes well (until clumps are no longer visible to the naked eye) with a pair of clean forceps.
5. Add 2 mL cold 1× NPB and mix well by pipetting the solution up and down with a 10 mL plastic pipet.
6. Squirt the sperm suspension through four thicknesses of cheesecloth placed into a funnel, and collect the solution in a 14-mL tube (Falcon 2059).
7. Rinse the dish with an additional 3 mL of cold 1× NPB, and force this through the cheesecloth into the 14-mL tube.
8. Add 5 mL of cold 1× NPB and squeeze the cheesecloth by hand, wearing gloves, to get any remaining liquid through the funnel into the 14-mL tube.
9. Pellet the sperm by centrifugation at 3,000 rpm for 10 min at 4°C (Sorvall HB-4 or similar swinging bucket rotor with the appropriate adapters). During spin, allow 1 mL of 1× NPB to equilibrate to room temperature.
10. Decant the supernatant and resuspend the sperm in 9 mL 1× NPB using a 10-mL plastic pipet and repellet by centrifugation at 3,000 rpm for 10 min at 4°C. During this spin, dissolve 1 mg of Lysolecithin in 100 μL of water (10 mg/mL) at room temperature.

11. Decant the supernatant and resuspend the pellet with a 1 mL blue tip in the 1 mL 1× NPB that has equilibrated at room temperature.
12. Dilute 20 μL of 10 mg/mL lysolecithin with 980 μL of water (0.2 mg/mL) and add 50 μL of that diluted lysolecithin to the sperm. Mix gently and incubate for 5 min at room temperature.
13. Add 10 mL of cold 3% BSA/1× NPB to the suspension to stop the reaction, and centrifuge at 3,000 rpm for 10 min at 4°C.
14. Decant the supernatant and resuspend the pellet in 5 mL cold 0.3% BSA/1× NPB. Mix well by pipetting with 5-mL plastic pipet and centrifuge at 3,000 rpm for 10 min at 4°C.
15. Take supernatant carefully and resuspend the pellet in 100 μL of sperm storage buffer, and transfer suspension into a 1.5-mL Eppendorf tube.
16. Count of the sperm nuclei: For a 1:100 dilution of our sperm stock, we typically obtain counts of ~80 (×10^4 nuclei/mL) in 1-mm × 1-mm × 0.1-mm square of an improved Neubauer hemacytometer. At this concentration, the undiluted stock contains 8×10^4 nuclei/μL, and we use 10 μL of that for a reaction. If the sperm stock is substantially less concentrated, let the sperm settle for a few hours or overnight, and remove some of supernatant. If the count is still low, you can use up to 30 μL of the sperm nuclei per reaction. Sperm nuclei can be frozen in aliquots at −80°C.

3.2.3. Nuclear Transplantations (*Xenopus tropicalis*)

1. Prime two females the day before the nuclear transplanations with 15 U of PMSG. Inject the females 3–4 h before the nuclear transplantation with 75 U of HCG and leave frogs at 22–24°C. We use two females for two rounds of the nuclear transplantation a day. If more rounds of transgenesis are needed in the same day, inject another two females 1 or 2 h after the first pair. This will allow for four rounds of transgenics, each around 1 h apart.
2. Allow an aliquot of MOH buffer to equilibrate to room temperature, and make up 2% Cysteine in 1× MMR, pH 8.0.
3. Turn on the switch of the Harvard Apparatus and start the infusion pump.
4. Set up a reaction: Take 1–2 μL of linearized DNA into a 1.5-mL tube. Using a clipped yellow tip, mix sperm nuclei stock and take 10 μL of sperm nuclei (~8×10^5 nuclei) to mix with the DNA.
5. Incubate the mixture for 5 min at room temperature.
6. Add 2 μL of egg extract to the sperm/DNA and mix well by gentle pipetting using a clipped yellow tip. Incubate for 15 min at room temperature.

7. During the reaction, collect eggs in a beaker by squeezing frogs and dejelly them in 2% Cysteine/0.1× MMR, pH 8.0.
8. Wash the dejellied eggs with 0.1× MMR three times and transfer the eggs to injection dishes containing 0.1% MMR/6% Ficoll. We generally fill the half of square space with eggs, because we need to inject eggs within 30 min.
9. After 15 min incubation, dilute all the reaction mix with 130 μL of MOH that has equilibrated at room temperature. Mix well and backfill a needle using a clipped yellow tip with a piece of tube. If you use more than 10 μL of the sperm (up to 30 μL), amount of MOH buffer should be reduced. The concentration of nuclei must be considerably higher with *X. tropicalis*, since the volume injected into each egg is about 1/5 the amount used for *X. laevis*.
10. Attach the needle to the injection system and inject eggs checking the flow regularly. Transplantation is performed at room temperature (22–24°C).
11. Incubate embryos at 22–24°C.
12. When embryos reach at the 4-cell stage, collect normally dividing embryos and transfer them to agarose-coated 24 well dishes (1–2 embryo/well) containing 0.01× MMR/6% Ficoll. Embryos reach the 4-cell stage approximately 1.5 h after injection at 22–24°C, therefore it may be necessary for one person to sort the embryos while others continue later rounds of injections (i.e., third and fourth rounds of injections). Culture embryos overnight at 22°C.
13. The next day transfer healthy embryos to an agarose-coated 90 mm Petri dish containing 0.01× MMR, and culture embryos at 22°C.

4. Notes

1. Generally the same restriction enzyme used to linearize the transgene is used in the transgenic reaction before the nuclear transplantation. However, a different enzyme, which was not used to linearize the transgene, can still be used for the reaction, as long as it does not cut within the transgene. We usually use enzymes, NotI, SalI, or SfiI purchased from Roche. Asp700 (XmnI) and SwaI can also be used. Asp700 digests once within Ampicilin resistant gene. We often use SwaI when the construct is generated by Tol2 kit (6) or pTransgenesis (7), which digests once within Tol2 transposon element. SwaI sites are generally quite rare in constructs. Some calibration may be required to determine the optimal amount of enzyme to add

to each reaction, as too much enzyme may adversely affect the development of embryos derived from nuclear transplantations. To co-integrate two different constructs into the genome, both DNAs can be digested with either the same enzyme or with different enzymes, without loss of co-integration efficiency. Although it is possible to omit restriction enzyme from the transgenic reaction, this usually results in a lower efficiency of transgenesis. This is desirable for some applications such as establishment of transgenic lines, as embryos generated from reactions that do not include restriction enzyme tend to develop to postmetamorphic and adult stages with fewer anomalies.

2. Egg quality is one of the factors that can affect transgenesis efficiency. Needles for nuclear transplantation are much larger than those used for RNA injections, in order to not damage swollen nuclei. Therefore eggs need to be of high quality to recover after injection. Eggs used for nuclear transplantation should, in particular, have a firm cortex and maintain a regular round shape.

3. One can determine whether the sperm dilution and the flow rate used for injections were appropriate by watching the first cleavage of the transplanted eggs. If few of the eggs received a nucleus, the frequency of cleavage will be low; one-third of our transplantations typically result in normally cleaving embryos. Eggs that were injected with more than one nucleus will divide at the time of first cleavage abnormally into three or four (or more) cells. Many of these embryos will develop to blastula stages, but most fail during gastrulation; in some, a region of the embryo will fail to cellularize and die. Eggs injected with multiple nuclei that do gastrulate usually do so abnormally; typically, blastopore closure is incomplete, resulting in embryos that form two wings of somites and neural tissue on each side of the exposed yolky tissue lying in the center of the trunk. This type of gastrulation failure is also common to stressed or unhealthy embryos (particularly embryos derived from "soft" eggs).

4. One of the main factors affecting the efficiency of transgenesis is the amount of chromosomal damage sustained by sperm nuclei prior to transplantation. Chromosomal damage can occur during sperm nuclear preparation and is also deliberately induced by addition of restriction enzyme to the sperm, extract, and plasmid reaction mixture used for transplantations. Nuclei with very little chromosomal damage will promote a much higher frequency of normal embryonic development to late stages. However, nuclei that have sustained more chromosomal damage give rise to transgenic embryos at higher frequencies. Thus a balance must be achieved between attaining

normal development of embryos vs. promoting high frequency transgenesis. We generally use a procedure that results in a lower (10–30%) efficiency of transgenesis, but with improved overall survival of the embryos.

References

1. Kroll KL, Amaya E (1996) Transgenic Xenopus embryos from sperm nuclear transplantations reveal FGF signaling requirements during gastrulation. Development 122(10):3173–3183
2. Hartley KO, Nutt SL, Amaya E (2002) Targeted gene expression in transgenic Xenopus using the binary Gal4-UAS system. Proc Natl Acad Sci U S A 99(3):1377–1382
3. Offield MF, Hirsch N, Grainger RM (2000) The development of *Xenopus tropicalis* transgenic lines and their use in studying lens developmental timing in living embryos. Development 127(9):1789–1797
4. Murray AW (1991) Cell cycle extracts. Methods Cell Biol 36:581–605
5. Amaya E, Offield MF, Grainger RM (1998) Frog genetics: *Xenopus tropicalis* jumps into the future. Trends Genet 14(7):253–255
6. Kwan KM, Fujimoto E, Grabher C, Mangum BD, Hardy ME, Campbell DS, Parant JM, Yost HJ, Kanki JP, Chien CB (2007) The Tol2kit: a multisite gateway-based construction kit for Tol2 transposon transgenesis constructs. Dev Dyn 236(11):3088–3099
7. Love NR, Thuret R, Chen Y, Ishibashi S, Sabherwal N, Paredes R, Alves-Silva J, Dorey K, Noble AM, Guille MJ, Sasai Y, Papalopulu N, Amaya E (2011) pTransgenesis: a cross-species, modular transgenesis resource. Development 138(24):5451–5458

Chapter 12

A Simple Method of Transgenesis Using I-*SceI* Meganuclease in *Xenopus*

Shoko Ishibashi, Nick R. Love, and Enrique Amaya

Abstract

Here we present a protocol for generating transgenic embryos in *Xenopus* using I-*SceI* meganuclease. This method relies on integration of DNA constructs, containing one or two I-*SceI* meganuclease sites. It is a simpler method than the REMI method of transgenesis, and it is ideally suited for generating transgenic lines in *Xenopus laevis* and *Xenopus tropicalis*. In addition to it being simpler than the REMI method, this protocol also results in single copy integration events rather than tandem concatemers. Although the protocol we describe is for *X. tropicalis*, the method can also be used to generate transgenic lines in *X. laevis*. We also describe a convenient method for designing and generating complex constructs for transgenesis, named pTransgenesis, based on the Multisite Gateway® cloning, which include I-*SceI* sites and Tol2 elements to facilitate genome integration.

Key words: I-*SceI* meganuclease, Transgenesis, Gateway

1. Introduction

1.1. Overview of Transgenesis Procedure Using I-SceI Meganuclease

Restriction enzymes have been used for more than a decade for producing transgenic animals in several species (1–6). Restriction enzymes stimulate recombination by creating double-strand breaks, facilitating integration of DNA constructs into the genome. The amount of the enzyme must be precisely controlled as the restriction enzymes may cause deletions, translocations, and inversions within the host genome.

I-*SceI* is an endonuclease from *Saccharomyces cerevisiae* that recognizes a 18-bp non-palindromic sequence (7, 8). Because of its rare recognition site, Thermes and colleague utilized I-*SceI* meganuclease for generating transgenic medaka fish, by injecting a DNA construct flanked by I-*SceI* recognition sites together with I-*SceI* meganuclease into early medaka embryos (9). Although the

Stefan Hoppler and Peter D. Vize (eds.), *Xenopus Protocols: Post-Genomic Approaches, Second Edition*,
Methods in Molecular Biology, vol. 917, DOI 10.1007/978-1-61779-992-1_12, © Springer Science+Business Media, LLC 2012

mechanism by which I-*Sce*I facilitates the integration of transgenes into the genome is not clear, it has subsequently been shown to facilitate the generation of transgenic animals in various species, including *Xenopus* (10–13). Due to its size, I-*Sce*I sites are expected to be found only once every 7×10^{10} bps of random sequence. As yet, no endogenous I-*Sce*I recognition site has been found in any vertebrate genome. Therefore it is unlikely that I-*Sce*I cuts the genome of embryos injected with the enzyme. Given that the I-*Sce*I enzyme has been found to remain bound to the ends of the linear DNA fragments, after cleavage, perhaps this interaction protects the ends of the DNA from degradation, thus facilitates integration of transgenes into the genome (14). This is consistent with the finding that I-*Sce*I-mediated integration events tend to be single copy rather than long concatemers (9). An additional feature of this enzyme, which may help explain its stimulatory effect on genomic integration, is that the enzyme is able to tolerate some sequence degeneracy within the recognition sequence (8). Therefore it is possible that I-*Sce*I is able to create double-stranded breaks in the genome without the exact recognition site, and in this way it may stimulate recombination.

To optimize the transgenesis protocol for *Xenopus tropicalis*, several constructs containing I-*Sce*I sites were tested. We found that a single I-*Sce*I site is sufficient for the transgenes to integrate into the genome, and, in fact, we found that constructs containing one site integrated more efficiently than constructs containing two inverted I-*Sce*I sites (Fig. 1).

1.2. Construction of I-SceI Transgenesis Compatible Transgenes Using Gateway® Multisite Cloning

Designing and constructing of transgene constructs are essential and the first task in generating transgenic animals. The traditional cloning method comprises digestion of DNA using restriction enzymes, purification of DNA, and ligation of DNAs. Although PCR-based techniques have significantly facilitated the cloning strategies available, each step remains time consuming, especially as the size of constructs increases. Generally transgenesis constructs consist of several elements. Minimally constructs contain a promoter and a protein coding gene. To distinguish transgenic animals from non-transgenic animals, it is useful to include a selection marker such as γ1-crystallin–GFP on the transgene constructs. Furthermore, since the expression of the transgene may be affected by the site of integration, additional elements, such as insulators, may prove valuable. Finally sites that facilitate genome integration, such as I-*Sce*I recognition sequences (and/or Tol2 elements, see Chap. 14), are essential in creating transgenic animals. Such complicated multi-element constructs are time consuming to generate using conventional cloning methods. Therefore we have developed a recombinational cloning method based on the Gateway® system, which we have named pTransgenesis, to facilitate the production of complex trasngenesis vectors (15). The Gateway® system utilizes

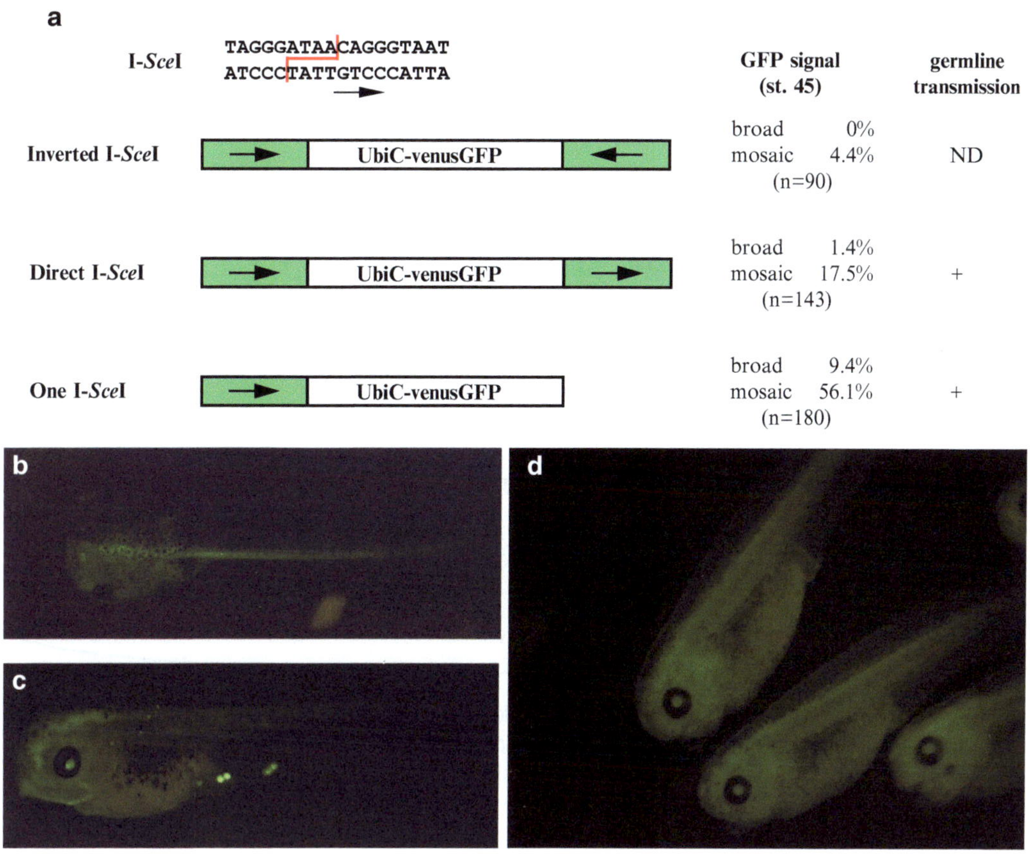

Fig. 1. One I-*Sce*I site is enough for the integration of transgenes into the genome. (**a**) A schematic diagram of I-*Sce*I constructs carrying UbiC promoter-driving venus GFP tested and frequencies of the GFP expression observed in surviving embryos after co-injection with I-*Sce*I. (**b**, **c**) An example embryo showing a broad expression of GFP after injection with Direct I-*Sce*I construct (**b**) and with One I-*Sce*I construct (**c**). (**d**) F1 embryos obtained by crossing a F0 founder animal injected with One I-*Sce*I construct with wild type, showing germline transmission.

in vitro site-specific recombination, which offers a rapid and convenient way for constructing complex, multi-element vectors without the need for restriction enzymes and ligases (16). Therefore we incorporated MultiSite Gateway Three Fragment Vector Construction Kit® (Invitrogen, La Jolla, USA) which is commercially available and allows for assembly of three different constructs (entry clones) into a destination vector. As shown in Fig. 2, we refer to three entry vectors as position 1 (p1), containing a selection marker flanked by *att*L4-*att*R1 Gateway recombination sites; position 2 (p2), containing a promoter flanked by *att*L1-*att*L2 sites; position 3 (p3), containing a reporter or coding sequence flanked by *att*R2-*att*L3. The destination vector, containing an I-*Sce*I site and two Tol2 elements (17) (see Chap. 14) flanking the recombination cassettes, is referred to as position 4 (p4).

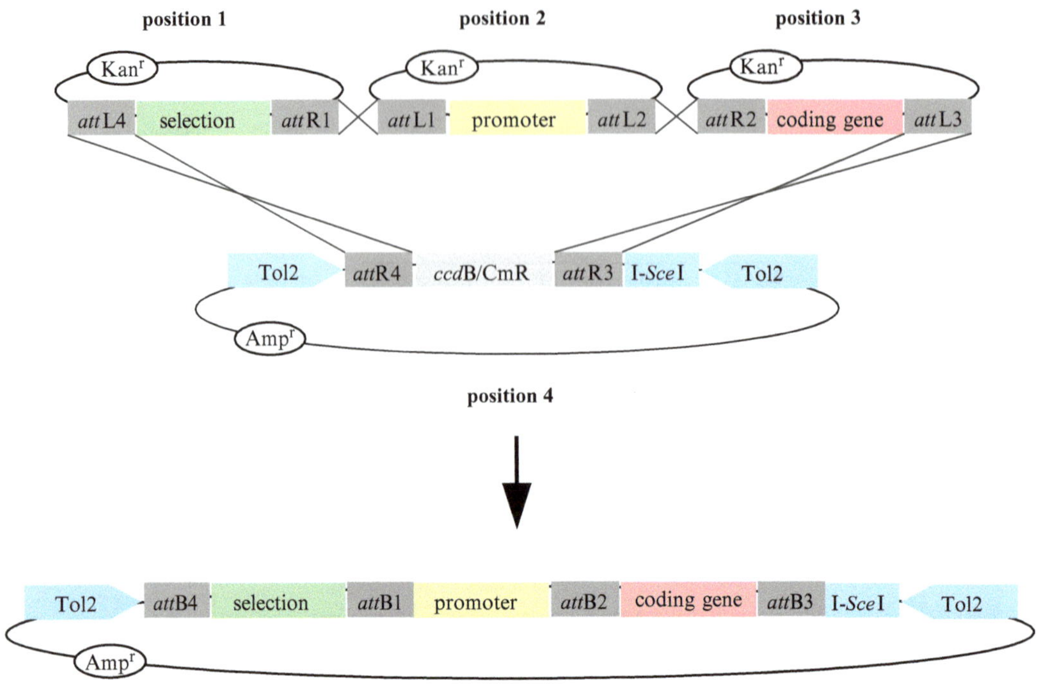

Fig. 2. Schematic representation of the pTransgenesis MultiSite Gateway recombination system for generating transgenic constructs. Position 1 (p1) contains the selection marker, Position 2 promoter, Position 3 coding sequence of transgene and Position 4 containing an I-*Sce*I site and two *Tol2* elements flanking the entire transgene construct.

2. Materials

1. Flaming/Brown Micropipette Puller (e.g., Model P-87, Sutter Instruments Co.).
2. Borosilicate glass capillaries (e.g., GC100TF-10, 1.0 mm OD × 0.78 mm ID, Harvard Apparatus), for pulling needles using these settings, $p = 50$, $v = 100$, and $t = 5$.
3. Microinjector (e.g., Picospritzer II, General Valve Corporation).
4. Dissection microscope (e.g., MZ6, Leica).
5. Micromanipulator.
6. Fluorescence dissection microscope (e.g., MZFLIII, Leica) with filters.
7. 60 mm Petri dish with mesh.
8. Microloader pipette tip (Eppendorf, 930001007).
9. Dissection tools and forceps.
10. I-*Sce*I meganuclease (NEB, R0694S or R0694L, see Note 1).
11. Plasmid purification kit, Qiagen kit or an equivalent.

12. Plasmid DNA containing the I-*Sce*I site (see Note 2).
13. 10× MMR (Marc's Modified Ringer) stock solution (1 M NaCl, 20 mM KCl, 10 mM MgCl2, 20 mM $CaCl_2$, 50 mM HEPES, pH 7.5).
14. 0.1× MMR.
15. Injection buffer: 0.1× MMR/2% Ficoll 400.
16. 0.01× MMR.
17. 90 mm petri dish with 1% Agarose/0.1× MMR.
18. 2% L-Cysteine, pH 8.0 (Sigma, C-7352).
19. Pregnant mare serum gonadotropin (PMSG, 100 U/mL (e.g., P.G.600®, Intervet, Inc., 021825). Dissolve in water and stored at −20°C).
20. Human chorionic gonadotropin (hCG, 500 U/mL (e.g., CHORULON®, Intervet, Inc., 057176). Dissolve in water and stored at 4°C).
21. LR Clonase® II Plus enzyme (Invitrogen, 12538–120).
22. Three entry clones and a destination vector.
23. DNA analysis software (e.g., ApE and details).

3. Method

Here we describe first the I-*Sce*I-mediated transgenesis protocol in *X. tropicalis*. The protocol involves coinjection of plasmid DNA containing I-*Sce*I sites and I-*Sce*I meganuclease. The generation of transgene constructs using MultiSite Gateway® kit is described in Subheading 3.2.

3.1. I-SceI Transgenesis Protocol

1. Inject females and males with 15 and 10 units of PMSG, respectively, on the day before the experiment. We routinely use four females and two males for four rounds (four fertilizations) of injections per day.
2. Four hours before the experiment, inject two females and one male with 75 and 50 units of hCG, respectively, 1–1.5 h later, inject the other females and male with hCG.
3. Set up the reaction mix without I-*Sce*I restriction enzyme as shown in Table 1 (see Note 3).
4. When the females start laying eggs, isolate testes from the male. It is very critical not to puncture the testes during the dissection, as this may decrease the efficiency of fertilization. One testis is used for the first fertilization and the other can be stored in a 1.5 mL tube filled with L-15/10% FCS solution on ice until it is used in the second fertilization (18).

Table 1
I-*Sce*I reaction mix

Stock solution	Volume (10 μL in total)	Final concentration
10× I-*Sce*I buffer (NEB)	1 μL	1×
1/10 BSA (NEB)	1 μL	1× (100 μg/ml)
DNA (100 ng/μL)	0.5–1 μL	5–10 ng/μL (Note 4)
Water	6.5–6 μL	
I-*Sce*I (5 units/μL, NEB)	1 μL	5 units

5. Manually extrude eggs from the females onto an empty 90 mm petri dish. It is important not to allow water to transfer to the dish during the egg extrusion.
6. Macerate and rub one of the testes throughout the freshly collected eggs with forceps. Leave the sperm/egg mixture for 5 min.
7. Flood the eggs with 0.1× MMR.
8. Around the same time, add I-*Sce*I to the DNA reaction mix and incubate at 37°C for 10 min (see Note 5).
9. 10 minutes after flooding the embryos with 0.1× MMR, dejelly the fertilized eggs with 2% Cysteine pH.8.0. This usually takes about 5 min.
10. During dejellying, backfill the needles with the reaction mixture using a pipetman with a microloader tip, and set needle into a micromanipulator. Clip the tip of the needle and calibrate the volume of injection to 2 nL per embryo, by adjusting the time and pressure of the injector.
11. Wash the dejellied fertilized eggs with 0.1× MMR three times and transfer to the petri dish with mesh containing injection buffer (0.1× MMR/2% Ficoll 400 (Fig. 3)).
12. Inject into one-cell embryos. Since early integration makes less mosaic animals, we inject only 10–20 embryos per fertilization.
13. During the preparation of the next reaction and fertilization, the first round of injected embryos can be transferred to a petri dish coated with 1% agarose/0.1× MMR, filled with 0.01× MMR.
14. The injected embryos should be kept at 22°C, the lowest permissible incubation temperature for *X. tropicalis*. The time flow of a typical experiment is shown in Table 2.
15. Screen injected embryos for expression of either or both, transgene and screenable marker. We choose embryos showing broad expression patterns and raise only these to the next generation (see Note 6).

Fig. 3. *X. tropicalis* embryos on nylon mesh positioned in a 60 mm petri-dish, ready for injection.

Table 2
Time flow of method

	Make a reaction mix without I-*Sce*I enzyme Isolate testes Manually extrude eggs from female Add testis onto the eggs Add I-*Sce*I enzyme	Add buffer to eggs
10 min	Incubate at 37°C	Incubate at room temperature
11 min~	Dejelly embryos	Load a needle/clip the needle
15–25 min	Transfer ~20 embryos into the injection dish with mesh Inject with 2 nL I-*Sce*I reaction mix	

3.2. Construction of Transgene Constructs Using the MultiSite Gateway Kit®

3.2.1. Design and Construction of Individual Vectors

We have incorporated a substantial number of DNA sequences commonly used in transgenesis into the pTransgenesis system, which, notably, supports I-*Sce*I-mediated transgenesis using the "p4 Tol2 I-*Sce*I" and similar vectors (see Table 3 and Note 7; (15)). These vectors were engineered to contain a single I-*Sce*I site. Moreover, we routinely utilize constructs containing the p1 γ1-crystallin or p1 αCE2x6 constructs, which drive various colors of fluorescent proteins in the eye. We have generated numerous transgenic *X. tropicalis* lines using such pTransgenesis derived constructs, and some of these lines are now in the F2 generation.

Table 3
Currently available constructs in the pTransgenesis system

P1 selection		
p1 γ-crystallin GFP	*X. laevis* γ1-crystallin promoter driving GFP	(19)
p1 γ-crystallin RFP	*X. laevis* γ1-crystallin promoter driving RFP	
p1 elastase GFP	Rat elastase promoter driving GFP	(20)
p1 αCE2X6 eCFP	Chicken α-crystallin gene enhancer and β-globin promoter driving eCFP	(21)
P2 promoter		
p2 CMV	CMV promoter (pCS2+)	(22)
p2 CAG	CAG promoter	(23)
p2 UbiC	Human ubiqutin C promoter	(24)
p2 Rosa26	Mouse Rosa26	(25)
p2 NBT	*X. laevis* neural specific β-tubulin promoter	(1)
p2 Pax6	*X. laevis* pax6 promoter	(26)
p2 CarA	*X. laevis* cardiac actin promoter	(1)
p2 Flk-1	*X. laevis* Flk1 promoter	(27)
p2 Lurp-1	*X. laevis* Lurp-1 promoter	(28)
p2 14XUAS	14XUAS E1b	(29)
p2 Nectin-2	*X. tropicalis* Nectin-2 promoter	(15)
p2 Vimentin	*X. tropicalis* Vimentin promoter	(15)
P3 coding gene		
p3 mCerulean	mCerulean	(30)
p3 eCFP	eCFP	(31)
p3 mTFP	mTFP	(32)
p3 Midori-ishi cyan	Midori-ishi cyan (MiCy)	(33)
p3 HyPer	Indicator of hydrogen peroxide	(34)
p3 KikGR	KikGR	(35)
p3 Kaede	Kaede	(36)
p3 eYFP	eYFP	(37)
p3 Venus	Venus	(38)
p3 Kusabira-Orange (KO)	Kusabira-Orange (KO)	(33)
p3 tdTomato	tdTomato	(39)
p3 DsRed2-nuc	Nuclear localised DsRed2	Clontech
p3 mCherry	mCherry	(39)
p3 mKate	Monomeric KatushkaRFP	(40)
p3 myc-KatushkaRFP	myc-tagged KatushkaRFP	(40)
p3 Gal4	Gal4	(41)
p3 MerCreMer	Cre recombinases carrying a mutated hormone binding domain at both the N and C-termini	(42)

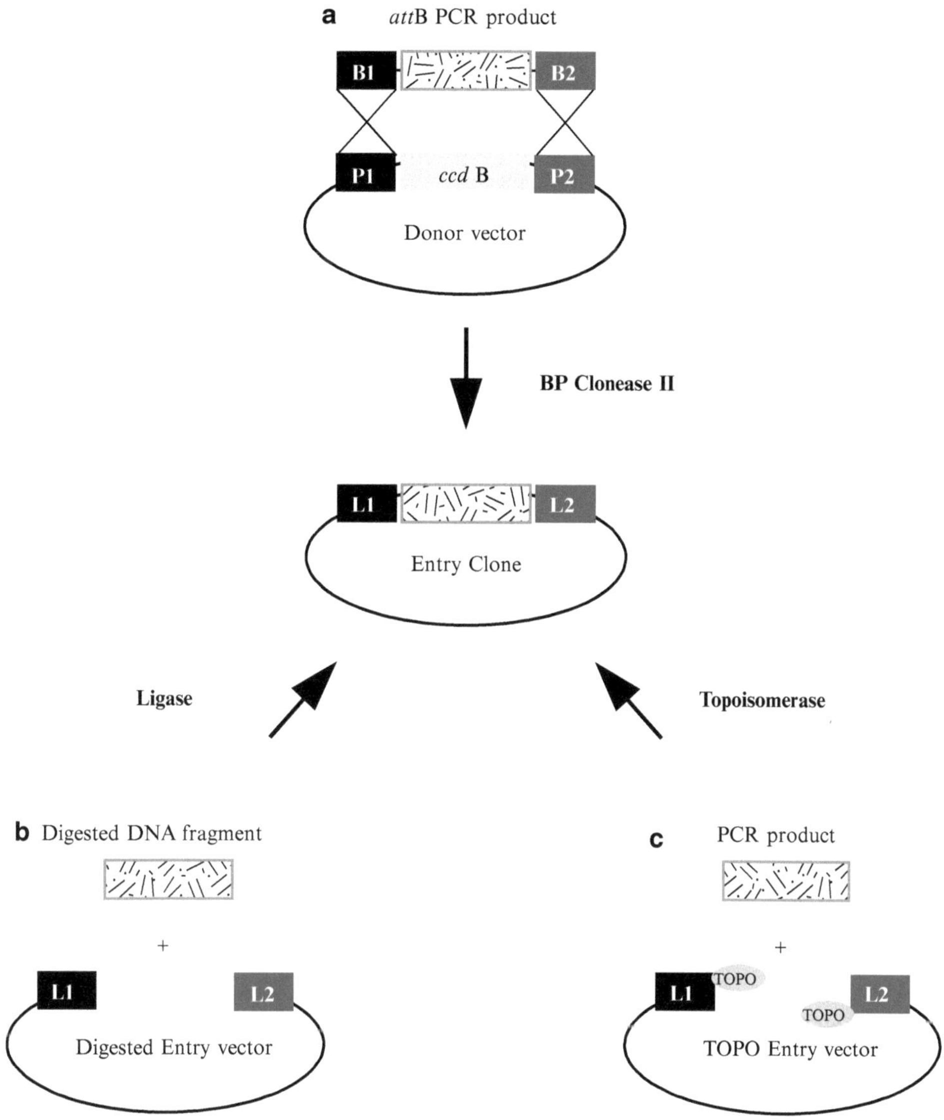

Fig. 4. Schematic representation of the methods to generate entry clones. (**a**) PCR product flanked by *att* recombination sites can be cloned into a donor vector by BP reaction, generating an Entry clone. (**b**) Cloning by restriction enzyme and ligation can also be used. (**c**) PCR product can be directly cloned to an Entry vector by TOPO® TA cloning. In this figure, it shows position 2 as an example.

Depending on the experimental design, one can either chose available entry clones or make a new entry clone (see Table 3 and Fig. 4). To make a p1, p2, p3 entry clone, one can amplify a sequence of interest via PCR with primers, which contain sequences compatible in a subsequent "BP" recombination (another recombinase, BP Clonase II, see Fig. 4a) with a p1, p2, p3 "DONR" vector (Invitrogen). Sometimes, however, it is easier to use traditional subcloning methods to rearrange or redesign an existing p1, p2, p3 clone (see Fig. 4b). In the case of the p2 position, a PCR product with TA overhangs can be cloned directly using a TOPO-based kit from Invitrogen (see Fig. 4c; pCR®8/GW/TOPO® TA Cloning Kit). This kit renders promoter cloning and testing highly efficient as it allows recombination with transgene coding sequences found in p3 vectors.

In the next section, we provide the protocol for performing pTransgenesis LR recombination reactions.

3.2.2. LR Recombination Reactions

The most critical point for the reaction is adding the correct molar amounts of the four vectors. We make "working stocks" of each vector such that 1 μL of a p1, p2, and p3 working stock contains 10 fmol, and for p4 vectors, 20 fmol. These dilutions are performed in TE pH 8.0 (NB we also dissolve the originally prepped DNA pellets in TE pH 8.0).

Working stocks can be calculated from the following formula (remember that the average molecular weight of 1-bp DNA is 660 g/mol and hence 660 fg/fmol):

$$\text{ng} = (x\,\text{fmol})(N)(660\,\text{fg/fmol}) / (1\,\text{ng}/10 \times 6\,\text{fg})$$

where N= bp length of a given plasmid
x= fmol required (10 for p1, p2, p3, 20 for p4)

For example, to make a p1 working stock of a 3333 bp plasmid:

$$(10\,\text{fmol})(3333)(660\,\text{fg/fmol}) / (1\,\text{ng} / 10 \times 6\,\text{fg}) =\sim 22\,\text{ng}$$

Hence, a ~22 ng/μL solution equals 10 fmol/μL for this plasmid.

1. Add the following to a 1.5-mL microcentrifuge tube:
   p1, p2, p3, p4 working stocks 1 μL each = 4 μL total
   TE pH 8.0 4 μL
2. Mix and then take 4ul of this 8ul mixture into a 1.5-mL microcentrifuge tube and add 1 μL of LR Clonase II Plus enzyme, vortex, spin down, and incubate the reaction at ~23°C overnight (NB: we store aliquots of the enzyme at –80°C).
3. Add 0.5 μL of the Proteinase K solution to the reaction. Incubate for 10 min at 37°C.
4. Transform a suitable *E. coli* host and plate (usually on ampicillin). We routinely use 2.75 μL of the 5.5 μL reaction for transformation into chemically competent OneShot TOP10 or DH5α cells (Invitrogen).
5. For plasmids ~5–15 kb, pick 2–4 colonies for culture and digestion analysis (usually a ~75–100% success rate). For plasmids larger than 20 kb, a greater number of colonies (4–8) should be picked (expect a ~25–75% success rate) and more care should be taken during culture (avoid prolonged incubation after bacterial growth plateau).

3.2.3. Analysis of Clones

Using DNA software one can easily predict the resulting recombinant DNA. We use a shareware called ApE (a plasmid editor, http://biologylabs.utah.edu/jorgensen/wayned/ape/). Using this elegant program, p1, p2, p3, and p4 clones can be recombined using the "Recombination Tool" from Tool tab. Chose the recombination type as Multisite LR reaction, and select your entry clones in order. By referring to the sequence of the recombinants, one can analyze the LR generated clones using appropriate restriction enzymes. For large plasmids, we generally use BamHI or XbaI for digestion analysis.

4. Notes

1. We have successfully generated transgenic *Xenopus* using I-*Sce*I meganuclease from both Roche and New England Biolabs. We use the reaction buffer following their instructions with 10 units of I-*Sce*I, but in the case of Roche, without the enhancer solution. We normally purchase I-*Sce*I enzyme from NEB, which provides smaller volume aliquots of enzyme. We believe that the activity of I-*Sce*I is the most critical factor for the efficiency of the transgenesis. Therefore I-*Sce*I enzyme should be stored in −80°C in aliquots sufficient for one day's experiment.
2. Plasmid DNAs for injections are generally prepared using a Midi or Maxi Kit from Qiagen. However, we have also found that DNA isolated by miniprep is generally not toxic to embryos.
3. Since the injection should be carried out within 20–30 min immediately after fertilization, the timing of the incubation of the reaction is important. The mixture except I-*Sce*I enzyme is setup before artificial fertilization (Table 1).
4. The amount of DNA in the reaction may vary depending on the construct. We usually inject 10 pg for plasmids ~10 kb or smaller, but use 20 pg for bigger construct such as recombinants generated by the Gateway system.
5. In our experience long incubation of the reaction mixture does not improve the efficiency of transgenesis.
6. To raise transgenic animals efficiently, it is important to screen the transgenic animals and identify the ones most likely to transmit the transgene to the next generation. The main purpose of injecting the I-*Sce*I—DNA mixture very soon after fertilization is to increase the likelihood of early integration, and thus non-mosaic integration of the transgene. If this has occurred, one should expect embryos to express the transgene throughout the expression domain of the promoter in question.

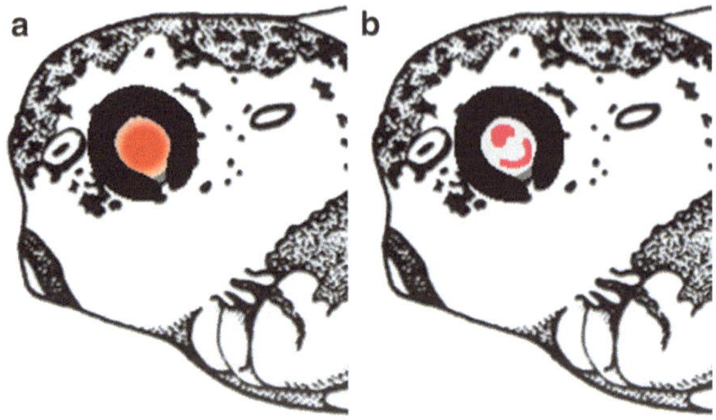

Fig. 5. Diagram of embryos showing broad or mosaic expression from the γ-crystallin promoter-driving RFP as a selection marker. (**a**) Broad expression of RFP in the lens will be observed if the transgene is integrated in the genome. (**b**) If the expression of transgene is transient, only a few cells in the lens shows strong expression of RFP.

Thus, the primary screening method is to identify embryos, which express the transgene and/or screenable marker broadly and homogenously throughout the expression domain of the promoter. Unintegrated DNA, by contrast, expresses the transgene in a highly mosaic and spotty pattern. We generally use the γ1-crystallin or αCE2x6 driving fluorescent protein as a screenable marker. Both promoters drive expression in the lens, and the signal can be detected even in adult frogs (19). We screen stage 45 tadpoles or older for fluorescence signal in the eye. If the transgene integrates into the genome at early stages, the expression is broad, throughout the lens (Fig. 5a). However, late integration or non-integrated DNA expression is seen mosaically within the lens (Fig. 5b). Tadpoles showing broad expression are selected, raised to adulthood, and mated with wild type animals to obtain F1 transgenic animals. From F0 selected embryos, which reach adulthood, from 18 to 50% transmit the transgene through the germline into the next generation.

7. The I-*Sce*I compatible and other pTransgenesis vectors are available from the European Xenopus Resource Centre: http://www.port.ac.uk/research/exrc/.

References

1. Kroll KL, Amaya E (1996) Transgenic Xenopus embryos from sperm nuclear transplantations reveal FGF signaling requirements during gastrulation. Development 122(10):3173–3183
2. Shemesh M, Gurevich M, Harel-Markowitz E, Benvenisti L, Shore LS, Stram Y (2000) Gene integration into bovine sperm genome and its expression in transgenic offspring. Mol Reprod Dev 56(2 Suppl):306–308
3. Harel-Markowitz E, Gurevich M, Shore LS, Katz A, Stram Y, Shemesh M (2009) Use of sperm plasmid DNA lipofection combined with REMI (restriction enzyme-mediated insertion) for production of transgenic chickens expressing eGFP (enhanced green fluorescent protein) or human follicle-stimulating hormone. Biol Reprod 80(5):1046–1052
4. Noh W, Kim SW, Dong-Won B, Kim JY, Ro HS (2010) Genetic introduction of foreign genes to Pleurotus eryngii by restriction enzyme-mediated integration. J Microbiol 48(2): 253–256

5. Yan W, Liu X, Shi T, Hao L, Tomley FM, Suo X (2009) Stable transfection of Eimeria tenella: constitutive expression of the YFP-YFP molecule throughout the life cycle. Int J Parasitol 39(1):109–117
6. Kuspa A, Loomis WF (1992) Tagging developmental genes in Dictyostelium by restriction enzyme-mediated integration of plasmid DNA. Proc Natl Acad Sci U S A 89(18):8803–8807
7. Dujon B (1980) Sequence of the intron and flanking exons of the mitochondrial 21S rRNA gene of yeast strains having different alleles at the omega and rib-1 loci. Cell 20(1):185–197
8. Colleaux L, D'Auriol L, Galibert F, Dujon B (1988) Recognition and cleavage site of the intron-encoded omega transposase. Proc Natl Acad Sci U S A 85(16):6022–6026
9. Thermes V, Grabher C, Ristoratore F, Bourrat F, Choulika A, Wittbrodt J, Joly JS (2002) I-SceI meganuclease mediates highly efficient transgenesis in fish. Mech Dev 118(1–2):91–98
10. Deschet K, Nakatani Y, Smith WC (2003) Generation of Ci-Brachyury-GFP stable transgenic lines in the ascidian Ciona savignyi. Genesis 35(4):248–259
11. Pan FC, Chen Y, Loeber J, Henningfeld K, Pieler T (2006) I-SceI meganuclease-mediated transgenesis in Xenopus. Dev Dyn 235(1):247–252
12. Ogino H, McConnell WB, Grainger RM (2006) Highly efficient transgenesis in *Xenopus tropicalis* using I-*Sce*I meganuclease. Mech Dev 123(2):103–113
13. Ochiai H, Sakamoto N, Suzuki K, Akasaka K, Yamamoto T (2008) The Ars insulator facilitates I-SceI meganuclease-mediated transgenesis in the sea urchin embryo. Dev Dyn 237(9): 2475–2482
14. Perrin A, Buckle M, Dujon B (1993) Asymmetrical recognition and activity of the I-SceI endonuclease on its site and on intron-exon junctions. EMBO J 12(7):2939–2947
15. Love NR, Thuret R, Chen Y, Ishibashi S, Sabherwal N, Paredes R, Dorey K, Noble AM, Guille MJ, Sasai Y, Papalopulu N, Amaya E (2011) pTransgenesis: A cross-species, modular transgenesis resource. Development 138(24): 5451–5458
16. Hartley JL, Temple GF, Brasch MA (2000) DNA cloning using in vitro site-specific recombination. Genome Res 10(11):1788–1795
17. Kawakami K, Takeda H, Kawakami N, Kobayashi M, Matsuda N, Mishina M (2004) A transposon-mediated gene trap approach identifies developmentally regulated genes in zebrafish. Dev Cell 7(1):133–144
18. Khokha MK, Chung C, Bustamante EL, Gaw LW, Trott KA, Yeh J, Lim N, Lin JC, Taverner N, Amaya E, Papalopulu N, Smith JC, Zorn AM, Harland RM, Grammer TC (2002) Techniques and probes for the study of *Xenopus tropicalis* development. Dev Dyn 225(4):499–510
19. Bronchain OJ, Hartley KO, Amaya E (1999) A gene trap approach in Xenopus. Curr Biol 9(20):1195–1198
20. Beck CW, Slack JM (1999) Gut specific expression using mammalian promoters in transgenic *Xenopus laevis*. Mech Dev 88(2):221–227
21. Yoshida T, Ohkumo T, Ishibashi S, Yasuda K (2005) The 5′-AT-rich half-site of Maf recognition element: a functional target for bZIP transcription factor Maf. Nucleic Acids Res 33(11):3465–3478
22. Turner DL, Weintraub H (1994) Expression of achaete-scute homolog 3 in Xenopus embryos converts ectodermal cells to a neural fate. Genes Dev 8(12):1434–1447
23. Niwa H, Yamamura K, Miyazaki J (1991) Efficient selection for high-expression transfectants with a novel eukaryotic vector. Gene 108(2):193–199
24. Schorpp M, Jager R, Schellander K, Schenkel J, Wagner EF, Weiher H, Angel P (1996) The human ubiquitin C promoter directs high ubiquitous expression of transgenes in mice. Nucleic Acids Res 24(9):1787–1788
25. Friedrich G, Soriano P (1991) Promoter traps in embryonic stem cells: a genetic screen to identify and mutate developmental genes in mice. Genes Dev 5(9):1513–1523
26. Hartley KO, Hardcastle Z, Friday RV, Amaya E, Papalopulu N (2001) Transgenic Xenopus embryos reveal that anterior neural development requires continued suppression of BMP signaling after gastrulation. Dev Biol 238(1): 168–184
27. Meadows SM, Salanga MC, Krieg PA (2009) Kruppel-like factor 2 cooperates with the ETS family protein ERG to activate Flk1 expression during vascular development. Development 136(7):1115–1125
28. Smith SJ, Kotecha S, Towers N, Latinkic BV, Mohun TJ (2002) XPOX2-peroxidase expression and the XLURP-1 promoter reveal the site of embryonic myeloid cell development in Xenopus. Mech Dev 117(1–2):173–186
29. Koster RW, Fraser SE (2001) Tracing transgene expression in living zebrafish embryos. Dev Biol 233(2):329–346
30. Rizzo MA, Springer GH, Granada B, Piston DW (2004) An improved cyan fluorescent protein variant useful for FRET. Nat Biotechnol 22(4):445–449
31. Heim R, Tsien RY (1996) Engineering green fluorescent protein for improved brightness, longer wavelengths and fluorescence resonance energy transfer. Curr Biol 6(2):178–182

32. Ai HW, Henderson JN, Remington SJ, Campbell RE (2006) Directed evolution of a monomeric, bright and photostable version of Clavularia cyan fluorescent protein: structural characterization and applications in fluorescence imaging. Biochem J 400(3):531–540
33. Karasawa S, Araki T, Nagai T, Mizuno H, Miyawaki A (2004) Cyan-emitting and orange-emitting fluorescent proteins as a donor/acceptor pair for fluorescence resonance energy transfer. Biochem J 381(Pt 1):307–312
34. Belousov VV, Fradkov AF, Lukyanov KA, Staroverov DB, Shakhbazov KS, Terskikh AV, Lukyanov S (2006) Genetically encoded fluorescent indicator for intracellular hydrogen peroxide. Nat Methods 3(4):281–286
35. Tsutsui H, Karasawa S, Shimizu H, Nukina N, Miyawaki A (2005) Semi-rational engineering of a coral fluorescent protein into an efficient highlighter. EMBO Rep 6(3):233–238
36. Ando R, Hama H, Yamamoto-Hino M, Mizuno H, Miyawaki A (2002) An optical marker based on the UV-induced green-to-red photoconversion of a fluorescent protein. Proc Natl Acad Sci U S A 99(20):12651–12656
37. Ormo M, Cubitt AB, Kallio K, Gross LA, Tsien RY, Remington SJ (1996) Crystal structure of the Aequorea victoria green fluorescent protein. Science 273(5280):1392–1395
38. Nagai T, Ibata K, Park ES, Kubota M, Mikoshiba K, Miyawaki A (2002) A variant of yellow fluorescent protein with fast and efficient maturation for cell-biological applications. Nat Biotechnol 20(1):87–90
39. Shaner NC, Campbell RE, Steinbach PA, Giepmans BN, Palmer AE, Tsien RY (2004) Improved monomeric red, orange and yellow fluorescent proteins derived from Discosoma sp. red fluorescent protein. Nat Biotechnol 22(12):1567–1572
40. Shcherbo D, Merzlyak EM, Chepurnykh TV, Fradkov AF, Ermakova GV, Solovieva EA, Lukyanov KA, Bogdanova EA, Zaraisky AG, Lukyanov S, Chudakov DM (2007) Bright far-red fluorescent protein for whole-body imaging. Nat Methods 4(9):741–746
41. Brand AH, Perrimon N (1993) Targeted gene expression as a means of altering cell fates and generating dominant phenotypes. Development 118(2):401–415
42. Verrou C, Zhang Y, Zurn C, Schamel WW, Reth M (1999) Comparison of the tamoxifen regulated chimeric Cre recombinases MerCreMer and CreMer. Biol Chem 380(12):1435–1438

Chapter 13

Using ΦC31 Integrase to Mediate Insertion of DNA in *Xenopus* Embryos

You E. Li, Bryan G. Allen, and Daniel L. Weeks

Abstract

The two most common methods used to generate transgenic *Xenopus* embryos, restriction enzyme-mediated insertion, and I-SceI meganuclease take advantage of relatively common but spatially unpredictable double-stranded breaks in sperm, egg, or early embryo genomes. These methods also tend to insert multimeric copies of the transgene. An alternative is to use bacteriophage- or transposon-derived integrase or recombinase to mediate more site-specific insertion of the transgene. The use of phiC31 integrase requires a defined sequence for insertion and is compatible with insertion of a single copy of the transgene. We describe the protocol we use to facilitate phiC31 integrase transgene insertion including the use of insulator sequences to reduce position effect disruption of transgene activity.

Key words: ΦC31 integrase, Embryo injection, Transgenesis, *Xenopus laevis*

1. Introduction

Select viruses and bacteriophage are capable of inserting their DNA into the genomes of their hosts. Bacteriophage ΦC31 is equipped to insert its DNA into the genome of *Streptomyces* species (1, 2). This process has many features that fit well with the goals for making transgenic *Xenopus* embryos. First, a single serine recombinase encoded by the phage, ΦC31 integrase, is sufficient to catalyze and complete integration (1, 3). Second, the DNA sequences that define the donor and acceptor sites been defined, and once integration occurs the excision of the inserted DNA cannot be accomplished by ΦC31 integrase by itself (1, 4). Third, relatively large pieces of DNA can be integrated, as is evident from the ability to insert the 41,491 basepairs of the ΦC31 phage into the *Streptomyces* genome.

The practical demonstration that ΦC31 integrase can act in a variety of backgrounds was first described by Calos and colleagues (5).

Stefan Hoppler and Peter D. Vize (eds.), *Xenopus Protocols: Post-Genomic Approaches, Second Edition*,
Methods in Molecular Biology, vol. 917, DOI 10.1007/978-1-61779-992-1_13, © Springer Science+Business Media, LLC 2012

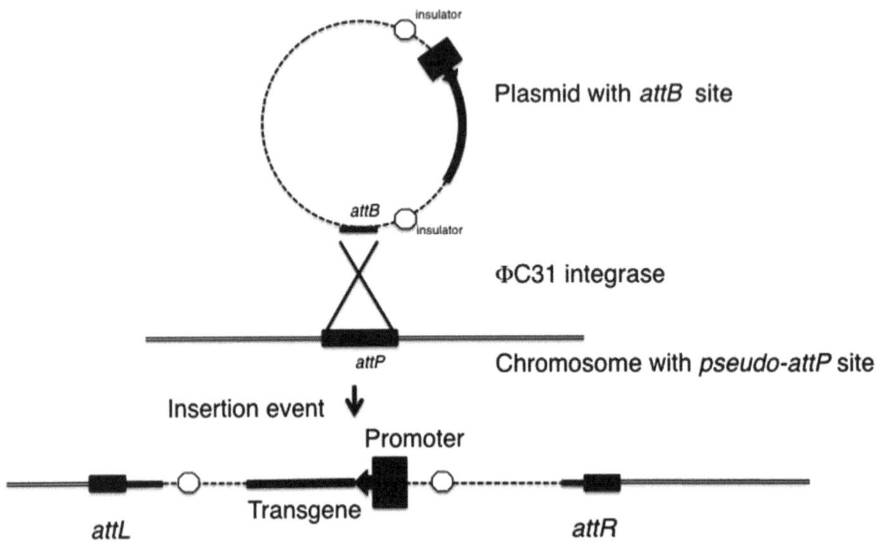

Fig. 1. Schematic of ΦC31 integrase-mediated insertion of plasmid DNA into a chromosome. ΦC31 integrase recognizes and binds to the *attB* site in the reporter plasmid. Typically the regions to be expressed would be flanked by insulators. Insulators are represented by octagons. The chromosome (*double black lines*) with a pseudo-*attP* site is bound in juxtaposition with the *attB* site in the plasmid. The *attB* site in the reporter plasmid and a pseudo-*attP* site in the embryo genome undergo recombination to form hybrid sites *attR* and *attL* that flank the integrated reporter plasmid.

Their work and that of others have used ΦC31 integrase to insert genetic material in yeast, plants, fruit flies, fish, *Xenopus*, and mammals (6–14).

In order to mediate phage integration, ΦC31 integrase must bind to a 34 bp sequence in the bacterial genome (the *attB*) site and a 39 bp sequence in the phage (the *attP* site) (4, 15). During integration neither site remains complete, rather ΦC31 integrase creates two new hybrid sites as depicted in Fig. 1. It is useful to note here for readers unfamiliar with this type of nomenclature that *attB* and *attP* are used to identify the DNA binding in a variety of phage/host systems; however, the sequences are specific for each phage and host. Although ΦC31 integrase-mediated integration efficiency is adversely affected when changes in *attB* or *attP* occur, integrase is capable of recognizing *attP* sites, called pseudo-*attP* sites that have as little as 29–54 % similarity to a consensus *attP* site (5). Tolerance for mismatches does not imply that integration at a pseudo-*attP* site will occur with the efficiency of a *bone fide attP* site. In studies in human cells, fruit fly, and zebrafish, the presence of a true *attP* site dramatically increases the efficiency of ΦC31 integrase (12, 16).

Although there are no consensus *attB* or *attP* sites in the *Xenopus* genome, integrase tolerance for pseudo-*attP* sites allows recombination to occur. As mentioned above, pseudo-*attP* sites work with reduced efficiency and frog to frog variation in pseudo-*attP* site sequence may explain the wide variation (from 1 % to as high as almost 50 %) in integration efficiency using this method.

Our analysis of transgenic embryos using this method suggests that a single transgene is normally inserted, though we have yet to find a particular genomic location favored for the insertion event. We have observed that expression of genes integrated using this method may be affected by position effects, presumably related to the specific site of integration. By flanking genes to be integrated with insulator elements, like the HS4 insulator found in the chicken β-globin gene, position effects can be significantly reduced (13, 17–19).

We note that this chapter is an update of previous descriptions of using ΦC31 integrase (13, 20–22).

2. Materials

2.1. Plasmids and Plasmid Preparation Reagents

1. Plasmids: The integrase plasmid pET11-ΦC31 poly(A) can be obtained from Michelle P. Calos (calos@stanford.edu) and the *attB* reporter plasmids, CMV-EGFP-DI-*attB* and CL-EGFP-DI-*attB* can be obtained from Daniel Weeks (daniel-weeks@uiowa.edu). A modified version of the ΦC31 integrase expression vector is (available from DLW, but an MTA for the use of the integrase should still be obtained from Dr. Calos) transcribed using T3 polymerase and containing the β-globin 5′ and 3′ UTRs in a pBluescript (S/K) vector has also been generated. Reporter plasmids, with either the *Xenopus* γ-crystalline lens promoter or the CMV promoter driving GFP flanked by the chicken β-globin HS4 insulators and a ΦC31 attB site can be obtained from the authors (DLW). However, permission to use HS4 insulator sequences and HS4 insulator constructs needs to be obtained from Gary Felsenfeld (gary.felsenfeld@nih.gov) (and see Notes 1 and 2).
2. Bacterial plate and culture reagents: Bacto-Yeast Extract, Bacto-Tryptone powder, and Bacto-agar (BD Biosciences, Sparks, MD).
3. Antibiotics: ampicillin and kanamycin (Sigma Aldrich, St. Louis, MO).
4. Plasmid purification kits: Qiaprep Spin Miniprep Kit, HiSpeed Plasmid Maxi Kit (Qiagen, Valencia, CA).
5. In vitro RNA transcription: T7 mMessage machine (Ambion, Austin, TX).
6. Agarose (RPI, Mt. Prospect, IL).
7. 10× TAE: 0.4 M Tris–acetate, 0.01 M EDTA.
8. 10× non-denaturing DNA loading buffer: 50 % glycerol, 60 mM EDTA, 1 % SDS, and 0.05 % bromophenol blue.
9. 1 % agarose, 2.2 M formaldehyde MOPS gel.

10. Denaturing RNA loading buffer (NorthernMax formaldehyde loading buffer, Ambion, Austin, TX or make loading buffer using 5 parts deionized formamide, 1.5 parts formaldehyde, and 1 part 10× MOPS buffer. Use 3 parts loading buffer to 1 part sample.).
11. Gel electrophoresis apparatus for flat bed agarose gel electrophoresis (Owl Separation Systems, Portsmouth, NH).
12. Spectrophotometer: Nanodrop ND1000 (Nanodrop Technologies, Wilmington, DE).
13. UV-light trans-illuminator.

2.2. Xenopus Injection

1. *Xenopus laevis*: May be obtained from either *Xenopus I* (Dexter, MI) or Nasco (Fort Atkinson, WI) and see Note 3.
2. Human Chorionic Gonadotropin (Sigma Aldrich).
3. Tricaine (3-aminobenzoic acid ethyl ester) (Sigma Aldrich).
4. 10× Marc's Modified Ringers Solution (MMR): 1 M NaCl, 20 mM KCl, 20 mM $CaCl_2$, 10 mM $MgCl_2$, 50 mM HEPES at pH 7.4.
5. Injection buffer: 88 mM NaCl, 10 mM HEPES.
6. 0.3× MMR with 3 % Ficoll Type 400 (Sigma Aldrich).
7. 2 % Cysteine (RPI) made in dH_2O, pH to 7.8–7.9 with NaOH. This solution should be made the day of injection.
8. Microinjection needle puller.
9. Micromanipulator: Singer MK-1 (Singer instrument company, Somerset, UK) or similar instrument.
10. Microinjector: inject+matic or similar instrument.
11. Glass capillary tubes of appropriate size for microinjector (Singer). Microinjection needles used in this technique have an injection tip of approximately 10 μm.
12. 18 °C Incubator.

2.3. Screening for Transgenics

1. Dissecting microscope: Nikon SMZ (Nikon Instruments, Melville, NY) and Zeiss Stemi SV 11 fluorescent dissecting microscope (Zeiss MicroImaging, Thornwood, NY).
2. Compound microscope with fluorescence: Zeiss Axioplan 2 (Carl Zeiss MicroImaging) or similar instrument.
3. Camera for photo documentation: SPOT camera (Diagnostic Instruments, Sterling Heights, MI), Zeiss Axiocam (Carl Zeiss MicroImaging) or similar instrument.
4. Genomic DNA purification method. Qiagen DNeasy (Qiagen) for example.
5. PCR primers that will amplify the plasmid across the *attB* site.
6. PCR primers that will amplify within the coding region of the gene to be inserted.

7. A control primer set to compare copy number of an endogenous gene with the transgene.
8. PCR thermocycler.
9. RNA purification using an RNA Aqueous kit (Ambion, Austin TX.) or similar protocol.
10. Reverse transcriptase polymerase chain reaction using OneStep RT-PCR (Qiagen,) or similar protocol.

3. Methods

3.1. Preparation of Plasmids and ΦC31 Integrase mRNA

1. Transform the pET11-ΦC31 poly(A) plasmid and the *attB* reporter plasmids (CMV-EGFP-DI-*attB* and CL-EGFP-DI-*attB*) into *Escherichia coli* and plate onto antibiotic selective plates using standard techniques. The pET11-ΦC31 poly(A) plasmid contains an ampicillin resistance gene and the *attB* reporter plasmids contain a kanamycin resistance gene. We have found that insulated *attB* reporter plasmids frequently undergo rearrangements during growth in *E. coli* and thus recommend propagating these plasmids in *E. coli* Stbl2 cells (Invitrogen) or similar *E. coli* strain that are designed to prevent recombination (see Note 2).
2. Select individual colonies on each plate and grow 3 mL cultures at 37 °C overnight with agitation. Use 1 mL for mini-preps (Qiaprep mini-prep kit) and save the other 2 mL of culture at 4 °C. Using the mini-prep generated DNA, confirm plasmid size and insulator orientation by restriction enzyme digestion and gel analysis. Once plasmid size and insulator orientation are confirmed, inoculate 1 L of LB broth containing the appropriate antibiotic with the remaining 2 mL of culture. Grow at 37 °C with agitation overnight.
3. Perform maxi-preps on the 1 L cultures following the manufactures instructions (Hi Speed Plasmid Maxi Kit). Do not add RNase to buffer P1.
4. Confirm maxi-prep-generated plasmid identity by restriction enzyme digest and gel analysis and by DNA-sequencing reactions. Store the maxi-prep generated DNA at –20 °C.
5. Linearize 5 μg of pET11-ΦC31 poly(A) maxi-prep DNA with either BamHI or EcoRI restriction enzymes. Run approximately one-tenth of the digestion on a 1 % agarose gel to ensure linearization. Heat inactivate the restriction enzyme in the remaining portion of digestion by heating reaction to 65 °C for 20 min. Precipitate the DNA by adding one-tenth the volume of 5 M NH_4 acetate and 2 volumes of ethanol. Place solution at –20 °C for 15 min and centrifuge at 10,000 × *g*

for 15 min. Remove the supernatant and resuspend the pellet in 10 μL of RNase free TE buffer.

6. Synthesize the ΦC31 integrase mRNA using the T7 mMessage machine following the manufactures instructions. Because the protocol includes a DNase treatment step, the DNase needs to be removed or inactivated. We routinely follow the LiCl precipitation protocol described in the manufactures instructions. Resuspend the integrase mRNA in RNase free water at a concentration of 1 mg/mL.
7. Run 1 μg of ΦC31 integrase mRNA on a formaldehyde MOPS gel in 1× MOPS buffer to ensure that the transcript is approximately 1.9 kb long.
8. Store ΦC31 integrase mRNA (for up to 1 month) at −80 °C until it is needed.

3.2. Injection of Xenopus Embryos

1. Any experiments with *Xenopus* should first start with approval of your plans to house and care for *Xenopus* adult frogs and embryos from your host institution. If you have limited experience using *Xenopus* other chapters in this volume, or the book, Early Development of *X. laevis*—a laboratory manual (23), may be a useful reference to consult.
2. Induce *Xenopus* females to lay eggs by injecting 1 mL (1,000 IU) of Human Chorionic Gonadotropin into the dorsal lymph sac the night before desired day of egg collection.
3. The next morning, the cloaca on the injected females should be swollen and there may be eggs in the water tank holding the frogs.
4. After eggs are present in the tank, inject a lethal dose of Tricaine (1 mL of a 10 % solution) into the dorsal lymph sac of a *Xenopus* male and then surgically remove the testes. Store the testes in 1× MMR.
5. Induce egg laying into a dry Petri dish by spreading apart the female's legs and gently putting pressure on the pelvic region. Immediately fertilize the eggs by rubbing a small piece of testis (~1/6th of a testis) through the eggs. Alternatively, the testis can be crushed in ~1 mL of 0.3× MMR and this solution spread over the eggs. After fertilization (allow about 1 min) flood the eggs with 0.1× MMR.
6. Successful fertilization can be determined if the eggs align with the animal pole (pigmented half) facing up. Embryos should be spherical, uniform size, and have smooth even pigmentation of the animal hemisphere. Thirty minutes after fertilization, remove the 0.1× MMR and place the embryos in 2 % cysteine, pH 7.8–7.9 for 2 min to remove the jelly coats.
7. Remove the cysteine and wash the embryos in three 5 mL washes of 0.3× MMR.

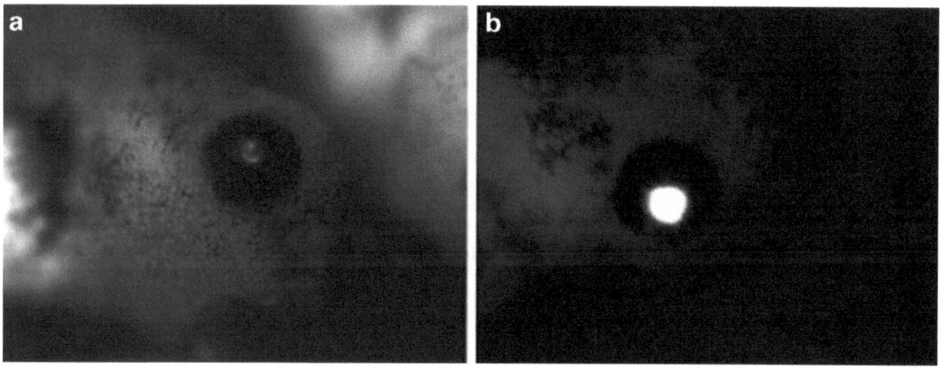

Fig. 2. ΦC31integrase-mediated transgenesis of insulated reporter plasmids in stage 42 (28) *Xenopus* embryos. Expression of GFP controlled by the γ-crystalline lens promoter. (**a**) Weak mosaic expression derived from extrachromasomal reporter plasmid CL-EGFP-DI-*attB*. (**b**) Full lens expression of GFP from integrated pCL-GFP dlattB (see Note 5).

8. Place the embryos in 0.3× MMR, 3 % ficoll.
9. Transfer embryos to an injection slide. A variety of injection slide formats can be used. We routinely fasten two 5×7.5 cm microscope slides together with a 0.5 cm offset, forming a "stairstep" like platform. Approximately 50 embryos can be placed in a single line in a cushion of the 3 % ficoll 0.3× MMR solution in the step, holding the embryos in place.
10. Inject single cell embryos into the center of the animal hemisphere with either: 10 nL of injection solution, 10 nL of injection solution containing 5 pg of reporter plasmid, or 10 nL of injection solution containing 5 pg of reporter plasmid + 1 ng of ΦC31 integrase mRNA (see Note 4).

3.3. Monitoring Developing Embryos for Transgenesis

1. Allow injected embryos to develop at 18 °C in 0.3× MMR, 0.3 % Ficoll for approximately 6–8 h after injection and then transfer to 0.3× MMR.
2. Remove delaminating or dead embryos and provide fresh 0.3× MMR at least twice a day for the first 3 days.
3. On the second day after fertilization, begin to monitor the embryos for GFP expression using a fluorescent microscope optimized to detect green fluorescence. Embryos injected with 1 ng integrase mRNA and 5 pg of CMV-EGFP-DI-*attB* should express GFP uniformly while embryos injected with 1 ng integrase mRNA and 5 pg of CL-EGFP-DI-*attB* should express GFP only in the lens of the eye (Fig. 2). Embryos injected with 5 pg of reporter plasmid alone rarely give GFP expression however we occasionally detect mosaic expression that is less intense than expression from that seen in transgenics (Fig. 2a, b show an example of this phenomena in embryos injected with the γ-crystalline lens:GFP plasmid). Digital photography with long exposure times may detect fluorescence before it is visible by eye.

4. Once fluorescence is detected, photograph the embryos using a fluorescent microscope with a digital camera. If embryo movement prevents photography, consider anesthetizing the embryos with 0.02 % Tricaine. Embryos may be exposed to the Tricaine solution for approximately 5 min; longer exposure times may lead to the demise of the embryos. Embryos should be placed back into 0.3× MMR after photography.

3.4. Molecular Analysis to Confirm Integration

1. The molecular confirmation of transgenesis depends, in part, on the experimental goal. Genomic DNA can be analyzed by Southern blot analysis as has been described previously (13, 20, 21). Alternatively, genomic DNA or mRNA can be used for PCR analysis of transgene insertion or expression.
2. Isolate DNA from single embryos using the Qiagen DNeasy kit following the manufacturer's instructions. In some studies whole embryos can be used for this analysis, while in others the DNA can be isolated from excised tails can be removed. For whole embryos, one stage 46 embryo should yield approximately 2–5 μg of DNA, the tails from a stage 48–50 can easily yield a similar amount of genomic DNA. DNA should be collected from non-injected embryos, embryos injected with reporter plasmid alone, and transgenic embryos determined by green fluorescence.
3. Starting with 250 ng of genomic DNA, and primer pairs designed to amplify the inserted plasmid or a single-copy gene from *Xenopus* (in Fig. 3a the control used was for the *Xenopus NKX2-10* gene with primers 5′CGCTACCTCTAC CCCCTTCT3′ and 5′TGATCAGCTGTGGAGGACTC 3′, the integrated plasmid being tested was being used to express an epitope tagged version of Nkx2-5 and the primers were 5′CATGACAGGAGGACAGCAGA3′ and 5′TGTGTGGAGG TGAGCTTCAG3′). If you wish to estimate copy number it is useful to pick control primers that will give the same size of amplified product as the primers used to identify the transgene. Twenty-eight cycles of PCR should easily show a single copy gene. To control for persistent extra-chromosomal (non-integrated plasmid) the isolated genomic DNA should be used for an amplification reaction using primers that generate a product across the *attB* site. Because the *attB* site is split apart during insertion, no product will be amplified in integrated copies of the plasmid. If information on insertion copy number is needed identify cycle numbers consistent with a doubling of product production in successive cycles.
4. Run amplified DNA on a 1 % agarose gel. Stain the gel with ethidium bromide and then photograph the gel while on a UV transiluminator.

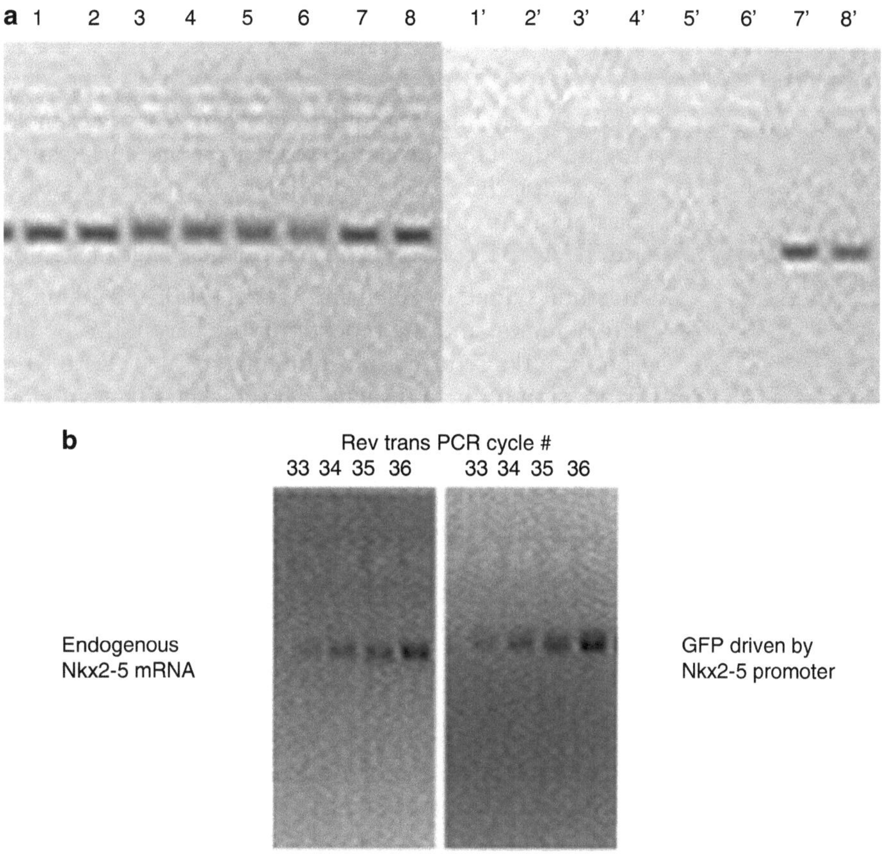

Fig. 3. Confirming ΦC31 integrase-mediated transgenesis and expression. (**a**) PCR amplification of an endogenous gene (*NKX2.10*) compared to the amplification of a transgene encoding a T7epitope tagged version of Nkx2-5. 200 ng of DNA from individual stage 47 (28) embryos was analyzed. Lanes 1–8 is the amplification of DNA from eight different embryos using the primers for an endogenous gene, lanes 1′–8′ the amplification of the DNA from the same eight embryos using primers that query the presence of the transgene. Embryos 7 and 8 were transgenic. (**b**) Analysis of the endogenous expression of Nkx2-5 compared to the transcription controlled by a transgene expressing GFP under the control of the *NKX2-5* promoter. Analysis used 200 ng of mRNA isolated from stage 46 embryos and RT-PCR and sampled sequential cycles from cycle 33–36.

5. To estimate the expression of a transgene compared to an endogenous gene, isolate mRNA from the embryos (including non-injected and plasmid only injected controls). When using the Qiagen OneStep RT-PCR, between 1 ng and 2 μg of RNA is needed, with PCR cycle numbers adjusted according to input RNA and target. In the example shown in Fig. 3b, the endogenous expression of the *NKX2-5* gene was compared to a transgene expression GFP under the control of 4.5 kb *NKX2-5* promoter described by Mohun and colleagues. RNA was isolated from stage 46 embryos and 200 ng of RNA was used in the RT-PCR reaction. Primers used for this amplification were 5′-ACTGACAGGAGGACAGCAGA-3′ and 5′-GCCGTTTG CATTTGTACCTT-3′ (*NKX2-5*) 5′-GACATGAAGCAGCAC GACTT-3′ and 5′-TGCTCAGGTAGTGGTTGTCG-3′(GFP).

4. Notes

1. MTA agreements are required from Dr. Gary Felsenfeld (NIH) to use the HS4 insulator sequences and Dr. Daniel Weeks (Iowa) for the CMV-EGFP-DI-*attB* and CL-EGFP-DI-*attB* plasmids. An MTA is also required to obtain the plasmid pET11ΦC31 poly(A) from Dr. Michele Calos (Stanford).
2. In studies done by Allen and Weeks (2005), flanking the regulatory region and the reporter in the plasmid to be integrated significantly reduced position effects that lead to reduced expression of the transgene. The chicken β-globin HS4 insulator sequence is effective both as a block to gene silencing and enhancers. Thus positioning the transgene between insulators should help reduce position effects. The Felsenfeld laboratory suggests that tandem copies of the core HS4 insulator, oriented all in the same direction work best. When configured this way the HS4 insulators form direct sequence repeats that can be unstable in many bacterial backgrounds.
3. We have historically used these suppliers, but other vendors of *X. laevis* may also provide useable animals. In fact, this method was successful using *X. laevis* from different breeding facilities (Rennes, France; Labor-Atoire Evolution et Développement, Université Paris Sud, France) (22). It is also important to note that in our hands we see significant variation in integration efficiency that seems to be animal dependent, perhaps due to the degree of mismatches in the *pseudo attP* sites. In the future we hope determine if animals that carry *bona fide attP* sites improve the ability to broadly use this method. If animals with *bona fide attP* sites improve efficacy of this approach they will be made available through the *Xenopus* stock centers.
4. The rapid production of ΦC31 integrase seems essential and care should be taken to ensure that the mRNA encoding ΦC31 integrase is stable and capped. Protocols that include RNaseA in the purification of plasmid can prove problematic if that plasmid is used for injection with ΦC31 integrase mRNA. We have also generated insertions using ΦC31 integrase protein made in vitro using a Promega TNT (transcription: translation) extract. However, to date, we have not seen significant increases in transgene generation using in vitro generated ΦC31 integrase protein.
5. The amount of fluorescence produced from a single-copy transgene (ΦC31 integrase most commonly gives a single insertion) may be significantly less than an embryo containing multiple copies of a transgene generated from the restriction enzyme-mediated insertion (24, 25) approach or the meganuclease approach (26, 27).

Acknowledgements

The authors would like to thank Michele Calos for providing the pET11ΦC31 poly(A) plasmid, Gary Felsenfeld and colleagues for providing the HS4 insulator sequences, Paul Krieg for providing the γ-crystallin lens promoter, and Tim Mohun for providing the Nkx2-5 promoter. This work was supported by funding from the NIH (GM069944 and DC007481). Bryan Allen was a student in the Medical Scientist Training Program at the Roy J. and Lucille A Carver College of Medicine, University of Iowa.

References

1. Thorpe HM, Smith MC (1998) In vitro site-specific integration of bacteriophage DNA catalyzed by a recombinase of the resolvase/invertase family. Proc Natl Acad Sci USA 95: 5505–5510
2. Smith MC, Burns RN, Wilson SE, Gregory MA (1999) The complete genome sequence of the Streptomyces temperate phage straight phiC31: evolutionary relationships to other viruses. Nucleic Acids Res 27:2145–2155
3. Groth AC, Calos MP (2004) Phage integrases: biology and applications. J Mol Biol 335: 667–678
4. Thorpe HM, Wilson SE, Smith MC (2000) Control of directionality in the site-specific recombination system of the Streptomyces phage phiC31. Mol Microbiol 38:232–241
5. Groth AC, Olivares EC, Thyagarajan B, Calos MP (2000) A phage integrase directs efficient site-specific integration in human cells. Proc Natl Acad Sci USA 97:5995–6000
6. Thyagarajan B, Guimaraes MJ, Groth AC, Calos MP (2000) Mammalian genomes contain active recombinase recognition sites. Gene 244:47–54
7. Thomason LC, Calendar R, Ow DW (2001) Gene insertion and replacement in Schizosaccharomyces pombe mediated by the Streptomyces bacteriophage phiC31 site-specific recombination system. Mol Genet Genomics 265:1031–1038
8. Thyagarajan B, Olivares EC, Hollis RP, Ginsburg DS, Calos MP (2001) Site-specific genomic integration in mammalian cells mediated by phage phiC31 integrase. Mol Cell Biol 21:3926–3934
9. Combes P, Till R, Bee S, Smith MC (2002) The streptomyces genome contains multiple pseudo-attB sites for the (phi)C31-encoded site-specific recombination system. J Bacteriol 184:5746–5752
10. Olivares EC, Hollis RP, Chalberg TW, Meuse L, Kay MA, Calos MP (2002) Site-specific genomic integration produces therapeutic Factor IX levels in mice. Nat Biotechnol 20:1124–1128
11. Ortiz-Urda S, Thyagarajan B, Keene DR, Lin Q, Fang M, Calos MP, Khavari PA (2002) Stable nonviral genetic correction of inherited human skin disease. Nat Med 8:1166–1170
12. Groth AC, Fish M, Nusse R, Calos MP (2004) Construction of transgenic Drosophila by using the site-specific integrase from phage phiC31. Genetics 166:1775–1782
13. Allen BG, Weeks DL (2005) Transgenic *Xenopus laevis* embryos can be generated using phiC31 integrase. Nat Methods 2:975–979
14. Held PK, Olivares EC, Aguilar CP, Finegold M, Calos MP, Grompe M (2005) In vivo correction of murine hereditary tyrosinemia type I by phiC31 integrase mediated gene delivery. Mol Ther 11:399–408
15. Smith MC, Till R, Brady K, Soultanas P, Thorpe H (2004) Synapsis and DNA cleavage in phiC31 integrase-mediated site-specific recombination. Nucleic Acids Res 32: 2607–2617
16. Lister JA (2010) Transgene excision in zebrafish using the phiC31 integrase. Genesis 48:137–143
17. Saitoh N, Bell AC, Recillas-Targa F, West AG, Simpson M, Pikaart M, Felsenfeld G (2000) Structural and functional conservation at the boundaries of the chicken beta-globin domain. EMBO J 19:2315–2322
18. Bell AC, West AG, Felsenfeld G (2001) Insulators and boundaries: versatile regulatory elements in the eukaryotic genome. Science 291:447–450
19. West AG, Gaszner M, Felsenfeld G (2002) Insulators: many functions, many mechanisms. Genes Dev 16:271–288

20. Allen BG, Weeks DL (2006) Using phiC31 integrase to make transgenic *Xenopus laevis* embryos. Nat Protoc 1:1248–1257
21. Allen BG, Weeks DL (2009) Bacteriophage phiC31 integrase mediated transgenesis in *Xenopus laevis* for protein expression at endogenous levels. Methods Mol Biol 518: 113–122
22. Chesneau A, Sachs LM, Chai N, Chen Y, Du Pasquier L, Loeber J, Pollet N, Reilly M, Weeks DL, Bronchain OJ (2008) Transgenesis procedures in Xenopus. Biol Cell 100: 503–521
23. Sive HL, Grainger RM, Harland RM (2000) Early development of *Xenopus laevis*: a laboratory manual. Cold Spring Harbor Laboratory Press, Cold Spring Harbor
24. Kroll KL, Amaya E (1996) Transgenic Xenopus embryos from sperm nuclear transplantations reveal FGF signaling requirements during gastrulation. Development 122:3173–3183
25. Sparrow DB, Latinkic B, Mohun TJ (2000) A simplified method of generating transgenic Xenopus. Nucleic Acids Res 28:E12
26. Ogino H, McConnell WB, Grainger RM (2006) Highly efficient transgenesis in Xenopus tropicalis using I-SceI meganuclease. Mech Dev 123:103–113
27. Pan FC, Chen Y, Loeber J, Henningfeld K, Pieler T (2006) I-SceI meganuclease-mediated transgenesis in Xenopus. Dev Dyn 235: 247–252
28. Nieuwkoop PD, Faber J (1994) Normal table of *Xenopus laevis* (Daudin). Garland, New York

Chapter 14

Xenopus Transgenics: Methods Using Transposons

Clair M. Kelley, Donald A. Yergeau, Haiqing Zhu, Emin Kuliyev, and Paul E. Mead

Abstract

The generation of transgenic animals is an essential tool for many genetic strategies. DNA "cut-and-paste" transposon systems can be used to efficiently modify the *Xenopus* genome. The DNA transposon substrate, harbored on a circularized plasmid, is co-injected into fertilized *Xenopus* embryos at the one-cell stage together with mRNA encoding the cognate transposase enzyme. The cellular machinery rapidly translates the exogenous mRNA to produce active transposase enzyme that catalyzes excision of the transposon substrate from the plasmid and stable integration into the genomic DNA.

Key words: Transposon, *Sleeping beauty*, *Tol2*, Transgenesis, *Xenopus*

1. Introduction

Frogs of the *Xenopus* species are particularly well suited for study of early vertebrate development as the large, externally developing embryos are produced in vast numbers throughout the year. Used for decades as models for cell and developmental biology, the South African clawed frog, *Xenopus laevis*, has not been viewed as a practical genetic model due to its long generation time (~1 year) and paleotetraploid genome. The closely related species, *Xenopus tropicalis*, offers the same experimental features as *X. laevis*, and is better suited for genetic manipulation due to a diploid genome and relatively short generation time (~6 months) (1). Due to the large size of the egg and external development, microinjection techniques are widely used in *Xenopus* laboratories. There are several strategies available for generating transgenic *Xenopus* that use microinjection to deliver exogenous DNA to the host genome either pre- or post-fertilization. Restriction endonuclease-mediated integration (REMI) uses injection of in vitro-manipulated sperm

Stefan Hoppler and Peter D. Vize (eds.), *Xenopus Protocols: Post-Genomic Approaches, Second Edition*, Methods in Molecular Biology, vol. 917, DOI 10.1007/978-1-61779-992-1_14, © Springer Science+Business Media, LLC 2012

nuclei into mature oocytes (2). The isolated sperm nuclei are incubated prior to injection with a cocktail of DNA encoding the transgene of interest and a restriction enzyme. The presence of the restriction enzyme is thought to increase integration efficiency by producing nicks in the host genomic DNA. The microinjection of the manipulated sperm nucleus activates the egg and results in normal cell division and development. The REMI technique has been used successfully to generate transgenic frogs; the efficiency of the technique is often around 15% but can be higher (2–4). In REMI the transgene frequently integrates as a concatamer and the in vitro manipulation of the sperm nuclei sometimes results in genetic damage to the DNA and thus the resulting transgenic lines produced may have underlying genetic lesions. The REMI procedure is technically more demanding than a simple microinjection procedure, and it requires a skilled hand for isolation and manipulation of the sperm nuclei. A simpler approach is to deliver the exogenous DNA post-fertilization. This offers two significant advantages; first, healthy zygotes can be readily sorted and selected for microinjection and second, as many *Xenopus* laboratories routinely use this technique for manipulating gene expression during early development, any laboratory equipped to do injections can readily generate transgenic lines. There are two basic strategies for generating transgenic *Xenopus* by injecting DNA post-fertilization; direct DNA injection and mobile element-mediated transgenesis.

The first strategy simply uses a linearized DNA template that is injected into the one-cell embryo (5). Integration of the linear exogenous fragment is mediated by the cellular machinery at sites of random nicks in the host genome. This method frequently results in the integration of the transgene as a concatamer, and the overall efficiency of the technique is low; in our hands ~1% of injected embryos will pass the transgene onto their progeny. As integration may occur at early cleavage stages, the transgenic founders are also likely to be mosaic for the transgene and will result in non-Mendelian inheritance in the first outcross generation. A modified version of this simple linear DNA injection method is to flank the transgene with rare-cutter restriction endonuclease sites, such as I-*Sce*I that recognizes an 18-bp sequence, and to co-inject the DNA fragment together with a small amount of the enzyme (Chapter 12) (6, 7). Linear DNA injected into the fertilized *Xenopus* embryo is rapidly joined end-to-end by the cellular DNA repair machinery, and the extended concatamer is thought to be a less efficient substrate for integration. The presence of the rare-cutter restriction enzyme maintains the injected DNA in a linear monomeric state. The host genome is unaffected by the presence of the I-*Sce*I enzyme as the long recognition sequence of the enzyme results in few, if any, target sites in vertebrate genomes.

The second strategy uses mobile DNA elements and their cognate enzymes to facilitate integration of the transgene into the host genome. DNA cut-and-paste transposon systems are ideally suited to this application (8–11). These systems are comprised of two parts, a substrate DNA molecule that is flanked by transposon repeat elements and a transposase enzyme that recognizes the repeat sequences, excises the transposon from the plasmid and inserts it into the host genome. One-cell embryos are injected with plasmid DNA harboring a transposon and synthetic mRNA encoding the transposase enzyme. In addition to DNA transposons, phage integrase systems such as *phiC31* have been used to modify the *Xenopus* genome (12, 13) (see *also* Chapter 13). We have focused our studies on the use of DNA transposable elements for generating transgenic *Xenopus* (14–16).

Transposons offer several advantages for germline transgenesis. First, as the transposon substrate is excised from the plasmid vector prior to integration into the host genome, the integrated exogenous DNA does not include plasmid sequences that have been shown to lead to epigenetic silencing of transgenes in higher vertebrates (17). Second, the presence of the repeat elements that flank the transposon allows rapid identification of the genomic integration site using simple PCR-based strategies. Third, DNA cut-and-paste transposons efficiently integrate into the frog genome, and transgenesis rates of 30–50% have been achieved using the *Sleeping Beauty* and *Tol2* systems in both *X. laevis* and *X. tropicalis* (14, 15, 18). Fourth, multiple independently segregating alleles can be recovered from a single founder. For example, using the *Tol2* system in *X. tropicalis* frequently results in up to six independent integration events in a single founder. Individual alleles can be segregated by serial outcross of the founder lines. Fifth, integration into the host genome is essentially random. The *Sleeping Beauty* system, for example, integrates at TA dinucleotides in the target genome. Finally, once integrated into the genome, the DNA transposon is now a substrate for subsequent excision and re-integration following re-expression of the transposase enzyme (16). This feature allows for two significant manipulations; if the integration event has caused a mutation, due to, for example, gene trapping and premature termination of the trapped open reading frame, this mutation can be corrected by excising the transposon out of endogenous locus by simply re-expressing the transposase enzyme. This approach can be used to verify that the transposon integration event is indeed responsible for the observed mutation. The ability to excise and re-integrate stably inherited transposons also provides a mechanism for large-scale mutagenesis screens, where transposons resident in the genome are hopped to new loci by providing the transposase enzyme *in trans*. In the accompanying chapter (see Chapter 6), we will outline transposon remobilization strategies for forward genetic screens in the *X. tropicalis*. Here, we will describe the protocols for generating transgenic *X. tropicalis* using two DNA transposons, *Sleeping Beauty* and *Tol2*. We have used the same method to efficiently generate transgenic *X. laevis* using both the *Sleeping Beauty* and *Tol2* transposon systems.

Sleeping Beauty and *Tol2* offer unique advantages for transgenesis in *Xenopus*. *Sleeping Beauty* (*SB*), a member of the *Tc1/mariner* family, was molecularly reconstructed from an ancient inactive transposon found in salmon (19). *SB* integrates at TA dinucleotides in the target genome (20), and thus promotes essentially random transgenesis. *SB*-mediated integration results in duplication of the TA dinucleotide target site. We, and others, have shown that when co-injected into *Xenopus*, the *SB* system frequently results in integration of small-order concatamers at discrete loci in the frog genome (15, 18, 21). These multimeric units can include parts of the plasmid vector and likely arise due to a non-canonical transposition. The genomic sequences flanking these complex integration products can be difficult to clone as the transgenic animal may contain multiple copies of each transposon arm, and sections of plasmid vector sequences, which can complicate conventional PCR-based cloning methods. Nevertheless, *SB* efficiently integrates into the frog genome and transgenes are stable in the subsequent generations. We routinely observe transgenesis rates of up to 30%, or more, using the *SB* system in *Xenopus*. The non-canonical transposition mechanism observed with the co-injection strategy but is not a general feature of *SB* in *Xenopus*. Re-expression of *SB* transposase in transgenic frogs that harbor an *SB* transposon results in precise excision and re-integration of discrete transposons via canonical transposition. The remobilized *SB* transposons integrate precisely at boundaries of the transposon arms, and the integrations are flanked by the expected target site duplication for *SB*, a TA dinucleotide (see accompanying Chapter 6).

Tol2 is a member of the *hAT* family of DNA transposons (22). Cloned from an albino line of Medaka fish, *Tol2* was identified as the first naturally occurring, active DNA cut-and-paste transposon in a vertebrate species (23). Like *SB*, *Tol2* efficiently integrates into the frog genome (14). *Tol2*, however, integrates using a conventional transposition mechanism. We have shown that *Tol2* mediates integration of discrete transposons, with no contaminating plasmid vector sequences, and the boundaries of the integrated *Tol2* transposons are flanked by the characteristic 8-bp target site duplications. Transgenesis rates achieved with *Tol2* in the frog are similar to those with *SB*; 30% or more. Unlike *SB*-mediated transgenics that usually contain one or, at most, two independently segregating alleles, we frequently identify *Tol2* founders with up to six transposon integration events. The independently segregating *Tol2* integrations can be isolated by serial outcross of the founders; however, given the fact that it may take several generations to segregate each of the transposon transgenes to individual alleles, this can add to the overall time it takes to derive discrete transgenic lines with the *Tol2* system.

Both *Tol2* and *SB* have high cargo capacities (>10 kB), although the transposition efficiency has been shown to decrease in the *SB* system with larger DNA inserts between the transposon arms (24). We have cloned a 10 kB tissue-specific transgene into both *Tol2* and *SB*, and achieved efficient germline transgenesis with each system in both *X. laevis* and *X. tropicalis*.

SB or *Tol2*? Both systems offer unique advantages. Injection-mediated delivery of the *Tol2* system results in canonical transposition of the substrate transposon into the frog genome. Each founder may have six, or more, independently segregating transposon integration sites, requiring serial outcross of the founder to isolate transgenic lines with single transposition events. With *SB*, the co-injection approach results in small-order complex concatamers stably integrated into the genome. The genomic sequences flanking the *SB* concatamers can be difficult to clone as the concatamer may contain several copies of the transposon arms that are used as anchors for PCR-cloning strategies. Furthermore, the concatamers do not integrate precisely at the end of the transposon arm, and frequently include parts of the plasmid vector (14). The formation of concatamers is an advantage when subsequent remobilization of the resident transposon is planned, as the concatamer provides more potential substrates for remobilization than a single-copy integration event. While both systems work well for making transgenic frogs, if the goal is to generate a single-copy transgenic line, then the *Tol2* system will be superior to *SB*.

Other DNA transposon systems, including *piggyBac* (25), *Frog Prince* (26), *minos* (27), and *Tol1* (28, 29), have been used for transgenesis in a variety of model systems, and are likely to function in *Xenopus*. We have recently demonstrated that the *Tol1* system is effective for transgenesis in *X. tropicalis* (unpublished observations). These mobile elements will add depth to the molecular toolbox available for manipulating the frog genome.

Finally, some general considerations for the design of transgenic constructs. Wherever possible, we try and use *Xenopus* sequences to generate transgene constructs rather than using those developed for other vertebrate species. The rationale being that the DNA regulatory elements are likely to function better in the same species and thus produce a more robust expression of the transgene. To determine the DNA sequences needed for a particular tissue-restricted expression pattern, we use published data from other species as a first guide to identify potential regulatory elements. Cross-species sequence comparisons are used to identify highly conserved non-coding regions and CpG islands using online resources such as VISTA (28) and metazome (http://www.metazome.net). It is important to include a functional intron into the transgene construct to allow normal processing of the nascent transcript. Whenever possible we include an intron from the genomic locus of the selected gene. In general, we use the first intron of the gene and engineer the reporter in frame with the sequences in the second exon. This ensures that the endogenous transcription and translation start sites, and any regulatory sequences harbored in the first intron, are intact. A functional polyadenylation signal is also required for efficient processing of the transgene. We generally use a strong polyadenylation signal such as those from SV40 or the rabbit β-globin gene.

2. Materials

Prepare all solutions with ultrapure nuclease-free water and analytical grade reagents. DNA and RNA stocks should be stored in small single-use aliquots to avoid repeated freeze–thaw cycles. All animal work should be done in accordance with local regulations and guidelines.

2.1. Components to Induce Egg Laying

1. Adult male frogs are readily available from commercial sources such as Nasco (http://www.enasco.com/page/xenopus), and cost ~ $35US per animal.
2. Human chorionic gonadotrophin (HCG; 1,000 units/mL, Novarel) resuspended in sterile water and stored at 4°C.
3. 1 mL syringes.
4. 27-gauge needles.
5. Plastic HDPE 5 gallon buckets (Green or Black; see Note 1) with lids (Letica, MI).
6. Frog System Water (FSW). Fresh water drawn directly from the adult *X. tropicalis* aquarium. The water conditions are maintained at a salinity of 1,000 μS and pH 7.0 at 25°C.
7. Polycarbonate tank (45 × 30 × 15 cm; 13.25 L) with lid (Rubbermaid Commercial Products, VA).
8. Mesh disc (~80 mm diameter) for harvesting eggs from mating tank (Spectramesh PP500, polypropylene mesh with 500 μm openings; Spectrum Laboratories, CA).
9. Glass crystallizing dishes (Kimax, 90 × 50 mm).
10. 3% (w/v) cysteine solution to de-jelly embryos. 15 g of l-cysteine (free-base, non-animal source, Sigma Aldrich, MO) dissolved in 500 mL of FSW. Adjust pH to pH 8.0 using 10 M NaOH.
11. 3% (w/v) Ficoll (PM 400, Sigma Aldrich, MO) in 0.05× MMR for microinjections. Dissolve 20 g of Ficoll in approximately 400 mL of sterile water in a sterile 1 L bottle. Microwave on full for ~2 min then carefully swirl to completely dissolve the Ficoll. Cool to room temperature. Add 3 mL of 10× MMR

(Marc's Modified Ringers; 1 M NaCl, 20 mM KCl, 10 mM $MgSO_4$, 20 mM $CaCl_2$, 50 mM HEPES pH 7.8 and 1 mM EDTA) and adjust final volume to 600 mL with water.

2.2. Preparing Plasmid DNA and mRNA for Microinjections

1. Plasmid DNA harboring a transposon substrate is isolated using a Qiagen midi-prep column and further purified by phenol:chloroform:isoamyl alcohol extraction, ethanol precipitation, and finally washed with 70% ethanol prior to resuspending the DNA pellet in nuclease-free water (see Note 2). The DNA concentration is estimated by measuring the absorbance at 260 nm of duplicate dilutions. The integrity of the sample is determined by digesting an aliquot of the plasmid DNA with a restriction endonuclease, and running the products on a standard agarose gel. If the sample appears smeary then do not use for injections and repeat the DNA isolation from a fresh overnight bacterial culture. Store aliquots of the DNA at −20°C.
2. Synthetic messenger RNA (mRNA) is prepared using an Ambion mMessage mMachine kit in accordance with the manufacturer's instructions (see Note 3). Determine the yield of RNA spectrophotometrically and check the integrity of the sample on a denaturing agarose gel. Store aliquots of the transposase mRNA at −80°C.
3. Immediately prior to injecting, prepare cocktail of plasmid DNA (50–100 pg in 3 nL) and mRNA (50–500 pg in 3 nL) diluted to the appropriate concentration with sterile nuclease free water (see Note 4). Centrifuge at full speed in a microfuge to pellet any particulate material that may clog the injection needle. Keep on ice.

2.3. Components Needed for Microinjections

1. Siliconized glass microscope slides.
2. Plastic disposable transfer pipettes.
3. Microinjection needles. These will vary depending on the microinjection step up in your laboratory. Borosilicate glass needles (10 mm, OD 1.0 mm, ID 0.50 mm) are pulled on a commercial needle puller (such as a Model P97 Flaming/Brown Pipette Puller, Sutter Instrument Co.), trimmed with a razor blade, and beveled at ~25° on a horizontal grinder (EG-40, Narishige). Needles for *X. tropicalis* injections need to be of a smaller diameter than those used for *X. laevis* eggs. For *X. tropicalis*, the injection volume is 3 nL. Needles are calibrated by measuring the numbers of pulses required to expel 1 μL of sterile water from the needle at an injection pressure of 20 psi and an injection time of 500 ms. We routinely use needles of 30 or higher for *X. tropicalis* embryo injections, and 20 or higher for *X. laevis* injections.

4. Watchmakers forceps to maneuver the embryos on the glass slide (Roboz 55, Dumont, Switzerland).
5. Agarose-coated 150 mm plastic dishes (Fisher 08-757-148). Dissolve 0.5% (w/v) of agarose (Roche 11 685 678 001) in FSW by heating in microwave oven, pipette 10 mL into each dish, swirl to evenly distribute the agarose solution, and allow to solidify at room temperature. Store at 4°C for up to 2 weeks (see Note 5).
6. Large glass crystallizing dishes (Kimax 190 × 100 mm).

2.4. Food for Tadpoles and Juvenile Frogs

1. Sera Micron food (Sera).
2. FSW is used to prepare a suspension of Sera micron powder for tadpole feeding. Fresh water is drawn directly from the adult *X. tropicalis* aquarium. The water conditions are maintained at a salinity of 1,000 μS and pH 7.0 at 25°C.
3. JurassiDiet Newt & Frog Food (JurassiPet).

3. Methods

3.1. Preparing Xenopus tropicalis Embryos for Microinjection

There are two methods available to harvest fertilized embryos for microinjection; in vitro fertilization and natural matings. The first method requires manually extracting eggs from primed females and mixing the eggs with macerated testes harvested from a euthanized male frog. Although this method has its advantages, including being able to time the production of eggs for injection experiments, it requires euthanizing at least one male frog each day that injections are planned. As we perform injection experiments at least 4 days a week throughout the year, we opt for natural matings for embryo harvest due to the cost of sacrificing male frogs for these experiments. Male frogs can be re-used for natural matings every 2 weeks. Female *X. tropicalis* adults can be used for egg collection every 8–10 weeks.

1. Pre-prime female frogs with HCG the night before embryo harvest. Two to three female frogs are used each day for embryo harvest. The female frogs are injected with 20U of HCG into the dorsal lymph sac using a 1 mL syringe fitted with a 27-gauge needle, and placed in a bucket of FSW overnight in the vivarium. A lid is placed loosely on the bucket to prevent escape of the frogs.
2. Prime female frogs with 200U of HCG in the morning and place the animals into a 45 × 30 × 15 cm tank containing FSW to a depth of ~8 cm (~7 L). Place the tank on a cart lined with a black plastic sheet large enough to wrap over and cover the tank (see Note 6).

3. Prime an equal number of male frogs as females by injecting with 100U HCG, and place them in the tank with the females. Place a lid on the tank and drape with the black plastic sheet. Place the tank in a quiet area of the laboratory and leave undisturbed for ~3 h. The frogs will initiate amplexus and begin laying eggs in 3–4 h after induction with HCG.
4. Harvest eggs from the tank using a disc of nylon mesh cut to fit loosely into a 90 mm crystallizing dish. The jelly coat surrounding the eggs will adhere to the mesh and sufficient eggs can be collected in minutes.
5. Place the mesh disc into a small crystallizing dish, cover with ~100 mL of 3% (w/v) cysteine solution and swirl gently. The mild reducing activity of the cysteine solution will dissociate the jelly coat surrounding the eggs in approximately 3 min. As the jelly coat dissolves, the eggs will release from the mesh disc and the mesh circle can be removed from the dish.
6. After 3 min in the cysteine solution, carefully pour off the liquid and wash the embryos four to five times with ~100 mL of FSW.
7. Finally, wash twice with ~100 mL of 3% (w/v) Ficoll solution, leaving the embryos that are now ready for injection, in the second Ficoll wash.
8. The mating pairs will continue to produce fertilized eggs for several hours during the afternoon, and embryos can be collected for microinjection throughout this time period.

3.2. Microinjection of *Xenopus tropicalis* Embryos

1. Using a transfer pipette, place a large drop of 3% (w/v) Ficoll solution on a siliconized slide on the injection apparatus.
2. Add approximately 50 embryos to the drop. The surface tension of the liquid on the siliconized slide will keep the liquid in a discrete drop. Observe the embryos under low-power through the binocular microscope, and gently move any defective embryos to one side of the slide with a pair of watchmakers forceps. Likewise, any embryos that have already progressed beyond the first cell division can be removed as injections are ideally performed at the one-cell stage.
3. Inject each fertilized embryo with 3 nL of DNA/mRNA mixture. Support the embryo with the forceps to prevent it slipping across siliconized slide when injecting. Move the injected embryos to one side until all have been injected.
4. With a transfer pipette, gently move the injected embryos to a 150 mm agarose-coated Petri dish containing the 3% ficoll solution. Place the dish of embryos in an incubator overnight at 26–28°C.
5. Repeat the injection procedure until all embryos have been injected and added to the agarose-coated dish containing ficoll

solution. If the embryos have begun to divide, discard the batch and collect fresh embryos from the mating tank as described above.

3.3. Care of Injected Embryos

1. Leave the injected embryos in the ficoll solution overnight. By the next morning the embryos will have begun to elongate and hatch (~stage 26). Healthy embryos are transferred to a large crystallizing dish containing ~2 L of FSW and maintained in a warm room at 28°C.
2. The embryos can be observed for transgene expression (GFP, RFP etc.) at various stages of development. We do not routinely sort injected embryos for transgene expression, but rather grow all the embryos and determine germline transmission of the transgene by outcrossing the adults (see Note 7).
3. As the embryos grow, they are transferred to larger polycarbonate tanks and raised to adulthood. Each week, approximately one-half of the water in each tank is removed and replaced with fresh FSW. An airstone, with very gentle bubbling, is added to each tadpole tank. Tadpoles are fed three times per day with a suspension of Sera-micron food. After metamorphosis, the juvenile frogs are kept in the same tank for at least 2 weeks before being moved to the general population in the adult aquarium (see Note 8). Juvenile frogs are fed once a day with Jurassi-Diet Newt And Frog Food (Jurassipet).

4. Notes

1. We use color-coded buckets to prevent cross contamination between our *X. laevis* (black buckets) and *X. tropicalis* (green buckets) aquaria. We avoid using light colored buckets as *Xenopus* frogs are less stressed when placed on a dark background.
2. Do not use mini-prep plasmid DNA as there will be contaminating nucleases that may interfere with the transgenesis protocol.
3. Precipitate mRNA using the ammonium acetate method provided with the kit. Do not use LiCl precipitation as residual lithium ions may interfere with normal development of the embryo. Fresh aliquots are used each day and the mRNA is never refrozen for use on subsequent days.
4. The ratio of DNA and mRNA can be changed, however, do not exceed the upper limits of DNA and RNA quoted above. We have had excellent transgenesis results with ratios over the

range of 1:1 (50 pg each of DNA substrate and enzyme mRNA in 3 nL) to 1:5 (100 pg of DNA and 500 pg of mRNA in 3 nL).

5. Coating the dishes with a thin layer of agarose prevents the injected embryos from adhering to the dish and enhances survival.
6. As noted above, frogs appear less stressed when they are able to hide on a dark background. Thus, a black plastic sheet is used under the mating tank. The sheet of plastic is large enough to wrap over the covered tank to provide an environment to promote natural mating and shield the animals from movements in the laboratory.
7. The co-injection method described here results in the formation of chimeric founder animals that may be highly mosaic for the transgene. This is due to fact that the injected enzyme mRNA needs to be translated by the host cellular machinery before it can catalyze the transposition reaction. Thus, early cleavage events may have occurred before sufficient transposase protein has been produced to initiate transposition. Furthermore, the injected nucleic acids may not evenly distribute throughout the fertilized egg, adding to the potential for mosaicism. The result is that different blastomeres of the developing embryo may have different integration events. Only blastomeres that contribute to the germline will pass the transposition event onto the progeny. Thus, observing transgene expression in the developing injected embryo may not provide a clear indication of germline transgenesis. Somatic expression of the transgene may not accurately reflect the germline status of the injected animal. As the rate of transgenesis with the transposon systems is relatively high; we generally do not presort the injected embryos based on transgene expression, but grow all of the injected animals and determine germline transmission by observing the outcross population.
8. Moving frogs to the aquarium system too soon after metamorphosis can result in significant morbidity. We usually keep the young frogs in the static tanks for at least 2–3 weeks postmetamorphosis to begin feeding on pellet food and to gain weight before transfer to the flow-through aquarium.

Acknowledgments

This work was supported by NIH grants HD042994 and MH079381 and by the American Lebanese and Syrian Associated Charities (ALSAC) to PEM.

References

1. Amaya E, Offield MF, Grainger RM (1998) Frog genetics: *Xenopus tropicalis* jumps into the future. Trends Genet 14:253–255
2. Amaya E, Kroll KL (1999) A method for generating transgenic frog embryos. Methods Mol Biol 97:393–414
3. Sparrow DB, Latinkic B, Mohun TJ (2000) A simplified method of generating transgenic *Xenopus*. Nucleic Acids Res 28:E12
4. Chesneau A, Sachs LM, Chai N, Chen Y, Du Pasquier L, Loeber J, Pollet N, Reilly M, Weeks DL, Bronchain OJ (2008) Transgenesis procedures in *Xenopus*. Biol Cell 100: 503–521
5. Etkin LD, Pearman B (1987) Distribution, expression and germ line transmission of exogenous DNA sequences following microinjection into *Xenopus laevis* eggs. Development 99:15–23
6. Ogino H, McConnell WB, Grainger RM (2006) Highly efficient transgenesis in *Xenopus tropicalis* using I-*Sce*I meganuclease. Mech Dev 123:103–113
7. Ogino H, McConnell WB, Grainger RM (2006) High-throughput transgenesis in *Xenopus* using I-*Sce*I meganuclease. Nat Protoc 1:1703–1710
8. Johnson Hamlet MR, Mead PE (2003) *Sleeping Beauty* and *Xenopus*: transposons as genetic tools. Curr Genomics 4:687–697
9. Miskey C, Izsvak Z, Kawakami K, Ivics Z (2005) DNA transposons in vertebrate functional genomics. Cell Mol Life Sci 62: 629–641
10. Ivics Z, Izsvak Z (2004) Transposable elements for transgenesis and insertional mutagenesis in vertebrates: a contemporary review of experimental strategies. Methods Mol Biol 260: 255–276
11. Yergeau DA, Mead PE (2007) Manipulating the *Xenopus* genome with transposable elements. Genome Biol 8(suppl 1):S11
12. Allen BG, Weeks DL (2005) Transgenic *Xenopus laevis* embryos can be generated using *phiC31* integrase. Nat Methods 2:975–979
13. Allen BG, Weeks DL (2006) Using *phiC31* integrase to make transgenic *Xenopus laevis* embryos. Nat Protoc 1:1248–1257
14. Johnson Hamlet MR, Yergeau DA, Kuliyev E, Takeda M, Taira M, Kawakami K, Mead PE (2006) *Tol2* transposon-mediated transgenesis in *Xenopus tropicalis*. Genesis 44:438–445
15. Yergeau DA, Johnson Hamlet MR, Kuliyev E, Zhu H, Doherty JR, Archer TD, Subhawong AP, Valentine MB, Kelley CM, Mead PE (2009) Transgenesis in *Xenopus* using the *Sleeping Beauty* transposon system. Dev Dyn 238: 1727–1743
16. Yergeau DA, Kelley CM, Kuliyev E, Zhu H, Sater AK, Wells DE, Mead PE (2010) Remobilization of *Tol2* transposons in *Xenopus tropicalis*. BMC Dev Biol 10:11
17. Whitelaw E, Sutherland H, Kearns M, Morgan H, Weaving L, Garrick D (2001) Epigenetic effects on transgene expression. Methods Mol Biol 158:351–368
18. Doherty JR, Johnson Hamlet MR, Kuliyev E, Mead PE (2007) A flk-1 promoter/enhancer reporter transgenic *Xenopus laevis* generated using the *Sleeping Beauty* transposon system: an in vivo model for vascular studies. Dev Dyn 236:2808–2817
19. Ivics Z, Hackett PB, Plasterk RH, Izsvak Z (1997) Molecular reconstruction of *Sleeping Beauty*, a *Tc1*-like transposon from fish, and its transposition in human cells. Cell 91: 501–510
20. Vigdal TJ, Kaufman CD, Izsvak Z, Voytas DF, Ivics Z (2002) Common physical properties of DNA affecting target site selection of *sleeping beauty* and other *Tc1/mariner* transposable elements. J Mol Biol 323:441–452
21. Sinzelle L, Vallin J, Coen L, Chesneau A, Du Pasquier D, Pollet N, Demeneix B, Mazabraud A (2006) Generation of trangenic *Xenopus laevis* using the *Sleeping Beauty* transposon system. Transgenic Res 15: 751–760
22. Koga A, Suzuki M, Inagaki H, Bessho Y, Hori H (1996) Transposable element in fish. Nature 383:30
23. Kawakami K (2007) *Tol2*: a versatile gene transfer vector in vertebrates. Genome Biol 8(suppl 1):S7
24. Karsi A, Moav B, Hackett P, Liu Z (2001) Effects of insert size on transposition efficiency of the *sleeping beauty* transposon in mouse cells. Mar Biotechnol (NY) 3:241–245
25. Ding S, Wu X, Li G, Han M, Zhuang Y, Xu T (2005) Efficient transposition of the *piggyBac* (*PB*) transposon in mammalian cells and mice. Cell 122:473–483
26. Miskey C, Izsvak Z, Plasterk RH, Ivics Z (2003) The *Frog Prince*: a reconstructed transposon

from *Rana pipiens* with high transpositional activity in vertebrate cells. Nucleic Acids Res 31:6873–6881

27. Franz G, Loukeris TG, Dialektaki G, Thompson CR, Savakis C (1994) Mobile *Minos* elements from *Drosophila hydei* encode a two-exon transposase with similarity to the paired DNA-binding domain. Proc Natl Acad Sci U S A 91: 4746–4750
28. Koga A, Cheah FS, Hamaguchi S, Yeo GH, Chong SS (2008) Germline transgenesis of zebrafish using the medaka *Tol1* transposon system. Dev Dyn 237:2466–2474
29. Koga A, Shimada A, Kuroki T, Hori H, Kusumi J, Kyono-Hamaguchi Y, Hamaguchi S (2007) The *Tol1* transposable element of the medaka fish moves in human and mouse cells. J Hum Genet 52:628–635

Chapter 15

Comparative Genomics-Based Identification and Analysis of Cis-Regulatory Elements

Hajime Ogino, Haruki Ochi, Chihiro Uchiyama, Sarah Louie, and Robert M. Grainger

Abstract

Identification of cis-regulatory elements, such as enhancers and promoters, is very important not only for analysis of gene regulatory networks but also as a tool for targeted gene expression experiments. In this chapter, we introduce an easy but reliable approach to predict enhancers of a gene of interest by comparing mammalian and *Xenopus* genome sequences, and to examine their activity using a co-transgenesis technique in *Xenopus* embryos. Since the bioinformatics analysis utilizes publically available web tools, bench biologists can easily perform it without any need for special computing capability. The co-transgenesis assay, which directly uses polymerase chain reaction products, quickly screens for the activity of the candidate elements in a cloning-free manner.

Key words: Cis-regulatory elements, Enhancer, Comparative genomics, Multiple alignment, Phylogenetic footprinting, Transcription factor-binding motif, *Xenopus* transgenesis

1. Introduction

Identification and analysis of cis-regulatory elements with conventional approaches have generally been slow and laborious work. Although promoters with basal transcriptional activity may be easily identified as the short upstream sequences adjacent to transcription start sites, enhancers that control spatio-temporal specificity and strength level of gene expression are often located far from the basal promoter. Even if enhancers are found, it is not easy to identify transcription factors that bind to the enhancer by recognizing ambiguous sequences (1). However, recent progress in comparative genomics has markedly improved the situation. Comparison of

Stefan Hoppler and Peter D. Vize (eds.), *Xenopus Protocols: Post-Genomic Approaches, Second Edition*,
Methods in Molecular Biology, vol. 917, DOI 10.1007/978-1-61779-992-1_15, © Springer Science+Business Media, LLC 2012

orthologous genomic sequences from different vertebrate species identifies conserved non-coding elements (CNEs) as reliable candidates for enhancers, and their phylogenetic footprinting analysis predicts transcription factor-binding motifs (TFBMs) that are highly conserved through evolution (2). Analysis of the genome-wide distribution of CNEs showed that they are especially enriched around developmental regulatory genes where there are clearly conserved regulatory gene circuits involved (3). The power of the comparative genomics-based approach and biological significance of the conserved enhancers have already been demonstrated by a number of studies, including the analyses of a *Shh* limb enhancer and a *Sox9* mandibular enhancer whose mutations are responsible for congenital polydactyly and cleft palate, respectively (4–6). There are also reports that show lack of evolutionary sequence conservation of vertebrate enhancers (7). However, if one focuses primarily on classic-style promoter bashing so as not to miss non-conserved elements, one may miss critical conserved enhancers located far away from basal promoters. In the above cases of *Shh* and *Sox9*, the enhancers are located more than 1 Mb away from their transcription start sites. If comprehensive identification of enhancers is necessary, it may be performed by more elaborate methods, for example, as with the ChIP-seq analysis using anti-p300 antibody followed by a large-scale transgenesis screening (8) (see Note 1).

In search of conserved enhancers, though fish genomic sequence is often used for comparison with mammalian sequences, the *Xenopus tropicalis* genome sequence may be more generally suitable for this purpose. This is because, while the *Xenopus* genome is evolutionarily distant enough from mammalian genomes for identifying functionally active CNEs, its structure is much closer to that of mammalian genomes than is the fish genome, where many duplications and rearrangements followed by subfunctionalization of cis-regulatory elements occurred after they had diverged from other vertebrates (9–11).

The power of combining the mammalian–*Xenopus* genome comparison with an efficient transgenesis technique in *Xenopus* was demonstrated by a cis-regulatory analysis of *foxe3*, a transcription factor gene essential for lens development (12). This study demonstrated that the classic-style promoter analysis and the comparative genomics-based approach gave consistent results by identifying the same enhancer. The phylogenetic footprinting analysis of this enhancer identified nine conserved TFBMs in its 0.4 kb length, and their mutational analyses in founder transgenics led to the discovery of Notch signaling as a new lens-inducing signal.

The approach described in this chapter includes the following steps: (a) retrieving large orthologous genome sequences of a gene of interest from comparative genome browsers; (b) aligning the retrieved multiple sequences; (c) extracting orthologous CNE sequences; (d) generating a phylogenetic footprint of these CNEs; (e) searching for conserved TFBMs in the CNEs; and (f) performing the co-transgenesis assay to assess possible enhancer activity of the CNEs in vivo (12). The comparative genome browsers, such as the ECR Browser (13), show pre-made, multiple pair-wise alignments of vertebrate genome sequences. While such alignments are quite useful for quickly looking at sequence conservation associated with a gene of interest, their parameter settings are typically adjusted for whole genome comparisons and are often too stringent for precise analysis of a specific gene locus. Thus, we use the comparative genome browsers as a source of large orthologous genome sequences, then re-align them using a sensitive genome aligner, PipMaker, for identifying CNEs (14). PipMaker often discovers CNEs that are not found by the comparative genome browsers due to their relatively low sequence conservation or to local inversions in compared genomes. Furthermore, while the comparative genome browsers show just pair-wise alignments at the nucleotide level view, PipMaker can generate a multiple alignment that contains more than three sequences. After the PipMaker analysis, the CNE sequences can be extracted from the output files and further re-aligned using a more sensitive, short sequence aligner, ClustalW, for phylogenetic footprinting (15). In the co-transgenesis assay, a polymerase chain reaction (PCR)-amplified CNE fragment is injected into *Xenopus laevis* eggs along with a basal promoter-GFP-polyA cassette. Since these two DNA fragments are integrated together into the host chromosome at the same site at the one-cell stage, GFP exhibits non-mosaic expression in the embryos if the injected CNE has enhancer activity. Since the co-transgenesis assay does not involve any cloning, one can finish enhancer screening of ~10 CNEs within 2 weeks, which appears to be a typical number of CNEs associated with one gene.

In the postgenomic era, one of the central aims of developmental biology is, undoubtedly, the global understanding of the gene regulatory networks (GRNs) that are hardwired in the genome. While pioneering GRN studies have achieved remarkable success in sea urchin, fruit fly, and nematode, this type of study has been very limited in vertebrates (16). This is mostly because the vertebrates have much larger genomes than these invertebrates, which makes it difficult to perform the genome-wide cis-regulatory analysis that is essential for GRN studies. However, the approach described in this chapter effectively combines the comparative genomics tools and the high-throughput transgenesis assay in *Xenopus* to overcome this problem. The resulting collection of cis-regulatory elements will be a powerful starting point for a genome-wide analysis of regulatory interactions between transcription factors and their target genes, and provide us a framework for elucidating the GRNs that are conserved in vertebrates.

2. Materials

2.1. Bioinformatics Tools

1. Comparative genome browser: ECR Browser (http://ecrbrowser.dcode.org/).
2. Genome aligner software: (basic) PipMaker, Advanced PipMaker, and MultiPipMaker (http://pipmaker.bx.psu.edu/pipmaker/).

   PipMaker and Advanced PipMaker generate pairwise alignments, and MultiPipMaker is used to align three or more sequences. MultiPipMaker is generally used because of the utility of multiple genome alignments. Links to detailed instructions of these programs are found in the website.
3. Supporting software tool for PipMaker: PipHelper (http://pipmaker.bx.psu.edu/piphelper/).
4. Simple sequence manipulation software tool: Range Extractor DNA (http://www.bioinformatics.org/sms2/range_extract_dna.html).
5. Short sequence aligner software: ClustalW (http://www.ebi.ac.uk/Tools/msa/clustalw2/).
6. TFBM search programs: rVista (http://rvista.dcode.org/) (17), ConSite (http://asp.ii.uib.no:8090/cgi-bin/CONSITE/consite) (18).
7. TFBM database: Transfac (http://www.gene-regulation.com/index.html, a free version is available from http://www.gene-regulation.com/cgi-bin/pub/databases/transfac/search.cgi), JASPAR (http://jaspar.cgb.ki.se/cgi-bin/jaspar_db.pl) (19).

It is helpful if one has a standard DNA analysis software tool, such as VectorNTI (Invitrogen) or equivalent.

2.2. Reagents and Equipment for Preparation of DNA Fragments for Xenopus Co-transgenesis Assay

Other reagents and equipments for obtaining *Xenopus* eggs and transgenesis are found in Chapter 11 of this book, and that for in situ hybridization analysis are found in reference (20).

- *X. tropicalis* genomic DNA: 50 ng/μL (see Note 2).
- pBSSK+βGFP (12): a pBluescript plasmid carrying the chicken β-*actin* basal promoter (21) adjacent to a GFP-polyA cassette. This plasmid is available from H. Ogino. A plasmid carrying the *gata2* basal promoter linked to the GFP-polyA (22), which is available from R. M. Grainger, may provide more sensitivity for the co-transgenesis assay since it appears to impart higher levels of transcriptional activity to enhancers without causing background expression on its own.
- QIAquick PCR Purification Kit (Qiagen).
- QIAquick Gel Extraction kit (Qiagen).

- Primer pairs for amplification of CNEs: 10 pmol/μL in TE (see Subheading 3 for details), stored at −20°C.
- Platinum *Pfx* DNA polymerase: 2.5 U/μL (Invitrogen), or an equivalent proof-reading DNA polymerase for high-fidelity PCR.

For the following, generic reagents and equipment are likely to suffice:

- 50 mM $MgSO_4$ (included in a set of the Platinum *Pfx* DNA polymerase).
- dNTP mixture, 2.5 mM each (Invitrogen).
- *Bam*HI: 10–20 U/μL (New England Biolabs).
- *Xho*I: 10–20 U/μL (New England Biolabs).
- 10× NEB3 buffer (for double-digestion with *Bam*HI and *Xho*I, New England Biolabs).
- 1× TAE buffer: 40 mM Tris-acetate, 1 mM EDTA, pH 8.0.
- Agarose (molecular biology grade).
- Ethidium bromide solution: 10 μg/mL (Sigma).
- TE buffer: 10 mM Tris–HCl, 1 mM EDTA, pH 8.0.
- 10 mM KCl.
- Ethanol (96–100%, molecular biology grade).
- 70% Ethanol (prepare by mixing 96–100% ethanol and distilled water).
- Distilled water.
- Heating block or water bath (for restriction enzyme digestion at 37°C, and for melting agarose gel at 50°C).
- GeneAmp PCR System 2400 (PerkinElmer) or an equivalent standard thermal cycler.
- UV transilluminator (for visualizing ethidium bromide-stained DNA after gel electrophoresis).
- Microcentrifuge.

3. Methods

The following sections describe how to identify and analyze enhancers in *Xenopus*, using as an example our analysis of the 5′ flanking region of a paired domain homeobox gene, *pax6* (23). This is only a small part of the large cis-regulatory region surrounding *pax6*, which spans at least 180 kb in mouse genome and contains more than nine enhancers (23, 24). The "C" region identified in the following sections was originally reported to be an enhancer for expression in the photoreceptors of the developing mouse retina (24). However, our analysis in *Xenopus* revealed its early embryonic activity in the developing optic vesicle and overlying presumptive lens ectoderm (H. Ogino and R. M. Grainger, unpublished).

3.1. Identification of CNEs Associated with a Gene of Interest

1. Extract genomic sequences from genome assembly data
   The first step in generating an alignment is identifying orthologous genomic regions of a gene of interest in multiple species and creating their text files in FASTA format (see Note 3). This can be done easily with the ECR Browser, where one can choose a base genome for the alignment from 13 vertebrate species including *X. tropicalis*, and then view a stacked pairwise genome alignment of any gene or location that was generated between the base genome and other vertebrate genomes (see Note 4). Figure 1 is a screen shot of the ECR Browser window, where the 5′ flanking region of human *PAX6* (hg18 chr11: 31786832–31806234; position 31786832–31806234 of chromosome 11 in human genome assembly 18) is chosen

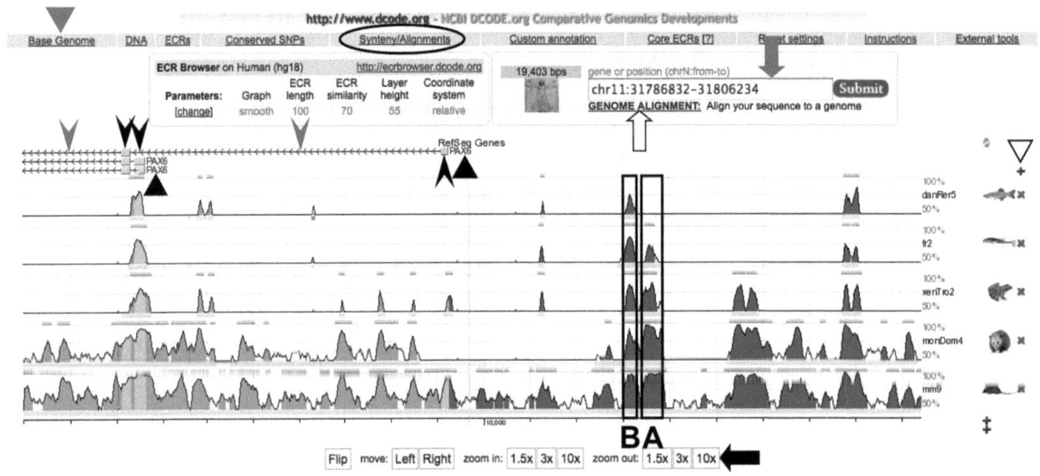

Fig. 1. A screen shot of the ECR Browser window. Above the schematic representation of the alignment, the gene names (*PAX6*, indicated by *black triangles*), exons (*boxes* indicated by *black arrowheads*), and introns (*horizontal lines* indicated by *gray arrowheads*) in the base genome are shown. *Arrows* associated with the *horizontal line* representing introns indicate transcriptional orientation of the gene. The conserved regions shown as peaks include not only exons but also CNEs located in introns and intergenic regions. Among these CNEs, those indicated as A and B (*boxed*) were shown to have enhancer activity in the pancreas and lens/cornea in transgenic mouse analyses, respectively (32). One can change the base genome species by clicking the "Base Genome" button (indicated by a *gray triangle*) at the *upper left*, but it is not necessary to choose *Xenopus* at this step, even if it will be used as the base sequence in a subsequent PipMaker analysis. On the *right* can be seen a + or an × next to the genomes shown (*white triangle*), permitting addition or removal of genomes from the alignment. The length of the alignment is adjusted using buttons such as "10×" at the *bottom* of the window (*black arrow*), or by directly inputting a desired genomic position of the base genome into the *rectangular box* at the *upper right* (*gray arrow*). One may also input just a gene name into the *rectangular box* to move to another gene locus. If one needs to align user-identified sequence to the genomes in this browser, click the "GENOME ALIGNMENT" button below the *rectangular box* (*white arrow*), and input the sequence as a FASTA format. For details, go to the instruction page of ECR Browser (http://ecrbrowser.dcode.org/ecrInstructions/ecrInstructions.html).

as the base genome sequence and is aligned with orthologous regions from mouse, opossum, *Xenopus*, *Fugu*, and zebrafish genome sequences. The regions with high sequence similarity are graphically displayed as peaks, whose height and width represent their sequence conservation level and length, respectively. One can change the base genome species, add or remove genomes from the alignment, and adjust the length of the alignment by clicking the buttons on the browser window (see the legend for Fig. 1). To retrieve the sequences aligned in the window, click the "Synteny/Alignments" button (circled in Fig. 1), which will open a new window showing a list of the sequences (see Note 5). In this window, click check boxes of the sequences that are needed for the subsequent PipMaker analysis, and then click the "Mulan" button at the left bottom corner (see Note 6). A link list will be visible for the sequences in FASTA format (ex. seq1.fa, seq2.fa). Download all the necessary sequences.

2. Build accompanying files
   The second step is to build Exons, Mask, and Underlay files for the sequence that is chosen to be the base sequence in the PipMaker analysis. The Exons and Mask files indicate the location of exons and repetitive sequences in the base sequence, respectively. The Underlay file instructs the program to shade specific gene structures (ex. exons and introns) with colors in the output. To build these files, open the FASTA file of the sequence that will be used as the base sequence in the PipMaker analysis below (ex. the FASTA file of human *PAX6* downloaded in the previous step, using TextEdit on a Macintosh or Notepad on a Windows computer to open it), and find the name of genome assembly and position of the sequence in the assembly in the header line of the text (ex. hg18 chr11:31786832–31806234). Next, go to the PipHelper website (http://pipmaker.bx.psu.edu/piphelper/), choose the genome assembly (ex. Mammal, Human, Mar. 2006 (NCBI36/hg18)), enter the position (ex. chr11:31786832–31806234), choose an annotation source for exons (ex. RefSeq Genes), and click the "submit" button. This will generate a link list for downloading the Exons, Mask, and Underlay files. If these files for the chosen base sequence are not available from PipHelper, one has to manually create them (see Note 7).

3. Generate a multiple genomic sequence alignment
   At the PipMaker website, click on MultiPipMaker, and enter the number of sequences to be aligned. On the next page, select "Generate nucleotide level view (PDF)" option to receive a raw sequence alignment data in addition to a schematic "Pip (Percent identity plot)" view as the output files. In the "First sequence" section, give a desired label for the base sequence,

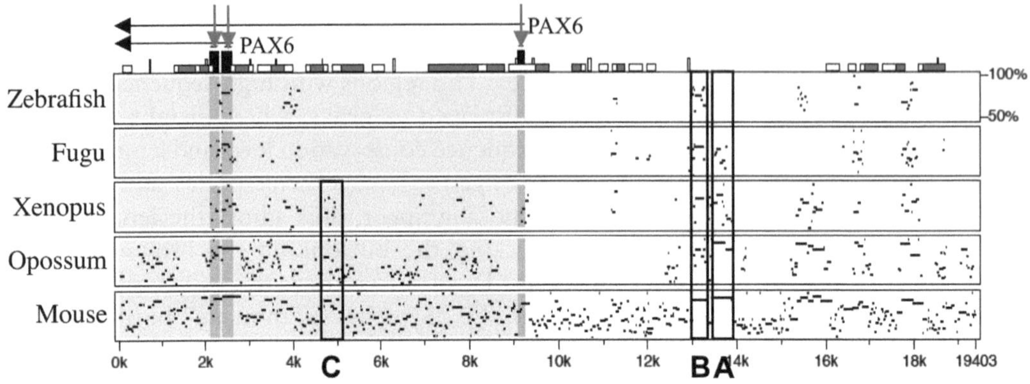

Fig. 2. A pip view output of MultiPipMaker analysis that was generated with sequences downloaded from the ECR Browser window shown in Fig. 1. The alignments between the base sequence (human) and zebrafish, *Fugu*, *Xenopus*, opossum, and mouse sequences are shown from the *top* to the *bottom*. Above the alignment, the gene name (*PAX6*) and its transcriptional orientation (*horizontal black arrows*) are shown. The *black boxes* on the *top* of the Pip view (indicated by *vertical gray arrows*) indicate the exons, and other *boxes* indicate repeats, transposons and other sequence features in the base sequence (go to PipMaker Instructions for details, http://pipmaker.bx.psu.edu/pipmaker/pip-instr.html). The regions of the zebrafish, *Fugu*, *Xenopus*, opossum, and mouse sequences that correspond to exons and introns of the base sequence are also shaded in the alignments with *dark gray* and *light gray*, respectively. In addition to the pancreas enhancer (A) and lens/cornea enhancer (B), the C region, which is conserved only in tetrapods, is *boxed*.

upload the base sequence in FASTA format, the Mask, Exons, and Underlay files for the base sequence, and check the "use as default in pip" box. An annotation file is not necessary for the analysis. Below the "First sequence" section, enter labels and upload sequence files for the additional orthologous sequences in FASTA format, and check the "Search both strands," "Show all matches" and "High sensitivity and low time limit" options (see Note 8). Underlay files are not necessary for these sequences. After entering the necessary information, click the "submit" button.

One will receive by e-mail a "Pip view" file that schematically shows the alignment (Fig. 2), and a "nucleotide level view" file that shows an actual multiple alignment of the genomic sequences (Fig. 3). In the Pip view, the height and width of the Pip represent the sequence conservation level and length, respectively. In our example, comparison of Figs. 1 and 2 reveals that PipMaker discovered not only the CNEs shown in the ECR Browser window (including the previously identified enhancers shown as A and B in Fig. 1), but also some additional CNEs. One of them, which is marked as "C" in Fig. 2, is conserved in tetrapods, but not in fish. In the nucleotide level view of the C region (Fig. 3), the *Xenopus*, opossum, and mouse nucleotides that are identical to the human nucleotides are shown as dots in the multiple alignment. This multiple alignment is useful for cloning of the sequences and for preliminary search for conserved TFBMs.

```
                4734          4741          4751          4761          4771
                   :             :             :             :             :
4725: CATTTTTACAACC---TCTCAATTAGCGACCGAGTGGATTAGCACCTTCT   Human
4467:   .....C...G.AAG...........A..T.TTG.......T..GAAA.   Xenopus
5963: .............---...........C.TTCG.................   Opossum
4644: .............---...........A...C..............C...   Mouse

                4781          4791                        4809          4819
                   :             :                           :             :
4772: TAAGCTTTGCCTCGGAAAGAAGCAGCCG--TAATCTTGTTGCTATTTCCC   Human
4515: A.T..C..TG...C..G.------...TC--.....A............T.   Xenopus
6010: .....AGAAG...T...G.....GC...CC....................   Opossum
4691: ..............AA......G......A--..................   Mouse

                4829          4836          4846          4856          4866
                   :             :             :             :             :
4820: CCCAGGGCAC---TGGCTCTTTGAATCAGCTCCTTCTCCCACCCCTCCAA   Human
4558: ..T......GAGTG....T.G.-----------------------A..C   Xenopus
6060: ..........A--G..T......TG..CA....C..CTT...TT.-T..C   Opossum
4739: ..........---..........C.......C...............--A...   Mouse

                4876          4886          4896          4905
                   :             :             :             :
4867: CCCCCATCTCGCTGCGGAGAAAAGTTTTAACAAAAAAT-CAGAAAAGGC-   Human
4584: ..T..---CGA.AAAAA.A...........GAG.....A-A.A....AT.A   Xenopus
6107: T.T..C.TC.C..C.--.A...T..........C.C.CA-...TC..AATC   Opossum
4784: ..........A.CT.A......................AT.....G.A..-   Mouse

                4923          4929          4939                        4953
                   :             :             :                           :
4915: -AAGGAGCGAGGAG----TGAATGCCACTGACGTCATT------GGGTGG   Human
4630: T....G.GAG....ATGA..T..A.AG............TTAATG.A.A..   Xenopus
6154: AT.A.G.GAG...ATCTC.C......-..........C.-----G......   Opossum
4833: -T......CT..CA----.........G...........------......   Mouse

                4959                                      4977
                   :                                         :
4954: G----GAAGGGGGCTCCG------------GGAAAGACTCCTGGAAA---   Human
4680: .----.GG.......G..                                   Xenopus
6198: .TAAA.GTT.....AAA.AGGGGAGAGAAA........ATT.A....ACA   Opossum
4872: .----..T.......T..------------..TGG....GT.......---   Mouse
```

Fig. 3. A nucleotide level view of MultiPipMaker analysis. Only part of the alignment that corresponds to the C region is shown.

3.2. Phylogenetic Footprinting Analysis of a CNE

1. Re-align orthologous CNE sequences using ClustalW

   It is very helpful to re-align the CNE sequences with a more accurate aligner for short sequences, ClustalW, and then shade conserved nucleotides with a multiple alignment editor, Jalview (25), for better visualization of the alignment and more efficient search of TFBMs. First, open the genome sequence files that were used for the PipMaker analysis, find the orthologous CNE sequences identified by PipMaker (in our example, the human, mouse, opossum, and *Xenopus* sequences for the C region of *Pax6*), and generate new text files in FASTA format for just the CNE sequences. This step can be easily done if you use Range Extractor DNA or a standard DNA analysis tool, such as VectorNTI. Next, go to ClustalW website, select the "DNA" mode, enter a set of the orthologous CNE sequences in a Multi-FASTA format (see Note 3), and click the "submit" button. One sees a raw ClustalW alignment of the input sequences in the next window.

2. Shade conserved nucleotides in a ClustalW alignment using Jalview
In the ClustalW alignment window, click "Result Summary," and then click the "Start Jalview" button. A new window appears, showing the sequence alignment. In this Jalview window, the alignment format can be modified as desired. For example, first click the "Format" menu, and then click "Wrap" to change the default "single column view" to a "multi-column view" so that the alignment length fits to the window width. Next, click "Color," and then click "Percent identity" to shade conserved nucleotides. Although the default setting shows a schematic representation of the sequence conservation under the alignment column, it can be removed by clicking "View" menu and then deselecting "Show Annotations." After the modification, the alignment image can be saved using the "screen shot" function (push "cmd" + "shift" + "4" + "space" keys, or use the packaged program "Grab" on a Macintosh, and "Print Screen" key on a Windows computer). Unfortunately, at this ClustalW website Jalview does not have a menu to export the alignment image to one's computer (see Note 9).
Figure 4 is an example of the Jalview output that shows a multiple alignment of the C region identified by the PipMaker analysis (compare with Fig. 3). Methods used for identifying the TFBMs shown in this figure are described in the next section.

3. Searching for TFBMs conserved in CNEs
Program tools for searching TFBMs are classified into three groups: pattern search programs, weight matrix search programs, and Hidden Markov Model (HMM)-based programs (1). Because HMM-based programs require some special mathematical knowledge, pattern and weight matrix search programs are more widely used for conventional analysis. We often use rVista for pattern search, and ConSite for a weight matrix search. This is because these two programs search TFBMs conserved in a pair of orthologous sequences, whereas most other programs work only on a single sequence (see Note 10). Although rVista and ConSite have default sets of TFBMs, they are limited in number and poorly defined TFBMs that may not be relevant to in vivo binding. Hence we strongly recommend creating a user-defined set by collecting TFBMs that

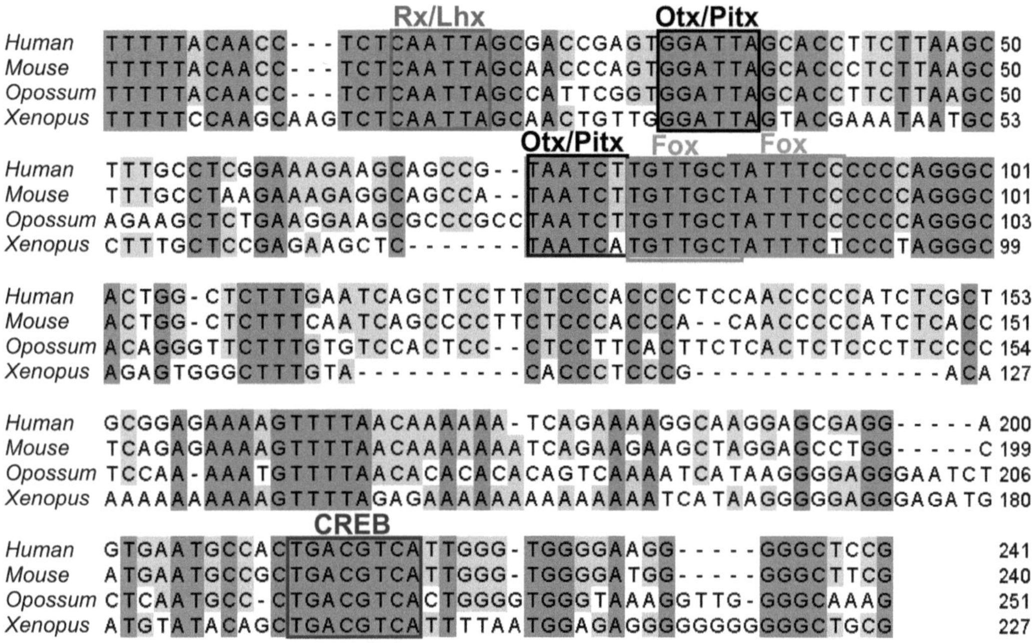

Fig. 4. Phylogenetic footprinting analysis of *Pax6* C region. The ClustalW alignment was shaded with Jalview, in which conserved TFBMs that were identified by rVista analysis are mapped.

are likely involved in regulation of a CNE of interest from the literature and TFBM databases for use in these programs (see Note 11).

Go to the rVista website, input two orthologous CNE sequences in FASTA format (in our example, the human and *Xenopus* sequences of the C region used for the ClustalW analysis), and click "Submit." Next, select "User-defined consensus sequences," enter the TFBMs (in our example, CREB/ATF tgacgtca, Otx/Pitx taakcy, Rx/Lhx taatkr, Fox trttkvy), and click "Submit." The "Results" window will then show TFBMs conserved between the human and *Xenopus* sequences. Since rVista shows the TFBMs only one by one in the resulting pair-wise alignment, one has to draw the identified TFBMs in the Jalview alignment for making a visually effective, final output, as shown in Fig. 4. Conservation of the TFBMs in other aligned sequences (the mouse and opossum sequences in Fig. 4) is also checked at this step.

3.3. Analysis of Enhancer Activity of CNEs by Co-transgenesis Assay in Xenopus

1. Amplify CNEs by genomic PCR

   First, amplify a *Xenopus* CNE sequence from *X. tropicalis* genomic DNA by PCR. PCR primers should have *Bam*HI sites and an extra 3-base overhang at the 5′ ends for efficient digestion of the resulting PCR product with *Bam*HI. The PCR product with *Bam*HI-protruding ends will be efficiently ligated to the reporter cassette with a compatible end, after their injection

Table 1
The reaction mixture and cycle setting used for amplification of the C region of *Xenopus Pax6*

Reagents	Volume (μL)
Distilled water	57
10 × *Pfx* buffer	20
50 mM $MgSO_4$	2
dNTP mix (2.5 mM each)	12
C-1 primer (10 pmol/μL)	3
C-2 primer (10 pmol/μL)	3
X. tropicalis genomic DNA (50 ng/μL)	2
Platinum *Pfx* DNA polymerase (2.5 U/μL)	1

The cycle setting: 94°C 5 min→(94°C 30 sec→58°C 1 min→68°C 1 min) × 38 cycles→68°C 7 min

into *Xenopus* eggs (see below). If the CNE sequence under study has *Bam*HI sites internally, *Bgl*II sites may be added to the primer ends in place of the *Bam*HI sites so that the PCR product has the protruding ends compatible with the *Bam*HI end of the reporter cassette (see Note 12). In our example, the primers used for amplification of the C region of *Xenopus pax6* are as follows (*Bam*HI sites are underlined with the extra three bases): C-1, GACGGATCCTTCCAAGCCTTGAATTGACCATCTG; C-2, GACGGATCCACTGATCGCTTCCAAAAACCCAG. They were designed to bind to the 5′ and 3′ flanking regions of the C element in *Xenopus* genome (see Note 13). The PCR condition used with these primers is shown in Table 1.

2. Purify DNA fragments (CNEs and a reporter cassette) for the co-transgenesis assay
   After the PCR reaction, take 1/10 volume of the resulting reaction for evaluation by agarose gel electrophoresis (0.7–1.5% agarose in 1× TAE buffer) and stain the gel with ethidium bromide (0.5 μg/mL in 1× TAE buffer) to check if the expected product is amplified as a single band (see Note 14). In our example, the primer sets for the *Xenopus* C element are expected to amplify a fragment of about 0.4 kb that contains a conserved 227 base sequence in the middle of the fragment. If the product has the expected size, purify the remaining product using the QIAquick PCR purification kit, and then digest with *Bam*HI at 37°C overnight to expose their *Bam*HI protruding ends. In parallel, digest pBSSK+βGFP with *Bam*HI and *Xho*I to cut out a reporter cassette (β-actin basal promoter-GFP-polyA) at 37°C overnight (see Note 15).

After the digestion, separate the DNA fragments by agarose gel electrophoresis, stain with ethidium bromide, cut out the desired DNA bands from the gel, and then purify the DNA using the QIAquick Gel Extraction Kit. The expected length of the reporter cassette is about 1.0 kb. After the purification with the QIAquick Gel Extraction Kit, precipitate the DNA with ethanol once, and dissolve in 10 mM KCl solution so that the final concentration will be 50–100 ng/μL (see Note 16). The purified DNA solution is stored at -20°C until use in transgenesis.

3. Co-inject a purified CNE fragment together with the reporter cassette into *X. laevis* eggs

A standard protocol for the sperm nuclear transplantation method of *X. laevis* transgenesis is found in Chapter 11 (see Note 17). For the co-transgenesis, mix the purified CNE fragment and the reporter cassette in a molar ratio of 2:1, and inject this mixture with the sperm nuclei into the eggs (Fig. 5a). After the injection and subsequent development, fix the resulting embryos at desired developmental stages, and subject them to in situ hybridization with antisense GFP probe. The in situ hybridization analysis is necessary for detecting GFP expression in embryos using this transgenesis protocol, because the expression from a β-*actin* basal promoter of the reporter cassette is not strong enough for epifluorescence microscopy in most cases. However, because the background expression from this promoter is very weak, it makes it possible to clearly distinguish the enhancer-driven expression from the promoter background in an in situ hybridization analysis. The GFP probe can be generated by transcribing the *Hind*III-digested pBSSK+βGFP with T7 RNA polymerase. For in situ hybridization and probe synthesis, follow the standard protocol for *Xenopus* (20). The *gata2* promoter discussed in the "Materials" section may elicit more robust expression of CNEs by co-transgenesis and therefore direct detection of expression by GFP fluorescence, though there may be a significant delay between the transcription of GFP mRNA and the accumulation and/or maturation of GFP protein for fluorescence during development.

If the injected CNE has enhancer activity, approximately 5–20% of the developing embryos typically express GFP, though the fraction of expressing embryos depends on a number of factors, including the strength of the enhancer. If the enhancer activity is relatively weak, the color reaction of the in situ hybridization experiment may take 1–2 days. Figures 5b and c show the example of GFP expression in the transgenic embryo generated with the C region and the reporter cassette. GFP expression was detected in this case in the eye (see Note 18).

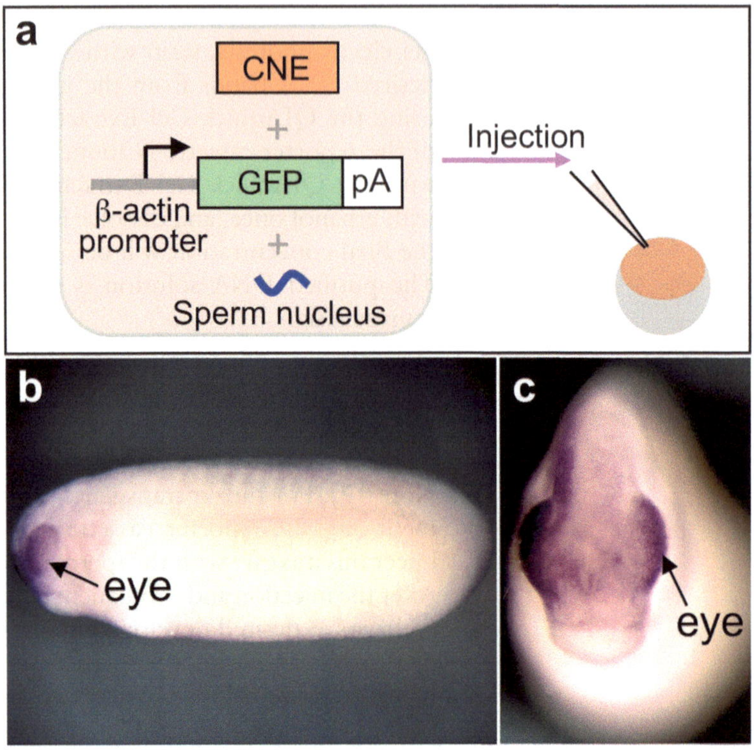

Fig. 5. The co-transgenesis assay and GFP expression in the resulting embryo detected by in situ hybridization. (**a**) A schematic representation of the co-injection of a CNE fragment, the reporter cassette, and the sperm nucleus into *Xenopus* eggs. (**b**, **c**) GFP expression in the embryo co-injected with the (**c**) element of *Xenopus Pax6* and the GFP reporter cassette. *Arrows* indicate eye-specific expression of GFP ((**b**) lateral view; (**c**) frontal view).

4. Notes

1. p300 is a transcriptional coactivator that is a near-ubiquitously expressed component of enhancer-associated protein assemblies. Since ChIP-seq analysis showed that p300 is significantly enriched within 10 kb up- and downstream of promoters, it may be effective to search non-conserved enhancers in the −10 to +10 kb region in addition to searching for enhancer activity in CNEs (8). It is also noteworthy that another ChIP-seq analysis revealed that enhancers of early developmental genes display three overlapping chromatin signatures, monomethylation of histone H3 at lysine 4 (H3K4me1), trimethylation of histone H3 at lysine 27, and p300-binding, in embryonic stem cells (26). Incorporation of such information about genome-wide epigenetic modifications may give the comparative genomics approach even more promise.

2. One may follow the *Xenopus* standard protocols for collecting blood cells from the heart, and purification of the genomic DNA from the blood cells or embryos (20). The DNeasy Blood & Tissue Kit (Qiagen) also works well.
3. Definitions of FASTA and Multi-FASTA format are found in Oxford University CBRG website (http://www.compbio.ox.ac.uk/bioinformatics_faq/format_examples.shtml).
4. The UCSC Genome Browser (http://genome.ucsc.edu/) and Vista Browser (http://genome.lbl.gov/vista/index.shtml) provide alternate options for retrieval of orthologous genomic sequences of a gene of interest. In the genome alignment window of the UCSC Genome Browser, click the "Convert" menu, and follow the directions. In the alignment window of the Vista Browser, click "Show Alignment(s)," and then click "Sequence (softmasked)" in the newly opened page.
5. The "Synteny/Alignments" window sometimes show two or more sequences as orthologs for a single base sequence. In such a case, one has to check whether the genome browser identified true multiple orthologs generated by a species-specific genome duplication (this often occurs in a mammalian–fish comparison), or misidentified closely related genes as the orthologs. For an example of the latter case, the genome browsers often misidentify both mouse *Sox2* and *Sox3* as the orthologs of human *SOX2*. A very useful database, Metazome, can help one to check orthologous relationships between genes from different species by showing their synteny conservation (http://www.metazome.net/).
6. This webpage is designed to forward the sequences to Mulan, a genome alignment tool that searches sequence conservation with higher sensitivity than ECR Browser in a limited length of orthologous genomic sequences. Since the sensitivity of Mulan is still lower than that of PipMaker, we usually choose to use Pipmaker for identifying CNEs. In our experience, Mulan often misses small CNEs that are found by PipMaker in mammalan–*Xenopus* genome comparison. Furthermore the nucleotide level output of Mulan is not a typical multiple sequence alignment and less accessible than that of PipMaker for bench biologists to extract CNE sequences for subsequent phylogenetic footprinting analysis, though the schematic view of the alignment generated by Mulan is better than that of PipMaker.
7. One can easily generate a Mask file using the RepeatMasker tool at the ISB website (http://www.repeatmasker.org/cgi-bin/WEBRepeatMasker) from a genome sequence of interest. For Exons and Underlay files, one has to generate these by analyzing gene structures using a standard DNA analysis software tool, such as Blast (http://blast.ncbi.nlm.nih.gov/Blast.cgi) or Vector NTI.

8. The "Show all matches" setting makes it possible for one region of the base sequence to align with several regions of the second sequence because of partial sequence duplications, or because of incomplete masking of low-complexity regions. The "Chaining" setting allows identifying and plotting only matches that appear in the same relative order in the base and second sequences. With the "Single coverage" setting, PipMaker selects a highest scoring set of alignments such that any position in the base sequence can appear in at most one alignment. We often use "Show all matches," which leads to "multi-hits" of low complexity regions, such as repetitive elements, of the base sequence to the aligned sequences and distinguishes them from CNEs with enhancer activity, which are generally unique in the genome and show a "single-hit" between the base and aligned sequences. The "High sensitivity and low time limit" option increases the MultiPipMaker sensitivity so that it can identify weak conservation between a sequence pair at a great evolutionary distance, such as human and *Fugu* sequences. The default setting without this option is tuned for comparison of two mammalian sequences. Further detailed instructions for these settings are found on the Advanced PipMaker instructions website (http://pipmaker.bx.psu.edu/pipmaker/pip-instr3.html).
9. Although Jalview can be easily used at this ClustalW website, its function is in part restricted. If its full function is needed, go to the Jalview website of University of Dundee (http://www.jalview.org/index.html), where one can use ClustalW, fully functional Jalview, and easily save the resulting alignment image to a local computer.
10. For searching TFBMs in a single sequence, Patch, Match, and P-Match are available for public use (http://www.gene-regulation.com/pub/programs.html). Patch and Match are pattern search and weight matrix search programs, respectively. P-Match combines pattern matching and weight matrix approaches thus providing higher accuracy of recognition than either of the methods alone. One can choose Transfac or a user-defined set of TFBMs for running these programs.
11. One can develop a custom set by selecting TFBMs from a TFBM database, such as Transfac or JASPAR (http://jaspar.cgb.ki.se/cgi-bin/jaspar_db.pl) (19). A collection of binding motifs of homeodomain proteins is also useful (27). It is also strongly recommended to search the literature for specific TFBMs that are likely involved in regulation of a CNE of interest and validated by stringent tests defining them as sites used in vivo. Many of the TFBMs in databases such as Transfac were generated by in vitro binding experiments and may include weak sites that do not work or are not relevant in vivo. Thus,

one may easily be misled by the multitude of false positives if the TFBM collection from the database alone is used for the search.

12. In the co-transgenesis assay, the addition of compatible protruding ends to both an enhancer fragment and the reporter cassette increases the number of transgenic embryos with enhancer-driven reporter expression by approximately twofold (C. Uchiyama, H. Ochi and H. Ogino, unpublished observation). The ligation reaction between these co-injected DNA fragments may be mediated by the homologous end-joining activity of *Xenopus* eggs (28).
13. A web-tool, Primer3 (http://primer3.sourceforge.net/) is convenient for designing PCR primers.
14. One may follow standard molecular cloning protocols for the restriction enzyme digestion, agarose gel electrophoresis, and ethanol precipitation of DNA (29).
15. Digestion of 10 μg of pBSSK+βGFP would be sufficient for preparing the reporter cassette for 10–20 rounds of co-transgenesis.
16. The last ethanol precipitation step is for removal of residual carryover of the gel extraction buffer and the DNA elution buffer from the kit. The DNA elution buffer in the kit contains Tris buffer, which is toxic for *Xenopus* embryos and sometimes causes ectopic transgene expression if introduced into the embryos with the DNA.
17. The co-transgenesis assay uses the sperm nuclear transplantation method (see Chapter 11). Because *X. laevis* eggs are larger and more tolerant to the sperm nuclear transplantation than *X. tropicalis* eggs, we usually choose *X. laevis* for this experiment. The sperm nuclear transplantation method may be somewhat technically demanding for those who are not familiar with the *Xenopus* system. In this case, one may clone a CNE fragment into a basal promoter–GFP reporter plasmid and perform the *Xenopus* transgenic assay using a more convenient technique, such as the *I-SceI* meganuclease method or others (see Chapters 12, 13, and 14) (30, 31). It is slower than the co-transgenesis assay, but still faster and more reliable than any enhancer assay in other vertebrates, and can be performed with *X. tropicalis* if necessary.
18. One may also use another eye enhancer of *pax6*, which is located in intron 7, as a positive control for the co-transgenesis assay (30). This enhancer may exhibit stronger activity than that of the C region in tailbud embryos. The primers for amplification of this intronic enhancer of *X. tropicalis pax6* are as follows (*Bam*HI sites are underlined with the extra 3 bases): int7-1, <u>GACGGATCC</u>ACTGGGTGGGGGTAATTCCT;

int7-2, GACGGATCCGGGAGATAAATACAGGGGGTC. The PCR condition for this primer pair is nearly the same as that shown in Table 1, except that the annealing temperature should be 54°C. The expected length of the amplicon is about 0.6 kb.

Acknowledgements

This work was supported by the Global COE Program in NAIST (Frontier Biosciences: strategies for survival and adaptation in a changing global environment), Grant-in-Aid for Scientific Research (C) (20579002 and 23570256) from JSPS and Grant-in-Aid for Scientific Research on Innovative Areas (21200064) from MEXT, Japan, to H. Ogino, by Grant-in-Aid for Young Scientists (B) (21770234) from JSPS and Research for Promoting Technological Seeds (A) (10-099) from JST, Japan, to H. Ochi, and by CREST (JST). This work was also supported by NIH grants EY00675, EY017400, EY018000, and RR013221 to R. Grainger.

References

1. Wasserman WW, Sandelin A (2004) Applied bioinformatics for the identification of regulatory elements. Nat Rev Genet 5:276–287
2. Pennacchio LA et al (2006) In vivo enhancer analysis of human conserved non-coding sequences. Nature 444:499–502
3. Woolfe A et al (2005) Highly conserved non-coding sequences are associated with vertebrate development. PLoS Biol 3:e7
4. Lettice LA et al (2003) A long-range *Shh* enhancer regulates expression in the developing limb and fin and is associated with preaxial polydactyly. Hum Mol Genet 12:1725–1735
5. Sagai T et al (2004) Phylogenetic conservation of a limb-specific, cis-acting regulator of *Sonic hedgehog* (*Shh*). Mamm Genome 15:23–34
6. Benko S et al (2009) Highly conserved non-coding elements on either side of *SOX9* associated with Pierre Robin sequence. Nat Genet 41:359–364
7. Fisher S et al (2006) Conservation of *RET* regulatory function from human to zebrafish without sequence similarity. Science 312:276–279
8. Visel A et al (2009) ChIP-seq accurately predicts tissue-specific activity of enhancers. Nature 457:854–858
9. Hellsten U et al (2010) The genome of the Western clawed frog *Xenopus tropicalis*. Science 328:633–636
10. Kasahara M et al (2007) The medaka draft genome and insights into vertebrate genome evolution. Nature 447:714–719
11. Canestro C, Yokoi H, Postlethwait JH (2007) Evolutionary developmental biology and genomics. Nat Rev Genet 8:932–942
12. Ogino H, Fisher M, Grainger RM (2008) Convergence of a head-field selector Otx2 and Notch signaling: a mechanism for lens specification. Development 135:249–258
13. Ovcharenko I et al (2004) ECR Browser: a tool for visualizing and accessing data from comparisons of multiple vertebrate genomes. Nucleic Acids Res 32:W280–W286
14. Schwartz S et al (2003) MultiPipMaker and supporting tools: alignments and analysis of multiple genomic DNA sequences. Nucleic Acids Res 31:3518–3524
15. Larkin MA et al (2007) Clustal W and Clustal X version 2.0. Bioinformatics 23:2947–2948
16. Stathopoulos A, Levine M (2005) Genomic regulatory networks and animal development. Dev Cell 9:449–462
17. Loots GG, Ovcharenko I (2004) rVISTA 2.0: evolutionary analysis of transcription factor binding sites. Nucleic Acids Res 32: W217–W221
18. Sandelin A, Wasserman WW, Lenhard B (2004) ConSite: web-based prediction of regulatory

elements using cross-species comparison. Nucleic Acids Res 32:W249–W252

19. Portales-Casamar E et al (2010) JASPAR 2010: the greatly expanded open-access database of transcription factor binding profiles. Nucleic Acids Res 38:D105–D110
20. Sive H, Grainger R, Harland R (2000) Early development of *Xenopus laevis*—a laboratory manual. Cold Spring Harbor Laboratory Press, New York
21. Kost TA, Theodorakis N, Hughes SH (1983) The nucleotide sequence of the chick cytoplasmic β-actin gene. Nucleic Acids Res 11:8287–8301
22. Navratilova P et al (2009) Systematic human/zebrafish comparative identification of cis-regulatory activity around vertebrate developmental transcription factor genes. Dev Biol 327:526–540
23. Kleinjan DA et al (2006) Long-range downstream enhancers are essential for *Pax6* expression. Dev Biol 299:563–581
24. Xu PX et al (1999) Regulation of *Pax6* expression is conserved between mice and flies. Development 126:383–395
25. Waterhouse AM et al (2009) Jalview version 2—a multiple sequence alignment editor and analysis workbench. Bioinformatics 25: 1189–1191
26. Rada-Iglesias A et al (2011) A unique chromatin signature uncovers early developmental enhancers in humans. Nature 470:279–283
27. Noyes MB et al (2008) Analysis of homeodomain specificities allows the family-wide prediction of preferred recognition sites. Cell 133:1277–1289
28. Lehman CW, Trautman JK, Carroll D (1994) Illegitimate recombination in *Xenopus*: characterization of end-joined junctions. Nucleic Acids Res 22:434–442
29. Sambrook J, Russell D (2001) Molecular cloning: a laboratory manual, 3rd edn. Cold Spring Harbor Laboratory Press, New York
30. Ogino H, Ochi H (2009) Resources and transgenesis techniques for functional genomics in *Xenopus*. Dev Growth Differ 51:387–401
31. Ogino H, McConnell WB, Grainger RM (2006) High-throughput transgenesis in *Xenopus* using *I-SceI* meganuclease. Nat Protoc 1:1703–1710
32. Kammandel B et al (1999) Distinct cis-essential modules direct the time-space pattern of the *Pax6* gene activity. Dev Biol 205:79–97

Chapter 16

Tet-On Binary Systems for Tissue-Specific and Inducible Transgene Expression

Daniel R. Buchholz

Abstract

Tissue-specific and inducible control of transgene expression is a cornerstone of modern studies in developmental biology. Even though such control of transgene expression has been accomplished in *Xenopus*, no general or widely available set of transgenic lines have been produced akin to those found in mouse and zebrafish. Here, I describe the design and characterization of transgenic lines in *Xenopus* constituting the Tet-On binary transgene expression system comprising two components: (1) rtTA transgenic lines, i.e., lines harboring the doxycycline- (Dox-) dependent transgenic transcription factor rtTA under control of a tissue-specific promoter and (2) transgenic promoter (TRE) transgenic lines, i.e., lines harboring a gene of interest (hereafter called the transgene) under control of a promoter (TRE). In double transgenic animals, i.e., embryos or tadpoles harboring both the rtTA and TRE components, transgene expression remains off the absence of Dox. Addition of Dox to the rearing water causes a conformational change in rtTA allowing it to bind the TRE promoter and induce transgene expression. Tissue specificity of transgene expression is determined by the promoter regulating rtTA expression, and inducibility is determined by the addition of Dox to the rearing water. Deposition of rtTA and TRE transgenic lines enabling tissue-specific inducible control of transgene expression into the *Xenopus* stock center will provide a powerful and flexible resource for studies in developmental biology.

Key words: Frog transgenesis, *Xenopus*, rtTA, Tetracycline response element, Doxycyline

1. Introduction

The simplest and most common transgene expression system involves inserting foreign DNA, consisting of a promoter controlling a transgene, into the genome (1, 2). Using tissue-specific promoters, investigators can examine the role of a transgene in a particular tissue. Commonly, early misexpression of a transgene alters development, precluding study of the role of the transgene later in development and leading to the need to control the timing of transgene expression. Inducible expression of transgenes using a heat shock (3, 4) or metallothionine promoter (5) allows researchers

Stefan Hoppler and Peter D. Vize (eds.), *Xenopus Protocols: Post-Genomic Approaches, Second Edition*, Methods in Molecular Biology, vol. 917, DOI 10.1007/978-1-61779-992-1_16, © Springer Science+Business Media, LLC 2012

to avoid early deleterious effects of a transgene. However, these inducible promoters result in ubiquitous transgene expression throughout the organism, dissallowing inducibility in a tissue-specific manner. Binary transgene expression systems enable tissue-specific and inducible control of transgene expression (6).

Binary transgene expression systems utilize (1) a ubiquitous or tissue-specific promoter controlling a transgenic transcription factor, which binds and regulates (2) a transgenic promoter controlling the transgene. Two such binary methods have been shown to work in frogs, the GAL4/UAS and Tet-On systems (6, 7). A third binary strategy using Cre/lox has been established in frogs, where Cre expressed from one component acts on a separate transgene with appropriately placed lox sites to activate another transgene (8, 9). We focus here on the Tet-On system because of its success to delimit the timing of thyroid hormone influence in hind limb innervation (6), to examine genes important for limb muscle development (10), to reveal gene switching during liver metamorphosis (11), and transdifferentiation of tadpole pancreatic acinar cells to duct cells (12). In contrast, the tissue-specific inducible versions of the Gal4/UAS system in frogs have not gone beyond the proof of principle (7, 13), and tissue-specific and inducible control has not yet been done with the Cre/lox system.

In the Tet-On system (commercialized by Clontech), two components work together, the transgenic transcription factor, rtTA, and the transgenic tetracycline-inducible promoter, TRE. The rtTA is derived from TetR, a tetracycline-inhibited transcriptional repressor from *Escherichia coli*. This protein, when fused to three copies of the minimal viral transactivation domain of VP16 from Herpes simplex virus, forms tTA (tetracycline-inhibited transcriptional activator). Mutations in tTA converted it to rtTA, reverse tetracycline-controlled transcriptional activator, which activates transcription upon addition of tetracycline or the more stable tetracycline mimic doxycycline (Dox). In the absence of Dox, rtTA does not bind DNA or regulate transgenes. In the presence of Dox, a conformational change allows rtTA to bind DNA and induce transgene expression from the TRE. The TRE is composed of a minimal CMV promoter (from cytomegalovirus which does not support transcription by itself) and seven copies of the bacterial tetO operator to which TetR and derivatives bind.

In addition to enabling tissue-specific inducible control of transgene expression, another advantage of Tet-On and other binary expression systems stems from the combinatorial possibilities of having separate rtTA and TRE transgenic lines. As more rtTA lines and more TRE lines are produced, the binary system approach rapidly multiplies the combinations of tissue specificity and transgenes of interest readily available to researchers. Transgenic lines from previous studies using the Tet-On system (6, 10) are of limited future use because the rtTA and TRE components are not separable, i.e., these components co-integrated into the same chromosomal position. Thus, few transgenic lines using binary expression systems are available that can be flexibly applied to a wide variety of research questions (14), highlighting the need to characterize new Dox-inducible transgenic *Xenopus* lines. The current protocol describes methods to design and characterize Tet-On transgenic rtTA and TRE lines with minimal leakiness and maximal levels of Dox-induced transgene expression.

2. Materials

1. Transgenesis plasmids. An rtTA and a TRE plasmid (Fig. 1) are available to researchers as a template to facilitate engineering their own transgenesis plasmids (see Note 1).
2. Materials for engineering transgenesis plasmids using standard cloning procedures (15).
3. Materials for breeding and rearing *X. laevis* (16) (see Note 6).
4. Transgenic lines harboring rtTA or TRE transgenesis constructs (Fig. 1) for characterizing Tet-On transgenic lines in this protocol (available from the *Xenopus* Stock Center (14)).
5. Materials for transgenesis (see Chapters 11, 12, 13 and 14) (see Note 7).
6. Doxycycline hyclate: 1,000× stock solution of 50 mg/mL in water stored at −20 °C.
7. 60 mm Petri dishes with dechlorinated fresh water ("frog water").
8. Foil for Petri dishes to block light (doxycycline is light sensitive).
9. Fluorescence dissecting microscope with filter sets for visualizing green and red fluorescent proteins.
10. 10%v/v Benzocaine in 100% EtOH for anesthesia.
11. Material and equipment for quantitative reverse transcriptase PCR (17).
12. Material and equipment for tissue sectioning and immunohistochemistry (18).

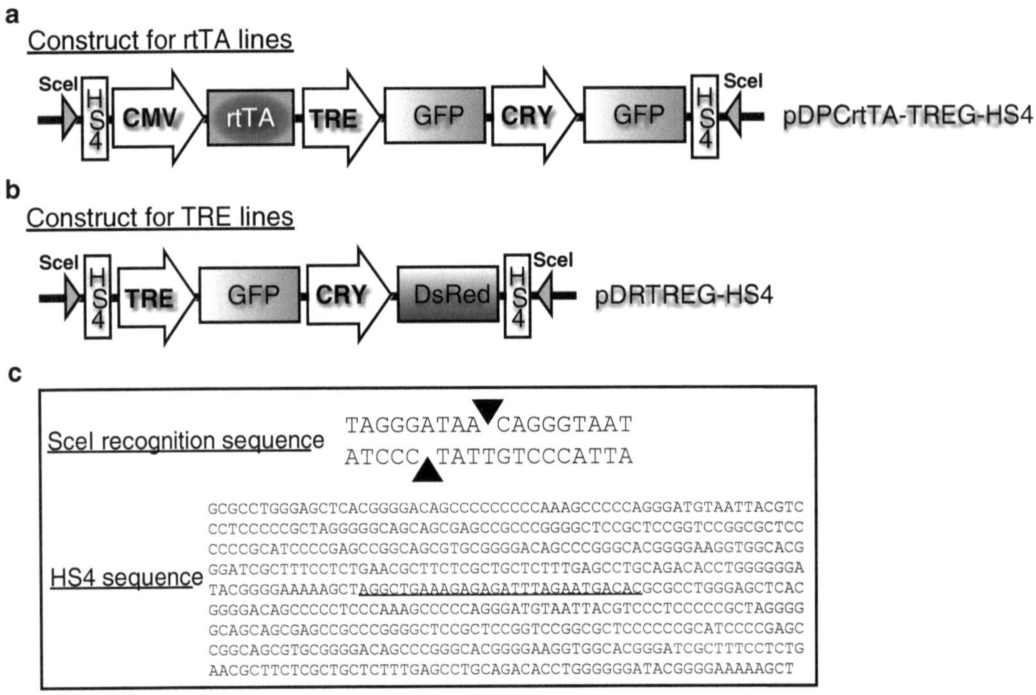

Fig. 1. Diagram of transgenesis plasmids. (**a**) Transgenic lines with this construct have ubiquitous expression of the Dox-activated transgenic transcription factor rtTA via the full length CMV (cytomegalovirus) promoter. Only in the presence of Dox does rtTA bind the transgenic tetracycline responsive promoter, TRE, and induce expression of GFP (green fluorescent protein) all over the body due to the TRE:GFP cassette (14) (see Notes 1 and 2). CRY is 300 base pairs of the *Xenopus* crystallin promoter driving constitutive GFP expression in the lens of the eye enabling identification of transgenic animals in the absence of Dox (25) (see Note 3). (**b**) Transgenic lines with this construct have GFP under control of TRE, such that GFP will be expressed only when Dox is present and only in tissues that express rtTA. The CRY:DsRed cassette drives DsRed expression in the eyes enabling identification of transgenic animals via DsRed expression in the eyes. I propose the convention that GFP be the marker for rtTA lines and DsRed be the marker for TRE lines so that Tet-On transgenic lines can be exchanged by researchers using Stock Centers. (**c**) Sequences for SceI and HS4. For both transgenesis constructs, the SceI recognition sites are required for transgenesis (see Note 4). Also, each HS4 represents two direct repeats of chicken 5' HS4 250 base pair core sequences (see Note 5).

3. Methods

3.1. Engineer Transgenesis Constructs

To suit individual research goals, the available transgenesis constructs (Fig. 1) can be modified using standard cloning techniques by replacing the promoter regulating rtTA expression in construct A and/or by replacing the transgene regulated by the TRE in construct B (15)(see Note 2).

3.2. Production of rtTA and TRE Transgenic Founders

1. Transgenesis (see Chapters 11, 12, 13, and 14), use method of choice (see Note 7).
2. Sort embryos and remove dead and deformed ones daily. Maintain healthy embryos in 60 mm Petri dishes (<50 per dish) in dechlorinated fresh water changed daily.

3. After 5–7 days at room temperature (about NF stage 38–42), select founders by choosing tadpoles that have lens-specific expression of GFP or DsRed in one or both eyes (see Note 8). GFP under control of the lens crystallin promoter is a marker for rtTA plasmids, and DsRed under control of the lens crystallin promoter is a marker for TRE plasmids. For small numbers of tadpoles, unanesthetized individuals may be observed in a droplet of water in a depression slide. For large numbers of tadpoles, 1–2 drops of benzocaine in ethanol added to the 60 mm Petri dish with 30–50 mL of frog water 5 min before viewing may be used.
4. Rear transgenic animals, i.e., tadpoles with GFP or DsRed expression in one or both eyes, to adulthood, about 12 months for *X. laevis* (see Note 9).

3.3. Characterization of rtTA Founders

1. Cross rtTA founders with pDRTREG-HS4 (Fig. 1) transgenic animals by natural mating (16) (see Note 10).
2. Examine eye fluorescence in embryos as above (see Subheading 3.2, step 3). Identify adult founders that have germ-line transgene expression by observing lens-specific GFP fluorescence in the F1 offspring using the GFP filter set. The lens-specific DsRed expression from the pDRTREG-HS4 parent (present in half of the F1 offspring) will not be visible if lens-specific GFP is expressed.
3. For all rtTA germ-line founders, sort F1 double transgenics harboring the rtTA and pDRTREG-HS4 transgenes by identifying tadpoles simultaneously expressing GFP in the eyes using the GFP filter set and DsRed in the eyes using the RFP filter set.
4. Treat F1 double transgenic tadpoles (both green and red fluorescence in eyes) with 50 μg/mL Dox for 1d (or 3d for higher Dox-induced GFP induction) at a stage the promoter controlling rtTA is expected to be strong.
5. Select rtTA germline founders which produce F1 offspring with the highest expressing, promoter-appropriate GFP fluorescence (see Note 11). In cases where Dox-induced GFP is not externally visible, dissection or histological sectioning and staining of tadpoles to check fluorescence will be necessary.
6. Rear F1 offspring from selected rtTA founders to adulthood and deposit some in the *Xenopus* Stock Center (see Note 12).
7. F1 double transgenic tadpoles from this cross can be used for GFP-dependent lineage tracing (Kerney and Buchholz, in prep).

3.4. Characterization of TRE Founders

1. Cross TRE founders with pDPCrtTA-TREG-HS4 (Fig. 1) transgenic animals by natural mating (16) (see Note 13).

2. Examine eye fluorescence in embryos as above (see Subheading 3.2, step 3). Identify adult founders that have germ-line transgene expression by observing lens-specific DsRed fluorescence in the F1 offspring using the red filter set. The lens-specific GFP expression from the pDPCrtTA-TREG-HS4 parent (present in half of the F1 offspring) will not be visible in the red filter set.
3. Treat F1 tadpoles double transgenic for the TRE and pDP-CrtTA-TREG-HS4 transgenes (as indicated by both green and red fluorescence in the eyes) with 0 or 50 μg/mL Dox for 1d at NF40-45.
4. Carry out quantitative reverse transcriptase PCR from whole bodies using TaqMan primer/probe set for the transgene. Follow external protocols (17) for isolating total RNA, synthesizing cDNA, designing and ordering TaqMan primer/probe sets for the gene of interest and a house keeping gene such as rpL8 (19), and carrying out and analyzing the quantitative PCR reaction. It is important to use DNAfree (Ambion) on the RNA sample prior to cDNA synthesis to remove contaminating DNA to avoid amplifying it in the quantitative PCR reaction.
5. Select TRE founders with the lowest leakiness and highest maximal expression by comparing results from quantitative reverse transcriptase PCR in double transgenic animals treated with and without Dox.
6. Rear single transgenic F1 offspring to adulthood from selected TRE founders and deposit some in the *Xenopus* Stock Center (see Note 12).
7. F1 double transgenic tadpoles from this cross can be used for studies where ubiquitous expression of the Dox-induced transgene is desired.

3.5. Crossing the rtTA and TRE Lines of Interest

Because there are so few rtTA and TRE transgenic lines as of now, it is likely that a line expressing rtTA in a particular tissue of interest and TRE regulating a particular transgene of interest will not be available. Thus, researchers will need to produce and characterize their own rtTA and TRE transgenic lines as detailed above. However, once available, any rtTA line can be crossed to any TRE line for specific research purposes.

1. Cross a singly transgenic rtTA line of choice to a singly transgenic TRE line of choice by natural mating (see Note 14). The male or female parent does not matter.
2. F1 double transgenic tadpoles can be identified as above (see Subheading 3.2, step 3) with both red and green fluorescence in the eyes. The parents will be heterozygous for their respective transgenes and thus at most one quarter of the offspring

will be double transgenic. The other offspring can be used as controls, but the best comparison will be between double transgenic animals treated with and without Dox.

3. Treat embryos or tadpoles with 50 μg/mL Dox for 1–3 days, minimum 3–6 h (14). The appropriate stage for Dox treatment will depend on the promoter controlling rtTA expression. Longer Dox treatments will increase the expression level of the Dox-induced transgene.
4. Analyze the experiment for morphological phenotype with or without sectioning and/or examine the effects on cell signaling or gene expression.

Theoretically with the Tet-On system, any transgene can be inducibly overexpressed in any tissue. Three limitations curb the widespread popularity for using this approach. First, few rtTA or TRE lines are currently available, requiring researchers to produce their own, a 2-year commitment at this time. Second, characterization of promoters used to drive rtTA into the tissue of interest is in itself a significant prerequisite for the production of rtTA transgenic lines (see Note 15). However, there are quite a few tissue-specific promoters available that have already been tested in frogs, e.g., expression in lens, neurons, muscle, tadpole skin, intestinal epithelium, exocrine and endocrine pancreas, limbs. Third, transgene expression levels remain a concern requiring careful characterization of transgenic lines to achieve sufficient transgene overexpression to observe a strong phenotype. Nevertheless, the availability of more rtTA and TRE lines will greatly increase the ease with which inducible tissue-specific transgene overexpression can be performed and will provide a significant additional resource to probe the mechanisms of vertebrate development using frogs as a model system.

4. Notes

1. The versions of rtTA and TRE used in frogs are rtTA-2S-M2 (20) and pTREtight (Clontech). Improved rtTA (Tet-On 3G) and TRE (pTRE3G) have been introduced by Clontech but have not been tested in frogs.
2. The TRE:GFP cassette in construct A (Fig. 1) is not necessary for the purposes of this protocol and may be removed from future constructs. The TRE:GFP cassette in construct A was used to characterize the first rtTA line in the absence of existing TRE lines (14).
3. Fluorescent proteins expressed in the lens of the eye are used to identify transgenic animals. If the eye is the organ of interest,

an alternate externally visible marker of transgenesis can be used, e.g., the muscle-specific promoter pCAR (21) can replace CRY to have constitutive GFP expression in the muscle to identify transgenic animals.

4. SceI sites are required for the meganuclease method of transgenesis (see Chapter 12) and are also suitable for the REMI method (see Chapter 11).

5. These insulators reduce the effects of chromosome position effects (14, 22, 23). To avoid enhancer-like carry over effects for uncharacterized promoters, I recommend including an additional HS4 insulator element to flank the promoter:rtTA cassette on the 5′ and 3′ ends. Improved insulator ability has been demonstrated with a 400 base pair sequence (24) but has not been tested in frogs.

6. This protocol uses established *X. laevis* rtTA and TRE transgenic lines (Fig. 1) for ease of characterizing additional rtTA and TRE lines via this protocol. The methods in this protocol are expected to be equivalent for *X. tropicalis* but no rtTA or TRE transgenic lines are thus far available. To begin the process of building a set of Tet-On transgenic lines in *X. tropicalis*, I recommend characterizing transgenic lines using the same transgenesis plasmids as initially used in *X. laevis* (14), which can then be used to characterize the additional lines.

7. The method of choice may depend on the required level of transgene expression. REMI tends to result in higher transgene expression levels compared to the methods using SceI/integrase/transposons because the copy number is much greater, up to 30 copies, compared to 1–8 copies for the other methods. However, the ideal transgene expression level varies with application. Thus, REMI is preferred when using a dominant negative, because expression level of the dominant negative must be high to overcome the activity of the endogenous protein, and even moderate levels of leakiness are unlikely to have an effect. On the other hand, SceI may be better when using cell signaling proteins which typically act at very low concentrations, such that the degree of leakiness may be more of an issue than maximal expression level.

8. A fully green or red fluorescent eye, as opposed to streaks or subsets of fluorescent lens cells, improves but does not guarantee germ-line expression, due to mosaicism particularly in the SceI/integrase/transposon methods. In addition, animals with one solid fluorescent eye should be kept, as it may be indicative of a "half transgenic" where transgene integration occurred at the two-cell stage. Selection of potential founders based on lens-specific GFP expression will eliminate transgenic animals that have germ-line expression but lack fluorescence in the eyes. However, the frequency that this occurs is not known.

9. Due to mosaicism with SceI/integrase/transposon transgenesis methods, the frequency of transgenic offspring is unlikely to be 50%, but more commonly, 5–25% transgenic offspring. Thus, a large number of potential founders will be needed, likely >20 animals, because germ-line transmission based solely on fluorescent protein expression in the eye is around 25%. On the other hand, the REMI method typically results in non-mosaic F0 founders requiring fewer founders to rear for testing because most will be germline.
10. Because females require 2–3 months of recovery time after breeding for re-breeding, it is time efficient to cross female founders directly with pDRTREG-HS4 males and test offspring for germ line expression and a Dox-induced response simultaneously. When females of the pDRTREG-HS4 line are limiting, the rtTA founder males, which can be induced to breed every 1–2 weeks, can first be checked for germline using wild type animals and then soon thereafter crossed with the pDRTREG-HS4 line.
11. Higher Dox-induced GFP expression levels should correlate with the greatest degree of Dox-induced transgene expression when crossed with other TRE lines.
12. Heterozygous transgenic individuals are sufficient for most applications, though homozygous individuals available in the F2 generation from crossing F1 transgenic siblings would enable a twofold higher transgene expression level in cases where increased expression would be useful. However, identifying homozygous individuals will depend on the potentially unreliable ability to detect a twofold difference in lens-specific GFP or DsRed expression.
13. Because females require 2–3 months of recovery time after breeding for re-breeding, it is time efficient to cross female founders directly with pDPCrtTA-TREG-HS4 males and test offspring for germline expression and a Dox-induced response simultaneously. When females of the pDPCrtTA-TREG-HS4 line are limiting, the TRE founder males, which can be induced to breed every 1–2 weeks, can first be checked for germline using wild type animals then soon thereafter crossed with the pDPCrtTA-TREG-HS4 line.
14. Transgenic founders have germline mosaicism (i.e., transgenic or not) explaining the <50% transgenic offspring. Founders also likely have mosaicism among transgenic gametes consisting of different germ cells having different insertion sites and copy numbers. Thus, if transgenic founders are used in this cross, be prepared for low frequencies of F1 double transgenic animals and expect a range of Dox-induced transgene expression levels and corresponding variation in the resulting phenotypes.

Use of F1 transgenic animals as parents in this cross will increase but not guarantee the homogeneity of Dox-induced transgene expression levels and phenotypic results.

15. If the future use of an uncharacterized promoter sequence is for tissue-specific inducible control of transgene expression, I recommend using the protocol outlined in Subheading 3.3 (Characterization of rtTA Founders) rather than first making promoter:GFP transgenic animals. Outside the use of promoter:GFP to mark tissue types, use of promoter:GFP transgenic animals has disadvantages for promoter characterization. Though straightforward, use of promoter:GFP constructs as a way to identify transgenic promoter expression domains limits the resolution of observed promoter activity because the highly stable GFP protein may be visible in cells days or weeks after the promoter has ceased to be active. In addition, transgenic lines from the protocol in Subheading 3.2 but not promoter:GFP are immediately available to carry out tissue-specific inducible control of transgenes by simply crossing to a TRE line.

Acknowledgements

Funding sources for this work were NIH R03HD059995 and NSF IOS 0950538.

References

1. Kroll KL, Amaya E (1996) Transgenic xenopus embryos from sperm nuclear transplantations reveal FGF signaling requirements during gastrulation. Development 122:3173–3183
2. Huang H, Marsh-Armstrong N, Brown DD (1999) Metamorphosis is inhibited in transgenic xenopus laevis tadpoles that overexpress type III deiodinase. Proc Natl Acad Sci U S A 96:962–967
3. Buchholz DR, Hsia SC, Fu L, Shi YB (2003) A dominant-negative thyroid hormone receptor blocks amphibian metamorphosis by retaining corepressors at target genes. Mol Cell Biol 23:6750–6758
4. Wheeler GN, Hamilton FS, Hoppler S (2000) Inducible gene expression in transgenic xenopus embryos. Curr Biol 10:849–852
5. Oofusa K, Tooia O, Kashiwagie A, Kashiwagie K, Kondoe Y, Obara M, Yoshizato K (2003) Metal ion-responsive transgenic xenopus laevis as an environmental monitoring animal. Environ Toxicol Pharmacol 13:153–159
6. Das B, Brown DD (2004) Controlling transgene expression to study xenopus laevis metamorphosis. Proc Natl Acad Sci U S A 101:4839–4842
7. Hartley KO, Nutt SL, Amaya E (2002) Targeted gene expression in transgenic xenopus using the binary Gal4-UAS system. Proc Natl Acad Sci U S A 99:1377–1382
8. Roose M, Sauert K, Turan G, Solomentsew N, Werdien D, Pramanik K, Senkel S, Ryffel GU, Waldner C (2009) Heat-shock inducible Cre strains to study organogenesis in transgenic xenopus laevis. Transgenic Res 18:595–605
9. Waldner C, Sakamaki K, Ueno N, Turan G, Ryffel GU (2006) Transgenic xenopus laevis strain expressing cre recombinase in muscle cells. Dev Dyn 235:2220–2228
10. Cai L, Das B, Brown DD (2007) Changing a limb muscle growth program into a resorption program. Dev Biol 304:260–271
11. Mukhi S, Cai L, Brown DD (2010) Gene switching at xenopus laevis metamorphosis. Dev Biol 338:117–126

12. Mukhi S, Brown DD (2011) Transdifferentiation of tadpole pancreatic acinar cells to duct cells mediated by notch and stromelysin-3. Dev Biol 351:311–317
13. Chae J, Zimmerman LB, Grainger RM (2002) Inducible control of tissue-specific transgene expression in xenopus tropicalis transgenic lines. Mech Dev 117:235–241
14. Rankin SA, Zorn AM, Buchholz DR (2011) New doxycycline-inducible transgenic lines in xenopus. Dev Dyn 240(6):1467–1474
15. Sambrook J, Russell D (2001) Molecular cloning: a laboratory manual, 3rd edn. Cold Spring Harbor Laboratory Press, New York
16. Sive HL, Grainger RM, Harland RM (2000) Early development of xenopus laevis: a laboratory manual. Cold Spring Harbor Laboratory Press, New York
17. Applied Biosystems (2010) TaqMan gene expression assays protocol. http://www.appliedbiosystems.com
18. Presnell JK, Schreibman MP (1997) Humason's animal tissue techniques. The Johns Hopkins University Press, Baltimore
19. Shi YB, Liang VC (1994) Cloning and characterization of the ribosomal protein L8 gene from xenopus laevis. Biochim Biophys Acta 1217:227–228
20. Urlinger S, Baron U, Thellmann M, Hasan MT, Bujard H, Hillen W (2000) Exploring the sequence space for tetracycline-dependent transcriptional activators: novel mutations yield expanded range and sensitivity. Proc Natl Acad Sci U S A 97:7963–7968
21. Mohun TJ, Garrett N, Gurdon JB (1986) Upstream sequences required for tissue-specific activation of the cardiac actin gene in xenopus laevis embryos. EMBO J 5:3185–3193
22. Allen BG, Weeks DL (2005) Transgenic xenopus laevis embryos can be generated using phiC31 integrase. Nat Methods 2:975–979
23. Sekkali B, Tran HT, Crabbe E, De Beule C, Van Roy F, Vleminckx K (2008) Chicken beta-globin insulator overcomes variegation of transgenes in xenopus embryos. FASEB J 22: 2534–2540
24. Aker M, Tubb J, Groth AC, Bukovsky AA, Bell AC, Felsenfeld G, Kiem HP, Stamatoyannopoulos G, Emery DW (2007) Extended core sequences from the cHS4 insulator are necessary for protecting retroviral vectors from silencing position effects. Hum Gene Ther 18:333–343
25. Smolich BD, Tarkington SK, Saha MS, Stathakis DG, Grainger RM (1993) Characterization of xenopus laevis gamma-crystallin-encoding genes. Gene 128:189–195

Part IV

Gene Expression: From Gene to Transcript to Protein

Chapter 17

Chromatin Immunoprecipitation Analysis of *Xenopus* Embryos

Robert C. Akkers, Ulrike G. Jacobi, and Gert Jan C. Veenstra

Abstract

Chromatin immunoprecipitation (ChIP) is a powerful technique to study epigenetic regulation and transcription factor binding events in the nucleus. It is based on immune-affinity capture of epitopes that have been cross-linked to genomic DNA in vivo. A readout of the extent to which the epitope is associated with particular genomic regions can be obtained by quantitative PCR (ChIP-qPCR), microarray hybridization (ChIP-chip), or deep sequencing (ChIP-seq). ChIP can be used for molecular and quantitative analyses of histone modifications, transcription factors, and elongating RNA polymerase II at specific loci. It can also be applied to assess the cellular state of transcriptional activation or repression as a predictor of the cells' capabilities and potential. Another possibility is to employ ChIP to characterize genomes, as histone modifications and binding events occur at specific and highly characteristic genomic elements and locations. This chapter provides a step-by-step protocol of ChIP using early *Xenopus* embryos and discusses potential pitfalls and other issues relevant for successful probing of protein–genome interactions by ChIP-qPCR and ChIP-seq.

Key words: Chromatin immunoprecipitation, ChIP, Genome-wide location analysis, Epigenome, Transcription, *Xenopus*

1. Introduction

The developmental program of gene expression involves highly regulated binding of sequence-specific DNA binding factors and other chromosomal proteins to *cis*-regulatory sequences such as promoters and enhancers. The accessibility of these sequences in the nucleus is dependent on chromatin structure. The basic unit of chromatin is formed by the nucleosome: DNA is wrapped around an octamer of histone proteins. The highly conserved N-terminal tails of the histone proteins protrude from the nucleosome and are subjected to extensive post-translational modifications. These epigenetic modifications affect most DNA-dependent processes,

Stefan Hoppler and Peter D. Vize (eds.), *Xenopus Protocols: Post-Genomic Approaches, Second Edition*,
Methods in Molecular Biology, vol. 917, DOI 10.1007/978-1-61779-992-1_17, © Springer Science+Business Media, LLC 2012

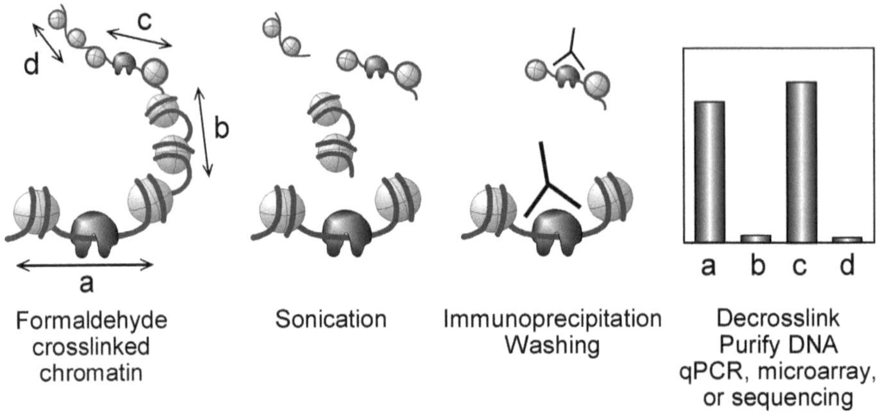

Fig. 1. Overview of chromatin immunoprecipitation (ChIP). Embryos are treated with formaldehyde, resulting in cross-linking of proteins to DNA. The embryos are subsequently homogenized and chromatin is fragmented by sonication or by enzymatic means. Epitopes of interest (transcription factors, modified histones, RNA polymerase) are enriched using immunoprecipitation, the cross-links are reversed, and the DNA is purified and analyzed using quantitative PCR (ChIP-qPCR), microarray hybridization (ChIP-chip), or deep sequencing (ChIP-seq).

including the binding of proteins involved in transcriptional regulation.

Chromatin immunoprecipitation (ChIP) was developed in the late 1980s to probe protein–DNA interactions in vivo (1, 2). Its development in combination with the availability of genomic sequence information has fueled the analysis of regulatory interactions in the nucleus (3, 4).

The ChIP procedure uses formaldehyde to cross-link proteins covalently to genomic DNA in vivo (Fig. 1). The chromatin is fragmented by sonication and the fragments that are associated with epitopes of interest are captured using antibodies. The cross-links are reversed and the captured DNA is purified. This DNA can be assessed by quantitative PCR for target gene enrichment (ChIP-qPCR), by microarray hybridization (ChIP-chip), or massive parallel sequencing (ChIP-seq). Cross-linking and sonication conditions are vital for the success of any ChIP procedure (5). For application to *Xenopus* embryos it is particularly important to avoid high salt or ionic detergents when homogenizing the embryos to prevent solubilizing the yolk. Several protocols for ChIP analysis in early *Xenopus* embryos have been published (6–8). Here, we describe a detailed protocol for ChIP experiments in *Xenopus*. The protocol can be used with both *Xenopus laevis* and *Xenopus tropicalis* embryos. The genome of *X. tropicalis* has been sequenced (9), which enables advanced genomic exploration in this species, including ChIP-chip and ChIP-seq applications of early development (10–12). Sequencing of the *X. laevis* genome will further enable ChIP analyses in this species. Integrating powerful genomic analyses with the unique advantages of the *Xenopus* model systems enhances our understanding of early vertebrate development. This chapter provides detailed instructions for ChIP and discusses experimental design and troubleshooting options.

2. Materials

All solutions are made with sterile water in autoclaved cylinders and stored at room temperature unless indicated otherwise.

2.1. Solutions

1. 1 M Tris–HCl pH 8.0: Weigh 121.1 g Tris and transfer to a cylinder. Add 800 mL water (milliQ). Adjust the pH and add water to 1 L.
2. 5 M Sodium chloride: Weigh 292.2 g NaCl and add water to a volume of 1 L.
3. 5 M Lithium chloride: Weigh 212 g LiCl and add water to 1 L.
4. 3 M Potassium chloride: Transfer 223.7 g KCl to a cylinder and add water to 1 L.
5. 0.5 M Ethylenediaminetetraacetic acid (EDTA) pH 8: Weigh 186.1 g Na_2-EDTA·$2H_2O$ transfer to a cylinder and add 800 mL water. Adjust the pH and add water to 1 L.
6. 1 M Dithiothreitol (DTT): Weigh 154.2 mg DTT and add 1 mL water. Store at −20 °C.
7. 10% Nonidet P40: Dilute Nonidet P40 ten times in water.
8. 10% Deoxycholate: Weigh 10 g deoxycholic acid sodium salt and adjust to 100 mL with water.
9. 1 M Glycine: Weigh 0.75 g glycine and adjust to 10 mL (see Note 1).
10. 12 mM Glycine/25% MMR: Mix 1 mL 1 M glycine, 6.8 mL water, and 0.2 mL 10× MMR (see Note 1).
11. 5% Bovine serum albumin: Weigh 0.5 g BSA and dissolve in 10 mL water. Store in aliquots at −20 °C.
12. 10% Sodium dodecyl sulfate: Weigh 10 g SDS and adjust to 100 mL.
13. 1 M Sodium bicarbonate pH 8.8: Weigh 84 g $NaHCO_3$ and transfer to a cylinder and add 800 mL water. Adjust the pH and add water to 1 L.
14. Protease inhibitor cocktail tablets (PIC): Dissolve one tablet of PIC (cat. no. 04 693 132 001, Roche) in 2 mL water (50× stock) and store at −20 °C.
15. 10× Marc's Modified Ringer (MMR) solution pH 7.4: Weigh 51.4 g NaCl, 2.94 g $CaCl_2$·$2H_2O$, 2 g $MgCl_2$·$6H_2O$ and 11.9 g HEPES. Transfer to a 1 L cylinder, add 6.67 mL 3M KCl and add water to a volume of 800 mL. Adjust the pH with HCl and add water to 1 L.

16. Sonication buffer (low-salt whole cell extract buffer; WCE-LS): 20 mM Tris–HCl (pH 8.0), 70 mM KCl, 1 mM EDTA, 10% glycerol, 5 mM DTT, 0.125% Nonidet P40 and PIC.
17. ChIP incubation buffer: 50 mM Tris–HCl (pH 8.0), 100 mM NaCl, 2 mM EDTA, 1 mM DTT, 1% Nonidet P40, and PIC.
18. Wash buffer 1: 50 mM Tris–HCl (pH 8.0), 100 mM NaCl, 2 mM EDTA, 1 mM DTT, 1% Nonidet P40, 0.1% deoxycholate, and PIC.
19. Wash buffer 2: 50 mM Tris–HCl (pH 8.0), 500 mM NaCl, 2 mM EDTA, 1 mM DTT, 1% Nonidet P40, 0.1% deoxycholate, and PIC.
20. Wash buffer 3: 50 mM Tris–HCl (pH 8.0), 100 mM NaCl, 250 mM LiCl, 2 mM EDTA, 1 mM DTT, 1% Nonidet P40, 0.1% deoxycholate, and PIC.
21. TE buffer: 10 mM Tris–HCl (pH 8.0), 1 mM EDTA.
22. Elution buffer (0.1 M $NaHCO_3$ pH 8.8 and 1% SDS): Mix 1 mL 1 M $NaHCO_3$ pH 8.8 and 1 mL 10% SDS with 8 mL milliQ and keep at room temperature.
23. Glycogen: Make a stock solution of 10 μg/μL and store at −20 °C.

2.2. Other Reagents and Disposables

1. Protein A/G beads (catalog number sc-2003, Santa Cruz).
2. BD Falcon tubes (catalog number 352096, BD Biosciences) are recommended in combination with the Bioruptor.
3. Safe-Lock Eppendorf tubes (catalog number 0030 123.328, Eppendorf), recommended for phenol extraction after reversal of cross-links.
4. The following antibodies are recommended as positive controls in ChIP applications:
   a. Antibody recognizing the Histone H3 lysine 4 trimethylation (H3K4me3) epitope (supplied by Abcam, cat. no. ab8580).
   b. Antibody recognizing RNA polymerase II (supplied by Diagenode, cat. no. AC-055-100).
5. The PCR primers listed in Table 1 can be used as positive and negative controls in combination with the antibodies listed as positive controls.

2.3. Equipment

1. Bioruptor sonication device (UCD-200 TO, Diagenode). This machine is suited for shearing genomic DNA in relatively small samples. Other devices can also be used. Sonication conditions for ChIP should be optimized for different devices (see Note 2).
2. Thermomixer (e.g., Eppendorf Thermomixer Compact).
3. PCR cycler for quantitative (real-time) PCR (e.g., Biorad MyiQ system).

Table 1
PCR primers used for positive and negative control experiments

X. tropicalis primers	Locus, xenbase gene (control)
5′- GCAGAGGTGAGTATGGGATGG 5′- AAAAGCAGGCCAGTAAGCCA	scaffold_73: 698,538-698,588 eef1a1o (positive control)
5′- TCGTCACCAAGAATGGAAGGT 5′- GGGTAGAATAAGTGGAGAGACCCTG	scaffold_432: 292,957-293,007 t (positive control)
5′- TGATTTTTTCACATATGCAGGCA 5′- GTGTACATGCAGGTGGCAGTG	scaffold_1: 6,458,583-6,458,633 N/A (negative control with minor RNAPII enrichment)
5′- CATTCCCTACTGGGCTGGGT 5′- CACACTGCTGGCCATCGTT	scaffold_919: 126,357-126,407 N/A (negative control)

X. laevis primers	Gene (control)
5′-TCTTCCGCGATTGTGACTAA 5′-GCCAATCAAGGCACAGTACA	Promoter of hsp70 (positive control)
5′-GGTTTGAGGAAGGACATGGA 5′-GTCCTGGTCAGTGGCTGAAT	First exon of foxb1 (FKH5) (positive control)
5′-TCCACATGCACAACCCTTTA 5′-TCTTTGCCCACAAATCTGGT	BAC locus (AC236459) (negative control)

3. Methods

3.1. Chromatin Isolation

1. Collect between 50 and 300 embryos in a 6-well plate (50 embryos per well, see Note 3). For each well, remove the buffer or water the embryos were placed in, and replace with 5 mL of 10% MMR premixed with 135 μL 37% formaldehyde (final concentration 1%). Incubate for 30 min at room temperature (see Note 4) while the embryos are gently shaken. The incubation in formaldehyde cross-links interacting proteins to the DNA.
2. Remove the 1% formaldehyde solution.
3. To each well, add 5 mL of 125 mM glycine in 25% MMR to the embryos (see Note 1) and incubate for 30 min at room temperature and shake the embryos gently on a shaking platform. Glycine quenches cross-linking by formaldehyde.
4. Remove the 125 mM glycine solution.
5. Wash the embryos twice at room temperature with 5 mL 25% MMR for 15 min (shake gently). Remove the 25% MMR between the washing steps.

6. Transfer the embryos (50–300) to a 15 mL high-clarity polypropylene conical Falcon tube and remove excess buffer (see Note 5).
7. Add 2 mL sonication buffer (see Note 6) and homogenize the embryos by pipetting up and down in a 1 mL (Gilson) pipette tip.

3.2. Sonication

1. Shear the chromatin into fragments by sonicating 15 min with cycles of 15 s high-intensity sonication and intervals of 30 s in ice-cold water using a Bioruptor (Diagenode, see Note 7).
2. Remove the cooling water and add fresh ice-cold water to the bucket and sonicate again for 15 min at identical settings.
3. Split the sonicated extract (2 mL) into two Eppendorf tubes. Centrifuge for 5 min at 13,000 rpm (all centrifugation steps in a top table Eppendorf centrifuge) at 4 °C to remove cell debris, pigment, and yolk. Snap freeze aliquots of the supernatant (chromatin extract) in liquid nitrogen and store at –80 °C.

3.3. Immuno-precipitation

All steps are performed in the cold room (4 °C) unless indicated otherwise.

1. For each ChIP reaction, mix 100 μL chromatin extract and 100 μL ChIP incubation buffer.
2. Keep 20 μL aliquot of chromatin extract separate as Input control (20%). Use this aliquot when arriving at step 11 of Subheading 3.4.
3. Add 1–5 μg of antibody to all ChIP reactions (see Note 8).
4. Incubate for 2 h using a rotating wheel. Proceed with steps 4–7 during this time.
5. Prepare 50 μL 25% protein A/G beads per reaction by incubation in IP buffer + 0.1% BSA for 1 h (see Note 9).
6. Spin down beads at 6,000 rpm for 1 min and wash the beads for 5 min (4 °C, rotating wheel) with 1 mLChIP buffer.
7. Repeat step 5 once.
8. Spin down and remove supernatant.
9. Add chromatin (from step 3) to freshly prepared beads.
10. Incubate overnight at 4 °C while rotating.

3.4. Washing

1. Centrifuge beads at 6,000 rpm for 1 min and remove the supernatant (non-bound chromatin).
2. Wash the beads with 1 mL ice-cold wash buffer 1 for 5 min at 4 °C on a rotating wheel.
3. Centrifuge beads at 6,000 rpm for 1 min and remove the supernatant.

4. Repeat steps 2 and 3 with:
   - 1 mL Wash buffer 2.
   - 1 mL Wash buffer 3.
   - 1 mL Wash buffer 1.
   - 500 μL TE buffer.
5. Centrifuge beads and remove supernatant.
6. Add 200 μL of elution buffer (see Note 10) and incubate for 15 min at room temperature while rotating.
7. Centrifuge beads at room temperature at 6,000 rpm for 1 min.
8. Collect supernatant and transfer to a Safe-Lock Eppendorf tube (see Note 11).
9. Repeat steps 6 and 7 and add supernatant to tube from step 8 (400 μL).
10. Add 16 μL 5 M NaCl.
11. Take the chromatin input control (Subheading 3.3, step 2) and add 380 μL elution buffer and 16 μL 5 M NaCl. Process the input chromatin along with the ChIP samples from this point onward.
12. Reverse cross-links by incubating for 5 h at 65 °C in a thermomixer, shaking at 1,000 rpm.

3.5. DNA Purification

1. Add 400 μL phenol/chloroform/isoamylalcohol (25:24:1) pH 8.0 and vortex 5 min at high speed.
2. Centrifuge at 13,000 rpm for 5 min.
3. Transfer aqueous upper layer to a new Eppendorf tube.
4. Add 400 μL chloroform/isoamylalcohol (24:1) and vortex 5 min at high speed.
5. Centrifuge at 13,000 rpm for 5 min.
6. Transfer upper layer to a new Eppendorf tube.
7. Add 10 μg glycogen, 40 μL3M NaAc pH 5.2, and 1,200 μL ethanol (96%).
8. Precipitate DNA overnight at −20 °C.
9. Centrifuge at 13,000 rpm for 25 min at 4 °C.
10. Remove the supernatant (see Note 12).
11. Wash the pellet carefully with 500 μL 70% ice-cold ethanol.
12. Centrifuge at 13,000 rpm for 10 min at 4°C.
13. Remove the supernatant and air-dry the pellet.
14. Dissolve the pellet in 200 μL milliQ water. Dilute the 20% input sample to obtain 10%, 1%, 0.1%, and 0.01% input dilutions. Proceed with qPCR analysis.

3.6. Data Analysis and Interpretation

3.6.1. Positive and Negative Controls

For proper evaluation and interpretation of an experiment, it is important that controls are included in the analysis. When establishing the ChIP procedure, it may be helpful to perform ChIP experiments with positive control antibodies (Subheading 2.2). Sometimes a "beads-only" (no antibody) or a non-specific IgG antibody is used as a negative control. In most experimental setups, however, the background signal of a ChIP experiment is most appropriately measured with the same antibody (ChIP sample) on a negative locus, i.e., with a primer pair that amplifies DNA where the epitope of interest is not found. The background of an antibody on a negative locus may be higher than the signal observed with a beads-only or IgG control. For uncharacterized epitopes, it is not known what loci would be suitable for background measurement; therefore it is a good practice to check a number of loci.

3.6.2. Determining Recovery and Enrichment Values

Quantitative (real-time) PCR is used to assess the success of ChIP experiments. For all primer pairs (positive control loci, negative controls loci and other loci of interest), the amount of template present in the ChIP sample needs to be determined and compared to the amount of template present in the input. This is most reliably done by testing multiple input dilutions (for example corresponding to 10%, 1%, 0.1%, and 0.01% of input) to generate a standard curve of qPCR threshold (Ct) values vs. % input for each primer pair. The standard curve is then used to convert the Ct value of the ChIP sample to a recovery value (% of input). It is best practice to use a standard curve rather than a simple calculation based on Ct values of the ChIP sample and a single input sample, as small differences in amplification efficiency (dependent on the primer pair, for example 1.8 or twofold amplification per cycle) may result in relatively large differences in the calculation (compare 1.8^10 and 2^10).

Once the recoveries relative to input are known, enrichment values can be calculated relative to the background based on the negative control loci. Expect background recovery values in the order of 0.01–0.1% and peak recoveries in the order of 0.5–3% for transcription factors and 10–40% for histone modifications (Fig. 2). Antibodies of suboptimal quality will produce lower peak recoveries or a higher background, which results in lower enrichment values. Enrichment values can also be lower on account of cellular heterogeneity, depending on how ubiquitous or cell type-specific the association of the epitope with genomic DNA is. See Note 13 for troubleshooting options if low enrichment values are obtained.

3.6.3. Microarray and NextGen Sequencing Readout of ChIP

Depending on the nature and scope of the research question, a genome-wide readout of ChIP enrichment may be desired. Genome-tiling microarray designs are available for *X. tropicalis* (10). In addition, it is possible to subject the ChIP-purified DNA to Next Generation sequencing (also called deep or massive parallel sequencing)

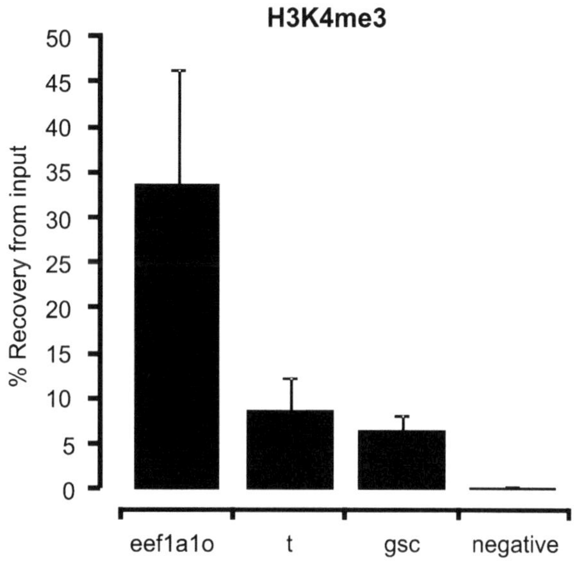

Fig. 2. ChIP-qPCR experiment showing enrichment of tri-methylation of histone H3 lysine 4 (H3K4me3) on the EF1α (eef1a1o), brachyury (t), and Goosecoid (gsc) loci, relative to a negative control.

and map the reads back to the reference genome (11, 12). A genome-wide analysis of epitope enrichment starts with characterizing the antibody and the enrichments that can be obtained at different loci. This is important because it allows monitoring, optimizing, and controlling the quality of the samples subjected to genome-wide analyses. The sample preparation procedures currently used for either microarray hybridization or NextGen sequencing typically involve repairing the ends of the ChIP fragments, followed by linker ligation and (limited) amplification. The details depend on the analysis platform and will not be reviewed in this chapter. Each of these steps, however, can be inefficient or introduce a bias. It is therefore important to monitor whether quantitative differences between loci are preserved during sample preparation. For this purpose, qPCR Ct values of a number of loci (with high and low recoveries) are plotted before and after sample preparation (Fig. 3). The resulting Quality Control plot needs to show both a good correlation between data points and a slope that is close to one. Once the genome-wide profile is generated, it can be used to visually inspect loci of interest in a suitable genome browser (Fig. 4). Obtaining a comprehensive picture of the genome-wide profile may include peak-calling, comparing the data to known genomic features such as genes, unsupervised clustering, sequence motif analyses, and supervised hypothesis-driven meta-analyses integrating other (published) datasets. These aspects of genome-wide ChIP analysis have been reviewed elsewhere (13, 14). In many cases, these analyses lead to new and immediate

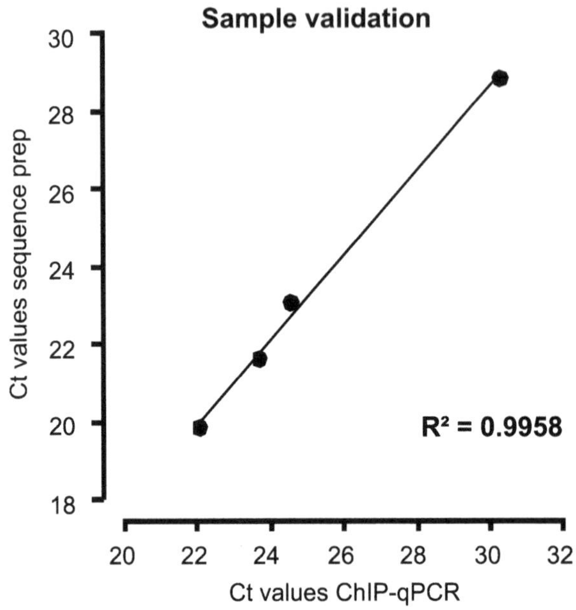

Fig. 3. Quality control qPCR analysis of ChIP samples prepared for Solexa sequencing. Ct values of samples before and after sample preparation are plotted.

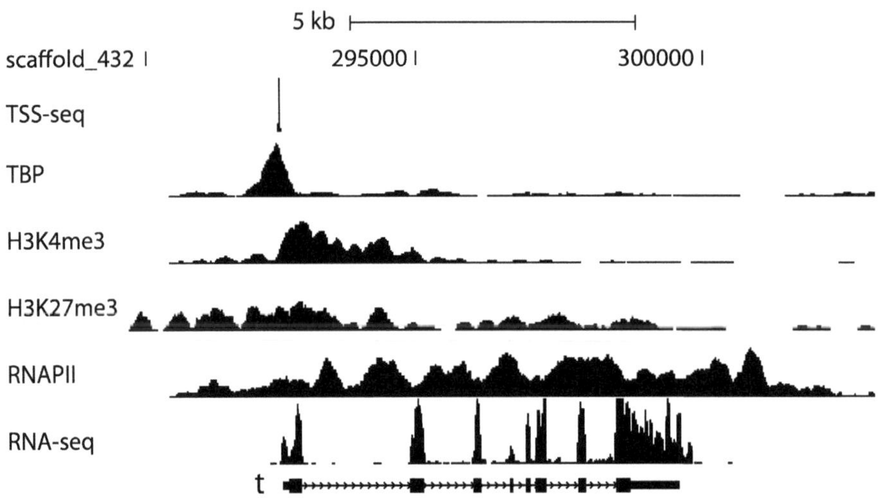

Fig. 4. Genome browser view of the brachyury (t) locus with several deep sequencing data sets. ChIP-sequencing tracks are shown for TATA binding protein (TBP), two histone modifications (H3K4me3 and H3K27me3), and RNA Polymerase II (RNAPII). In addition tracks are shown for polyA + RNA (RNA-seq) and transcript 5′ ends (TSS-seq to determine transcription start sites at high resolution).

questions for follow-up experiments. In addition, applying these techniques to *Xenopus* embryos will generate new levels of understanding of molecular mechanisms of gene regulation, improve genomic annotation of genes and regulatory elements, and enable systems level analyses of development.

4. Notes

1. Prepare glycine solutions fresh.
2. The extent of sonication should be checked when establishing the conditions for ChIP (new equipment, different sonication time or intensity) and whenever the recovery or enrichment at positive control loci is not satisfactory. Over-sonication results in very small fragments that cannot be PCR-amplified efficiently (low recoveries in ChIP), whereas too little sonication leads to large fragments that cause a high background (low enrichment over genomic background). Ideally, the DNA fragments in the sonicated material should range from 150 to 1,000 bp in size, with the majority of fragments around 500 bp. To analyze the size distribution of DNA, take an aliquot of the sonicated chromatin extract (four gastrula-stage embryo equivalents) and prepare it for agarose gel electrophoresis:
   (a) Add one volume of Stop buffer (containing 30 mM EDTA, 20 mM Tris–HCl pH 7.5, 1% SDS, and 0.5 mg/mL proteinase K) and incubate for at least 1 h at 37 °C.
   (b) Perform two phenol/chloroform extractions, precipitate with ethanol.
   (c) Resuspend in TE containing 100 μg/mL RNAse A and incubate for 1 h at 37 °C.
   (d) Load >2 μg of DNA on a 1% agarose gel. Expect to see a smear. Samples with bands around 18–20 kb (intact genomic and/or mitochondrial DNA) are severely undersonicated. Very small products may correspond to incompletely degraded RNA.
3. ChIP is most robust with large amounts of starting material. This is due to (a) a low cross-linking efficiency and recovery of genomic DNA, and (b) the optimal volumes for sonication and immunoprecipitation. The protocol in this chapter can be used with between 50 and 600 embryos without further modifications. For *X. laevis* and *X. tropicalis* the same number of embryos is used for preparation of extracts and immunoprecipitation. Even though the size and volume of the embryos differ, the number of cells at the same stage is comparable. When using 50 embryos, use 1 well of a 6-well plate, cross-link and wash as described, and sonicate in 2 mL. When using fewer embryos, reduce the volume of sonication buffer added in step 7 of Subheading 3.1. Optimize sonication conditions when changing the sonication volume. It is possible to perform ChIP-qPCR with as few as 1,000 cells and ChIP-seq with only 10,000 cells (15–17). The number of embryos taken for ChIP analysis also depends on experimental design (manipulated

embryos or normal embryos), developmental stage, and the intended read-out of the experiment (ChIP-qPCR or ChIP-seq).

4. Make sure that all embryos are fully submerged; they tend to float to the surface after changing solutions. For untested antibodies, it is recommended to test different cross-linking times: 0, 15, 30, and 60 min.
5. Falcon tubes are recommended in combination with the Bioruptor.
6. Prepare fresh sonication buffer.
7. Alternatively, sonicate six times for 10 s at power 5 using a Branson sonifier (250) with at least 30 s intervals between the pulses. Cool with ice water during sonication. Protect your ears. Resume at step 3 of Subheading 3.2.

   Chromatin fragmentation can also be achieved by enzyme digestion (e.g., with micrococcal nuclease).
8. To establish a successful ChIP, the quality of the antibody is most important (cf. troubleshooting note). It is worth testing several concentrations of antibody to obtain the best signal-to-noise ratio.
9. Beads for multiple ChIP reactions can be blocked at the same time in a single Eppendorf tube.
10. Prepare fresh elution buffer.
11. Safe-Lock Eppendorf tubes are recommended for the long incubation at 65 °C and the subsequent phenol extraction.
12. Be careful not to disturb the pellet.
13. Troubleshooting suggestions:
    (a) *Nothing works—what is wrong?* When setting up ChIP it is recommended to use various controls to evaluate whether the basic procedure works. A number of ChIP-grade antibodies, as well as PCR primers for positive and negative loci are listed in Subheading 2.2. If the positive controls do not work, you may try the following: (1) Test multiple chromatin preparations derived from strong, healthy embryos; (2) check the efficiency of sonication (cf. see Note 2).

    (b) *Good results are obtained with the positive control antibodies, but my own antibody does not work. What is the problem?* It is possible that non-enriched loci are tested. This is a major problem with proteins for which the genomic enrichment profile is unknown. Mine the literature for possible targets and design more primer pairs. Of course, it is also possible to blindly sequence the DNA to find target loci, but this is generally not recommended. In case a new, uncharacterized antibody is used, test the antibody in regular immunoprecipitation (IP), for example using in vitro translated proteins containing the epitope. This will establish whether the antibody recognizes the *Xenopus* protein with sufficient affinity. However, even if the antibody performs well in IP, it is not necessarily "ChIP grade" as the epitope may be masked by other chromatin-associated proteins or by cross-linking. Several things can be tried to resolve epitope-masking, including different cross-linking times (cf. see Note 4) and different conditions for cross-linking and sample preparation. One could add up to 0.1–0.5% of SDS to the sonicated, cleared chromatin extract (Subheading 3.3, step 1) and test the effect. A more radical approach that works well for some epitopes is to homogenize the embryos first and then perform the cross-linking in solution (instead of the other way around). This can reduce epitope-masking. One potential concern with this approach is that irrelevant proteins, for example cytoplasmic factors, may be cross-linked to genomic DNA. In most cases this is not a serious concern because the dilution during homogenization renders non-specific interactions very inefficient. To perform this alternative cross-linking procedure, homogenize 300 embryos in 5 mL 10% MMR at step 1 of Subheading 3.1. Scale down proportionally if fewer embryos are used. Add formaldehyde to 1%, mix by pipetting up and down, and incubate for 30 min at room temperature on a rotating wheel. Add glycine to 125 mM final concentration, mix gently, and incubate for 30 min on a rotating wheel. Centrifuge for 5 min at 1,000 × *g*. Resuspend the pellet (which contains most of the high molecular weight genomic DNA) in 10 mL 25% MMR. Repeat the wash step once and resuspend in 2 mL sonication buffer. Resume the protocol at step 1 of Subheading 3.2. It should be noted that in this procedure the cross-linking is more extensive (consider reducing the cross-linking time), that sonication time should also be adjusted (longer), and that some DNA is lost in the washing steps.

(c) *My antibody is apparently not ChIP-grade. Can I use epitope-tagged proteins?* In this scenario it is generally recommended to test epitope tags at different positions in the protein of interest. Artificial epitopes can work as well as natural epitopes, but can also be masked, depending on the protein and the location where the epitope tag is placed (at the N terminus or C terminus). A concern with epitope-tagged proteins is the level of overexpression obtained; even moderate overexpression may cause additional binding events and reduced interactions with other proteins. Therefore it may be necessary to perform additional control experiments when using epitope tags to assess the validity of the results.

References

1. Solomon MJ, Larsen PL, Varshavsky A (1988) Mapping protein-DNA interactions in vivo with formaldehyde: evidence that histone H4 is retained on a highly transcribed gene. Cell 53:937
2. Hebbes TR, Thorne AW, Crane-Robinson C (1988) A direct link between core histone acetylation and transcriptionally active chromatin. EMBO J 7:1395
3. Mendenhall EM, Bernstein BE (2008) Chromatin state maps: new technologies, new insights. Curr Opin Genet Dev 18:109
4. Farnham PJ (2009) Insights from genomic profiling of transcription factors. Nat Rev Genet 10:605
5. Orlando V (2000) Mapping chromosomal proteins in vivo by formaldehyde-crosslinked-chromatin immunoprecipitation. Trends Biochem Sci 25:99
6. Sachs LM, Shi YB (2000) Targeted chromatin binding and histone acetylation in vivo by thyroid hormone receptor during amphibian development. Proc Natl Acad Sci U S A 97:13138
7. Jallow Z, Jacobi UG, Weeks DL, Dawid IB, Veenstra GJC (2004) Specialized and redundant roles of TBP and a vertebrate-specific TBP paralog in embryonic gene regulation in *Xenopus*. Proc Natl Acad Sci U S A 101:13525
8. Blythe SA, Reid CD, Kessler DS, Klein PS (2009) Chromatin immunoprecipitation in early *Xenopus laevis* embryos. Dev Dyn 238:1422
9. Hellsten U, Harland RM, Gilchrist MJ, Hendrix D, Jurka J, Kapitonov V, Ovcharenko I, Putnam NH, Shu S, Taher L, Blitz IL, Blumberg B, Dichmann DS, Dubchak I, Amaya E, Detter JC, Fletcher R, Gerhard DS, Goodstein D, Graves T, Grigoriev IV, Grimwood J, Kawashima T, Lindquist E, Lucas SM, Mead PE, Mitros T, Ogino H, Ohta Y, Poliakov AV, Pollet N, Robert J, Salamov A, Sater AK, Schmutz J, Terry A, Vize PD, Warren WC, Wells D, Wills A, Wilson RK, Zimmerman LB, Zorn AM, Grainger R, Grammer T, Khokha MK, Richardson PM, Rokhsar DS (2010) The genome of the Western clawed frog Xenopus tropicalis. Science 328:633
10. Akkers RC, vanHeeringen SJ, Manak JR, Green RD, Stunnenberg HG, Veenstra GJC (2010) ChIP-chip designs to interrogate the genome of *Xenopus* embryos for transcription factor binding and epigenetic regulation. PLoS One 5:e8820
11. Akkers RC, van Heeringen SJ, Jacobi UG, Janssen-Megens EM, Francoijs K-J, Stunnenberg HG, Veenstra GJC (2009) A hierarchy of H3K4me3 and H3K27me3 acquisition in spatial gene regulation in Xenopus embryos. Dev Cell 17:425
12. van Heeringen SJ, Akhtar W, Jacobi UG, Akkers RC, Suzuki Y, Veenstra GJC (2011) Nucleotide composition-linked divergence of vertebrate core promoter architecture. Genome Res 21:410–421. doi:10.1101/gr.111724.110
13. Hawkins RD, Hon GC, Ren B (2010) Next-generation genomics: an integrative approach. Nat Rev Genet 11:476
14. Barski A, Zhao K (2009) Genomic location analysis by ChIP-Seq. J Cell Biochem 107:11
15. O'Neill LP, VerMilyea MD, Turner BM (2006) Epigenetic characterization of the early embryo with a chromatin immunoprecipitation protocol applicable to small cell populations. Nat Genet 38:835
16. Dahl JA, Collas P (2008) A rapid micro chromatin immunoprecipitation assay (microChIP). Nat Protoc 3:1032
17. Adli M, Zhu J, Bernstein BE (2010) Genome-wide chromatin maps derived from limited numbers of hematopoietic progenitors. Nat Methods 7:615

Chapter 18

Transcriptomics Using Next Generation Sequencing Technologies

Dasfne Lee-Liu, Leonardo I. Almonacid, Fernando Faunes, Francisco Melo, and Juan Larrain

Abstract

Next generation sequencing technologies may now be applied to the study of transcriptomics. RNA-Seq or RNA sequencing employs high-throughput sequencing of complementary DNA fragments delivering a transcriptional profile. In this chapter, we aim to provide a starting point for *Xenopus* researchers planning on starting an RNA-Seq transcriptomics study. We begin by providing a section on template isolation and library preparation. The next section comprises the main bioinformatics procedures that need to be performed for raw data processing, normalization, and differential gene expression. Finally, we have included a section on studying deep sequencing results in *Xenopus*, which offers general guidance as to what can be done in this model.

Key words: *Xenopus tropicalis*, *Xenopus laevis*, RNA-Seq, Small RNA-Seq, Transcriptomics, Transcriptional profiling, High-throughput sequencing, Next generation sequencing, Massively parallel sequencing, Illumina

1. Introduction

Next generation sequencing technologies are a group of methods that employ high-throughput sequencing (also known as *deep sequencing*) of DNA fragments. Massively parallel sequencing is used, whereby billions of fragments may be sequenced simultaneously. As a result, vast amounts of data are obtained in a short period of time and at a much lower cost than with Sanger sequencing (1). These methodologies can be used for genome sequencing projects and for global gene expression analyses (1, 2). In this chapter we will focus on the use of these methodologies for transcriptomics analysis. In particular, we will summarize their use in *Xenopus*, one of the main models for studying vertebrate developmental biology, and more recently, regeneration (see Chapter 30).

Stefan Hoppler and Peter D. Vize (eds.), *Xenopus Protocols: Post-Genomic Approaches, Second Edition*,
Methods in Molecular Biology, vol. 917, DOI 10.1007/978-1-61779-992-1_18, © Springer Science+Business Media, LLC 2012

Currently, there are several platforms available to perform deep sequencing. Each platform employs a different method, albeit maintaining the following three main strategies: template preparation, sequencing and imaging followed by alignment and assembly of sequences. We will focus on Illumina, presently the most widely used platform. For more technical details and a comparison of the available platforms, readers may refer to an article by Metzker (1).

In the year 2008, Michael Snyder's group first published the "RNA-Seq" or RNA Sequencing method using the Illumina platform. They described the use of high-throughput sequencing of complementary DNA (cDNA) fragments as a quantitative method to obtain a high-resolution transcriptome map of the yeast genome. Their method followed these steps: isolation of polyadenylated RNA, generation of double-stranded cDNA by reverse transcription, fragmentation, high-throughput Illumina sequencing, and mapping to the reference genome (2). Illumina uses reversible terminator amplification, whereby fluorescent nucleotides are added to each sequence, followed by an imaging step that allows the identification of the incorporated nucleotide. Once the imaging is finished, the incorporated terminators are removed to allow the entrance of a new nucleotide (3). Following a determined number of cycles, a read length (according to the manufacturers) of approximately 35–150 nucleotides can be obtained. These reads can be mapped to the corresponding reference genome or transcript sequence database (e.g., UniGene collection). Although the availability of a reference genome is important for the alignment and assembly of sequences, this is not strictly necessary. Full-length transcript assembly can be performed without a reference genome using Trinity, a recently published software for de novo transcriptome assembly (4). We will not cover this method, but interested readers may refer to this article for details.

The use of RNA-Seq over DNA microarrays for the study of transcriptomics presents several advantages. First, it does not rely on previous knowledge about genome or transcript sequences. Second, sequencing methods lack the problem of cross-hybridization, reducing background levels and improving the dynamic range of detection, as signals are not as easily saturated (5). Furthermore, not all probes will hybridize with the same affinity, which may create biases towards certain transcripts (6). Finally, RNA-Seq delivers information that cannot be obtained using *Xenopus* commercially available microarrays, such as identification of new transcripts and differential mRNA processing events (7).

When compared to tag-based methods, such as serial analysis of gene expression (SAGE) (8), RNA-Seq also provides many advantages. In the latter, the template is attached to a solid surface, where it will be clonally amplified creating a cluster of identical templates, hence eliminating the need for bacterial cloning (1). In addition, in tag-based methods, only a portion of the transcript can be analyzed, and many short sequences will not be uniquely mapped to the reference genome (5).

In addition to global mRNA studies, these methodologies can also be employed for characterizing the population of small RNAs (small RNA-Seq) after modifications in the library preparation (9).

These methodologies have been used for transcriptomics analyses in several organisms (2, 6, 7, 10, 11). Recently, in *Xenopus* some groups have successfully used high-throughput sequencing. Veenstra's group carried out the first ChIP-Seq study (see also Chapter 17). They combined chromatin immunoprecipitation with Illumina deep sequencing to study epigenetic regulation of gene expression in *Xenopus* embryos (12). In addition to this ChIP-seq study, other groups have performed deep sequencing of small RNAs. Miska's group studied the expression of small RNA populations in the germline and soma of *Xenopus tropicalis* (13). Blower's group performed analysis of PIWI-interacting RNAs (piRNAs) in *X. tropicalis* eggs (14). Lai's group showed that primary piRNAs can be derived from 3′-UTR in *Drosophila*, mouse, and *Xenopus* (15). Finally, in our laboratory, we studied piRNAs specifically derived from transposable elements in the *X. tropicalis* gastrula (16). In all these cases, the mapping was performed to the *X. tropicalis* genome sequence (17). However, Xenbase, EST, and UniGene databases can also be used to map to transcripts for both *X. tropicalis* and *X. laevis* (18, 19).

Through this chapter, we aim to provide a starting point for researchers working in *Xenopus* who are planning to use RNA-Seq for the study of transcriptomics. We provide a section on template isolation and library preparation, followed by bioinformatics procedures that need to be performed for data analysis. Finally, we have included a section on studying deep sequencing results in *Xenopus*, which offers general guidance as to what can be done in this model.

2. Materials

2.1. Template Isolation: Total RNA

1. RNeasy Mini Kit (QIAGEN, #74104).
2. RNA*later* RNA Stabilization Reagent (QIAGEN, #76104).
3. RNase-Free DNase Set (QIAGEN, #79254).
4. Dithiothreitol (DTT) 2 M.
5. Ethanol 70%.
6. Rotor-stator homogenizer (or a drill with a 1.5 mL Eppendorf tube polypropylene pestle attached to it).
7. Nanodrop (Thermo Scientific) (or fluorimetric quantification equipment).
8. 2100 Bioanalyzer (Agilent Technologies).

2.2. Template Isolation: Small RNA

1. TRIzol reagent (Invitrogen #15596-026).
2. Chloroform.
3. Isopropyl alcohol.
4. Nanodrop (Thermo Scientific) (or fluorimetric quantification equipment).
5. 2100 Bioanalyzer (Agilent Technologies).

2.3. Library Preparation and Sequencing

1. TruSeq RNA Sample Prep Kit (Illumina # FC-122-1001).
2. TruSeq small RNA Sample Prep Kit (Illumina # RS-200-0012).
3. Illumina Genome Analyzer IIx or Illumina HiSeq 2000.
4. 2100 Bioanalyzer (Agilent Technologies).

2.4. Bioinformatics Analyses

For convenience, all computer tools and databases described in this chapter are freely available to download from our laboratory Web site (http://melolab.org/ngs). However, it is important to mention that most of these tools were developed by several independent research groups or private companies, and not by us. Additionally, we would like to emphasize that many of these tools are rapidly evolving, and that some of them might have license restrictions. Therefore, it is up to the user to check that all the required restraints are fulfilled. Our aim in compiling these tools at our Web site is simply to facilitate the implementation of the procedures described in this book chapter, but we cannot guarantee that these tools and databases will be continually updated in the future. The above mentioned software tools are the following:

1. *cutadapt* software (for adapter removal).
2. *FASTX Toolkit* software suite (sequence filtering).
3. Perl.
4. Bowtie (20).
5. edgeR package from Bioconductor.

3. Methods

3.1. RNA Template Isolation

We provide two different protocols for RNA isolation from experimental sample tissue: one for general applications to study mRNA expression (Subheading 3.1.1), and the other specifically designed for isolation and analysis of small RNA expression (Subheading 3.1.2).

3.1.1. Template Isolation: Total RNA

For total RNA isolation, we have successfully used the RNeasy Mini Kit, according to manufacturer's instructions (RNeasy Mini

Table 1
Recommended use of RNA*later*, buffer RLT, and estimated total RNA isolated in certain *Xenopus laevis* tissues/embryos

Sample	Amount of sample (maximum)	RNA*later* volume (μL)	Buffer RLT volume (μL)	Estimated total RNA (μg)
Oocytes	10 oocytes	500	600	~25
Stage 10–20 embryos	10 embryos	500	600	~25
Stage 21–40 embryos	5 embryos	500	600	~25
Stage 10 dorsal/ ventral explants	50 explants	500	600	~25
Stage 50 spinal cords (half)	50 spinal cords	200	1,000	~7–10
Stage 66 spinal cords (half)	10 spinal cords	200	1,000	~7–10

We provide here the estimated volumes of RNA*later* and Buffer RLT (tissue lysis buffer) we have used for different amounts of *X. laevis* embryos and tissues, together with the estimated yield of total RNA. We would like to note that we have not performed a detailed study regarding these amounts. They are only based upon our experience

Handbook, Fourth Edition, 2009, QIAGEN) for processing of animal tissues. Carefully consider good molecular biology practice for working with RNA (see Note 1).

1. Transfer tissue immediately after isolation into *at least* 10 volumes of RNA*later* solution to avoid RNA degradation (see Table 1, Note 2).
2. After isolating all samples, transfer each of them (up to 30 mg) into 600 μL of Buffer RLT (see Table 1) for tissue disruption and homogenization using a rotor-stator homogenizer, until it is uniformly homogenous (usually two or three cycles of 20 s each is enough, see Note 3).
3. Centrifuge the lysate for 10 min at full speed (see Note 4).

   For the following steps (4–9), *use the same collection tube until instructed to change it.*
4. Carefully transfer the supernatant into a new 1.5 mL tube, and add 1 volume of 70% ethanol. Mix by carefully inverting the tube. Do *not* centrifuge. Transfer up to 650 μL of the sample, including any precipitate that may have formed into an RNeasy spin column placed in a 2 mL collection tube (supplied). Close the lid (see Note 5) and centrifuge for 15 s at 8,000 × *g*. Discard the flow-through. As the sample volume will usually exceed 650 μL, centrifuge successive aliquots in the same RNeasy spin column and discard the flow-through after each centrifugation.
5. Add 350 μL of Buffer RW1 to the spin column. Close the lid and centrifuge for 15 s at 8,000 × *g* to wash the spin column membrane. Discard the flow-through.

6. Add 10 μL DNase I stock solution to 70 μL Buffer RDD. Mix by gently inverting the tube (*do not vortex*), and centrifuge briefly to collect residual liquid from the sides of the tube. Add this mix (80 μL) directly onto the middle of the RNeasy spin column membrane and incubate at room temperature (~20°C) for 15 min (see Note 6).
7. Add 350 μL Buffer RW1 to the spin column. Close the lid and centrifuge for 15 s at 8,000 × *g*. Discard the flow-through.
8. Add 500 μL of Buffer RPE to the spin column. Close the lid and centrifuge for 15 s at 8,000 × *g*. Discard the flow-through.
9. Add 500 μL of Buffer RPE to the spin column. Close the lid and centrifuge for 2 min at 8,000 × *g*. Discard the flow-through.
10. Transfer the spin column into a new collecting tube (supplied). Close the lid and centrifuge for 1 min at maximum speed.
11. Transfer the spin column into a new 1.5 mL labeled collection tube (supplied—this tube will contain your final RNA sample), and leave the spin column lid open for at least 1 min to ensure that all ethanol has evaporated.
12. Add 50 μL of RNase-free water (supplied) in the middle of the spin column (see Note 6), and incubate at room temperature (~20°C) for 10 min to ensure a higher RNA yield.
13. Close the lid and centrifuge for 1 min at 8,000 × *g* to elute the RNA. *Keep the flow-through*, as it contains the RNA.
14. A 2 μL aliquot of the eluted RNA may be used for quantification using Nanodrop. A260/A280 should be higher than 1.8 (see Note 7).
15. To evaluate RNA integrity, Agilent 2100 Bioanalyzer may be used. RIN (RNA integrity number) value should be higher than 8 (see Note 8).

3.1.2. Template Isolation: Small RNA (see Note 9)

1. Place the recommended amount of sample in 1 mL of TRIzol reagent (see Table 2, and Note 2). It is possible to store samples at −20°C for several months in this step if necessary.
2. For embryos, vortex samples for 1–2 min for homogenization. Use a rotor-stator homogenizer with a 1.5 mL Eppendorf tube polypropylene pestle for spinal cord or harder tissues.
3. Add 200 μL chloroform and vortex. Keep samples on ice. Centrifuge the samples at 4°C for 5 min at 16,000 × *g*.
4. Transfer the aqueous phase (colorless upper phase, usually ~450 μL) to a fresh new tube. Add 1 volume of chloroform, vortex, and centrifuge at 4°C for 5 min at 16,000 × *g*.
5. Transfer the aqueous phase (colorless upper phase, usually ~400 μL) to a fresh new tube. Add 1 volume of isopropyl

Table 2
Recommended use of sample amount per 1 mL TRIzol reagent, and estimated total RNA isolated in certain *Xenopus tropicalis* and *X. laevis* tissues/embryos

Sample	Amount of sample per mL TRIzol	Estimated total RNA (μg)
Oocytes (*X. tropicalis*)	10 oocytes	~10
Stage 10–20 embryos (*X. tropicalis*)	10 embryos	~10
Stage 10 dorsal/ventral explants (*X. tropicalis*)	30 explants	~1
Stage 50 spinal cords (half) (*X. laevis*)	10 spinal cords	~1–2
Stage 66 spinal cords (half) (*X. laevis*)	2–3 spinal cords	~1–2

We provide here the estimated amount of several *Xenopus* tissues and embryos that can be lysed in 1 mL of TRIzol reagent, and the estimated total RNA amount isolated. We would like to note that we have not performed a detailed study regarding these amounts. They are only based upon our experience

alcohol, vortex, and incubate the samples at –80°C for at least 30 min (see Note 10).

6. Centrifuge the samples at 4°C for 30 min at 16,000 × *g*.
7. Discard the supernatant. Optional: Wash the pellet with 500 μL 70% ethanol and centrifuge the samples at 4°C for 10 min at 16,000 × *g* (see Note 11).
8. Discard the supernatant and centrifuge again at 4°C for 2 min at 16,000 × *g* to discard the supernatant completely.
9. Resuspend the pellet in 20–50 μL RNAse-free water and determine the concentration of RNA using Nanodrop (see Note 12).

3.2. Transcriptome Library Preparation for Deep Sequencing

3.2.1 Library Preparation for mRNA-Seq

Illumina is continually updating their library preparation kits, reducing processing time, increasing efficiency, and making it possible to start with lower amounts of total RNA. For example, in the newest version of the sample preparation kit it is possible to construct a library starting with 0.1–4 μg of total RNA. Library preparation consists of the following main steps: fragmentation, reverse transcription followed by adapter ligation, finishing with template enrichment using PCR. As protocols change, we will give an overview with the main steps of the library preparation process, based on Illumina's TruSeq RNA Sample Preparation Guide (November 2010).

1. *Polyadenylated mRNA Isolation.* Polyadenylated mRNA (poly-A) molecules are separated from other RNAs (mainly ribosomal RNA, which comprises >90% of total RNA) through poly-T oligo-attached magnetic beads, using two rounds of purification. Poly-A mRNA is then eluted for the following step (see Note 13).

2. *Fragmentation*. Poly-A mRNA molecules are then fragmented using elevated temperature in the presence of divalent cations (see Note 14).
3. *Reverse Transcription*. First strand cDNA is then obtained using a reverse transcriptase (e.g., SuperScript II) and random hexamers. The RNA strand is then degraded using RNase H, and is followed by DNA Polymerase I second strand cDNA synthesis. Afterwards, an ethanol cleanup is performed.
4. *End-Repair*. This process removes 3′ overhangs and fills in 5′ overhangs, usually using T4 and Klenow DNA polymerases.
5. *Adenylation of 3′ Ends*. A single "A" nucleotide is added to the 3′ end of the blunt double-stranded cDNA, enabling adapter ligation in the next step. dATP and polymerase activity of Klenow fragment (3′ to 5′ exo minus) are used for this purpose.
6. *Adapter Ligation*. Adapters contain a single "T" nucleotide overhang on the 3′ end enabling the ligation of the adapter to the fragment and thereby lowering the rate of template concatenation. They are added to each reaction tube together with T4 DNA ligase, and incubated for ligation. Adapters at both ends of double-stranded cDNA will enable hybridization to the flow cell.
7. *Product Purification and PCR Enrichment*. PCR is used to enrich templates that have successfully acquired adapter molecules. Primer cocktails that anneal to the ends of adapters will preferably select templates that have successfully acquired adapter molecules. Only 15 PCR cycles are then used to enrich the correct templates, avoiding library construction biases. The product is then washed and purified, followed by quality and size validation using Agilent 2100 Bioanalyzer (see Note 15).

3.2.2. Library Preparation for Small RNA-Seq

Small RNA sequencing library preparation is similar to that for mRNA. We will describe only the main differences. Adapters are ligated to both ends of the small RNA before reverse transcription and library generation. In addition, the fragmentation step is unnecessary, and the process ends with size selection through gel purification. Like for mRNA, hereafter is an overview of the main steps of small RNA library preparation, based on Illumina's TruSeq Small RNA Sample Preparation Guide (March 2011).

1. *3′ Adapter Ligation*. 3′ RNA adapter ligation is performed by incubating the total RNA with the 3′ adapters, followed by addition of Truncated T4 RNA Ligase 2, which specifically ligates the pre-adenylated 5′ end of the RNA adapter to the 3′ end of the RNA in the sample. The enzyme does not require ATP for ligation, but does need the pre-adenylated substrate, optimizing adapter ligation.

2. *5′ Adapter Ligation.* 3′ RNA adapter ligation follows by incubating with the 5′ adapters, followed by addition of ATP and full-length T4 RNA Ligase 2 to the mix.
3. *RT-PCR Amplification.* As for mRNA library preparation, this step ensures enrichment of sequences that have successfully ligated both adapters. To accomplish this, primers that ligate to the ends of adapter molecules are used. Reverse transcription is performed to generate the cDNA library for the sequencing process, and is followed by PCR amplification, using only 11 cycles, to avoid any bias in library construction.
4. *Purification of Small RNA Library.* For purification, the PCR amplified library is loaded onto a gel, and size selection is performed by gel extraction and purification (see Note 16).

3.3. Sequencing and Imaging

Currently, with Illumina technology, there are three slightly different types of sequencing that can be carried out: (1) single-end sequencing (Fig. 1a), (2) paired-end sequencing (Fig. 1b), and (3) mate-pair sequencing (Fig. 1c). The first type corresponds to the sequencing of only one end of the cDNA molecule. The second and third types perform sequencing from both ends of the cDNA, differing only in the size of the sequenced molecule (see Note 17). Once sequencing has been completed, the enormous amount of data (see Note 18) obtained needs to be analyzed in order to extract the desired information from it. The handling and processing of data is one of the main challenges in using RNA-Seq. Subheadings 3.4, 3.5, and 3.6 provide the bioinformatics tools and databases needed for this purpose. It is important to note that due to the complexity of the required data analyses, these sections were aimed at an audience with knowledge on bioinformatics.

3.4. Raw Data Processing

One of the most important goals in transcriptomics analysis is the process of obtaining general and specific knowledge from the large amount of data that is retrieved. This task requires computer processing of hundreds of millions of short reads (~100 nucleotides). It usually involves filtering, alignment, assembly, clustering, counting, and normalization of data for each experimental condition or source, as well as analyses of differential expression across experimental conditions. Different deep sequencing platforms may deliver their output in a variety of data formats (see Note 19). However, the FASTQ format (see Note 20) has recently become the standard in the field. To illustrate the logical order of the steps involved in the processing of the raw data, we have designed a simple and general flow chart (Fig. 2).

3.4.1. Raw Data Processing: Removing Adapter Sequences

The *cutadapt* software can be used to remove the adapters from the raw sequence data. This software is implemented in Python, but has an extension module written in C language that implements

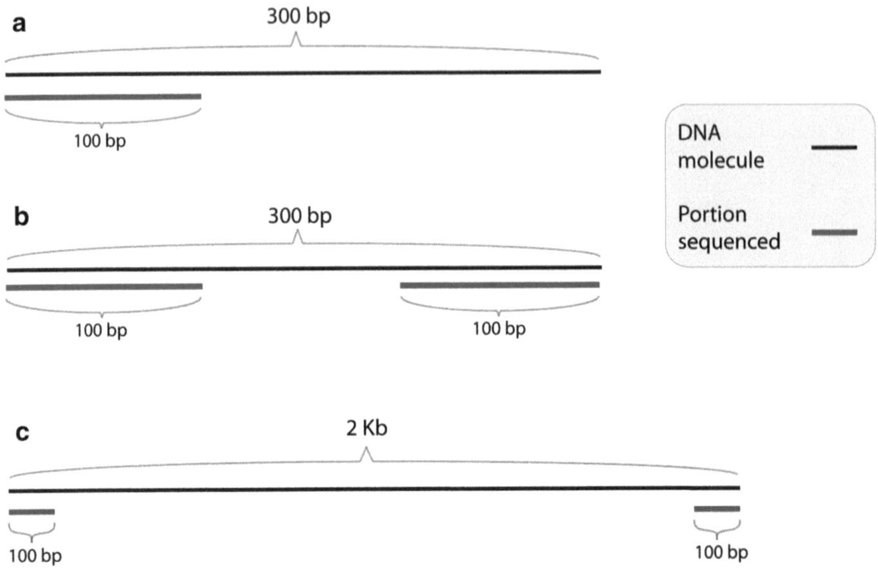

Fig. 1. Illumina sequencing categories. (**a**) Single-end sequencing. Only one end of the molecule is sequenced. (**b**) Paired-end sequencing. Both ends of the cDNA molecule are sequenced. This allows the unambiguous allocation of the cDNA segment in the assembly process, or during the task of mapping reads, by employing the distance between the two short sequence pairs (~200–400 bp). (**c**) Mate-pair sequencing. As in paired-end sequencing, both ends are sequenced. However, due to differences in the library preparation procedure, the distance that can exist between the two short sequences is much higher (~2–5 Kb).

the alignment algorithm. Assuming your sequencing data is available as a FASTQ file, an execution example of this software to achieve this task will be the following:

cutadapt -a ADAPTER_SEQUENCE input.fastq > output.fastq

In this example, the processed output sequence data without the adapters will be stored in the file "output.fastq" (see Note 21). Please note that the term ADAPTER_SEQUENCE must be replaced with the specific adapter sequence used (e.g., ATCGATCGTGTGACGAT).

3.4.2. Raw Data Processing: Filtering Sequences

The main filters applied to the data are the following: read length, read quality, sequence complexity, removal of singletons, and genome frequency. The *FASTX Toolkit* software suite can be used for this purpose, as it contains a collection of individual software applications for most of these tasks. The following list contains a description of each filter and an example of how to execute it in the program.

1. Length Filtering
   For small RNAs (miRNA, siRNA, piRNA, etc.), filtering of reads that are not between 19 and 32 bp is necessary. Unfortunately, FASTX Toolkit software does not provide a script for achieving this. However, the simple Perl script *FilterLength.pl* written by us can be used for this task.

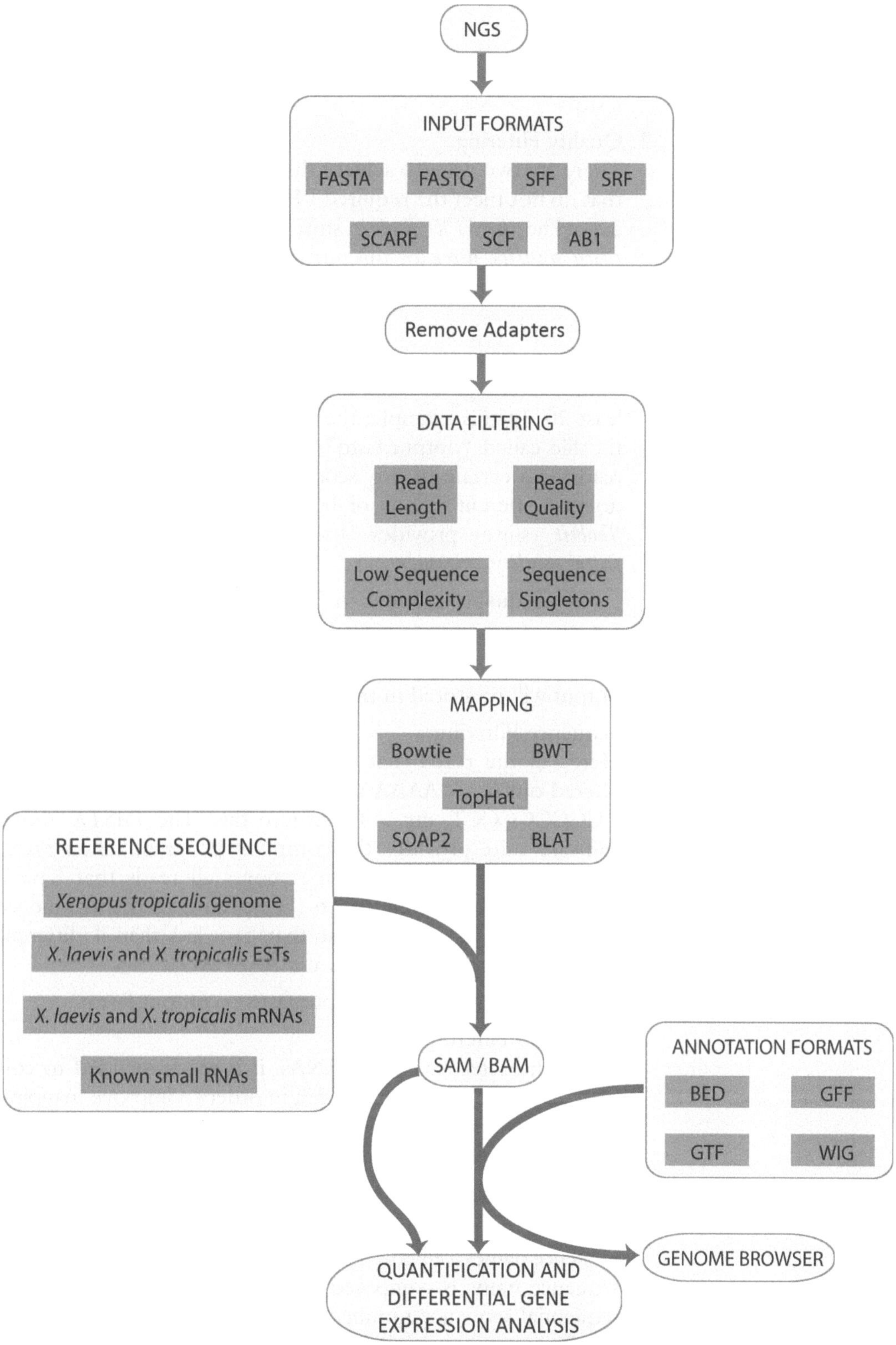

Fig. 2. RNA-Seq data analysis flowchart. Next generation sequencing results can be delivered in several input formats. However, FASTQ format has become a standard in the field. Adaptors are first removed from raw sequences, followed by data filtering. The next stage is the mapping of the filtered sequences to *Xenopus* reference sequences. Output files (SAM/BAM) can be used for quantification and differential gene expression analyses. It is important to mention that although deep sequencing technologies and their associated software are rapidly changing, the diagram displayed here is robust because it is general and not attached to any specific software or technology.

perl FilterLength.pl –m 19 –M 32 –i input.fastq –o output.fastq

2. Quality Filtering
   There are two ways to achieve this. The first is to remove reads that do not meet the required PHRED quality score (see Note 22). The *FASTX Toolkit* suite provides a program called *fastq_quality_filter* for this purpose.

   fastq_quality_filter –q 20 –p 80 –i input.fastq –o output.fastq

   Here, 20 is the minimum quality score and 80 is the minimum percentage of bases that must have a quality score of at least 20. In this example, the filtered output will be stored in the file called "output.fastq" The second one is to trim the reads by a certain quality score, which means that the read is going to be cut in one or both ends. For this task, *FASTX Toolkit* suite provides the computer program called *fastq_quality_trimmer.*

   fastq_quality_trimmer –t 20 –i input.fastq –o output.fastq

   Here, 20 is the quality score threshold (nucleotides with lower quality will be trimmed). In this example, the trimmed output will be stored in the file called "output.fastq."

3. Sequence Filtering
   Here, all the reads that have low sequence complexity are filtered out (e.g., AAAAAAAAAAA, GGGGGGGGG, CCCCCCCCCCCCCT, etc.). To achieve this, The *FASTX Toolkit* software suite provides the computer program called *fastx_artifacts_filter.* This program removes all reads that have a sequence length lower than 6 (irrespective of their sequence composition) as well as those that have less than 4 different nucleotides (e.g., AAAAAAAAAAAAAAAGT).

   fastx_artifacts_filter –i input.fastq –o output.fastq

4. Singletons Filtering
   When characterizing small RNAs, it is recommended to collapse the reads that are identical, in order to improve mapping time (see Note 23). *FASTX Toolkit* suite provides the program called *fastx_collapser f*or this task:

   fastx_collapser –i reads.fastq –o input-collapsed.fasta

   In the output file (i.e., "input-collapsed.fasta"), the original sequence names found in the input file are discarded. The output sequence name is composed of two numbers: the first is the sequential sequence number in the input file, and the second number is the multiplicity value (i.e., how many times the particular sequence was found in the input file). The collapsing of the reads allows the identification of those sequences that were observed only once. These "singletons" should be removed before mapping. The following script written by us, *RemoveSingletons.pl*, can achieve this task:

perl RemoveSingletons.pl –n 2 –i input-collapsed.fasta –o output.fasta

Here, option "n" sets the minimal allowed count, which in this case is two because we want to remove singletons (reads with just one count). "input-collapsed.fasta" is the input file that contains the sequences in FASTA format, and also the headers of each sequence are given in the format provided by the *fastx_collapser* program. "output.fasta" is the ouput file without the singletons sequences.

5. Genome Filtering
This filter can be applied only when a reference genome or transcriptome is available. It is usually used when the mapping of a particular read is intended to a single transcript or to a single position in the genome. Therefore, after the mapping is carried out, reads that are found in only one region of the reference sequences are selected (e.g., transcripts, ESTs, contigs, scaffolds, chromosomes). This filter can only be applied after the mapping has been carried out, not before (see below).

3.5. Mapping Reads Against Reference Sequence Data

Before mapping, two elements must be defined and available: (1) the reference sequences and (2) the mapping software that will be used.

1. The Reference Sequences
Currently, a draft of the *X. tropicalis* genome is available in public databases (17) (see Chapter 4). Albeit unassembled, 20,000 scaffolds can be retrieved from Biomart (21), containing the most updated annotation. For *X. laevis,* we have the mitochondrial genome, which has been completely sequenced and assembled (22). In addition, there are around 600,000 ESTs and 30,000 mRNAs currently available for *X. laevis* in UniGene (23), which have been collected from GenBank and dbEST (24). Xenbase (25) also has a large set of ESTs (about 677,000) and mRNAs (about 30,000) that can supplement the final count. Summarizing, we can use the genome, ESTs, mRNAs, and UniGene clusters for the analysis of *X. tropicalis* sequences (see Note 24), and for *X. laevis,* although its genome is yet unavailable, ESTs, mRNAs, and UniGene clusters provide an ample database to use as reference sequences (see Note 25).

2. Mapping Software
There are several softwares that can be used for mapping (see Note 26), but due to the large number of sequences that are obtained after sequencing with deep sequencing platforms, *Bowtie* computer program is a good choice because it is fast and efficient in terms of memory usage and mapping time (20). *Bowtie* is a computer software that works in two consecutive

steps: (1) an index of the reference sequences is created first and (2) the short reads are then aligned against the indexed reference sequences. The indexation of the reference sequences (e.g., *X. tropicalis* genome, *X. laevis* UniGene clusters) is carried out with *bowtie-build* tool as follows:

bowtie-build –f RefGenome.fasta RefGenome

where "-f" indicates to the program that the format of the input file is FASTA, and the last argument corresponds to the indexed output file (see Note 27).

The alignment of short reads against the indexed reference sequences is also carried out with *bowtie* program for single-end, paired-end, and mate-pair sequencing (Fig. 1).

(a) *For single-end sequencing*:

bowtie –a –v 2 RefGenome reads.fastq

where "-a" option indicates to report all alignments per read and "-v" option is the number of mismatches allowed (see Note 28).

(b) *For paired-end and mate-pair sequencing*:

bowtie --fr -S –I 200–X 400 –a –v 2 RefGenome -1 reads1.fastq -2 reads2.fastq Mapping.sam

where "-a" and "-v" options mean the same as above. "--fr" indicates the orientation of the mate -1 and -2 (Forward/Reverse, see Note 29). Options "-1" and "-2" correspond to the respective mates. "-I" and "-X" are the minimum and maximum insert size for paired-end alignment. All the alignments are written in the output file called "Mapping.sam" in SAM format (see Note 30) because "–S" option was given.

Read mapping in *Xenopus* can be performed using different references: *X. tropicalis* genome (Subheading 3.5.1), transcripts (Subheading 3.5.2) and ESTs (Subheading 3.5.3) (for both *X. tropicalis* and *X. laevis*), and small RNA (Subheading 3.5.4).

3.5.1. Mapping of Short Reads Against the X. tropicalis Genome

1. Download the *X. tropicalis* genome in FASTA format (see Note 31). After downloading, it should be uncompressed (see Note 32).
2. Index the genome with *bowtie-build* as follows:

   bowtie-build –f Xentr4.allmasked XtGenome
3. Align the short reads against the recently indexed file (this example assumes that you have paired-end sequencing):

   bowtie –-fr -S –I 250 –X 350 –a –v 2 –m 1 XtGenome -1 reads1.fastq -2 reads2.fastq map.sam

   It is important here that the flag "-m" is set to 1 (see Note 33). Also, for paired-end or mate-pair sequencing the distance between mates must be known (see Note 34).

3.5.2. Mapping of Short Reads Against Known X. tropicalis or X. laevis Transcripts (UniGene)

UniGene is a database that contains partitioned transcript sequences (including ESTs) from GenBank into a nonredundant set of clusters, each representing a potential gene locus.

As mentioned above, two things are needed to perform the mapping, and we will only change the reference sequence (i.e., after indexation takes place).

1. Download the FASTA files of the UniGene cluster for *X. tropicalis* or *X. laevis* (see Note 35). Downloaded files should have the names "Str.seq.uniq.gz" and "Xl.seq.uniq.gz" (or similar), respectively. After downloading, it should be uncompressed.
2. Repeat steps 2 and 3 of Subheading 3.5.1, but instead of using the genome, use one of the previously downloaded files. Also rename the output of *bowtie-build* to something coherent with the used file (e.g., XlUniGene or XtUniGene).

3.5.3. Mapping of Short Reads Against Known X. tropicalis or X. laevis ESTs

Mapping against expressed sequence tags can be a very demanding task computationally, because of the large and growing number of existing ESTs for *X. tropicalis* and *X. laevis* species (to date, 1,271,375 and 677,806 ESTs respectively). Additionally, the volume of short sequence reads delivered by current deep sequencing platforms make the access to an adequate computer infrastructure very important, in order to carry out the alignment task and also to store the input and output data.

As mentioned above, two things are needed for mapping, for which here we will only change the reference sequence.

1. Download the file containing the ESTs for *X. tropicalis* or *X. laevis* in FASTA format (see Note 36). The files to download should be named as xlaevisEST.fasta and xtropEST.fasta, respectively.
2. Repeat steps 2 and 3 of Subheading 3.5.1, but instead of using the genome, use one of the previously downloaded files. Also, rename the output of *bowtie-build* to something coherent with the used file (e.g., XlESTs or XtESTs).

3.5.4. Mapping of Small RNAs to Reference Sequences

Mapping of small RNAs is similar to the previously described alignments. First, the reference sequences need to be defined, and then the alignment with *bowtie* must be carried out. It is not recommended to allow mismatches in the alignment (i.e., to only use unambiguously mapped reads). If the aim is to characterize the expression level of known small RNAs, then the reads should first be aligned against some existing database of noncoding RNAs (see Note 37), and then to only use those sequences that match in the posterior mapping against the reference sequences.

For *X. tropicalis* genome there is an annotation available (see Note 38), thus after aligning the genome to the mapped reads, they can be crossed with this annotation to search for known sRNAs.

3.6. Data Analysis of Transcriptomics Analysis Results

3.6.1. Data Normalization

Normalization requires two initial definitions: (1) which subset of genes (also called the normalization baseline or reference population) will be used to calculate the normalization factor and (2) which normalization method will be employed.

1. *Choosing a Normalization Baseline*: In the absence of knowledge regarding invariant genes, all genes can be used to calculate the normalization factor. The advantage of this strategy is that no assumptions are made about the expression patterns of individual genes. This type of baseline assumes that the median or mean expression level across all of the genes is mostly unchanged (26). This assumption is sustained on the fact that RNA content is constant between different experimental samples, and that most genes do not significantly change their expression levels between them. The definition of this normalization baseline can be applied to both mRNA and small RNA data obtained from RNA-Seq experiments.
2. *Normalization Method*: Many studies have been performed comparing different normalization methods (27–33). The general conclusion is that there is no single "best" normalization method. However, the trimmed mean of M value (TMM) method (34) has been shown to constitute a fair normalization method and some authors have encouraged its use. Alternatively, upper-quartile normalization method is also recommended over RPKM normalization, which is another widely used method (35).

Please note that in the context of this subject, sensitivity is defined as the ability to detect changes between two conditions. This is reflected into how many transcripts show differential expression patterns when distinct normalization methods and statistical tests are employed. It has been shown that sensitivity variance is more prone to vary upon the normalization procedure adopted, rather than in the statistical test used (35). Therefore, the choice of the data normalization procedure is a key step that should be carefully selected when inferring differential expression patterns.

3.6.2. Assessing Differential Gene Expression Between Samples

Statistical Analyses: The package *edgeR* from Bioconductor, which adopts the TMM normalization method, is a good choice to assess differential gene expression. In addition to implementing the calculation of the normalization factor, this package also provides statistical functions for the assessment of differentially expressed genes. These functions are based on an over-dispersed Poisson model and an empirical Bayes procedure to moderate the degree of over-dispersion.

3.7. Experimental Validation of Results from Deep Sequencing Analyses in Xenopus

An important step following high-throughput sequencing is the study of specific sequences of interest identified in a deep sequencing experiment. These analyses allow the experimental validation of novel sequences, but also to verify the differential expression of transcripts when two or more conditions were compared during deep sequencing.

One of the main advantages of working with *Xenopus* (both *X. laevis* and *X. tropicalis*) is the availability of several methods to study gene expression, such as RT-PCR (see Note 39), in situ hybridization, and Northern blot.

qRT-PCR is the most adequate method to accurately validate transcript abundance results. This method has been extensively used for studying gene expression of transcripts. The primers can be designed using information contained in the same sequence identified by deep sequencing, by mapping to known transcripts or to the genome sequence (in the case of *X. tropicalis*). In *Xenopus*, several genes can be studied using relatively low amounts of material (see Tables 1 and 2 for approximate RNA yield obtained from samples). It is important to mention that qRT-PCR can also be used to study the expression of small RNAs (36–38). qRT-PCR for small RNAs includes a polyadenylation step of total or small RNA followed by a reverse transcription using a modified oligo-dT primer. The PCR reaction is then performed using the identified small RNA sequence and a sequence included in the modified oligo-dT as primers. PCR products can be cloned and sequenced to verify the specificity of the PCR reaction. This protocol has been successfully used in our laboratory (16).

On the other hand, standard protocols are available for *Xenopus* whole mount and section in situ hybridization (39). Aside from permitting validation of high differences in transcript contents across samples, it provides further information regarding their localization. This method has also been used for studying small RNAs (see Chapter 25), in particular, miRNAs (40). The use of labeled locked nucleic acids (LNA)—complementary to the sequence of interest—can be a useful tool for determining the expression pattern of small RNAs in *Xenopus*. However, it is important to consider a possible hybridization with precursors, especially in the case of siRNAs and piRNAs.

Finally, most *Xenopus* samples provide enough RNA to perform Northern blotting (see Tables 1 and 2), which proves especially useful when identifying new transcripts, as transcript length can be verified through this method. This valuable information cannot be provided by qRT-PCR. This method can also be used for small RNAs. A cDNA or RNA probe must be prepared. Please refer to published protocols for details on small RNA Northern blots (41). For Northern blotting of small RNAs, the use of LNA probes is recommended for increased sensibility.

In summary, all these methodologies can be used to confirm the expression of transcripts (long and small RNAs) in *Xenopus*. In addition, as obtaining embryos at different stages of development, or explants in *Xenopus*, is relatively easy, it is possible to extend the analyses of gene expression to samples from different embryonic stages and tissues that were not included in the original deep sequencing experiment.

4. Notes

1. Before starting carefully consider the following advice on working with RNA:
   - Work (quickly) at *room temperature* during the whole procedure (including centrifugation steps).
   - If using the RNeasy kit for the first time, add 4 volumes of 100% ethanol to Buffer RPE (e.g., if kit comes with 11 mL of buffer RPE, add 44 mL of 100% ethanol to make a total volume of 55 mL).
   - Prepare a 2 M Dithiothreitol (DTT) stock solution in RNase-free water, and store in single-use aliquots at -20°C.
   - Add 20 μL of 2 M DTT per 1 mL Buffer RLT before use. Only prepare the amount of buffer needed for the day's samples—do not add DTT directly to stock solution.
   - Prepare DNase I stock solution by dissolving the lyophilized DNase I (1,500 Kunitz units) in 550 μL of the RNase-free water provided in the kit. Do this by injecting the RNase-free water into the vial using an RNase-free needle and syringe (we use a 1 mL BD syringe for this purpose). Mix gently by inverting the vial. Do not vortex, as DNase I is especially sensitive to physical denaturation. Homogenize by gently inverting the tube only.

   For long-term storage of DNase I, store in single-use aliquots and store at -20°C for up to 9 months. Thawed aliquots can be stored at 2–8°C for up to 9 months. Do not refreeze the aliquots after thawing.
   - Clean a pair of forceps using 70% ethanol.
   - Prepare 70% ethanol adding 7 mL of 100% ethanol to 3 mL of RNase-free water, and mix by inverting the tube.
2. Table 1 provides recommended values that we have used at our laboratory. However, we have not performed a detailed study on the amount of buffer needed, nor the RNA yield. We have only provided approximate values according to our experience.
3. You may use a 1.5 mL Eppendorf tube polypropylene pestle attached to the rotating station. When the sample tissue requires a Buffer RLT volume exceeding 600 μL, first place all tissue into 500–600 μL of Buffer RLT and disrupt for two 20 s rounds, then add the rest of Buffer RLT to the same tube, mix by pipetting and separate the contents again into as many tubes as necessary, using up to 600 μL per tube, for two more 20 s rounds of tissue disruption and homogenization. This optimizes the RNA yield.

4. Although the kit instructs to centrifuge for 3 min, it has not been enough for some of the tissues we have worked on, which is why we centrifuge for 10 min.
5. Always close the lid of the spin column gently.
6. Be careful to avoid adding the incubation mix to the walls or the *O*-ring of the spin column, as the DNase I digestion may be incomplete.
7. In our experience an A260/A280 above 2 is desirable (it is rarely not achieved using this kit).
8. Prior to library preparation and sequencing, it is advisable to check the samples using RT-PCR of spatial and/or temporal gene expression markers. This will be especially useful when validating sequencing results.
9. The above protocol for total RNA isolation purifies all RNA fragments longer than 200 nt (according to the manufacturer), which is why we include the following protocol for small RNA isolation.
10. It is important to incubate for 30 min for an efficient isolation of small RNAs.
11. A fraction of small RNAs may be soluble in ethanol solutions. We have successfully performed the protocol without this ethanol wash. However, we have not compared the effect of including or excluding the ethanol wash.
12. This procedure can be used for the preparation of Illumina libraries, Northern blots, and for RT-PCR of both small RNAs and long RNAs. As with total RNA isolation, samples can be checked using RT-PCR for specific temporal or spatial gene expression markers.
13. In the latest Illumina kit, poly-A mRNA elution, fragmentation, and priming with random hexamers are performed at the same time, using the Elute, Prime, Fragment Mix.
14. Molecules may be fragmented at either the RNA or cDNA stage. The former is the standard procedure in Illumina sample preparation kits. It creates low bias towards the transcript body, but transcript ends tend to be depleted. The latter, on the other hand, creates a strong bias towards the 3′ ends of transcripts, which may prove useful by providing the precise identity of these ends (5).
15. The previous version of the Illumina sample preparation kit contained an extra size selection and gel purification step. They have eliminated this step from the newest version of the kit, for enhanced preparation time and robustness.
16. According to the manufacturer's instructions, once loaded onto the gel, there will be a 147 nt band corresponding mainly to mature microRNAs (originated from 22 nt small RNA fragments), and a 157 nt band containing PIWI-interacting

RNAs (and possible other microRNAs and other regulatory small RNA molecules).

17. The main difference between paired-end and mate-pair sequencing is the distance between the paired reads. While the former is from 200 to 400 bp, the latter goes from 2 to 5 Kb. The source of this difference is found in the library preparation procedures. We have not included library preparation for mate-pair sequencing, but the following kit may be used for this purpose (Mate Pair Library Prep Kit, Illumina # PE-112-2002).
18. Both sequencing and library preparation can be outsourced to services to which total RNA can be sent, and they are able to prepare the sequencing library using original Illumina kits. We provide the following details as a reference only. We obtained these from our own sequencing results using *Xenopus* samples.
    - Total RNA used for library construction: 1 μg of total RNA per sample.
    - Platform: Illumina HiSeq2000.
    - Sequencing: 100 bp paired-end.
    - Number of lanes: 2.
    - Number of samples: 8 (4 per lane).
    - Expected yield*: 4–5 GB of data per sample, equivalent to ~20 million reads per sample, at $Q>15–20$.
    - Actual yield: 10 GB of data per sample, equivalent to ~80 million reads per sample (~40 million reads per end), with 90% of sequences with $Q>30$.
    - *Refers to yield offered by sequencing service. Actual yield obtained was much higher than that stated by the sequencing service. However, we are not able to explain this.
19. Despite only a few deep sequencing platforms being currently widely used, there are several output data formats: FASTA, FASTQ (fastqsanger or fastqillumina variations), SFF (Standard Flowgram Format), SRF (Sequence Read Format also called Short Read Format), SCARF (Solexa Compact ASCII Read Format), SCF, and AB1.
20. FASTQ format is a text-based format for storing both a nucleotide sequence and its corresponding quality scores. Both the sequence letter and its quality score are simultaneously encoded within a single ASCII character for succinctness.
21. The procedure to remove the adapters requires previous knowledge of their sequence; otherwise it is impossible to achieve this task.
22. Illumina FASTQ score (sq) can be converted into Qphred score (Q) using the Perl code: $Q=10 * log (1+10 (ord($sq) – 64)/10.0)/log (10).

23. For single-end sequencing it is recommended to collapse the repeated reads because it speeds up the sequence mapping process.
24. We recommend using the raw *X. tropicalis* genome sequence first and then crossing the results with the known genome annotation.
25. We recommend using the UniGene clusters because they represent a nonredundant set of ESTs and mRNAs from *X. laevis.*
26. Bowtie (20), BWT (42), SOAP2 (43), MAQ (44), Zoom (45) are some alignment softwares for short sequences.
27. Bowtie-build software will add the extension "*.ebwt" to the output file.
28. It is difficult to establish a precise number of mismatches that allows obtaining a set that will be free of any bias. However, it is common practice to allow up to two mismatches for the mapping. It is important to remember that some differences with the reference sequence may also emerge because of biological differences between individuals of the same species (e.g., polymorphisms or posttranscriptional RNA sequence modifications such as RNA editing).
29. When --fr is specified and there is a candidate paired-end alignment, where mate1 appears upstream of the reverse complement of mate2 and the insert length constraints are fulfilled, the alignment is valid. Alternatively, if mate2 appears upstream of the reverse complement of mate1 and all other constraints are fulfilled, the alignment will also be valid. Most Illumina datasets have this orientation.
30. The Sequence Alignment/Map (SAM) format is a generic alignment format for storing the alignments of reads against reference sequences, supporting short and long reads produced by different sequencing platforms (46).
31. Assembly version 4.1 is currently available for *X. tropicalis* genome at http://genome.jgi-psf.org/Xentr4. Masked regions are represented with lowercase characters; gaps in the assembly are represented with Ns.
32. This is a regular file in multi FASTA format (i.e., with several sequences on it).
33. It is recommended to only use those sequences that map to a single region in the genome/transcriptome, because they can be unambiguously assigned to a specific locus or transcript. However, this choice depends on the aim of the experiment (i.e., those sequences that map to multiple loci may be the focus of interest). For example, when searching for small RNAs derived from transposons, all mapped sequences are considered in downstream analyses, given the nature of these elements (e.g., multiple copies in the genome). This will not be the

case when the aim is to characterize the expression of mRNAs (e.g., mRNA-seq). In this situation, paired-end or mate-pair sequencing help to increase the number of uniquely mapped sequences

34. When paired-end or mate-pair sequencing reads are used, there is a third piece of information that is added: the distance between pairs/mates. With this additional information, the odds of getting multiple matches against the reference sequences or a spurious path during the assembly process are diminished.
35. The UniGene clusters can be downloaded from: ftp://ftp.ncbi.nlm.nih.gov/repository/UniGene/Xenopus_laevis/ for *X. laevis* and ftp://ftp.ncbi.nlm.nih.gov/repository/UniGene/Xenopus_tropicalis/ for *X. tropicalis.*
36. ESTs of *X. tropicalis* and *X. laevis* can be obtained from: ftp://ftp.xenbase.org/pub/Genomics/Sequences
37. Some noncoding RNA databases are RNAdb (REF: RNAdb 2.0—an expanded database of mammalian noncoding RNAs), ncRNAdb (47), Rfam (48), and NONCODE (49).
38. BIOMART (http://www.ensembl.org/biomart) provides a list of annotated small RNAs for *X. tropicalis.*
39. qRT-PCR approaches are generally more adequate but it is possible to perform semiquantitative RT-PCR as well. However, it is important to mention that it is likely that only validation of genes that show high differences in mRNA levels will be possible.

Acknowledgements

This work was funded by research grants from FONDECYT (No. 1110400), ICM (No. P09-016-F) (LIA and FM), Center for Aging and Regeneration (CARE), and Millennium Nucleus in Regenerative Biology (MINREB) (DLL, FF, JL). We thank Dr. Mauricio Moreno for providing information on RNA yield from *Xenopus* embryos.

References

1. Metzker ML (2010) Sequencing technologies–the next generation. Nat Rev Genet 11:31–46
2. Nagalakshmi U, Wang Z, Waern K, Shou C, Raha D, Gerstein M, Snyder M (2008) The transcriptional landscape of the yeast genome defined by RNA sequencing. Science 320:1344–1349
3. Bentley DR, Balasubramanian S, Swerdlow HP, Smith GP, Milton J, Brown CG, Hall KP, Evers DJ, Barnes CL, Bignell HR, Boutell JM, Bryant J, Carter RJ, Keira Cheetham R, Cox AJ, Ellis DJ, Flatbush MR, Gormley NA, Humphray SJ, Irving LJ, Karbelashvili MS, Kirk SM, Li H, Liu X, Maisinger KS, Murray LJ, Obradovic B,

Ost T, Parkinson ML, Pratt MR, Rasolonjatovo IM, Reed MT, Rigatti R, Rodighiero C, Ross MT, Sabot A, Sankar SV, Scally A, Schroth GP, Smith ME, Smith VP, Spiridou A, Torrance PE, Tzonev SS, Vermaas EH, Walter K, Wu X, Zhang L, Alam MD, Anastasi C, Aniebo IC, Bailey DM, Bancarz IR, Banerjee S, Barbour SG, Baybayan PA, Benoit VA, Benson KF, Bevis C, Black PJ, Boodhun A, Brennan JS, Bridgham JA, Brown RC, Brown AA, Buermann DH, Bundu AA, Burrows JC, Carter NP, Castillo N, Chiara ECM, Chang S, Neil Cooley R, Crake NR, Dada OO, Diakoumakos KD, Dominguez-Fernandez B, Earnshaw DJ, Egbujor UC, Elmore DW, Etchin SS, Ewan MR, Fedurco M, Fraser LJ, Fuentes Fajardo KV, Scott Furey W, George D, Gietzen KJ, Goddard CP, Golda GS, Granieri PA, Green DE, Gustafson DL, Hansen NF, Harnish K, Haudenschild CD, Heyer NI, Hims MM, Ho JT, Horgan AM, Hoschler K, Hurwitz S, Ivanov DV, Johnson MQ, James T, Huw Jones TA, Kang GD, Kerelska TH, Kersey AD, Khrebtukova I, Kindwall AP, Kingsbury Z, Kokko-Gonzales PI, Kumar A, Laurent MA, Lawley CT, Lee SE, Lee X, Liao AK, Loch JA, Lok M, Luo S, Mammen RM, Martin JW, McCauley PG, McNitt P, Mehta P, Moon KW, Mullens JW, Newington T, Ning Z, Ling Ng B, Novo SM, O'Neill MJ, Osborne MA, Osnowski A, Ostadan O, Paraschos LL, Pickering L, Pike AC, Chris Pinkard D, Pliskin DP, Podhasky J, Quijano VJ, Raczy C, Rae VH, Rawlings SR, Chiva Rodriguez A, Roe PM, Rogers J, Rogert Bacigalupo MC, Romanov N, Romieu A, Roth RK, Rourke NJ, Ruediger ST, Rusman E, Sanches-Kuiper RM, Schenker MR, Seoane JM, Shaw RJ, Shiver MK, Short SW, Sizto NL, Sluis JP, Smith MA, Ernest Solma Solma J, Spence EJ, Stevens K, Sutton N, Szajkowski L, Tregidgo CL, Turcatti G, Vandevondele S, Verhovsky Y, Virk SM, Wakelin S, Walcott GC, Wang J, Worsley GJ, Yan J, Yau L, Zuerlein M, Mullikin JC, Hurles ME, McCooke NJ, West JS, Oaks FL, Lundberg PL, Klenerman D, Durbin R, Smith AJ (2008) Accurate whole human genome sequencing using reversible terminator chemistry. Nature 456:53–59

4. Grabherr MG, Haas BJ, Yassour M, Levin JZ, Thompson DA, Amit I, Adiconis X, Fan L, Raychowdhury R, Zeng Q, Chen Z, Mauceli E, Hacohen N, Gnirke A, Rhind N, Di Palma F, Birren BW, Nusbaum C, Lindblad-Toh K, Friedman N, Regev A (2011) Full-length transcriptome assembly from RNA-Seq data without a reference genome. Nat Biotechnol 29:644–652
5. Wang Z, Gerstein M, Snyder M (2009) RNA-Seq: a revolutionary tool for transcriptomics. Nat Rev Genet 10:57–63
6. Marioni JC, Mason CE, Mane SM, Stephens M, Gilad Y (2008) RNA-seq: an assessment of technical reproducibility and comparison with gene expression arrays. Genome Res 18: 1509–1517
7. Li P, Ponnala L, Gandotra N, Wang L, Si Y, Tausta SL, Kebrom TH, Provart N, Patel R, Myers CR, Reidel EJ, Turgeon R, Liu P, Sun Q, Nelson T, Brutnell TP (2010) The developmental dynamics of the maize leaf transcriptome. Nat Genet 42:1060–1067
8. Velculescu VE, Zhang L, Vogelstein B, Kinzler KW (1995) Serial analysis of gene expression. Science 270:484–487
9. Lu C, Meyers BC, Green PJ (2007) Construction of small RNA cDNA libraries for deep sequencing. Methods 43:110–117
10. Mortazavi A, Williams BA, Mccue K, Schaeffer L, Wold B (2008) Mapping and quantifying mammalian transcriptomes by RNA-Seq. Nat Methods 5:621–628
11. Morin R, Bainbridge M, Fejes A, Hirst M, Krzywinski M, Pugh T, McDonald H, Varhol R, Jones S, Marra M (2008) Profiling the HeLa S3 transcriptome using randomly primed cDNA and massively parallel short-read sequencing. Biotechniques 45:81–94
12. Akkers RC, van Heeringen SJ, Jacobi UG, Janssen-Megens EM, Françoijs K-J, Stunnenberg HG, Veenstra GJC (2009) A hierarchy of H3K4me3 and H3K27me3 acquisition in spatial gene regulation in Xenopus embryos. Dev Cell 17:425–434
13. Armisen J, Gilchrist MJ, Wilczynska A, Standart N, Miska EA (2009) Abundant and dynamically expressed miRNAs, piRNAs, and other small RNAs in the vertebrate Xenopus tropicalis. Genome Res 19:1766–1775
14. Lau NC, Ohsumi T, Borowsky M, Kingston RE, Blower MD (2009) Systematic and single cell analysis of Xenopus Piwi-interacting RNAs and Xiwi. EMBO J 28:2945–2958
15. Robine N, Lau NC, Balla S, Jin Z, Okamura K, Kuramochi-Miyagawa S, Blower MD, Lai EC (2009) A broadly conserved pathway generates 3'UTR-directed primary piRNAs. Curr Biol 19:2066–2076
16. Faunes, F., Sanchez, N., Moreno, M., Olivares, G. H., Lee-Liu, D., Almonacid, L., Slater, A. W., Norambuena, T., Taft, R. J., Mattick, J. S., Melo, F., and Larrain, J. (2011) Expression of transposable elements in neural tissues during Xenopus development, PLoS ONE 6, e22569
17. Hellsten U, Harland RM, Gilchrist MJ, Hendrix D, Jurka J, Kapitonov V, Ovcharenko I, Putnam NH, Shu S, Taher L, Blitz IL, Blumberg B, Dichmann DS, Dubchak I, Amaya E, Detter JC, Fletcher R, Gerhard DS,

Goodstein D, Graves T, Grigoriev IV, Grimwood J, Kawashima T, Lindquist E, Lucas SM, Mead PE, Mitros T, Ogino H, Ohta Y, Poliakov AV, Pollet N, Robert J, Salamov A, Sater AK, Schmutz J, Terry A, Vize PD, Warren WC, Wells D, Wills A, Wilson RK, Zimmerman LB, Zorn AM, Grainger R, Grammer T, Khokha MK, Richardson PM, Rokhsar DS (2010) The genome of the Western clawed frog Xenopus tropicalis. Science 328:633–636

18. Gilchrist MJ, Zorn AM, Voigt J, Smith JC, Papalopulu N, Amaya E (2004) Defining a large set of full-length clones from a Xenopus tropicalis EST project. Dev Biol 271:498–516
19. Wheeler DL, Barrett T, Benson DA, Bryant SH, Canese K, Church DM, DiCuccio M, Edgar R, Federhen S, Helmberg W, Kenton DL, Khovayko O, Lipman DJ, Madden TL, Maglott DR, Ostell J, Pontius JU, Pruitt KD, Schuler GD, Schriml LM, Sequeira E, Sherry ST, Sirotkin K, Starchenko G, Suzek TO, Tatusov R, Tatusova TA, Wagner L, Yaschenko E (2005) Database resources of the National Center for Biotechnology Information. Nucleic Acids Res 33:D39–D45
20. Langmead B, Trapnell C, Pop M, Salzberg SL (2009) Ultrafast and memory-efficient alignment of short DNA sequences to the human genome. Genome Biol 10:R25
21. Haider S, Ballester B, Smedley D, Zhang J, Rice P, Kasprzyk A (2009) BioMart Central Portal–unified access to biological data. Nucleic Acids Res 37:W23–W27
22. Roe BA, Ma DP, Wilson RK, Wong JF (1985) The complete nucleotide sequence of the Xenopus laevis mitochondrial genome. J Biol Chem 260:9759–9774
23. Schuler GD (1997) Pieces of the puzzle: expressed sequence tags and the catalog of human genes. J Mol Med (Berl) 75:694–698
24. Sayers EW, Barrett T, Benson DA, Bolton E, Bryant SH, Canese K, Chetvernin V, Church DM, DiCuccio M, Federhen S, Feolo M, Fingerman IM, Geer LY, Helmberg W, Kapustin Y, Landsman D, Lipman DJ, Lu Z, Madden TL, Madej T, Maglott DR, Marchler-Bauer A, Miller V, Mizrachi I, Ostell J, Panchenko A, Phan L, Pruitt KD, Schuler GD, Sequeira E, Sherry ST, Shumway M, Sirotkin K, Slotta D, Souvorov A, Starchenko G, Tatusova TA, Wagner L, Wang Y, Wilbur WJ, Yaschenko E, Ye J (2011) Database resources of the National Center for Biotechnology Information. Nucleic Acids Res 39:D38–D51
25. Bowes JB, Snyder KA, Segerdell E, Gibb R, Jarabek C, Noumen E, Pollet N, Vize PD (2008) Xenbase: a Xenopus biology and genomics resource. Nucleic Acids Res 36: D761–D767
26. McCormick KP, Willmann MR, Meyers BC (2011) Experimental design, preprocessing, normalization and differential expression analysis of small RNA sequencing experiments. Silence 2:2
27. Autio R, Kilpinen S, Saarela M, Kallioniemi O, Hautaniemi S, Astola J (2009) Comparison of Affymetrix data normalization methods using 6,926 experiments across five array generations. BMC Bioinformatics 10(Suppl 1):S24
28. Bolstad BM, Irizarry RA, Astrand M, Speed TP (2003) A comparison of normalization methods for high density oligonucleotide array data based on variance and bias. Bioinformatics 19: 185–193
29. Irizarry RA, Wu Z, Jaffee HA (2006) Comparison of Affymetrix GeneChip expression measures. Bioinformatics 22:789–794
30. Barbacioru CC, Wang Y, Canales RD, Sun YA, Keys DN, Chan F, Poulter KA, Samaha RR (2006) Effect of various normalization methods on Applied Biosystems expression array system data. BMC Bioinformatics 7:533
31. Binder H, Preibisch S, Berger H (2010) Calibration of microarray gene-expression data. Methods Mol Biol 576:375–407
32. Harr B, Schlotterer C (2006) Comparison of algorithms for the analysis of Affymetrix microarray data as evaluated by co-expression of genes in known operons. Nucleic Acids Res 34:e8
33. Millenaar FF, Okyere J, May ST, van Zanten M, Voesenek LA, Peeters AJ (2006) How to decide? Different methods of calculating gene expression from short oligonucleotide array data will give different results. BMC Bioinformatics 7:137
34. Robinson MD, Oshlack A (2010) A scaling normalization method for differential expression analysis of RNA-seq data. Genome Biol 11:R25
35. Bullard JH, Purdom E, Hansen KD, Dudoit S (2010) Evaluation of statistical methods for normalization and differential expression in mRNA-Seq experiments. BMC Bioinformatics 11:94
36. Ro S, Park C, Jin J, Sanders KM, Yan W (2006) A PCR-based method for detection and quantification of small RNAs. Biochem Biophys Res Commun 351:756–763
37. Ro S, Yan W (2010) Detection and quantitative analysis of small RNAs by PCR. Methods Mol Biol 629:295–305
38. Martello G, Zacchigna L, Inui M, Montagner M, Adorno M, Mamidi A, Morsut L, Soligo S, Tran U, Dupont S, Cordenonsi M, Wessely O, Piccolo S (2007) MicroRNA control of Nodal signalling. Nature 449:183–188

39. Sive HL, Grainger RM, Harland RM (2000) Early development of Xenopus laevis. A laboratory manual. Cold Spring Harbor Laboratory Press, New York
40. Agrawal R, Tran U, Wessely O (2009) The miR-30 miRNA family regulates Xenopus pronephros development and targets the transcription factor Xlim1/Lhx1. Development 136: 3927–3936
41. Watanabe T, Imai H, Minami N (2008) Identification and expression analysis of small RNAs during development. Methods Mol Biol 442:173–185
42. Li H, Durbin R (2009) Fast and accurate short read alignment with Burrows-Wheeler transform. Bioinformatics 25:1754–1760
43. Li R, Yu C, Li Y, Lam TW, Yiu SM, Kristiansen K, Wang J (2009) SOAP2: an improved ultrafast tool for short read alignment. Bioinformatics 25:1966–1967
44. Li H, Ruan J, Durbin R (2008) Mapping short DNA sequencing reads and calling variants using mapping quality scores. Genome Res 18:1851–1858
45. Lin H, Zhang Z, Zhang MQ, Ma B, Li M (2008) ZOOM!Zillions of oligos mapped. Bioinformatics 24:2431–2437
46. Li H, Handsaker B, Wysoker A, Fennell T, Ruan J, Homer N, Marth G, Abecasis G, Durbin R (2009) The Sequence Alignment/Map format and SAMtools. Bioinformatics 25:2078–2079
47. Szymanski M, Erdmann VA, Barciszewski J (2007) Noncoding RNAs database (ncRNAdb). Nucleic Acids Res 35:D162–D164
48. Griffiths-Jones S, Moxon S, Marshall M, Khanna A, Eddy SR, Bateman A (2005) Rfam: annotating non-coding RNAs in complete genomes. Nucleic Acids Res 33:D121–D124
49. He S, Liu C, Skogerbo G, Zhao H, Wang J, Liu T, Bai B, Zhao Y, Chen R (2008) NONCODE v2.0: decoding the non-coding. Nucleic Acids Res 36:D170–D172

Chapter 19

Databases of Gene Expression in *Xenopus* Development

Michael J. Gilchrist and Nicolas Pollet

Abstract

Gene expression data for *Xenopus* are collected and curated in diverse forms and locations. The intention of this chapter is to give the reader a guide to the publicly accessible databases where these data can be found and an idea of the current scope and limitations of the data in these resources. Instructions are given on how to access and interpret the data provided by the NCBI Gene database, Xenbase, and the *Xenopus* full-length EST, quickImage, and Xenmark databases.

Key words: EST, cDNA, In situ image, Expression pattern, Expression profile, Database

1. Introduction

Given a gene of interest that we might be working on, we would often wish to ask questions like: when is this gene expressed; in what tissues; in response to what internal or external cues or signals? Or, in a different context, we might invert these questions and ask: what are the genes that are expressed in some specific region of the embryo at a certain time? Such questions are amenable to experimental investigation, but the data may already exist, and we could then just look it up. There may be questions about the authenticity and reliability of data from other sources, and there may be questions about the ease with which we can extract the data we want. But if the data are there, and they are trustworthy, then there is much to be gained from using them.

There is another issue here, and that is the difference between *data* and *databases*. Just because someone has generated some useful data does not mean that we can get our hands on it. And even if we can get to the data, there may be too much of it for us to handle easily, or we may not be able to interpret it appropriately. To put primary research data usefully into the public domain usually means that a group of people have to design, build, and maintain

Stefan Hoppler and Peter D. Vize (eds.), *Xenopus Protocols: Post-Genomic Approaches, Second Edition*,
Methods in Molecular Biology, vol. 917, DOI 10.1007/978-1-61779-992-1_19,

databases to store and query the data, and web interfaces to allow members of the research community to make queries, view, and sort the data that is returned in response to a query.

Here, we use the term database rather widely to mean any pool of data accessible to non-technical users through some sort of public (usually web-based) interface. We also use the term *gene* loosely to mean a transcribed locus in the genome that produces one or more biologically functional molecules. We will focus primarily on protein-coding genes in *Xenopus*, but remain aware that some genes produce non-coding RNA molecules as their primary output (tRNAs, etc.), and some appear to produce non-coding RNAs as their secondary output (microRNAs), although this is to oversimplify a picture of growing complexity.

Gene expression data is often based on the detection of mRNA or protein gene products, and the first of these, for reasons of cost and simplicity, is used much more widely. It has been shown that mRNA and protein levels are not always correlated (1, 2), and therefore some issues arise concerning the match between the detection method used and the inferred behavior of a gene. This is discussed in Note 1 below.

2. Database Resources

A list of database resources and their different types of expression data is presented in Table 1. Although this is primarily designed to be a list of resources where useful data can be easily queried from user-friendly interfaces, some of the more intractable data stores are included for completeness. There is an indication of the scope and current usability of the data for each entry. If you look at any of these data sources and think they could be better, it can be very useful to give the database providers some feedback (see Note 2).

The different types of gene expression data available have very different characteristics, depending on how they are generated, and this can affect how you interpret them. For more detailed information about the different types of data and how to make the most effective use of them see below. Areas covered include: image data (see Note 3); sequence data (see Note 4); high-throughput sequencing data (see Note 5); and microarray data (see Note 6).

In the current chapter, emphasis has been placed on using a series of *Xenopus*-specific databases, apart from the NCBI Gene database (3)—one of the primary resources for the establishment of gene identity. We look in detail at databases from two sources. The first of these is the *Xenopus* model organism community database Xenbase (4) which collates many different types of *Xenopus*-specific data in a gene-centric manner. The second is a group of databases associated originally with the Gurdon Institute in

Table 1
A list of primary database resources for *Xenopus* gene expression data, with a description of the types of data they contain and an indication of their scope and usability

Database	Data	Notes
Xenbase (www.xenbase.org)	Sequence, image	*Xenopus* model organism community database. Coordinates gene annotation, gene naming, and other gene-centric data. Annotates and curates both direct gene expression data, like images, and indirect data in the form of peer-reviewed literature. Provides links to external data sets
Xenopus full-length EST database (genomics.nimr.mrc.ac.uk/online/xt-fl-db.html)	Sequence, digital expression profile	Database of all known cDNA sequences, both ESTs and full-length cDNAs. Groups of ESTs and cDNAs are assembled to predict the likely sequence of the transcript(s). One can visually check the sequence assembly to see whether it looks correct
NCBI-UNIGENE (compbio.dfci.harvard.edu/tgi)	Sequence, digital expression profile	Database of all known cDNA sequences, both ESTs and full-length cDNAs. Includes descriptions of the libraries from which the EST sequences were made. Digital expression profile both for the gene (what stages/tissues the gene is expressed in), and for the library sample (what genes are expressed in that stage/tissue). Data for gene expression profiles are roughly quantitative once you normalize for library size. Easy access to get a good idea of the most highly expressed gene in a tissue or at given stage of development
DFCI (www.ncbi.nlm.nih.gov/unigene)	Sequence, digital expression profile	Database of all known cDNA sequences, both ESTs and full-length cDNAs. Groups of ESTs and cDNAs are assembled to propose a Tentative Consensus (TC) sequence, the latter being annotated through a series of bioinformatic pipelines. Some unique annotations available in this database are interesting, such as Gene Ontology classification, Metabolic and Signaling Pathways, alternative splice forms and SNP report
NCBI-GEO (www.ncbi.nlm.nih.gov/geo)	Microarray, high throughput sequence	Most of the data currently in this database originates from micro-arrays, but this is now being augmented by HTS data, and this trend is likely to increase as more researchers use this newer platform. A significant challenge for a user is to try and get a sense of what is in there and whether it might be useful

(continued)

Table 1 (continued)

Database	Data	Notes
EBI-ArrayExpress (www.ebi.ac.uk/arrayexpress)	Microarray	Covers similar territory to the NCBI-GEO database described above. The difference between these may be in the species coverage, depth, or manner in which the data is subsequently re-analyzed and re-presented. A significant challenge for a user is to try and get a sense of what is in there and whether it might be useful
Axeldb (indigene.issb.genopole.fr/axeldb)	Image, sequence	The database has a few rough edges, but there is a lot of data there and it may be worth persevering. There are a number of ways to query the data, including a stage- and anatomy-based tool. Most but not all expression data are available through XenMARK and Xenbase, and therefore it makes sense to perform queries to look for additional images documenting in situ hybridization results. The database is no longer updated
XDB3 (xenopus.nibb.ac.jp)	Image, sequence	It can be a little frustrating trying to get to the image data from the sequence data, as it is not obvious which links might lead to images. The WISH Photo Browser does not have keyword search, and although it is a good way to look at large numbers of images from the same developmental stage, it is difficult to get to gene-specific images without knowing precise clone IDs. However, once you get a clone name, you can call up all the images from that clone with ease. So the most effective way may be via the sequence data (BLAST or keyword search) and then take the clone name to the WISH browser. The sequence search will usually yield a set of contig links, and the best strategy appears to be just click through the list of links to find any contigs that have associated images
XEXTDB (xextdb.nibb.ac.jp)	Image, sequence	As you are unlikely to know many actual clone IDs, the best way into this data is probably through gene names. Again, it is a bit hit and miss, and a list of all genes for which images were available would have helped, but we found some nice images
XenMARK/quickImage (genomics.nimr.mrc.ac.uk/apps/XenMARK)	Image, sequence	XenMARK organizes in situ gene expression images that were annotated, as well as providing an annotation platform for these images. See text for more details
VisiGene (genome.ucsc.edu/cgi-bin/hgVisiGene)	Image	The VisiGene image browser organizes in situ and other gene expression images into a searchable database, with the added value that the images can be inspected with a "virtual microscope". There is currently only the one set of *Xenopus* data, from the NIBB database

Harland/Berkeley (tropicalis.berkeley.edu/home/gene_expression/insitugallery)	Image	Contains around 500 images of expression patterns for key genes at relevant stages of developmental. If you know the name of gene you are looking for, and the Harland Lab have used that name for it, then you will find high-quality, useful images very quickly, or be reasonably sure there aren't any
XGEBASE (www.euregene.org/xgebase)	Image	*Xenopus* Gene Expression Database (XGEBase) is part of the larger European Renal Genome Project and contains over a thousand images from 200+ genes, with detailed annotation where appropriate, of expression within the developing kidney/pronephros from the lab of André Brändli (5). Searching on gene name is a bit hit or miss, but the anatomy driven query tool is impressively easy to use
BGEE (bgee.unil.ch/bgee/bgee)	Image, microarray, digital expression profile	The species covered are *Xenopus*, zebrafish, fly, mouse and human, although the coverage of different types of data varies between the species. A lot of attention has been paid to anatomical ontologies and the development of cross species homologous anatomical groupings, or Homologous Organ Groups (HOGs), enabling cross species expression comparison on an anatomical basis. Of special interest if you want to look at evolutionarily conserved gene expression in relation to anatomy
quickLit (genomics.nimr.mrc.ac.uk/apps/quickLit)	Literature	Sequence-based literature search
quickGene (genomics.nimr.mrc.ac.uk/apps/quickGene)	Gene name	Sequence-based gene name search. See text for more details

Cambridge, and moved to their current home with the setting up of the Gilchrist lab in the Systems Biology Division at the MRC National Institute of Medical Research in London (6–8). The authors of this chapter are strongly associated with the motivation for, and construction of, this latter group of databases. Following is a list of the main databases described in this chapter, and these will be useful places to search for gene expression data in *Xenopus laevis* and *Xenopus tropicalis*:

1. http://www.xenbase.org/: Xenbase, the *Xenopus* model organism database. Contains curated and systematic data for both *X. tropicalis* and *X. laevis.*
2. http://www.ncbi.nlm.nih.gov/gene/: the NCBI Gene database. Contains gene identity and related information across many species including *Xenopus.*
3. http://genomics.nimr.mrc.ac.uk/apps/quickGene/: the quickGene database. Can be used for establishing gene identity on the basis of sequence similarity.
4. http://genomics.nimr.mrc.ac.uk/apps/quickImage/and http://genomics.nimr.mrc.ac.uk/apps/quickImage/XenMARK/: the quickImage and XenMARK databases. Contain annotated in situ images which can be queried on the basis of gene identity or expression pattern.
5. http://genomics.nimr.mrc.ac.uk/online/xt-fl-db.html: the *Xenopus* full-length EST database. Contains assembled EST and cDNA data for both *X. tropicalis* and *X. laevis.*

3. Methods

3.1. Clarifying Your Gene of Interest and Retrieving Its Sequences

If you are looking for gene expression data for a specific gene or genes, or planning to do experimental work, then it is important to clarify the identity of your gene(s) first. This is not as straightforward as it should be, partly because many genes have changed their name over the last few years, partly because genes were not consistently named across species, partly because not all genes have (functional) names in any species, and partly because not all *Xenopus* genes have been assigned their correct names yet (see also Chapter 4 for *X. tropicalis*). Once you have clarified the identity of your gene, you will be able to retrieve its mRNA and protein sequences. These will be useful as you develop an investigation of its function, or wish to search for further data. There are several different databases where you can do this type of search: Xenbase, NCBI Gene, and quickGene, and it is certainly worth comparing the results from different databases if there is any uncertainty. You will probably start off with a text-based search using the name of your gene, but be aware that the name you are using may be an alias or synonym. Or if you are starting with just a sequence and an uninformative accession number some databases will allow you to search for genes with that sequence. This approach may be especially useful if you are starting from a gene in a different species, and it may also give you useful information about the family of related genes (see Note 7), not all of which will have obviously similar names.

3.1.1. Protocol to Verify Gene Identity and Retrieving Sequences in Xenbase

1. Connect to Xenbase by pointing an internet browser to http://www.xenbase.org/ and type your gene symbol into the search box just under the main menu bar at the top of the page.
2. You will see a list of alternative matching names as you type, select the one you want, and click the [search] button or hit return. This will take you to the *gene page* for your gene.
3. Inspect the list of gene names and synonyms to confirm that you have found the correct gene. Return to the search box if you are not sure, where you can inspect alternatives.
4. To retrieve the sequences for your gene, go down the page to the **Molecules** section, and click on the magnifying glass icon next to the RefSeq entry you are interested in (mRNA or protein, *X. tropicalis*, or *X. laevis*). This opens a sequence viewer from which you can copy and paste the sequence in fasta format into a text file for future reference. Alternatively click on any of the links, and this will take you to an external database where the sequence is stored, and follow instructions there. If this is an NCBI database, see Subheading 3.1.2, step 4 for more detail.

3.1.2. Protocol to Verify Gene Identity and Retrieving Sequences from the NCBI Gene Database

1. Connect to the NCBI Gene database by pointing an internet browser to http://www.ncbi.nlm.nih.gov/gene and type your gene symbol into the search box near the top of the page. It may be useful to add the text "AND Xenopus," as this will restrict the list of genes returned to *Xenopus* entries.
2. Inspect the list of gene entries for the one you are interested in. Be especially careful if your search term has come up as an alias. Click on the gene symbol link for your gene, which will take you to the NCBI *gene page*.
3. To retrieve the sequences for your gene, scroll down the gene page until you find the section ***mRNA and Protein(s)***, and click on the link to the sequence you require: NM_ or XM_ accession numbers will be mRNAs, and NP_ and XP_ will be proteins. This will take you to the GenBank sequence entry.
4. Click on the FASTA link near the top of the page and the left margin. This will re-display the sequence in fasta format, which you may then copy and paste into a text file for future reference.

3.1.3. Protocol to Search Gene Names in QuickGene Using a Sequence as Input

1. Connect to quickGene by pointing an internet browser to http://genomics.nimr.mrc.ac.uk/apps/quickGene/ and clicking on the [enter] button.
2. Copy and paste a protein or DNA sequence into the sequence field. You must select between protein or DNA search according to your query sequence. You may change the other parameters to narrow down the list of species searched, change the sensitivity of the search, or the maximum number of results. Click on the [search] button.
3. Wait a few seconds for the results. You will see a screen showing which reference sequences your sequence has matched, in an alignment diagram. You would normally be interested in the best match, which will be the top one (see Note 8).
4. Scroll down to see the gene name output. Depending on the species which your search sequence came from you may need to interpret the source species and percent identity value in the detailed results section (first two columns of the output) to ensure you are looking at the results for your gene or an orthologous gene, and not a paralog (see Note 9). The gene name is indicated in bold grey and links to the NCBI Gene database. You can use that name to query Xenbase or other databases of interest.

3.2. Finding In Situ Hybridization Expression Patterns for Your Gene of Interest

As you begin to investigate the function of your correctly identified gene of interest, you will probably want to study the anatomical distribution of its gene expression pattern in the relevant stages of development. You might choose to analyze gene expression with RNA in situ hybridization yourself, but it is worth searching collections of existing images first, in case this has already been done. The community has placed tens of thousands of such images into various public databases, and although there is some duplication, there are two databases well worth searching, as they give you very different ways to get to the data. Xenbase in situ images can be accessed via gene pages and text-based anatomical annotation of images, or by a sequence search. Those in the quickImage and XenMARK databases can be accessed via a sequence-based search (see Note 7), or through user-defined expression patterns drawn on the Nieuwkoop and Faber (9) schematic stage drawings. This means that you can do forward searches: "show me the images for a specific gene," and reverse searches: "show me the genes that are expressed in a specific region of the embryo," in both these databases.

3.2.1. Protocol to Search Xenbase for In Situ Images for Specific Genes

1. Connect to Xenbase by pointing an internet browser to http://www.xenbase.org/ and type your gene symbol into the search box as in Subheading 3.1.1, step 1 above.

2. When you reach the gene page you will see a series of tabs just under the search box. Click on the Expression tab, which is currently the second from the left.
3. You will now see a graphical display of the available images for this gene from both *X. tropicalis* and *X. laevis* (see Fig. 1). These are divided into groups: single images of RNA in situ data submitted by the community, usually at relatively high resolution, and which may or may not be published; and published images annotated from the literature by Xenbase curators, available as the published figure, and with the relevant citation. Individual images may be clicked on to view larger sized versions.

3.2.2. Protocol to Search quickImage for In Situ Images for Specific Genes and Related Gene Family Members

1. Acquire the mRNA or protein sequence of your gene of interest (see Subheading 3.1.1 and 3.1.2 above).
2. Connect to quickImage by pointing an internet browser to http://genomics.nimr.mrc.ac.uk/apps/quickImage/, click the [enter] button.
3. Paste your gene sequence into the "sequence" box. You may also set any additional parameters as described in Subheading 3.1.3, step 2 above; although this application will auto-detect the sequence type. Then click the [search] button; you may have to wait a few moments for the results, as the application is performing a BLAST search.
4. You will see an alignment diagram showing sequence matches (see Fig. 2). You would normally be interested in the best match, which will be the top one.
5. Scroll down to see the images attached to your sequence matches (see Fig. 2).If there are more than ten images available you can see the rest by clicking on the [ALL] button in the **Output format** section and re-running the search. You can view any image at full size by clicking on it.
6. If a [search] button is displayed below an image, it means that you can search for genes/images with similar expression patterns at this stage. Just click on the [search] button to do this. This is described in more detail in Subheading 3.3.2, step 4.

3.3. Finding Genes Expressed in Specific Regions of the Embryo by Searching Through Expression Patterns

This method reverse the search described in Subheading 3.2 above and tries to find genes expressed in regions of interest defined by the user, usually anatomical features of some sort. The first step is for the user to define the region of interest, and then images are retrieved according to how well their annotated gene expression matches the request. The classical method for large collection of images is anatomical text association by manual inspection, for example in zebrafish where the efforts of Bernard Thisse and others are visualized through the Zfin database (10). This approach

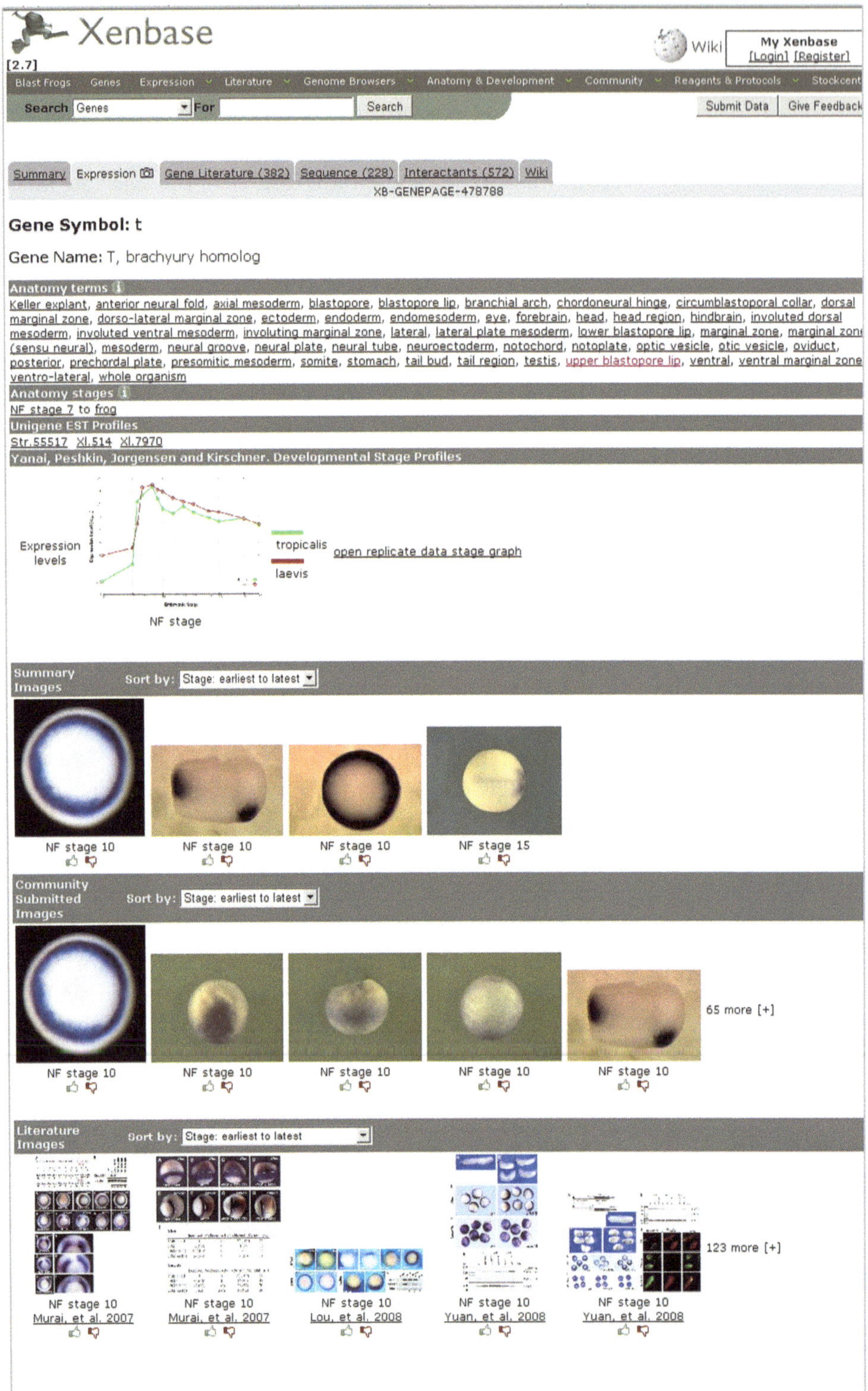

Fig. 1. Using Xenbase to retrieve in situ images for a gene of interest. Here we see the results of searching for expression data found from the Xenbase Gene Page for the gene **brachyury**. The lower part of the page contains in situ images collected for this gene, organized into three groups. A high-level Summary section, followed by Community Submitted individual images, followed by images scanned from the literature complete with publication context. The upper part of the results sections contains time-based expression profiles for this gene in both *Xenopus tropicalis* and *Xenopus laevis* sections between stages 2 and 33. These were generated from microarray data in the Kirschner laboratory (11).

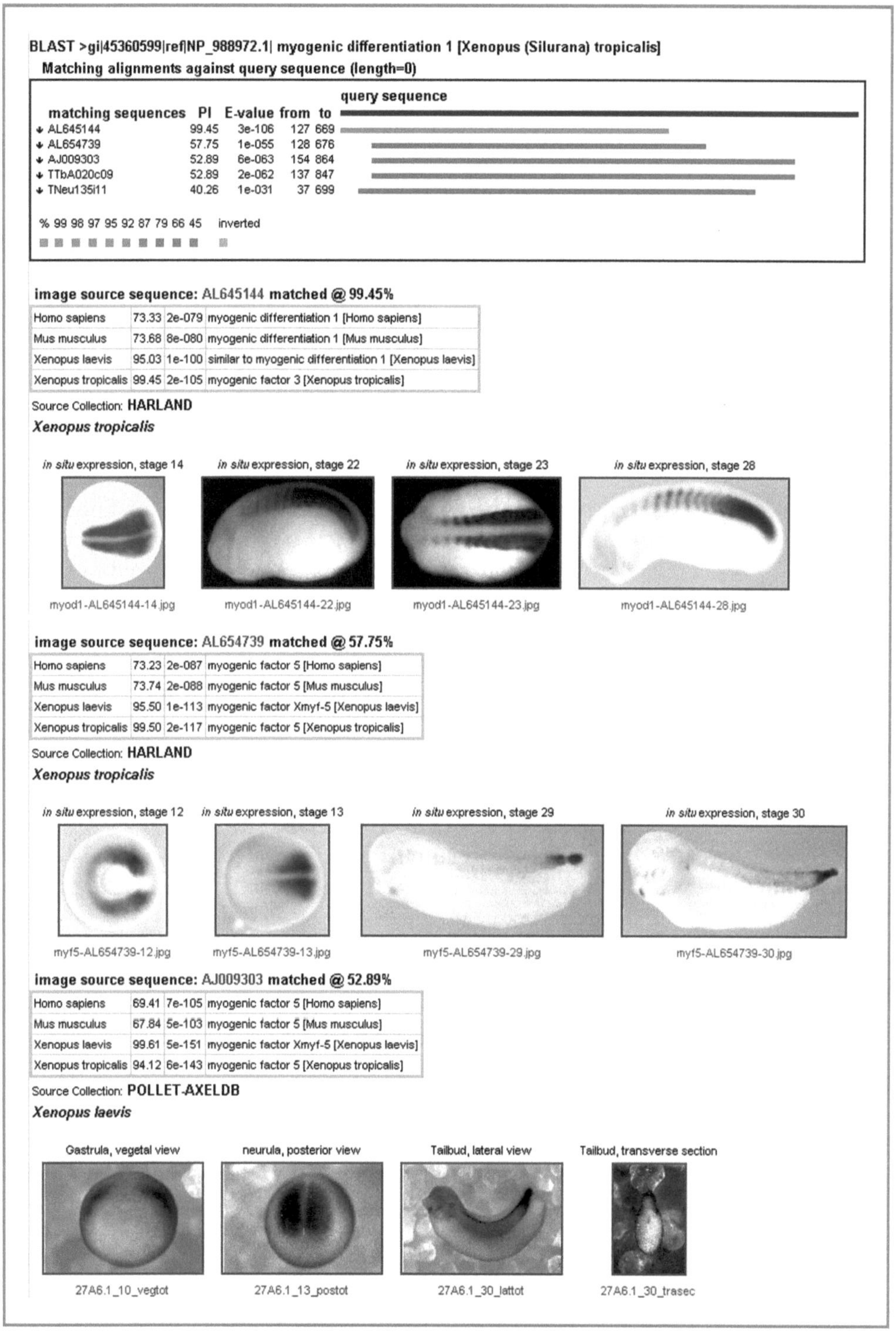

Fig. 2. Using quickImage to retrieve in situ images by sequence similarity. Here we see the results of searching for in situ images using the *X. tropicalis* **myod1** protein sequence as the query sequence. The top section shows the sequence alignments schematically, with the query sequence in blue and the matched sequences underneath, color coded according to the degree of similarity. The best match here is *X. tropicalis* **myod1** itself, but then we find images for the related gene **myf5**, first in *X. tropicalis* and then in *X. laevis*. We can make an observation that would be harder in a more gene-specific, text-based, search: that the expression patterns for these two genes are practically identical at around stage 13/14, and this has implications for the functional analysis of these genes. The screen shot has been edited for clarity. See ref. (7) for more details.

has been used by Xenbase. The XenMARK database (which evolved from the quickImage database, described above) uses a novel method of large-scale manual annotation of computational style *x*, *y* coordinate-based expression patterns registered to the framework of the schematic stage diagrams (8). Computer automated annotation of in situ expression patterns has long been known to be a difficult problem (12), and, with the wide range of embryo morphologies through the course of development in *Xenopus*, a universal solution may remain perpetually elusive. The XenMARK in situ expression patterns (8) provide sensitive and precise information about where and when a gene is expressed, and allow the image retrieval process to score the retrieved images according to how closely they match the query pattern. Although many images are shared between the two databases, we recommend using both databases to maximize the chances of finding useful information. In principle this method of image retrieval could, for example, give you a list of candidates genes involved in the development of a particular organ.

3.3.1. Protocol to Search Xenbase for In Situ Images Where Gene Expression Is Associated with Specific Anatomical Features

1. Connect to Xenbase by pointing an internet browser to http://www.xenbase.org/ and click on the Expression link in the top menu bar.
2. You will reach a general query page with quite a lot of options. Go to the **Anatomy Terms:** section and tick the boxes for anatomical regions you would like to see gene expression in. These should accumulate in the **Selected Search Terms:** box alongside. You may select for search results with annotated expression in any one of the selected tissues (OR), or in all the selected tissues at the same time (AND): the latter search will tend to give fewer results. If your tissue of interest is not listed, start typing the name of the desired target in the box **Search Entire Anatomy Ontology**. As you type, suggestions will appear, and when you see your target, click on it. Now click the [Search] button at the bottom of the page.
3. You will now see a page where genes with annotated expression in your regions of interest are listed alphabetically, and a single representative image is displayed. Click on the left hand link, which indicates the number of sources of expression data for this gene, to see other images. These may be a mixture of community submitted and literature sourced images.

3.3.2. Protocol to Search XenMARK for In Situ Images with Specific Expression Patterns

1. Connect to XenMARK by pointing an internet browser to http://genomics.nimr.mrc.ac.uk/apps/XenMARK/ and click the [enter] button. You should arrive at the "Search By Expression Pattern" page.
2. The entry point for this type of search is the set of heat maps generated from the sum of all the annotations against each

schematic diagram. Without altering any of the settings, click the [show] button in the main panel to display the set of Nieuwkoop and Faber schematic diagrams (9) overlaid with heat maps where red indicates a high density of annotated expression patters and blue a low density. This shows you both for which stages there are many annotated images, and within any schematic the apparent distribution of gene expression over the surface of the embryo.

3. Click on the [setup search] button under the schematic that represents the development stage you are interested in. A fresh version of the schematic will appear alongside the heat map. You can scribble on this schematic with the mouse pointer to define your region of interest. For example, if we fill in the region of the pronephros on the stage 29 schematic, and hit the [search] button we retrieve a number of images with gene expression clearly localized in the pronephros, which would be candidate genes for involvement in kidney development (see Fig. 3). The fit between the area you marked up and the annotation of the expression pattern is scored, and the ranking of the images by these scores allows you to see the expression patterns that best match your request.
4. In a further refinement of this search, you may now use one of your retrieved images as the query to search for images with expression patterns like *it* has. Simply click the [search] button immediately underneath any of the images you have just retrieved and a new search will be initiated with that image as the query (see Note 10).

Recently this approach has been extended to allow for the annotation of specialized tissue section images, which facilitates the investigation of gene expression patterns on a much finer scale. This has been prototyped in retinal section images and recently accepted for publication (13). The authors welcome suggestions for similar projects (see Note 11).

3.4. Finding Temporal Expression Profiles for Your Gene of Interest

In the same way that the spatial distribution of the gene expression pattern within the embryo will give an indication of gene function, the variation of expression over time during embryonic development will give other clues about the function: genes whose expression peaks quickly and then dies away quickly may be responsible for inducing a specific program of cell differentiation, whereas genes whose expression rises and then plateaus at a steady level may be responsible for the maintenance of a differentiated state. We can find useful example of these data in two databases: one in Xenbase from a single set of experiments by the Kirschner laboratory (11), and one in the Gurdon EST database (6) compiled from the many millions of ESTs from many sources, deposited in GenBank over the last 10 or 15 years.

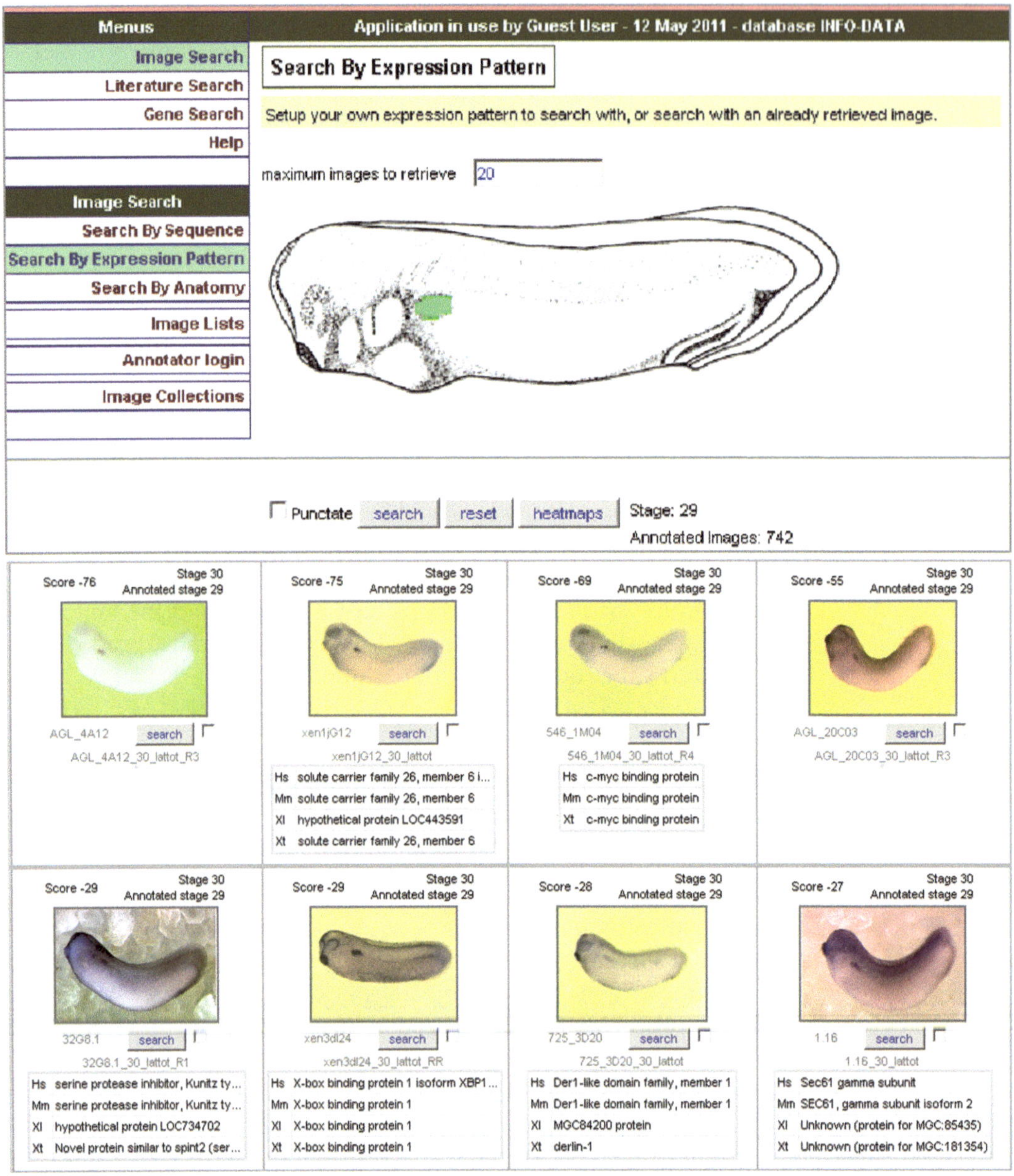

Fig. 3. Using XenMARK to retrieve in situ images from a user defined expression pattern. The user has outlined the pronephros in the stage 29 tailbud schematic, and the database has retrieved images whose annotated expression patterns most closely matches the user's request. We can see from the retrieved images the specificity of this search method, and that we can thus define a set of candidate genes for involvement in the development of the vertebrate kidney. The name of the clone sequence used to generate the in situ images, and an indirect annotation of the gene involved, are provided under the image. See ref. (8) for more details.

3.4.1. Protocol to Search Xenbase for Time-Based Developmental Expression Profiles of Given Genes

1. Connect to Xenbase by pointing an internet browser to http://www.xenbase.org/, find your gene of interest and navigate to the Expression tab by following Subheading 3.2.1, steps 1–2 above.

2. Once in the Expression tab, scroll down until you come to the section entitled **Yanai, Peshkin, Jorgensen and Kirschner. Developmental Stage Profiles**. If there are data for this gene, you will find a chart with two traces: one for *X. tropicalis* and one for *X. laevis* (see Fig. 1). These show the variation in gene expression from stages 2 to 33, with the period from stages 8 to 20 densely covered at every second stage. The vertical axis is log10 of the relative expression level of the gene, derived from microarray data (11). This gives you a very good idea of the approximate variation in gene expression over these important early development stages. Clicking on the chart will display a larger version.
3. To see expression profile data from NCBI scroll back up the Expression tab to the section **UniGene EST Profiles**, where you will find links for *X. tropicalis* and *X. laevis*. Clicking on these will take you to the UniGene page for this gene and display spatial and temporal distributions determined by frequency of EST reads in various cDNA libraries. The time scale of development stages is a little coarser than in the other databases.

3.4.2. Protocol to Search the Gurdon EST Database for Time-Based Developmental Expression Profiles of Given Genes

1. Acquire the mRNA or protein sequence of your gene of interest (see Subheading 3.1.1 and 3.1.2 above).
2. Connect to the *Xenopus* full-length EST database by pointing an internet browser to http://genomics.nimr.mrc.ac.uk/online/xt-fl-db.html. Click the [enter] button, and then choose a clustering project. These are listed in descending order of date of compilation, most recent first, so pick the top project for the frog you are interested in. Bear in mind that it is always likely to be interesting to look at the behavior of a given gene in both frogs. Hit the [find] button on the right of your selected project, which will take you directly to the BLAST search page.
3. Paste your sequence into the sequence box. Check the "amino acid sequence" box if your query is a protein sequence, and hit the [blast] button to compare your sequence to the current EST clustering data.
4. After a few seconds, you will see a screen showing which EST clusters your sequence has matched, in an alignment diagram (similar to that in Fig. 2). You would normally be interested in the best match, which will be the top one, and you are looking for an alignment very close to 100% identity. Once you have ascertained which cluster is likely to be the one you want, scroll down to the Matching Sequences detail box. The BLASTx descriptions of protein matches to each cluster should enable you to verify or crosscheck your gene identity at this point.

5. Click on the cluster name link on the left of the top line of the details box, which will take you to the cluster view page. This gives you a graphical view of the assembled ESTs that constitute the cluster and other related data. The top panel below the input form is labeled "Estimated expression levels relative to total library clones" and displays an approximate expression profile in green for this gene from stage 1 (Egg) through stage 65. A second panel shows similar expression profiles, in orange, but over a range of adult tissues rather than time of development. These profiles are derived from counting ESTs in the cluster, and knowing which library each EST comes from and how many ESTs there are in each library. The counts are normalized as a percentile of each library, and a blue bar at the left end of each set gives you a reference percentile to compare the relative expression levels of different genes (see Note 12).

3.5. Further Use of the EST Expression Data

The *Xenopus* community is fortunate to have access to large amounts of systematically generated EST data, sequenced from well-defined and well-documented cDNA libraries. In addition to the expression profiles mentioned above, the assembled EST clusters can provide crucial validation of sequences from other sources, and enable us to locate physical clones for experimental work.

Sequence validation is important because many of the *X. tropicalis* transcript sequences in the public databases come from computationally generated gene models, and a surprising proportion of these are probably incorrect. If you are designing (for example) a morpholino, you want to be sure you have the right section of a real mRNA.

3.5.1. Protocol to Validate an mRNA Sequence Taken from a Public Database

1. Take the mRNA sequence you have found for your gene of interest and find the open reading frame (ORF) by pointing your browser at (for example) the NCBI ORF Finder http://www.ncbi.nlm.nih.gov/gorf/. Paste your sequence into the input box and hit the [OrfFind] button. Click on the longest ORF. Your selected ORF should change color; then click on [Accept]. Change the View options to "Fasta nucleotide" and then click on [View]. You have now extracted the sequence of the apparent ORF, and can copy and paste this into your file of collected sequences.
2. Now take this sequence of the open reading frame of your gene and follow the steps in Subheading 3.4.2, steps 1–5 above. This will eventually take you to the cluster view page in the Gurdon EST database for your gene. Inspect the assembly, and if the depth and coverage of the ORF is convincing (see Note 13), then you have good data to validate your own sequence. If not, you may have to find other methods. The help link next to the [download] button in the top panel may help you understand the cluster display.

3. Validate the extent of the ORF by inspecting the position of the pale blue "blast aligned region," which should be visible just below the consensus sequences at the top of the cluster display panel. This should start exactly below the green ATG on the upper consensus sequence marking the start of the open reading frame, it should be labeled with a "1," and it should agree with the analyzed ORF, visible (purple) just below the protein alignments and above the ESTs alignments a little further down the display. The same should be true at the 3′ end of the ORF (although there is a 2 bp error in the position of the BLAST alignment at this end). Your sequence may be truncated if it is shorter than the ORF of the EST cluster.
4. Validate the sequence of your ORF by pair wise alignment with the EST consensus sequence. Click on the open reading frame link at the beginning of the purple analyzed ORF in the cluster display, and copy and paste the mRNA sequence into your data file. Now point your browser to the NCBI BLAST page at http://blast.ncbi.nlm.nih.gov/. Click on the nucleotide blast link, and then tick the "Align two or more sequences" box on the sequence input page, top panel. Paste your original found ORF into the top sequence input box, and the ORF from the EST cluster into the bottom input box. Then click on the blue [BLAST] button below. Inspect the resultant alignment, which should show substantially 100% agreement between the two sequences over their whole lengths. If there is a difference, you will want to satisfy yourself that you understand the cause, and take the sequence you believe to be more likely correct as your reference sequence.

3.5.2. Protocol to Identify a Full-Length Clone from the EST Data for Your Gene of Interest

1. Take the sequence of your gene of interest and follow the steps in Subheading 3.4.2, steps 1–5 above. This will take you to the cluster view page in the Gurdon EST database for your gene, where you should inspect and decide on the completeness of the open reading frame, as described in Subheading 3.5.1, step 2 above.
2. The alignments of the EST sequences relative to the ORF will tell you which clones are likely to be full length. Clones names are found in the left hand margin of the cluster view, in the last column before the sequence alignments. If the 5′ EST of a clone matches the consensus sequence of the cluster and starts before the initial ATG of the ORF, and if the 3′ EST of the same clone continues past the STOP codon, then there is reasonable likelihood that the clone is full length (see Note 14 and the published method (6)). Most clusters have been preanalyzed and candidate full-length clones are marked with a green label, two columns to the left of the clone name. Click on the clone of choice, which will take you to the NCBI nucleotide database where you should be able to work out the originator of the clone and possibly ask them for a copy.

4. Conclusion

There will probably never be a time when databases like these are mature and stable, but this is certainly a time of rapid change, both in our understanding of gene expression, and in the technologies with which we can generate data to investigate it. Although this chapter will not quite be out of date before it is printed, be aware that new databases, new variants of existing databases, and new ways of accessing the data are very likely to appear—so be prepared to be both adaptable and curious. And please feed back your experience of Web sites to the people who built them—this is often the only way they have of determining what people want, and allows the improvement of databases in the most useful way. The XenMARK database continues to be developed, and we encourage any *Xenopus* researchers with gene expression pattern images to contact us and get their images included in the database.

5. Notes

1. *Gene expression data may be based on detected mRNAs or proteins*
   The standard model for (protein coding) gene expression is very familiar: a gene locus is transcribed from one strand of genomic DNA to a single strand of primary RNA; this may be processed to remove unwanted sequence (introns) and produce messenger RNA; part of which (the open reading frame) is then translated by the ribosomal complex to yield a protein molecule. This paradigm has an important consequence for gene expression data, which is simultaneously both useful and non-useful, although in rather different ways. It is the protein that generally does the functional work of the gene, but the mRNA can be both easier to detect and to quantify. There are also some questions of timing and persistence; clearly mRNA must be produced (or present) before the protein can be made, and both types of gene product decay over time, either naturally or specifically through various targeting mechanisms. Also one copy of the mRNA may be used as the template to make many copies of the protein. As a result, detection of an mRNA does not guarantee that the protein is present, and equally the protein may (still) be present in the absence of the mRNA. We accept that this means a relatively imperfect correlation between mRNA expression levels and protein concentrations, but in general we are comfortable with the notions (a) that easily detectable amounts of messenger-RNA *are* a good indication of the presence of the protein product of the gene, and (b) that

sharp changes in mRNA expression levels imply regulation of the amount of protein in the affected cells.

2. *Providing feedback to database providers*
If you have tried to use an online database or other resource, and you feel it could be better, database providers generally welcome constructive feedback. Please make the effort to provide this, as it gives them the opportunity to improve their databases according to what their users (actually) want—as long as it is not too complicated, and would be generally useful to other people. Occasionally you will get an unenthusiastic response (or none at all), but do not be put off as this input is very valuable. The gene expression profiles in the Gurdon EST database were created in response to just such a suggestion.

3. *Characteristics of gene expression image data*
Images of whole embryos or tissue sections may be stained to show the precise location of gene expression at the time of imaging. By far the greater proportion of images are RNA in situ hybridization results using cRNA (complementary RNA) hybridization probes matching known mRNA sequences, showing location of these mRNAs in whole-mount (whole embryo) preparations or sectional in situ images. *Xenopus* is particularly suitable for whole mount images, as the embryos, especially in early development, are moderately translucent, and can be rendered more so when required. Protein localization may be imaged by using fluorescent tagged antibodies (see Chapter 23) in the process of *immunostaining* and the production of *immunofluorescence* images. These are less widely used, primarily because of the difficulty in obtaining antibodies for a wide range of genes, although they are clearly more informative if you want to know where a gene product is actually present, rather than where it may have been produced. The range of antibodies available is rapidly expanding and can be viewed in Xenbase (see http://www.xenbase.org/xenwiki/index.php/Antibodies_and_lectins).

   Gene expression images usually come with associated *meta data*, such as the gene name, the development stage or tissue, and possibly the view or section. This makes it relatively easy to construct databases to retrieve images based on these pieces of information, although generally the most useful of these is to retrieve the images for a particular gene. But it remains challenging to retrieve images on the basis of localization (see ref. (12) for a discussion of the issues).

   The same images may appear in several different databases. There is a tendency for researchers who have generated large sets of images to set up an initial database and Web site, but also to give permission for their use in the more community-centric databases. Even if you find a useful image in a community database, it may be worth tracking the image down in its original context, where there could be more information about it, and in any case, if the image is useful in your research you may wish to acknowledge its original creator in any subsequent publication.

4. *Characteristics of gene expression sequence data*
   Both messenger RNAs, and long and short non-coding RNAs may be cloned as cDNAs and sequenced. This is generally done in *libraries*, where the source of a given library is well defined, i.e. which tissue and/or development stage it was derived from. Analysis of (typically) 10,000–50,000 *Expressed Sequence Tags (ESTs)* from each library enables us to detect and quantify the main protein coding genes active in these materials. The *Xenopus* community made a series of strategic choices of libraries to make for both *X. tropicalis* and *X. laevis*, and in consequence we can construct approximate expression profiles from early- through mid-development stages, and across a wide range of adult tissues for genes detectable above about the 0.01% level (compared to all other mRNAs). This enables us to answer such questions as: at what stage of development (or adult tissue) is a gene expressed? The accuracy of this approach depends on various factors, such as the depth of the sequencing of a library, and the fidelity of its construction, but also the bioinformatic analysis of the subsequent EST data into *gene clusters*, and the correct identification of the genes involved. This approach is not very good at detecting genes functionally present at low levels, and is, at the time of writing, beginning to be superseded by high-throughput sequencing (see Note 5).

5. *Characteristics of high-throughput sequence data*
   New technologies capable of generating tens of millions of short (~50–100 bp) sequence reads for little more than the cost of a few hundred ESTs have appeared over the last 2 or 3 years. These technologies are generically termed *high throughput sequencing* (HTS) or *next generation sequencing* (NGS) technologies, and they are currently revolutionizing our approach to large-scale gene expression analysis (see Chapter 18). Although potentially very informative, it is likely to be a while before such data flow into queryable databases in a useful manner. However, individual datasets are beginning to become available in data repositories like NCBI's GEO (14), and, even with basic computational skills, it is possible to extract significant value from these datasets.

   The data are essentially similar to EST-based data, in that one is randomly sampling fragments of mRNAs from libraries prepared under defined conditions, and that by mapping the millions of reads to the transcript data, or gene loci, we can build a high-resolution picture of the transcriptional state of the active genes under those conditions. This is also a good type of data to investigate for alternative exon usage.

6. *Characteristics of microarray data*
Microarrays are a slightly older and less accurate technology for assessing the expression levels of many genes in parallel, and predate high-throughput sequencing-based approaches. They use hybridization of messenger-RNAs in an RNA sample to gene-specific oligos anchored to a substrate, typically a glass slide. The shortcomings are well known (15), and they are more appropriate for *comparing* two cellular or tissue states than determining absolute expression levels. Nevertheless, a good deal of data has been generated using this technology, and much of it has been submitted to NCBI's GEO (14) or the EBI's ArrayExpress (16) databases. They are somewhat going out of fashion for quantitative genomics application, such as ChIP and mRNA profiling, but they are still a valuable tool for the analysis of polymorphism variation between individuals. You will probably not find databases that allow you to easily query microarray data, although both GEO and ArrayExpress do provide some query interfaces. However, you should be able to download and—with the requisite bioinformatics skills—investigate any published micro-array data sets which have relevance to your own research. GEO profile data for *Xenopus* genes is linked from Xenbase gene pages.

7. *The quickDatabases and sequence-based search strategies*
Searches for non-sequence gene data (e.g. in situ images) is traditionally done by text searches on gene symbols, names, or fragments thereof. This can be unreliable for many reasons: genes change names, orthologous genes in different species may have quite different names, getting the precise spelling of longer, complex names may be tricky, and so on. Although it is obvious to search the expressed *sequence* data by sequence similarity, it could be seen that it was also advantageous to search for non-sequence data in the same way, as other types of data can be associated, in one way or another, with specific sequences (7). This novel search method allows databases to be set up with no additional curation effort, and one can thus find gene-related data without worrying about using the right gene name or version of it, and even be somewhat confident that the absence of data returned from a query means that that database likely has no data for that gene.

8. *Interpreting the alignment diagram for a sequence based search*
When searching any of the quickDatabases via a gene sequence, the first set of results on the page is a series of schematic alignments between the query and the target sequences. These are ordered either by descending percent identity of the sequence

match, or by e-value. This may lead to slightly unexpected ordering as longer matches at lower identity may have better e-values than shorter matches at higher identity. You are probably more concerned with the sequence identity values here, and the alignments are color coded from pure green at 100% identity, towards red. In general, for the data you have retrieved to have come from the same gene (and same species) that you are searching with, you will be looking for at least 99% match; although small gaps in the alignment of otherwise 100% identical sequences can degrade the reported PI value. In addition, if the e-value is too relaxed, then short, nearly identical matches may be reported, and these will probably either be chance matches or from repeat sequences. Orthologous gene sequences between the two frog species are ~90–95% similar, although we should perhaps more properly refer to the pair of orthologs in *X. laevis* as homeologs. If you use a search sequence from a non-frog vertebrate species, then the highest reported match, over most of the sequence, may well be the frog ortholog, but you should probably consider some independent check of the relationship between the genes in question.

9. *Orthologs vs. paralogs for sequence percent similarity*
Understanding the results of a sequence similarity based search across species in terms of orthologs or paralogs can be tricky. Imagine you were looking for in situ images to help understand the gene expression pattern of the frog ortholog of a mouse gene. Searching with (say) a mouse protein sequence, we would expect the frog ortholog(s) to be ~70% similar, and one would interpret the list of sequence alignments in that light, bearing in mind that there is no guarantee that the database contains images for any given gene. More obviously, the *second best* hit for *X. tropicalis* is unlikely to be the ortholog, although the top hit may not be either. The situation with *X. laevis* will be correspondingly more complex, although the two homeologs should be about equidistant in similarity from their joint ortholog in another species.

10. *Using an image in XenMARK to search for other images*
The existence of the underlying expression pattern annotations in the XenMARK database allows one to search a priori for images with expression patterns matching your idealized pattern, but it also enables one to search the image collection using one of the annotated images itself as the input. In any set of images returned from quickImage or XenMARK, each annotated image has a search button underneath it. When a search is initiated with this, the database simply uses the annotation attached to the selected image to search for similar annotations in the rest of the database, and then returns those that fit most closely.

11. *Sectional in situ images and XenMARK*
In a recent elaboration of its annotation approach, the XenMARK database has been extended to explore the use of sectional in situ patterns, where gene expression is examined on a tissue section rather than a whole mount embryo. This allows expression to be localized almost down to the cellular level, if required. The prototype system was based on work with retinal sections from stage 39 embryos, with a particular interest in stem cell markers in the ciliary marginal zone of the retina, and more general interest in markers for the other regions of retinal structure (13). In cases like this, the research group driving the project would take responsibility for providing a suitable sectional diagram, and for the image annotation of the expression patterns. A group of four or five post docs and students can annotate hundreds of images in just a few days. An example of the search method being used to discover region-specific markers can be seen in Fig. 4. In this study, known markers from the literature were included, where XenMARK was used to create the annotation of the expression pattern from published images. For copyright reasons, these published images cannot be displayed in XenMARK, so the annotation pattern is shown instead, and a link is provided to the original reference in PubMed.

12. *EST cluster information*
Once you have reached an assembled EST cluster, there is a lot of information available: relationships to other clusters, either longitudinal fragments of the same gene, or paralogs; approximate expression profiles through development and within all the adult tissues for which libraries have been made; the consensus sequence of the mRNA, and the presence of polymorphisms and alternative transcripts; and the precise alignments of each EST used to assemble the cluster. The most directly useful of these are the expression profiles generated from the cDNA source library data stored for each EST sequence. Simple inspection of these profiles, even though they are necessarily somewhat approximate, will give you a sense of whether and when the genes are required in development, or whether they figure prominently in the running of major adult tissues. Equally it may be clear that you are looking at a ubiquitously expressed, or housekeeping, gene, which you may not want to spend time on in a functional study. Even with the relatively coarse grain of the EST-based information these profiles can be surprisingly informative. This is illustrated in Fig. 5, where we see the EST assemblies for the two *X. laevis* homeologs (also called pseudo-alleles) for the gene **atp5e**. They are about 90% similar in sequence, which can be observed easily in the only slightly different patterns of ATG/STOP codons, and look to

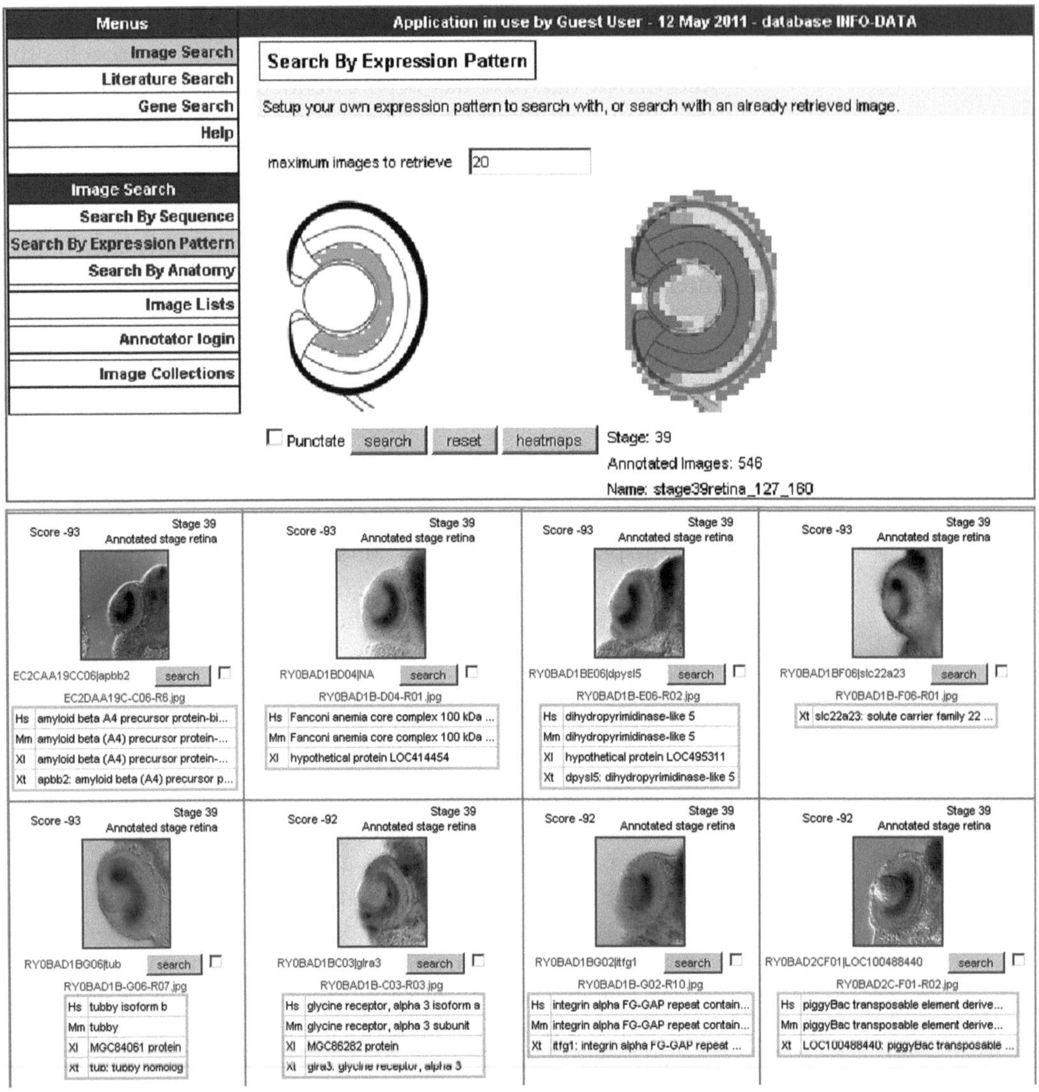

Fig. 4. Expression pattern search based on retinal section models. In the top panel, the green area represents the expression pattern query drawn by the user on a schematic diagram of a stage 39 retinal section, and the heat map on the right is a summation of the gene expression patterns available in Xenmark for that stage of development in the retina. The sectional in situ images retrieved by the requested expression pattern are presented below. The name of the clone sequence used to generate the in situ images, and an indirect annotation of the gene involved, are provided under the image. See ref. (13) for more details.

have detectably different expression profiles, with one peaking in early development during gastrulation and the other in the tailbud phase.

13. *Interpreting an EST cluster for quality and completeness*
    If you have reached an EST cluster display page in search of further information, it may pay to assess the quality of the assembled cluster to understand the reliability of any associated data. These EST clusters are constructed computationally and

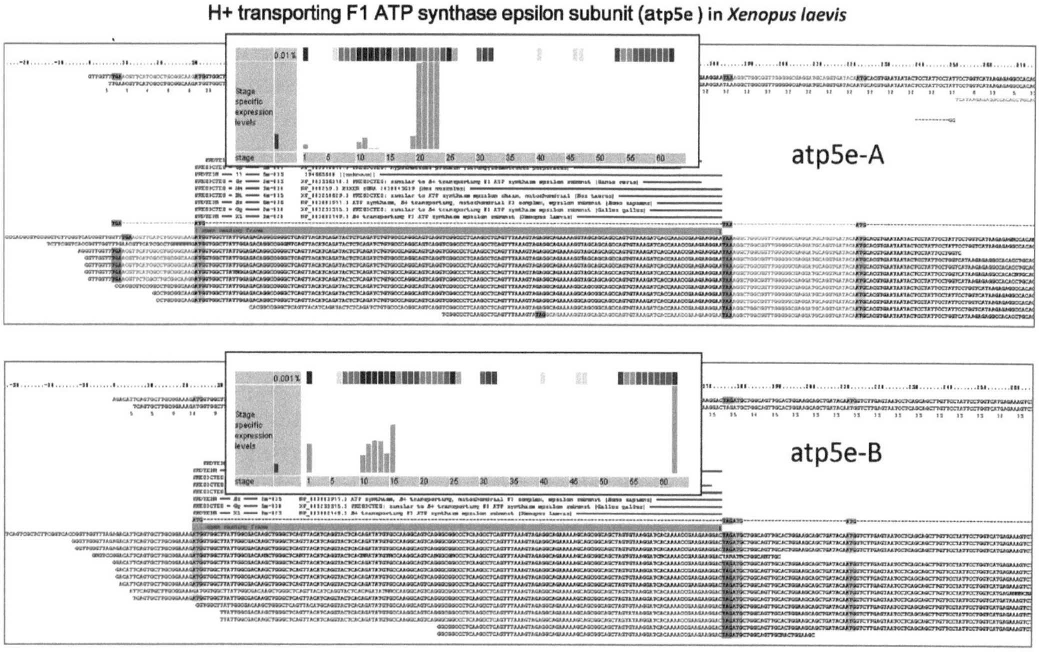

Fig. 5. Gene expression profile for *X. laevis* **atp5e** homeologs generated from existing EST data. Here we see that EST clustering is capable of resolving the closely similar homeologs of *X. laevis* H+ transporting F1 ATP synthase epsilon subunit. As a consequence, we can construct homeolog-specific expression profiles through early development from an analysis of the library source of the ESTs in each cluster, and thus discern the probable sub-functionalization of these two genes.

this process can give rise to errors. At the same time, genes for which we have few ESTs, or which have very long transcripts, are harder to build into coherent clusters, and may present a challenge to interpret. Fortunately, visual inspection of the cluster can give you a good sense of its quality and completeness, and therefore how confident you can be of using it. Look at the depth of the EST alignment data, and the quality of the alignments (easily discernible from the pattern of bases and marked codons in the display), and see if there are any weak (few ESTs) sections of the alignment. In addition, you will see two consensus sequences at the top of the cluster (just below the length scale); the lower (brown) one of these is a check sequence, and contains 'x's at any position where the consensus is not obvious. Also look at the open reading frame (purple box and codons just above the aligned ESTs) and check that it is unbroken and properly closed at either end. This is particularly important if you are trying to assess the completeness of a gene model or full-length cDNA sequence.

14. *Finding full-length clones in the Xenopus full-length EST database*

A common problem is to track down a full-length cDNA clone for a given gene, one which you can then order or retrieve from somewhere (your own freezer, another laboratory, or commercial clone services) and do experiments with. The best starting point for searching for expressed sequences for a given gene is the gene sequence itself. Less precise but quite usable is the text-based search of protein descriptions associated with each gene *cluster* (generic term for a group of assembled EST sequences), although you may have to explore the list of possible results to find the right gene. In the unlikely event you know the name of a clone whose EST or cDNA sequences are in the data, you can search directly with the clone name to find the cluster that its sequences are in.

If you are starting with a gene sequence from a species other than *Xenopus* you may get better results by starting with the protein sequence. If you are starting with a *Xenopus gene* sequence then the best starting point may be the mRNA sequence (cDNA), although you may want to try and use only the coding region, as repeat sequences in the UTRs can make results harder to interpret.

To find a full-length cDNA clone in the *Xenopus* full-length database, follow the protocol described on the Web site of the X-Omics project, which can be found at: http://indigene.issb.genopole.fr/spip.php?article64, and also accessed via the login page to the Gurdon *Xenopus* EST database, described above: http://genomics.nimr.mrc.ac.uk/online/xt-fl-db.html. You may find your clone from one of the existing sets of re-arrayed full-length clones, or you may try and obtain it from the original submitting authors.

At this point you should also be aware of some of the unavoidable limitations of the clone distribution process. This involves much physical manipulation, re-growing, etc. during which a small number of clones (at least 10%) will effectively lose their original nature. So if you have the option, it may well be worth ordering two clones at the same time, should either one prove not to contain the gene sequence you identified. In any case you should probably always end-sequence the clone you get to verify that it contains the correct gene sequence.

References

1. Gygi SP, Rochon Y, Franza BR, and Aebersold, R (1999) Correlation between protein and mRNA abundance in yeast. Mol Cell Biol 19:1720–1730
2. Maier T, Guell M, and Serrano L (2009) Correlation of mRNA and protein in complex biological samples. FEBS Lett 583:3966–3973
3. Maglott D, Ostell J, Pruitt KD, and Tatusova T (2007) Entrez Gene: gene-centered information at NCBI. Nucleic Acids Res 35:D26–31
4. Bowes JB, Snyder KA, Segerdell E, Gibb R, Jarabek C, Noumen E, Pollet N, and Vize PD (2008) Xenbase: a Xenopus biology and genomics resource. Nucleic Acids Res 36:D761–767
5. Raciti D, Reggiani L, Geffers L, Jiang Q, Bacchion F, Subrizi AE, Clements D, Tindal C, Davidson DR, Kaissling B, et al (2008) Organization of the pronephric kidney revealed by large-scale gene expression mapping. Genome Biol 9:R84

6. Gilchrist MJ, Zorn AM, Voigt J, Smith JC, Papalopulu N, and Amaya E (2004) Defining a large set of full-length clones from a Xenopus tropicalis EST project. Dev Biol 271:498–516
7. Gilchrist MJ, Christensen MB, Harland R, Pollet N, Smith JC, Ueno N, and Papalopulu N (2008) Evading the annotation bottleneck: using sequence similarity to search non-sequence gene data. BMC Bioinformatics 9:442
8. Gilchrist MJ, Christensen MB, Bronchain O, Brunet F, Chesneau A, Fenger U, Geach TJ, Ironfield HV, Kaya F, Kricha S, et al (2009) Database of queryable gene expression patterns for Xenopus. Dev Dyn 238:1379–1388
9. Nieuwkoop PD, and Faber J (1956) Normal Table of Xenopus laevis (Daudin). Garland Publishing Inc, New York ISBN 0-8153-1896-0
10. Sprague J, Clements D, Conlin T, Edwards P, Frazer K, Schaper K, Segerdell E, Song P, Sprunger B, and Westerfield M (2003) The Zebrafish Information Network (ZFIN): the zebrafish model organism database. Nucleic Acids Res 31:241–243
11. Yanai I, Peshkin L, Jorgensen P, and Kirschner MW (2011) Mapping gene expression in two Xenopus species: evolutionary constraints and developmental flexibility. Dev Cell 20:483–496
12. Peng H, Long F, Zhou J, Leung G, Eisen MB, and Myers EW (2007) Automatic image analysis for gene expression patterns of fly embryos. BMC Cell Biol 8 Suppl 1:S7
13. Parain K, Mazurier N, Bronchain O, Borday C, Cabochette P, Chesneau A, Colozza G, El Yakoubi W, Hamdache J, Locker M, et al (2011) A large scale screen for neural stem cell markers in Xenopus retina. Dev Neurobiol 72:491–506
14. Barrett T, Troup DB, Wilhite SE, Ledoux P, Rudnev D, Evangelista C, Kim IF, Soboleva A, Tomashevsky M, and Edgar R (2007) NCBI GEO: mining tens of millions of expression profiles–database and tools update. Nucleic Acids Res 35:D760–765
15. Walker MS, and Hughes TA (2008) Messenger RNA expression profiling using DNA microarray technology: diagnostic tool, scientific analysis or un-interpretable data? Int J Mol Med 21:13–17
16. Parkinson H, Kapushesky M, Shojatalab M, Abeygunawardena N, Coulson R, Farne A, Holloway E, Kolesnykov N, Lilja P, Lukk M, et al (2007) ArrayExpress–a public database of microarray experiments and gene expression profiles. Nucleic Acids Res 35:D747–750

Chapter 20

Investigating Alternative RNA Splicing in *Xenopus*

Agnès Mereau and Serge Hardy

Abstract

Alternative splicing, the process by which distinct mature mRNAs can be produced from a single primary transcript, is a key mechanism to increase the organism complexity. The generation of alternative splicing pattern is a means to expand the proteome diversity and also to control gene expression through the regulation of mRNA abundance. Alternative splicing is therefore particularly prevalent during development and accordingly numerous splicing events are regulated in a tissue or temporal manner. To study the roles of alternative splicing during developmental processes and decipher the molecular mechanisms that underlie temporal and spatial regulation, it is important to develop in vivo whole animal studies. In this chapter, we present the advantages of using the amphibian *Xenopus* as a fully in vivo model to study alternative splicing and we describe the experimental procedures that can be used with *Xenopus laevis* embryos and oocytes to define the *cis*-regulatory elements and identify the associated *trans*-acting factors.

Key words: *Xenopus*, Splicing, Alternative splicing, RNA, Embryo, Oocyte

1. Introduction

In vertebrates more than 90% of the genes are subject to alternative splicing (1), a process that can generate multiple mRNAs from a single primary transcript through the differential joining of exons. It is considered that alternative splicing contributes strongly to the complexity of metazoans by expanding the transcriptome and consequently the proteome and by participating in the regulatory network of gene expression (2, 3). One effect of alternative splicing is generation of specific isoforms with subtle changes in protein activity leading thereby to a fine-tuning of protein function. Alternative splicing is also commonly used to produce mRNAs that differ in their 3' untranslated region (UTR) with consequences for their translation, stability, localization, and miRNA regulation. Finally, through the generation of mRNAs that harbor a premature stop codon and are degraded by nonsense-mediated decay (NMD),

Stefan Hoppler and Peter D. Vize (eds.), *Xenopus Protocols: Post-Genomic Approaches, Second Edition*, Methods in Molecular Biology, vol. 917, DOI 10.1007/978-1-61779-992-1_20, © Springer Science+Business Media, LLC 2012

alternative splicing is also a means to regulate the abundance of mRNA transcripts (4).

General rules have emerged over the last few years to determine whether an alternative exon is spliced or not (5). Alternative exons are usually delimited by noncanonical splice sites and are poorly recognized by components of the basal splicing machinery. Therefore, their usage can be negatively or positively modulated by interplay of ubiquitous or tissue-specific RNA binding proteins (RNA-BP) that interact with *cis*-regulatory sequences present in both exons and introns. Positively acting elements within exons or introns are designated exonic splicing enhancer (ESE) and intronic splicing enhancer (ISE), respectively. While exonic and intronic repressor elements are called exonic splicing silencer (ESS) and intronic splicing silencer (ISS).

Alternative splicing may be particularly important during development and accordingly numerous splicing events are regulated in a tissue or temporal manner. The importance of alternative splicing during developmental processes can only be evaluated by in vivo whole animal studies. Whole animal studies are also required to evaluate the consequences of alternative splicing for cellular functions and to define the molecular mechanisms that underlie temporal and spatial regulation in vivo and in particular during development. *Xenopus* is particularly amenable for such studies (6) and in this chapter we describe the experimental procedures that can be used with *X. laevis* embryos and oocytes to define the *cis*-regulatory elements and identify the associated *trans*-acting factors.

1.1. Identifying Tissue-Specific Regulatory Element

The identification of tissue-specific elements that control the usage of an alternative exon requires establishing a gene model that recapitulates the splicing of the endogenous pre-mRNA. The first step involves the construction of a DNA reporter plasmid containing the genomic region encompassing the alternative exon positioned between a promoter and a poly(A) signal (Fig. 1) and reviewed in (7). In general the genomic region consists of the alternative exon surrounded by the flanking introns and constitutive exons. This construct is referred as a minigene since it contains all the signals required for its expression. In theory every eukaryotic expression vector may be used for such a construction but we routinely use the pcDNA3-1 plasmid as backbone for our constructs (Fig. 2). Several approaches can then be used to analyze the tissue-specific splicing pattern of the minigene. The first one consists of using tissue-specific promoters and to directly assay the splicing pattern by RT-PCR analysis. This approach has the advantage of simplicity, since no further handling of the embryos is required. Several *Xenopus* promoters that drive tissue-specific expression have been described and characterized. We developed such a strategy to identify the sequences that determine the usage of a terminal exon of the α-tropomyosin pre-mRNA in the epidermal and the

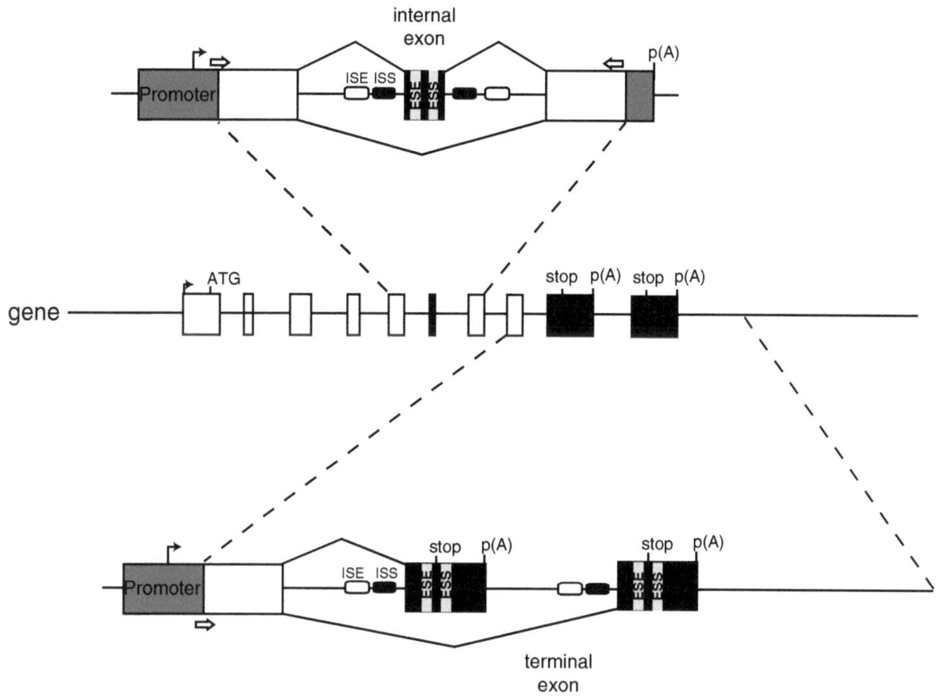

Fig. 1. Minigenes design to study the regulation of an alternative exon. *Middle*: schematic representation of a typical eukaryotic gene. Exons and introns are represented by *boxes* and *horizontal lines*, respectively. Alternative exons are represented by *solid boxes*. The transcription initiation site is represented by a *bent arrow*. The start codon (ATG), the stop codon (stop), and the polyadenylation signal (p(A)) are indicated. *Top* and *bottom*: diagrammed representation of minigenes constructs to study an internal alternative exon (*top*) and alternative terminal exons (*bottom*). Intronic and exonic regulatory elements are represented. *ESE* exonic splicing enhancer; *ESS* exonic splicing silencer; *ISE* intronic splicing enhancer; *ISS* intronic splicing silencer. The primers use to PCR amplify the produced RNA are indicated by *arrows*.

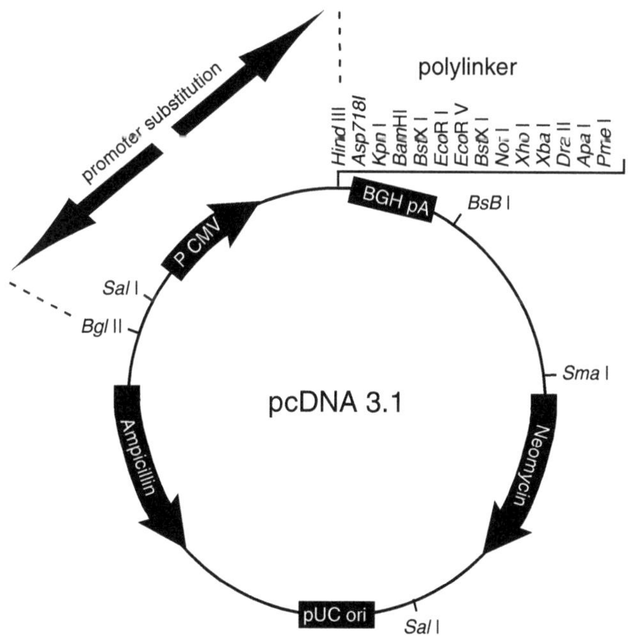

Fig. 2. Schematic representation of the pcDNA 3.1 plasmid. The region that can be removed to substitute a tissue-specific promoter is indicated. Several unique restriction sites within the polylinker can be used to insert the genomic fragment. *P CMV* cytomegalovirus promoter; *BGH pA* bovine growth hormone polyadenylation signal.

myotomal cells using the keratin and the cardiac actin promoters respectively (8, 9). In a second approach, the genomic region of interest is placed under the control of a ubiquitously or widely expressed promoter. The tissue-specific splicing pattern is then studied by RT-PCR after dissection of the appropriate tissues. An endogenous widely expressed promoter such as the EF1α promoter works well for such an approach and was successfully used to identify determinant sequences of the α-tropomyosin pre-mRNA (10). If the tissue of interest is not easily accessible to dissection, the animal cap explant assay can also be a useful tool for investigating a tissue-specific splicing pattern. The *Xenopus* animal cap cells retain pluripotency and in addition to differentiate to ectodermal tissue they can be driven to differentiate toward neural, mesodermal, and endodermal derivatives. Several in vitro differentiation protocols have been established to produce specific tissues such as cardiac, pancreatic, renal, and neural tissues (11). In this third approach, a minigene driven by a ubiquitous promoter is microinjected into the animal pole and at the blastula stage the animal pole tissue is excised and cultured in presence of the adequate inducer. The splicing pattern can then be assayed by RT-PCR in control and induced animal cap explants.

Although using a single gene construction to study alternative splicing in many tissues is labor-saving, the dissection of the tissues or the animal cap explant requires practical skills and is in itself time consuming. Hence, a ubiquitously expressed promoter is used when no tissue-specific promoters recapitulating the splicing pattern are available. However, this approach is also useful to demonstrate that the splicing pattern obtained with minigenes driven by tissue-specific promoters is not biased by the promoter structure.

Two strategies can then be used to express the minigene. The straightforward and simple one is the transient injection of a DNA plasmid into the embryo. With this procedure, DNA does not integrate in the genome and remains extrachromosomal throughout early development to late tailbud stages (12, 13). However, mosaic gene expression from the injected plasmid DNA is observed and ectopically activation may also occur, depending on the tissue-specific promoter utilized (14). Despite these limits, we have shown, for two sets of alternative exons (8–10), that pre-mRNA splicing from injected DNA templates is accurately regulated, therefore providing an extremely rapid assay to recapitulate splicing patterns. We therefore recommend this strategy to dissect the *cis*-regulatory sequences, in which case a number of mutant minigenes has to be tested. However, one important limitation of the transient transgenesis method is the time window during which the gene is expressed. Expression from injected minigenes generally becomes undetectable after the late tailbud stage (15). It is therefore impossible to use DNA injection to study splicing events that occur in tissues that differentiate after the late tailbud stage notably during organogenesis. This limitation can be overcome by the generation of stable transgenic embryos. The most common and efficient method to produce transgenic embryos is based on the REMI (restriction enzyme-mediated insertion) technique and is described in Chapter 11. It is noteworthy that if a unique restriction site is situated outside of the expressed sequence, then the minigene reporter plasmid used for DNA injection can also be employed to make stable transgenic embryos. For minigenes constructed with the pcDNA3.1 backbone we successfully used the restriction sites *Sma*I positioned upstream of the neomycin gene and the *Sal*I present upstream of the pUC origin (see Fig. 2). F0 embryos are directly accessible for RT-PCR analysis.

When a model that recapitulates the tissue-specific splicing pattern has been established, it can be used to seek regulatory sequences. As discussed above, regulatory elements can be present in either the regulated exons (ESE and ESS) or the adjacent introns (ISE and ISS). Because exons are relatively short (200 nt on average) the straightforward method to identify ESE or ESS sequences within an alternative exon is the linker scanning mutagenesis procedure. In a first approach, a ten bases substitution linker should be sufficient to identify an exonic regulatory element. This initial analysis could then be further refined by introducing point mutations. The identification of an intronic element is generally more difficult because the length of introns varies extensively in vertebrates ranging from one hundred to several thousand of nucleotides and the position of the branch point, a motif required for the first transesterification reaction may differ from one intron to the other. Indeed, the branch point is generally located 30–40 nt upstream of the 3' splice site but can be much more distant in a subset of alternative exons (16). In *Xenopus*, the branch point associated to an alternative terminal exon of the α-tropomyosin pre-mRNA is located 274 nt upstream of the 3' splice site (8). Interestingly, it was shown for several alternative exons that distant branch sites accommodate space for regulatory sequences (8, 9, 17, 18). Before seeking intronic *cis*-regulatory element, it is therefore necessary to precisely assign the position of the functional branch site within the intron. As a first approach, *in silico* mapping of branch sites can be realized with appropriate algorithms ((19) and see (http://regulatorygenomics.upf.edu/SVM_BP/)). However, the strong degeneracy of the motif in vertebrates (20) means this approach has no guaranty of success and we recommend mutating the potential branch site(s) in the minigene reporter construct and monitoring the effects upon splicing. Intronic elements can then be identified by overlapping deletions upstream and downstream of the branch site and further characterized by linker scanning mutagenesis. It is important to notice that silencer and enhancer elements frequently overlap, rendering the straightforward delineation of regulatory motifs difficult.

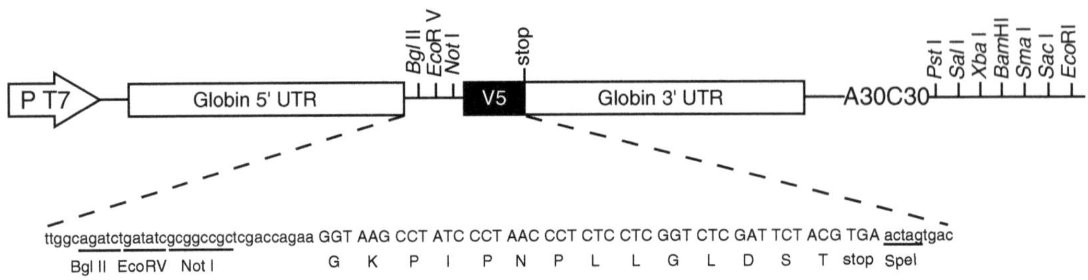

Fig. 3. pT7TS-V5 plasmide. The V5 epitope tag was positioned downstream of the three unique sites (BglII, EcoRV, NotI) available to insert the ORF of interest. The unique sites to linearize the transcription template downstream of the A30C30 sequence are indicated.

1.2. Studying the Role of trans-Acting Factors

Minigene injection associated with loss- or gain-of-function experiments in *Xenopus* embryos is a powerful strategy to identify *trans*-acting factors involved in the regulation of an alternative splicing event and to correlate their activity with a particular *cis*-regulatory element. Indeed, while the depletion or the overexpression of the protein can be effective in the entire embryo, the consequences of changing *trans*-acting factor expression upon splicing can be studied in distinct tissues depending on the tissue-specific expression of the minigene. With this strategy, it is also easy to verify the responsiveness of a specific *cis*-regulatory element by testing different mutant minigenes. Because of mosaic expression, DNA injection cannot be used to overexpress proteins and generally gain-of-function experiments of splicing factors are realized by injection of synthetic mRNAs that spread uniformly throughout the embryo (21). mRNA stability and translational efficiency are therefore the only limitation of this procedure which affects the stage of embryonic development that can be studied. To overcome this problem, we use pT7TS-V5, a derivative of pSP64 (22) as a transcription vector (Fig. 3). It contains the 5' and 3' UTR of *Xenopus* β-globin mRNA which enhances the stability and translational efficiency of the chimeric mRNA and the V5 tag which can be used to follow the expression of the chimeric protein. The effects of the candidate proteins upon splicing are assayed by co-injection of exogenous mRNA (encoding the *trans*-acting splice factor) together with the reporter minigene (containing the *cis*-regulatory splice sequences).

If overexpression of a specific protein cannot be attained in the targeted embryonic tissue because of the time window or because it causes development defects, the oocyte can be an interesting alternative. This cell is an active site of pre-mRNA processing and has been used by a number of laboratories to study constitutive splicing (23–25). In addition, the majority of exogenous mRNAs are translated very efficiently when injected directly into the cytoplasm and since the oocyte does not divide, overexpression should not affect its physiology, at least during the time course of the experiment. The screening of *trans*-acting factors is realized with a two-step microinjection protocol (26). First the mRNA encoding the V5 tagged protein is injected into the cytoplasm and its accumulation in the cytoplasm and nucleus can be monitored by Western blot. Second, after several hours to allow the accumulation of the exogenous protein, the minigene is injected into the nucleus. Ubiquitous cellular promoters (for example EF1α) or viral promoters such as the CMV or early SV40 promoters can be used to drive expression of the minigene.

Because numerous RNA binding protein have redundant activity in splicing and may bind with low affinity to nonspecific motifs present in the pre-mRNA (27), one needs to be cautious when analyzing overexpression data. Indeed, while overexpression is a powerful means to address the potential mechanistic activity of a protein, it does not demonstrate its physiological relevance. Loss-of-function experiments are therefore an important complement to overexpression analysis. In *Xenopus* embryo, the most efficient way to knockdown a specific endogenous protein is the injection of antisense morpholino oligomers (AMOs) that target the 5' UTR sequence or the start codon of the mRNA encoded by the protein of interest (see Chapter 8). The tissue-specific effects of the knockdown upon the splicing pattern are easily studied by co-injecting AMOs with the minigene. The knockdown effectiveness is measured by Western blot (if there is a suitable antibody to analyze the protein expression of the endogenous splice factor). Rescue experiments are required in order to ascertain the specificity of the knockdown. These can be easily carried out by co-injecting AMOs, the minigene and a synthetic mRNA encoding the tagged protein but modified so that it is no longer targeted by AMOs. This approach was used to demonstrate that the knockdown of the polypyrimidine tract binding protein (PTB) resulted in the derepression of a 3' terminal exon of the α-tropomyosin pre-mRNA (8, 26). In the oocyte, AMOs cannot be used to deplete maternally expressed splicing factors unless they are very unstable proteins. Inhibition of a splicing factor could be attempted by injection of neutralizing antibodies but in this case, a rescue experiment is normally excluded. A second alternative is the injection in the nucleus of competitor RNAs that present high affinity binding sites for the candidate RNA-BP. The specificity can be assayed by injection of control RNA that does not harbor the high affinity binding sites (8).

1.3. Investigating Endogenous Alternative Pre-mRNA Splicing

Minigene studies are very efficient to define *cis*-regulatory elements that determine the tissue-specific usage of an exon. However, they present several limitations that must be taken in consideration. First, minigenes are generally driven by heterologous promoters and it is known that some alternative splicing pattern can be modulated by the promoter (28). Indeed, the promoter structure can affect the rate of transcription elongation that in turn may affect the timing with which competing splices are recognized by the splicing machinery. In addition, promoter identity may also affect the recruitment of regulatory splicing factors to the transcribing RNA polymerase II and therefore their availability to the *cis*-regulatory sequences. Second, as the injected DNA plasmid remains in an episomal form it is not subjected to chromatin regulation and it has recently been shown that chromatin structure and histone modifications may participate in the regulation of alternative splicing through changes in RNA polymerase II processivity and modification in the recruitment of splicing factors (29). Third, only a small part of the pre-mRNA is present and it cannot be ignored that missing exonic or intronic sequences may influence the endogenous splicing pattern. It is therefore important to determine whether a *cis*-regulatory sequence characterized with a minigene model is physiologically relevant in the context of the full-length endogenous pre-mRNA expressed from its own promoter. An efficient way to test the activity of a candidate *cis*-regulatory element within an endogenous pre-mRNA is to design modified antisense oligonucleotides that bind to complementary sequences in the targeted pre-mRNA and sterically inhibit the binding of *trans*-acting proteins (30, 31). The *Xenopus* embryo is particularly amenable to such studies. First, microinjection of the embryos very efficiently delivers the oligonucleotides to all cell types. Second, oligonucleotides are present before the onset of zygotic transcription and therefore before pre-mRNA starts to be processed allowing a robust switching in splice sites selection. In *Xenopus,* among the different available chemistries, antisense AMO proved very efficient to inhibit translation or splicing while, as a class of molecules, they do not affect development (32, 33). At present they are the most effective tool to achieve the targeting of a *cis*-regulatory element (see Chapter 8). In addition to evaluate the relevance of a candidate motif, morpholino-mediated inhibition of a regulatory element can be used to redirect a splicing pattern and assay the function of a specific mRNA isoform in the developing embryo. Such a strategy was used to demonstrate the physiological relevance of an ISE required to activate the splicing of a 3′ terminal exon of the α-tropomyosin pre-mRNA in myotomal cells (9). Experimentally altering endogenous α-tropomyosin pre-mRNA splicing in this way also resulted in myotome contraction defects, indicating that the embryonic α-tropomyosin isoform displays specific properties in embryonic myofibrils contraction.

2. Materials

2.1. Identification of cis-Regulatory Elements

2.1.1. Minigene Construction

1. Expression vector (e.g., pCDNA 3.1 from Invitrogen, Carlsbad, CA).
2. Restriction enzymes, DNA ligase, and appropriate buffers.
3. Thermostable polymerase with proofreading activity (e.g., PFU from Promega, Madison, WI).

2.1.2. Microinjection of Minigenes into Xenopus Embryos

1. Adult male and female *Xenopus laevis.*
2. Human chorionic gonadotrophin (HCG) (e.g., from Schering-Plough; Courbevoie, France).
3. 10× stock F1 medium: 312 mM NaCl, 17.5 mM KCl, 590 μM $MgCl_2$, 20.2 mM $NaHCO_3$, 100 mM Hepes-NaOH (pH 7.6). Filter sterilize and store at 4 or −20°C. To prepare 1× F1 medium add 100 mL 10× stock solution and 250 μL 1 M $CaCl_2$ to 900 mL H_2O. Store at 4°C.
4. Freshly made 2% Cysteine-NaOH, pH 7.9 in 1× F1.
5. Ficoll injection buffer: 5% (w/v) Ficoll in 1× F1.
6. Thin-walled glass capillaries (e.g., Narishige G-100).
7. Pipet puller (e.g., Narishige PN-30).
8. Nanoject Microinjection apparatus (Drummond. Broomall, PA).
9. Micromanipulator (e.g., Sutter instruments MM-33).
10. Dissecting microscope (e.g., Olympus SZ40) and a cold light source (e.g., Olympus Highlight 2100).
11. Controlled temperature incubator (16–22°C).

2.1.3. RNA Analysis

1. Tri reagent solution (Molecular research center, Inc, Cincinnati, OH, cat. no. TR 118).
2. Turbo DNase from Ambion (Applied Biosytems, Carlsbad, CA).
3. MMLV reverse transcriptase (e.g., from Promega, Madison, WI) and appropriate buffer.
4. dNTP mix: dATP, dCTP, dGTP, and dTTP, each at 10 and 2.5 mM.
5. T4 polynucleotide kinase (Promega, Madison, WI) and appropriate buffer.
6. (γ ^{32}P) ATP 3,000 Ci/mmol, 10 μCi/μL (NEN Perkin Elmer, Waltham MA).
7. Microspin G25 columns (Roche, Indianapolis, IN).
8. Taq DNA polymerase (Promega, Madison, WI).

2.2. Investigating trans-Acting Splice Factors

1. pT7TS-V5 plasmid (available upon request).
2. mMESSAGE mMACHINE T7 Kit (Ambion, Applied Biosytems, Carlsbad, CA).

2.2.1. RNA Overexpression

1. AMO are ordered from Gene Tools, LLC, Philomath, OR.

2.2.2. Morpholino-Mediated Knockdown

2.2.3. Analysis of the Overexpression and Knockdown Assays

1. Suspension buffer: 20 mM Tris pH = 7.5; 4 mM EDTA.
2. Protein extraction buffer: 100 mM Tris–HCl pH 7.5, 4 mM EDTA pH 8, 1% Triton X100. 2.5 μL of the Sigma protease inhibitor cocktail P8340 is added freshly to 1 mL of buffer.
3. V5 monoclonal antibody (Invitrogen, Carlsbad, CA).

2.3. Microinjection of Oocytes

2.3.1. Preparation of the Oocytes

1. 10× stock OR2 medium: 825 nM NaCl, 25 nM KCl, 10 mM $MgCl_2$, 10 mM NA_2HPO_4, 50 mM Hepes-NaOH pH 7.8. Filter sterilize and store at 4°C. To prepare 1× OR2 minus Ca^{2+} make a 1:10 dilution of the stock solution in distilled water. To prepare 1× OR2 add 1 mL of 1 M $CaCl_2$ to 999 mL of 1× OR2 minus Ca^{2+}.
2. Collagenase type I (Sigma-Aldrich, St. Louis, MO).

2.3.2. Minigene Expression: Nuclear Injection

1. 24-well cell tissue plates.
2. Iron rod of 2 mm diameter.

2.4. Investigating and Modifying the Endogenous Pre-mRNA

1. Random primers stock solution (3 μg/μL) (Invitrogen, Carlsbad, CA).
2. Superscript II reverse transcriptase (Invitrogen, Carlsbad, CA).
3. RNasin and RNAse H (Promega, Madison, WI).
4. TaqMan universal PCR master mix (Applied Biosystem, Carlsbad, CA).

3. Methods

3.1. Identification of cis-Regulatory Elements

3.1.1. Minigene Construction

1. Amplify the tissue-specific promoter using primers containing restriction enzymes sites at their 5′ ends compatible with the *Bgl*II and *Hind*III restrictions sites of pcDNA3.1 plasmid.
2. Using standard procedures substitute the CMV promoter of pcDNA 3.1 with the tissue-specific promoter. The CMV promoter is removed by *Bgl*II/*Hind*III digestion and the promoter fragment is subsequently ligated to the remaining plasmid (Fig. 2).
3. Amplify the genomic region encompassing the alternative exon using a high fidelity polymerase so as not to introduce mutations

and primers containing different restriction sites at their 5′ end compatible with the polylinker of pcDNA3.1 (see Notes 1–3).

4. Using standard procedures subclone the PCR product into the polylinker of the modified pcDNA 3.1 that contains the tissue-specific promoter.

3.1.2. Microinjection of Minigenes into Xenopus Embryos

For fertilization we are using established methods that are precisely described in (34). Briefly our procedure is:

1. Induce ovulation by subcutaneous injection of 500 IU of human chorionic gonadotrophin 12 h before laying.
2. Squeeze out eggs onto a 14 cm glass petri dish and gently spread out them to form a monolayer.
3. Fertilize eggs with sperm of ¼ of testes minced in 1.5 mL of F1 medium. After 2 min cover the eggs with F1 medium and place at 21°C.
4. Following cortical rotation (20 min at 21°C) incubate the embryos for a few minutes in 2% cysteine-HCl in F1 medium pH 7.9 to remove the jelly coat. When embryos are completely dejellied they will pack together very closely. The precise incubation time is very important and is dependent on the quality of the laid eggs. If the embryos are only partially dejellied they will be very difficult to inject. If they are incubated for too long they will not survive injection. Transfer the embryos in a large volume of water at the room temperature and rinse twice with a large volume of water to insure that no more cystein remain.
5. Rinse the embryos twice in F1 medium, transfer them to the Ficoll injection buffer, and incubate one half at 16°C and the other half at 21°C. The embryo incubated at 21°C will cleave 90 min after the fertilization while the embryos incubated at 16°C will cleave 2 h after the fertilization. This will therefore give a 1 h time-window for the injection of the two-cell stage embryos.
6. Select embryos that have just started to cleave and line them up on a plastic grid placed in a 5 cm petri dish filled up with injection buffer. For each minigene to be tested we routinely inject one blastomere of two. Injection is performed with a nanoject injector (Drummond) and micropipets with a tip of 20 μm diameter.
7. Inject 125–250 pg of minigene DNA in 13.8 nL (see Note 4).
8. Keep the embryos at 21°C in injection buffer until they reach stage 7–8 (4–5 h) and then transfer them in 0.1× F1 medium for further development.
9. When embryos reach the appropriate stage (see Note 5), transfer them by pool of 5 into Eppendorf tubes. After removing the remaining medium, freeze them in liquid nitrogen and store at −80°C.

3.1.3. RNA Analysis

The RT-PCR procedure is the most appropriate method to analyze the splicing of a minigene. It permits low amount of processed mRNAs to be detected with a low background and it can selectively amplify only mRNA expressed from the minigene (see Note 6). The main concern with this assay is to obtain a ratio of the PCR products that reflects the ratio of the mRNA isoforms. It is therefore important to perform less than 22 cycles during the amplification procedure in order to be in the linear range of amplification. One way to increase the sensitivity of the PCR assay is to use a 5′-end ^{32}P labeled forward oligonucleotide, that also gives a means to accurately quantify the different PCR products (see Note 7).

1. Extract total RNA with the Tri reagent solution according to the manufacturer's procedure. We routinely homogenize pool of 5 embryos in 250 μL and obtained 25 μg of total RNA. Following extraction RNA is kept in Milli-Q water at −80°C for short-term storage or precipitated in ethanol at −20°C for long-term storage.
2. Remove contaminating DNA with Turbo DNase according to the manufacturer's procedure.
3. Reverse transcribe the mRNAs isoforms by priming with a reverse oligonucleotide that hybridizes to the downstream constitutive exon (see Note 8).
   - Assemble a 25 μL reaction containing 15 μL RNA (10 μg), 5 μL 5× Reverse transcriptase buffer, 2 μL of reverse primer (2.5 μM).
   - Heat at 85°C for 3 min.
   - Add 2.5 μL dNTPs (10 mM) and 0.5 μL MMLV Reverse transcriptase.
   - Incubate the reaction at 37°C for 1 h.
   - The reaction can then be stored at −20°C or used as template for PCR.
4. End-label the forward or reverse primers using (γ ^{32}P) ATP and T4 polynucleotide kinase.
   - Set up a 15 μL reaction containing 5 μL Milli-Q H_2O, 1 μL 30 μM oligonucleotide, 1.5 μL 10× polynucleotide kinase buffer, 5 μL (γ ^{32}P) ATP (3,000 Ci/mMole, 10 μCi/μL), and 2 μL T4 polynucleotide kinase.
   - Incubate 30 min at 37°C.
   - Add 35 μL of Milli-Q H_2O to the reaction.
   - Use a microspin G25 column (Roche) to remove unincorporated labeled nucleotides.
   - Store at −20°C.

5. Amplify the cDNA by PCR.
   - Assemble a 50 μL reaction containing 32 μL Milli-Q H_2O, 2–5 μL cDNA, 5 μL 10× PCR buffer with $MgCl_2$, 4 μL dNTPs (2.5 mM), 2 μL labeled forward primer, 0.5 μL forward primer (30 μM), 1 μL reverse primer (30 μM), and 0.5 μL Taq polymerase (5 U/μL).
   - The reactions are cycled 22 times.
6. Size fractionate the PCR products on a polyacrylamide gel (see Note 9).
   - Depending on their size, the PCR products are electrophoresed on 4–6% nondenaturating polyacrylamide gels and quantified in a PhosphorImager (GE Healthcare).

3.2. Investigating the Role of trans-Acting Splice Factors

3.2.1. RNA Overexpression

1. Amplify the ORF of the candidate factor without the stop codon using primers containing restrictions sites compatibles with the restriction sites of pT7TS-V5 plasmid. The reverse primer must also be designed to be in frame with the V5 tag (see Fig. 3).
2. Using standard procedures subclone the PCR product into the polylinker of pT7TS-V5.
3. Linearize 10 μg of the recombinant plasmid DNA with one of the restriction enzymes whose sites are present downstream of the A30C30 sequence (see Fig. 3).
4. Phenol-chloroform extract the reaction, ethanol precipitate, and resuspend in 10 μL of Milli-Q H_2O.
5. Synthesize capped mRNA with the T7 messenger machine kit (Ambion) according to the manufacturer's recommendations. Resuspend RNA in Milli-Q water and store at −80°C.
6. Follow the same procedure as described in Subheading 3.1.2, steps 1–6.
7. Inject 125–250 pg of minigene with 100–300 pg (0.5–2 fmol) of mRNA in 13.8 nL in one or two blastomeres (see Note 10).
8. Follow the same procedure as described in Subheading 3.1.2, step 8 and 9.

3.2.2. Morpholino-Mediated Knockdown

1. Design a 25 mer antisense AMO complementary to the 5' UTR or spanning the start codon (see Note 11).
2. Design a rescue mRNA that is not complementary to the morpholino and follow the procedure described in Subheading 3.2.1 to synthesize it.
3. Prepare the embryos by following the same procedure as described in Subheading 3.1.2, steps 1–6.
4. Inject 125 pg of minigene with 10–25 ng of morpholino in 13.8 nL in one or two blastomeres (see Note 12). For the rescue

experiment include in the mix of injection 100–300 pg (0.5–1.5 fmol) of the rescue mRNA encoding the splice transfactor under investigation.

5. Follow the same procedure as described above in Subheading 3.1.2, steps 8 and 9.

3.2.3. Analysis of the Overexpression and Knockdown Assays

Proteins and RNA are extracted from the same embryos using the following procedure.

1. Homogenize pool of 5 embryos in 60 μL of suspension buffer.
2. Transfer 2/3 of the suspension in 250 μL of Tri reagent solution and proceed according to the manufacturer's procedure to extract RNA.
3. Transfer 1/3 of the suspension in 50 μL of protein extraction buffer. Spin for 20 min at 13,000 × *g* at 4°C. Transfer the supernatant to a new tube and add one volume of 2× Laemmli buffer. Heat the samples 10 min at 90°C.

The proteins levels in both assays can be measured by Western blot using antibodies directed again the candidate protein. If these are not available, overexpression of the exogenous V5-tagged protein can be monitored by Western blot with a V5 antibody. We used the monoclonal V5 antibody from Invitrogen in numerous experiments and never observed cross-reaction with endogenous *Xenopus* proteins.

If specific antibodies for the protein encoded by the targeted mRNA are not available, the translation inhibitory potential of the AMO can be assessed by co-injecting it with the V5-tagged version of the mRNA. One limitation of this strategy is that the endogenous mRNA and the reporter one may not have the same accessibility to the AMO. The effect of the AMO on the translation of the exogenous RNA is therefore not proof of the endogenous knockdown. It is therefore advisable to test a second AMO directed to the same target to determine if both yield a similar splicing switch.

The analysis of the splicing pattern is realized by RT/PCR as described in Subheading 3.1.3.

3.3. Oocyte Microinjection Protocol

3.3.1. Preparation of the Oocytes

All steps are carried out at 18–22°C.

1. Anaesthetize one female frog in a solution of 0.2% phenoxyethanol.
2. Remove ovary fragments and place them in OR2 medium.
3. Open the ovarian lobes so that all oocytes are in contact with the medium and cut a fragment into small pieces (1–2 g), clean them from vessels and coagulated blood using forceps, and let them recover for at least 2 h in the OR2 medium at 16°C.

4. Dissolve 20–30 mg of collagenase (type I, sigma) corresponding to 5,000 IU in 15 mL OR2 minus Ca^{2+}.
5. Rinse ovarian fragments with OR2 minus Ca^{2+} medium by successively dipping them in 3 baths and finally incubate them in this buffer for 30 min under gentle agitation.
6. Change OR2 minus Ca^{2+} buffer to the collagenase solution and agitate gently for 25 min. After 25 min, stop the reaction by transferring the ovarian fragment in OR2 medium and check under a stereo microscope that the follicular layer surrounding the oocytes can be easily removed manually using fine forceps. If oocytes explode during this step, extend the incubation time for two additional minutes up to a maximum incubation time of 30 min. We noticed that beyond 30 min in the collagenase solution, oocytes are weakened and have a lower recovery rate from the centrifugation step and the two-step microinjection protocol.
7. Allow oocytes released from the follicule to equilibrate overnight at 16°C in OR2 medium before microinjection. Oocytes can be kept at this temperature for 2–3 days by changing medium at least once a day and discard dying or unhealthy oocytes.

3.3.2. RNA Overexpression: Cytoplasmic Injection

1. Select healthy oocytes and line them up on a plastic grid disposed in a 5 cm petri dish filled up with OR2.
2. Using a pair of fine forceps, orient the oocytes with the animal hemisphere (darkly pigmented) upward. For each condition to be tested we routinely inject 20 oocytes and injection is performed with a nanoject injector (Drummond) and micropipets with a tip of 20 μm diameter.
3. Inject 1–30 ng of mRNA in 18.4 nL into the cytoplasm at the level of the middle equatorial line (see Note 13).
4. Incubate overnight at 16°C to allow accumulation of the protein.

3.3.3. Minigene Expression: Nuclear Injection

The major problem with nuclear injection is to localize the nucleus within the pigmented animal hemisphere. To increase the success rate we centrifuge oocytes. This results in a movement of the nucleus toward the surface that is visualized by the apparition of a non-pigmented spot.

1. Prepare the plates for the centrifugation of oocytes: pour 2 mL of agarose gel (2%) in each well of a 24-well cell tissue plate and shape small cavities in agarose gel using a warmed iron rod (around 6 cavities per well and each cavity will receive one oocyte).
2. Place one oocyte in each cavity in the 24-well plate filled with OR2 medium. With a fine forceps orient the oocyte in the cavities so that the animal hemisphere is directed upward.

3. Spin the plate at 600 × *g* for 15 min at 15°C to allow the visualization of the nucleus (see Note14).
4. Line up the oocytes on a plastic grid disposed in a 5 cm petri dish filled up with OR2. Using a pair of fine forceps, orient the oocytes with the unpigmented spot toward the micropipette.
5. Inject 1–5 ng of minigene in 13.8 nL into the nucleus.
6. Keep the oocytes for 6 h at 16°C and then transfer them by pools of 5 into Eppendorf tubes. After removing the remaining medium, freeze them in liquid nitrogen and store at −80°C.

3.3.4. Analysis of the Overexpression Assay

Due to the large size of the nucleus (germinal vesicle) (500 μm diameter), it is easy to manually separate it from the cytoplasm. The expression of the protein encoded by the microinjected mRNA in both cellular compartments can therefore be monitored by Western blot as described in Subheading 3.2.3.

The analysis of the splicing pattern is realized by RT/PCR as described in Subheading 3.1.3.

3.4. Investigating and Modifying the Endogenous Pre-mRNA

1. Prepare embryos as described in Subheading 3.1.2 (steps 1–5).
2. For each AMO to be tested we routinely inject 50–100 two-cell stage embryos into both blastomeres (see Note 15). Select embryos that have just start to cleave and line them up on a plastic grid placed in a 5 cm petri dish filled up with injection buffer. With dull forceps orientate the embryos so that the cleavage furrow is turned toward the micropipet. This procedure requires some time but the injection in both blastomeres will be subsequently easier.
3. Inject 13.8 nL of AMO solution in each blastomere (see Notes 16 and 17).
4. Keep the embryos at 21°C in injection buffer until they reach stage 7–8 (4–5 h) and then transfer them in 0.1× F1 medium for further development (see Note 18).
5. When embryos reach the appropriate stage transfer half of them in groups of 5 into Eppendorf tubes. After removing the remaining medium freeze them in liquid nitrogen and store at −80°C. Keep the other half at 21°C for the description of phenotypes.

3.4.1. Analysis of the Alternative Splicing Patterns

Because of its sensitivity the endogenous splicing patterns are commonly analyzed with quantitative or semiquantitative RT-PCR assays. If the expression level is sufficient RNAse protection (26) or S1 nuclease essays (35) can be useful for detecting unspliced product or show unexpected splicing patterns. The semiquantitative RT-PCR and quantitative RT-PCR assay are described below.

1. RNAs are extracted as described in Subheading 3.1.3, step 1
2. Reverse transcribe RNA using random primers by assembling the following reaction:
   - Add 2 μg of total RNA to 1 μL of 300–500 ng random primers and 1 μL of dNTP mix (10 mM each) and Milli-Q water up to 12 μL.
   - Heat the reaction mixture to 65°C for 5 min and quick chill on ice.
   - Collect the contents of the tube by brief centrifugation and add 4 μL of 5× RT buffer (250 mM Tris–HCl pH 8.3; 375 mM KCl; 15 mM $MgCl_2$), 2 μL of 0,1 M DTT, and 1 μL of RNAsin solution and incubate at 25°C for 5 min.
   - Add 1 μL of RT (we usually use SuperScriptII RT from Invitrogen) and incubate at 42°C for 1 h.
   - Inactivate the reaction by heating at 70°C for 15 min.
   - Remove RNA complementary to the cDNA by addition of 1 μL (2 units) of *E. coli* RNase H and incubate at 37°C for 20 min.
3. For semiquantitative PCR assay use one tenth of the RT reaction and a ^{32}P-labelled forward primer that hybridizes to a constitutive exon and a reverse primer that specifically hybridizes to one specific alternative exon or to a downstream constitutive exon. Proceed as described in Subheading 3.1.3, steps 5–6.
4. For quantitative PCR reaction use primers that hybridize to a constitutive exon and an alternative exon. Assemble a 10 μL reaction containing 5 μL of TaqMan Universal PCR Master Mix (Applied Biosystem), 1 μL of a 1:100 dilution of the RT reaction and primers at a final concentration of 200 nM. Run in an appropriate thermal cycler with fluorescence detection such as the 7900 HT Real-time PCR system (Applied Biosystem).

4. Notes

1. Since intronic regulatory elements are generally present in the introns upstream or downstream of the alternative exon, we recommend amplification of a genomic fragment spanning from the beginning of the exon upstream of the regulated exon to the end of the exon downstream of the regulated exon. If the introns size is not compatible with the amplification or the subsequent cloning of such a fragment, one may remove the central part of the intron leaving 500 nt on each side associated with the constitutive and regulated exons.

2. The minigene design must be slightly adapted for the study of 3′-terminal alternative exons (see Fig. 1). It is important to extend the genomic fragment about 200 nt beyond the hexanucleotide (AAUAAA) of the poly(A) signal of the last exon to insure that the poly(A) signal is complete and functional. The poly(A) signal from the expression plasmid must also be removed to avoid competition with that of the regulated exons. Using the pcDNA3.1 plasmid the genomic fragment can be cloned between a restriction site of the polylinker and the *Bsb*I restriction site present downstream of the bovine poly(A) signal (see Fig. 2).
3. Except for specific purposes it is best not to have an open reading frame in the mRNAs produced by the minigene and therefore not to introduce a start codon in the minigene. Indeed, cryptic splicing pattern or activation of cryptic splice sites may produce nonproductive mRNAs with a premature stop codon or no in-frame stop codon that will be targeted by the translation-dependant surveillance mechanisms NMD (Nonsense-Mediated Decay) or NSD (No stop decay) respectively. The steady-state levels of these nonproductive RNA isoforms could thereby be biased, invalidating the quantification of the different RNA isoforms.
4. The quality of DNA is very important in order to obtain reproducible results and we routinely inject supercoiled DNA purified via Qiagen columns and diluted in water at the appropriate concentration. However it is very important to regularly monitor DNA quality since we have observed progressive misregulation when nicked DNA accumulates in the injected DNA. To overcome this problem one can store DNA either at 4°C to avoid freeze/thaw cycles that nick DNA or as aliquots at −20°C.
5. The stage at which the embryos are fixed is dependent of the tissue specificity, the strength and the activation timing of the promoter that drives the minigene expression, the major concern being to accumulate enough RNA for the RT-PCR analysis. For the cardiac actin and keratin promoters that begin expression at stage 12 (36) and 15 (37) respectively we fix the embryos at stage 26.
6. mRNAs produced from the minigene are specifically amplified using a forward primer that spans the junction between the promoter sequence and the 5’ constitutive exon or a reverse primer that spans the junction between the 3’ constitutive exon and the plasmid poly(A) signal (see Fig. 1).
7. Depending on the splicing pattern the results can be expressed as the percentage of each mRNA isoforms or as the ratio between two of them. While the second representation can emphasize small changes in splicing, we favor the first one in

which it is possible to estimate the relative amount of the different isoforms.

8. When studying the regulation of alternative terminal exons, there is no downstream constitutive exon. In this case we prime with an anchored oligo d(T) primer (5'-VTTTTTTTTTTTT TTTTCAGCTGTAGCTATGCGCACAG-3') and then amplification is performed with a PCR reverse anchor primer (5'-CAGCTGTAGCTATGCGCACAG-3') and a specific forward primer.
9. The analysis of mRNA isoforms containing mutually exclusive exons of the same size requires to preliminary discriminate them by using restriction enzyme that cleaves specifically in each exon. The digested products can then be analyzed by electrophoresis.
10. Injection of the minigene with exogenous mRNA in only one blastomere of a two-cell stage embryo is generally sufficient since mRNA will be translated in all cells expressing the minigene. It has the advantage that the noninjected blastomere may partially compensate the developmental defects that can occur in the injected blastomere upon the misexpression of an essential factor allowing the time window for analysis to be expanded.
11. *X. laevis* is a pseudotetraploid species meaning that very often two paralogs with small differences in their sequences are expressed during embryogenesis. It is therefore important to take this feature into consideration when designing AMOs as these must be effective against both paralogs. In general a close inspection of EST databases is sufficient to enable the conception of a single AMO that target all the expressed copies of a gene. If no common 25 nucleotides sequence are identified, then two different AMOs are required.
12. Injection of the minigene with AMOs in only one blastomere of a two-cell stage embryo is sufficient since translation will be inhibited in all cells expressing the minigene. It has the advantage that the noninjected blastomere may partially compensate the developmental defect that can occur in the injected blastomere upon the knockdown of an essential factor allowing the time window for analysis to be expanded. However, if specific antibodies are available the efficiency of the knockdown must be verified by injecting AMOs in both blastomeres. The amount of microinjected AMO that will generate an efficient knockdown must be empirically tested for each AMO. We are generally successful with amount ranging between 15 and 25 ng per blastomere. Depending on the morpholino efficiency and protein stability we observed 4–10-fold decrease in the protein levels.

13. Even with the 5' and 3' untranslated sequence of the β-globin mRNA we observed strong differences in the intrinsic translational efficiency of different chimeric mRNAs.
14. Since the quality of oocytes may vary between individual females, it is necessary to adjust the spin speed to each batch of oocytes. Practically, we test the initial speed (600 × *g* for 15 min) with 5 oocytes and depending on the result we reduce or increase the speed (between 400 and 800 × *g*).
15. Depending of the batch of embryos, approximately 30–60% of the embryos injected into both blastomeres will survive. We therefore consider that 50–100 injected embryos is a good start to obtain enough embryos for further biochemical procedures and phenotype descriptions.
16. AMO are diluted in sterile Milli-Q water from a 10 mg/mL stock solution. The appropriate amount of microinjected AMO that will generate a biological activity without causing nonspecific effects must be empirically tested for each AMO. We had success with amounts ranging between 5 and 15 ng per blastomere. We therefore recommend to test 5, 10, and 15 ng per blastomere and then to use the lowest possible amount that elicit a splicing switch.
17. To maximize the efficiency of AMOs several aspects must be taken in consideration. It is first required to test whether the targeted sequence is accessible or present within a secondary structure. As the thermodynamic strength of binding of an AMO to its target correlates strongly with its biological activity, it is also interesting to test *in silico* the theoretical binding strength of candidate AMOs using appropriate software (38). Despite these considerations the design of an efficient AMO remains essentially empirical and therefore to increase success we recommend designing two different oligomers for each targeted element.
18. Around 40–60% of the embryos generally die through necrosis. Since the necrotic embryos may harm their neighbors, it is important to remove them as soon as they are observed. We also keep embryos at a low density with no more than 20 embryos in a petri dish of 9 cm diameter, especially for over-night incubations.

Acknowledgments

Many thanks to Beverley Osborne for critical comments on the manuscript.

References

1. Pan Q, Shai O, Lee LJ, Frey BJ, Blencowe BJ (2008) Deep surveying of alternative splicing complexity in the human transcriptome by high-throughput sequencing. Nat Genet 40: 1413–1415
2. Nilsen TW, Graveley BR (2010) Expansion of the eukaryotic proteome by alternative splicing. Nature 463:457–463
3. Licatalosi DD, Darnell RB (2010) RNA processing and its regulation: global insights into biological networks. Nat Rev Genet 11: 75–87
4. McGlincy NJ, Smith CW (2008) Alternative splicing resulting in nonsense-mediated mRNA decay: what is the meaning of nonsense? Trends Biochem Sci 33:385–393
5. Hallegger M, Llorian M, Smith CW (2010) Alternative splicing: global insights. FEBS J 277:856–866
6. Mereau A, Le Sommer C, Lerivray H, Lesimple M, Hardy S (2007) *Xenopus* as a model to study alternative splicing *in vivo*. Biol Cell 99:55–65
7. Cooper TA (2005) Use of minigene systems to dissect alternative splicing elements. Methods 37:331–340
8. Hamon S, Le Sommer C, Mereau A, Allo MR, Hardy S (2004) Polypyrimidine tract-binding protein is involved *in vivo* in repression of a composite internal/3' -terminal exon of the *Xenopus* alpha-tropomyosin Pre-mRNA. J Biol Chem 279:22166–22175
9. Anquetil V, Le Sommer C, Mereau A, Hamon S, Lerivray H, Hardy S (2009) Polypyrimidine tract binding protein prevents activity of an intronic regulatory element that promotes usage of a composite 3'-terminal exon. J Biol Chem 284:32370–32383
10. Duriez P, Lesimple M, Allo MR, Hardy S (2000) Alternative splicing of *Xenopus* α_{fast}-tropomyosin pre-mRNA during development: identification of determining sequences. DNA Cell Biol 19:365–376
11. Asashima M., Ito, Y., Chan T., Michiue T., Nakanishi M., Suzuki K., Hitachi K., Okabayashi K., Kondow A. and Ariizumi T. *In vitro* organogenesis from undifferentiated cells in *Xenopus*. *Dev Dyn*, **238**, 1309–1320.
12. Forbes DJ, Kirschner MW, Newport JW (1983) Spontaneous formation of nucleus-like structures around bacteriophage DNA microinjected into *Xenopus* eggs. Cell 34:13–23
13. Etkin LD, Pearman B (1987) Distribution, expression and germ line transmission of exogenous DNA sequences following microinjection into *Xenopus laevis* eggs. Development 99:15–23
14. Kroll KL, Amaya E (1996) Transgenic *Xenopus* embryos from sperm nuclear transplantations reveal FGF signaling requirements during gastrulation. Development 122:3173–3183
15. Gurdon JB, Melton DA (1981) Gene transfer in amphibian eggs and oocytes. Annu Rev Genet 15:189–218
16. Gooding C, Clark F, Wollerton MC, Grellscheid SN, Groom H, Smith CW (2006) A class of human exons with predicted distant branch points revealed by analysis of AG dinucleotide exclusion zones. Genome Biol 7:R1
17. Southby J, Gooding C, Smith CW (1999) Polypyrimidine tract binding protein functions as a repressor to regulate alternative splicing of alpha-actinin mutally exclusive exons. Mol Cell Biol 19:2699–2711
18. Wollerton MC, Gooding C, Wagner EJ, Garcia-Blanco MA, Smith CW (2004) Autoregulation of polypyrimidine tract binding protein by alternative splicing leading to nonsense-mediated decay. Mol Cell 13:91–100
19. Corvelo A, Hallegger M, Smith CW, Eyras E (2010) Genome-wide association between branch point properties and alternative splicing. PLoS Comput Biol 6:e1001016
20. Gao K, Masuda A, Matsuura T, Ohno K (2008) Human branch point consensus sequence is yUnAy. Nucleic Acids Res 36:2257–2267
21. Hopwood ND, Gurdon JB (1990) Activation of muscle genes without myogenesis by ectopic expression of MyoD in frog embryo cells. Nature 347:197–200
22. Krieg PA, Melton DA (1984) Functional messenger RNAs are produced by SP6 in vitro transcription of cloned cDNAs. Nucleic Acids Res 12:7057–7070
23. Green MR, Maniatis T, Melton DA (1983) Human beta-globin pre-mRNA synthesized *in vitro* is accurately spliced in *Xenopus* oocyte nuclei. Cell 32:681–694
24. Pan ZQ, Prives C (1988) Assembly of functional U1 and U2 human-amphibian hybrid snRNPs in *Xenopus laevis* oocytes. Science 241:1328–1331
25. Ge J, Liu H, Yu YT (2010) Regulation of pre-mRNA splicing in *Xenopus oocytes* by targeted 2'-O-methylation. RNA 16:1078–1085
26. Le Sommer C, Lesimple M, Mereau A, Menoret S, Allo MR, Hardy S (2005) PTB regulates the processing of a 3'-terminal exon by repressing both splicing and polyadenylation. Mol Cell Biol 25:9595–9607

27. Singh R, Valcarcel J (2005) Building specificity with nonspecific RNA-binding proteins. Nat Struct Mol Biol 12:645–653
28. Kornblihtt AR (2005) Promoter usage and alternative splicing. Curr Opin Cell Biol 17: 262–268
29. Luco RF, Allo M, Schor IE, Kornblihtt AR, Misteli T (2011) Epigenetics in alternative pre-mRNA splicing. Cell 144:16–26
30. Bruno IG, Jin W, Cote GJ (2004) Correction of aberrant FGFR1 alternative RNA splicing through targeting of intronic regulatory elements. Hum Mol Genet 13:2409–2420
31. Hua Y, Vickers TA, Okunola HL, Bennett CF, Krainer AR (2008) Antisense masking of an hnRNP A1/A2 intronic splicing silencer corrects SMN2 splicing in transgenic mice. Am J Hum Genet 82:834–848
32. Heasman J (2002) Morpholino oligos: making sense of antisense? Dev Biol 243:209–214
33. Morcos PA (2007) Achieving targeted and quantifiable alteration of mRNA splicing with Morpholino oligos. Biochem Biophys Res Commun 358:521–527
34. Cleaver O, Krieg PA (1999) Expression from DNA injected into *Xenopus* embryos. Methods Mol Biol 127:133–153
35. Ryner LC, Baker BS (1991) Regulation of doublesex pre-mRNA processing occurs by 3'-splice site activation. Genes Dev 5: 2071–2085
36. Mohun TJ, Garrett N, Gurdon JB (1986) Upstream sequences required for tissue-specific activation of the cardiac actin gene in *Xenopus laevis* embryos. EMBO J 5:3185–3193
37. Jonas EA, Snape AM, Sargent TD (1989) Transcriptional regulation of a Xenopus embryonic epidermal keratin gene. Development 106:399–405
38. Popplewell LJ, Trollet C, Dickson G, Graham IR (2009) Design of phosphorodiamidate morpholino oligomers (PMOs) for the induction of exon skipping of the human DMD gene. Mol Ther 17:554–561

Chapter 21

Immunoisolation of Protein Complexes from *Xenopus*

Frank L. Conlon, Yana Miteva, Erin Kaltenbrun, Lauren Waldron, Todd M. Greco, and Ileana M. Cristea

Abstract

The immunoaffinity isolation of protein complexes is an essential technique for the purification and concentration of protein complexes from cells and tissues. In this chapter we present the methodologies for the purification of proteins and protein complexes from *Xenopus laevis* and *Xenopus tropicalis*. Specific to this protocol are the techniques for the cryolysis of *Xenopus* cells and tissues, a procedure that limits contamination from yolk proteins while preserving endogenous protein complexes, the methodologies for immunoaffinity purification of proteins using magnetic beads, and the protocols for western blot analysis. In addition, the procedures in this chapter can be extended to use with proteomic analysis of protein complexes as presented in the following chapter.

Key words: *Xenopus*, Immunoaffinity purification, Immunoprecipitation, Immunoisolation, Cryolysis, Tissue lysis, Protein extraction, Protein complex, Protein interactions, Western blot analysis

1. Introduction

It is becoming increasingly clear that many forms of human disease are associated with defects in genes that are required for early steps in embryonic development. Moreover, the molecular and cellular pathways through which these genes function can be elucidated using established model systems such as the African clawed frog, *Xenopus*. *Xenopus* has numerous advantages as a model system in which to identify and characterize cellular and developmental processes particularly in regards to proteomic-based approaches. Most critically, unlike the mouse, the *Xenopus* embryo develops externally and the embryo is relatively large and is amenable to surgical manipulations, allowing defined regions to be excised and cultured in simple salt solutions. These classical approaches are complemented

Stefan Hoppler and Peter D. Vize (eds.), *Xenopus Protocols: Post-Genomic Approaches, Second Edition*,
Methods in Molecular Biology, vol. 917, DOI 10.1007/978-1-61779-992-1_21, © Springer Science+Business Media, LLC 2012

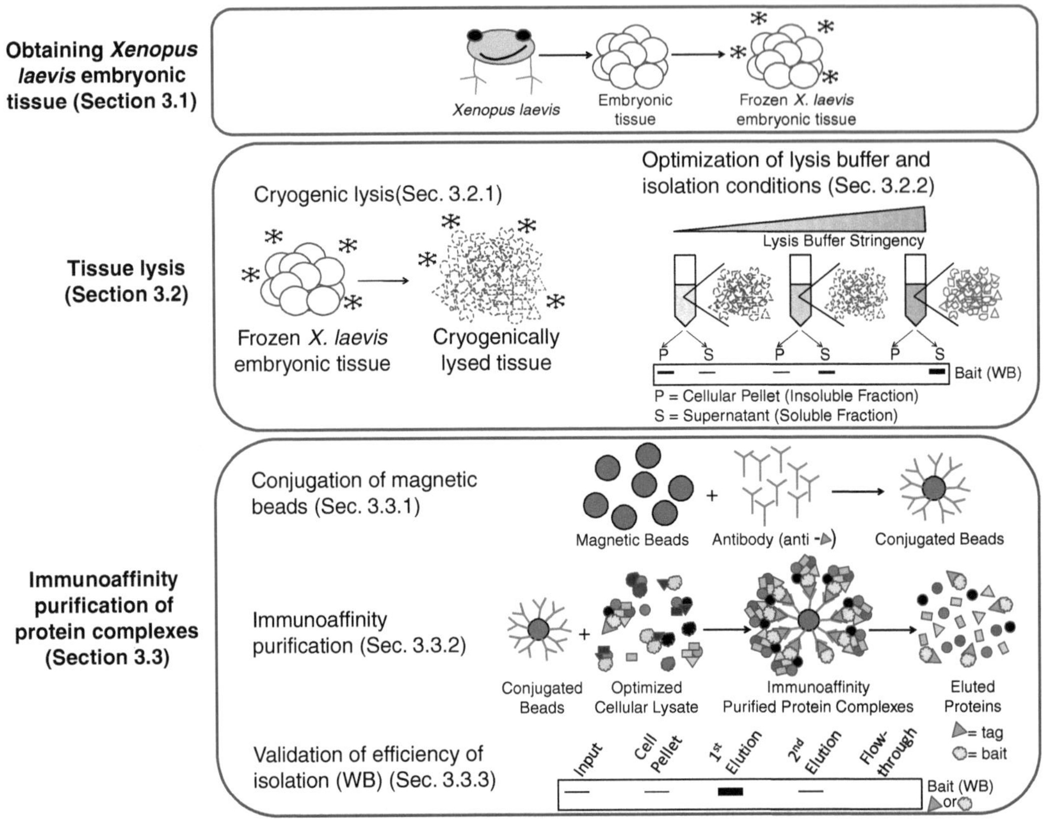

Fig. 1. Immunoisolation of protein complexes from *Xenopus laevis*.

by molecular techniques that allow the ectopic expression, overexpression, or knock-down of specific gene transcripts in the early embryo and transgenic technologies.

Complementary to these approaches are emerging biochemical approaches. In this regard, *Xenopus* offers a unique model system for the identification and characterization of protein complexes *in vivo*. However, the use of these approaches has been limited due to the lack of optimized protocols for isolation of early stage *Xenopus* tissues and the large abundance of yolk proteins. As shown in Fig. 1, this chapter describes methods for conducting immunoprecipitation of endogenous protein complexes in *Xenopus laevis* and *Xenopus tropicalis* which combines the cryogenic lysis of tissues with immunoisolation on magnetic beads. An overview of the approach is shown in Fig. 1. Collectively, these approaches function to preserve endogenous protein complexes, limit problems associated with yolk platelets, and provide a specific isolation of a given protein.

2. Methods and Equipment

2.1. Obtaining Xenopus laevis Embryonic Tissue

1. Fine watchmaker's forceps such as Dumont number 5 forceps.
2. *X. laevis* embryos cultured to desired stage of development (1)
3. 10× Modified Barth's Saline (MBS), pH 7.8: 880 mM NaCl, 10 mM KCl, 10 mM $MgSO_4$, 50 mM HEPES pH 7.8, 25 mM $NaHCO_3$. 1× MBS is made by mixing 100 mL of 10× stock solution with 700 μL 1 M $CaCl_2$ and adjusting the volume to 1 L with dH_2O. Store at room temperature.
4. 1% agarose plates for dissections: Weigh 1 g agarose and transfer to 250 mL Erlenmeyer flask containing 100 mL dH_2O. Heat flask in microwave until agarose has completely dissolved. Cool molten agarose until cool enough to hold flask. Pour a layer of agarose into small plastic Petri dishes (5 cm). Allow agarose to set. Store plates at 4°C.
5. Plastic transfer pipettes.
6. Liquid nitrogen.
7. Syringe needle (19G1½).
8. 50 mL conical tubes.
9. A dissecting microscope (e.g., Leica MZ6).

2.2. Tissue Lysis and Protein Extraction

1. Retsch MM 301 Mixer Mill with 2×25 mL jars and 2×20 mm (tungsten carbide or stainless steel) grinding balls (Retsch, Newtown, PA).
2. Liquid nitrogen, Styrofoam container, and a pair of long forceps.
3. Windex.
4. Methanol.
5. 50 mL conical tubes.
6. Ultrapure water.

2.3. Immunoaffinity Purification of Protein Complexes

Unless otherwise stated all solutions can be stored at room temperature

2.3.1. Conjugation of Magnetic Beads

1. Dynabeads M-270 Epoxy (Invitrogen). Store at 4°C.
2. Affinity purified antibodies against a protein of interest or tag (e.g., anti-GFP antibodies as shown below for the isolation of GFP-tagged proteins) or Immunoglobulin G (for isolation of Protein A-tagged proteins). Store at −80°C.
3. 0.1 M Sodium Phosphate buffer, pH 7.4: Prepare as 19 mM NaH_2PO_4, 81 mM Na_2HPO_4 in water and adjust pH to 7.4, if necessary. Filter sterilize (0.2 μm filter (Millipore)). Store at 4°C.

4. 3 M Ammonium Sulfate: Prepare in 0.1 M Sodium Phosphate buffer, pH 7.4. Filter sterilize (0.2 μm filter (Millipore)).
5. 100 mM Glycine-HCl, pH 2.5: Prepare in water. Adjust pH to 2.5 with HCl. Filter sterilize (0.2 μm filter (Millipore)). Store at 4°C.
6. 10 mM Tris, pH 8.8: Prepare in water. Adjust pH to 8.8 with HCl. Filter sterilize (0.2 μm filter (Millipore)).
7. 100 mM Triethylamine: Prepare fresh in water. CAUTION: toxic and extremely flammable. Must handle in a chemical hood and dispose of appropriately.
8. DPBS, pH 7.4 (Dulbecco's Phosphate-Buffered Saline (1×), liquid), (Invitrogen): Store at 4°C.
9. 0.5% Triton X-100: Prepare in DPBS. Store at 4°C.
10. 0.02% Sodium azide (NaN_3): Prepare in DPBS. Store at 4°C. CAUTION: NaN_3 is a toxic solid compound. Must handle in a chemical hood and dispose of appropriately.
11. Rotator (at 30°C).
12. Magnetic separation tube rack (Invitrogen).
13. Tube shaker (Tomy shaker).
14. Round bottom 2 mL Safe-Lock tubes (Eppendorf).
15. Ultrapure water (e.g., from a Milli-Q Integral Water Purification System).

2.3.2. Immunoaffinity Purification

1. Frozen tissue powder (see Subheading 3.1). Store at −80°C
2. Optimized lysis buffer (see Subheading 3.2) prepared fresh prior to each experiment. Store on ice.
3. Magnetic beads conjugated with antibodies (see Subheading 3.3). Store at 4°C.
4. 50 mL conical tubes.
5. Polytron for tissue homogenization (e.g., PT 10-35 Polytron from Kinematica).
6. Centrifuge and rotor, compatible with 50 mL conical tubes and capable of 8,000 × *g* at 4°C.
7. Tube rotator at 4°C.
8. Ultrapure dH_2O.
9. Round bottom eppendorf tubes (Fisher).
10. Axygen Maxymum Recovery microcentrifuge tubes, 1.5 mL (VWR).
11. Bar magnets (for conical tubes) and magnetic separation rack (for eppendorf tubes) (Invitrogen).
12. Ammonium hydroxide, 14.8 M (Sigma). Store at 4°C.

13. Base elution buffer: Mix 4.826 mL of ultrapure H_2O, 5 μL of 0.5 M EDTA, pH 8.0, and 169 μL of ammonium hydroxide. Prepare fresh before use.
14. 4× LDS elution buffer: Dissolve 0.666 g of Tris–HCl, 0.682 g of Tris-Base, 0.8 g of LDS, and 0.006 g of EDTA (free acid) in ultrapure dH_2O to a final volume of 10 mL. Aliquot and store at −20°C.
15. 10× Reducing Agent (Invitrogen). Store at 4°C.
16. 1 M iodoacetamide (IAA) (Sigma): Dissolve 0.185 g of iodoacetamide in 1 mL HPLC grade water. Dispense into 50 × 20 μL aliquots and store at −20°C.
17. Heat block at 70°C.

2.3.3. Assessment of Immunoaffinity Purification

Sample Preparation

1. Reserved fractions (from Subheading 3.3.2).
   (a) Cell pellet (step 7).
   (b) Input supernatant (step 7).
   (c) Flow-through (step 11).
   (d) Primary eluate (step 21).
   (e) Secondary eluate (step 19).
2. Ultrapure dH_2O.
3. Acetone (−20°C).
4. 1.7 mL eppendorf tubes (Fisher).
5. Microcentrifuge.
6. NuPAGE 4–12% Bis-Tris precast SDS-PAGE gel, 10 well (Invitrogen).
7. Xcell SureLock Mini-Cell electrophoresis system (Invitrogen).
8. 20× NuPAGE MOPS SDS Running Buffer (Invitrogen).
9. 4× NuPAGE LDS Sample Buffer (Invitrogen).
10. 10× Reducing Agent (Invitrogen). Store at 4°C
11. Heat block at 70°C.

SDS-PAGE and Western Blot Analysis

1. Prepared fractions (from Subheading 3.3.3).
   (a) Cell pellet (step 1).
   (b) Input supernatant (step 2).
   (c) Flow-through (step 3).
   (d) Secondary eluate (step 4).
   (e) Primary eluate (step 5).
2. NuPAGE 4–12% Bis-Tris gel, 10 well (Invitrogen).
3. Xcell SureLock Mini-Cell electrophoresis system (Invitrogen).
4. 20× NuPAGE MOPS SDS Running Buffer (Invitrogen).

5. 1× Running Buffer: Dilute 20× NuPAGE MOPS SDS Running Buffer in 700 mL of ultrapure water.
6. Precision Plus Protein Dual Color Molecular Weight Standards (BioRad).
7. 4× NuPAGE LDS Sample Buffer (Invitrogen).
8. 10× Reducing Reagent (Invitrogen). Store at 4°C.
9. PVDF membrane (BioRad).
10. Methanol.
11. Transfer apparatus (e.g., Mini Trans-blot Cell from BioRad).
12. 10× Transfer Buffer: Dissolve 144 g glycine and 30.3 g Tris base in final volume of 1 L dH_2O. Prepare 1 L of 1× Transfer Buffer containing 20% methanol. Chill at 4°C for 30 min before use.
13. 2× Whatman filter paper and 2× sponges for transfer.
14. 20× TBST: 200 mM Tris–HCl pH 8, 3 M NaCl, 2% Tween-20 in dH_2O. Dilute to 1× with dH_2O for use.
15. Blocking Buffer: 5% nonfat dry milk powder in 1× TBST.
16. Appropriate primary and secondary antibodies, diluted in Blocking Buffer.
17. Autoradiography cassette (FisherBiotech Cat# FBCA 57).
18. ECL chemiluminescent substrate kit (Thermo Scientific).
19. Autoradiography film (Kodak).

3. Methods and Procedures

3.1. Obtaining Xenopus laevis Embryonic Tissue

Of all the proteins in *X. laevis* embryonic tissue, yolk proteins are among the most abundant, especially at earlier developmental stages when the embryo is still dependent on yolk for nutrients. The abundance of yolk proteins can be problematic when performing immunoaffinity purifications, as these proteins can nonspecifically react with antibodies and mask a less abundant interaction. For this reason, it is desirable to remove as much of the yolk from the embryo as possible.

1. Fill a 1% agarose plate with cold 1× MBS. Transfer *X. laevis* embryos to MBS in agarose plate.
2. Using fine forceps and a dissecting microscope, remove as much of the yolk as possible from the rest of the embryo. Using a plastic transfer pipette, transfer the embryo to a new agarose plate containing fresh 1× MBS. Keep tissue on ice until all dissections are completed. Collect appropriate number of embryos for each immunopurification to be performed (see Note 1).

3. Using a syringe needle, poke four holes in the cap of a 50 mL conical tube. Remove cap and secure tube into a rack in a styrofoam cooler. Fill cooler and tube with liquid nitrogen.
4. Using a plastic transfer pipette, drop embryos one by one into liquid nitrogen in conical tube. When finished, replace the cap and screw on tightly. Remove the tube from the cooler (using a paper towel for protection) and invert to remove the liquid nitrogen. Store frozen tissue at −80°C.

3.2. Tissue Lysis and Protein Extraction

Tissue lysis can be carried out utilizing several approaches, including homogenization in a detergent-containing lysis buffer, passage through a needle (different needle gauges can be tested for efficiency of lysis), and cryogenic tissue disruption using traditional mortar and pestle or a Mixer Mill. While the procedures described below for immunoaffinity purification of protein complexes utilize as starting material tissue disrupted cryogenically using a Mixer Mill, the other types of tissue lysis can also be incorporated. We prefer the type of cryogenic disruption described below as it leads to an increased efficiency of extraction (i.e., isolation of the targeted protein) and decreased level of nonspecific associations. This method has provided us with a reliable and effective means of cell lysis for isolating varied protein complexes (2–7). In circumstances that require a mild tissue lysis, such as the maintenance of intact organelles or large structures, e.g., postsynaptic densities (8), cryogenic disruption may not be the method of choice.

3.2.1. Cryogenic Tissue Disruption

1. Clean one spatula, the Retsch Mixer Mill jars, and the grinding balls sequentially with ultrapure dH_2O, Windex, ultrapure dH_2O, and 100% methanol. Allow all parts to dry completely in a chemical hood.
2. Cool the jars and balls in liquid nitrogen (e.g., using a Styrofoam container filled with liquid nitrogen). Once cooled (i.e., liquid nitrogen no longer appears to be bubbling) remove them from the liquid nitrogen container using a pair of long forceps and place the frozen tissue into the jar. The tissue can fill up to a maximum of one-third of the total volume of the jar for optimal cryogenic grinding (e.g., ~7 g frozen tissue pellets per 25 mL jar). Add the chilled ball on top of the tissue (use one ball per jar), close the jar, and place it back into the liquid nitrogen container to cool.
3. Place the filled jars in the Retsch Mixer Mill holders. If only one jar contains frozen tissue for grinding, then use the other empty jar (without a ball) as a balance. Grind the tissue using 20 cycles of 2 min 30 s each at a frequency of 30 Hz. Place the jars in liquid nitrogen in between cycles to cool and ensure that the jars are still tightly closed.

4. Open the jar and use a chilled spatula to transfer the frozen tissue powder to a 50 mL conical tube kept on dry ice. Work as quickly as possible to avoid thawing of the ground sample. Periodically chill the spatula in liquid nitrogen. Store the powder at −80°C until immunopurification is to be performed.

3.2.2. Optimization of Lysis Buffer and Isolation Conditions

Successful isolation of a protein of interest and its interacting partners is dependent on several criteria including protein abundance and subcellular localization, sample amount, affinity of the antibody used for immunoaffinity purification, efficiency of bead conjugation, and lysis buffer conditions for immunoaffinity purification. During the cell lysis and protein isolation steps it is crucial to extract and preserve the targeted protein with its interactions in a soluble fraction. Therefore, the lysis buffer conditions utilized prior and during the affinity purification have to be optimized for each protein of interest before proceeding with larger scale immunoaffinity purifications for proteomics studies. This can be done by performing small-scale experiments (i.e., 20 embryos per immunopurification) that use western blotting to assess (1) the efficiency of protein solubilization (see procedure below) and (2) efficiency of isolation (see Subheading 3.3.3). It is recommended to compare at least three lysis buffer conditions with varied levels of stringency. Generally, the stringency of a lysis buffer is determined by the concentrations and combinations of detergents and salts. Table 1 provides examples of frequently used detergents, and Table 2 lists several lysis buffers that differ slightly in their compositions and were successfully utilized in immunoaffinity purifications of protein complexes from varied species.

1. Split cryogenically ground tissue into equal small aliquots (e.g., 0.1 g) (see Note 2). Ensure that the tissue powder does not thaw during the weighing.
2. Place the small aliquots on ice (4°C) and add a different lysis buffer (5 mL buffer per 1 g cells) to each sample.
3. Homogenize the tissue powder in the buffer by vortexing for 1 min with intermittent cooling. This step is different than the usual homogenization for immunoaffinity purifications (see Subheading 3.3.2), which uses a polytron and a larger volume for the starting material.
4. Separate the soluble and insoluble fractions by centrifugation at 8,000 × *g* at 4°C for 10 min. Recover soluble fraction and label "supernatant."
5. Wash the pellet in water and discard supernatant. Extract pellet by sequential sonication then boiling at 95°C for 5 min in 50 mM Tris–HCl, pH 7.4, containing 2% SDS. Centrifuge at 20,000 × *g* for 10 min. Recover supernatant and label "Pellet."

Table 1
Examples of detergents commonly used for cell lysis and their properties

Detergent	Properties	Notes
Triton X-100	Nonionic detergent pH 6.0–8.0 (5% aqueous solution) Critical micelle concentration (CMC): 0.22–24 mM	Depending on the utilized concentration, Triton is considered a relatively mild, nondenaturing detergent. Many enzymes remain active in 0.1–0.5% Triton X-100 solution (e.g., Proteinase K is still active in 1% solution)
	Soluble at 25°C in all proportions; soluble in water, benzene, toluene, xylene, trichloroethylene, ethylene glycol, ethyl ether, ethanol, isopropanol, ethylene dichloride	Can be used to preserve protein–lipid interactions
Sodium deoxycholate (DOC)	Anionic detergent pH 5.0–9.0 (1% aqueous solution) CMC: 2–6 mM (0.083–0.249%, w/v). Micelle molecular weight: 2,000 g (average), at concentrations above 2 mM Soluble at 20°C; soluble in water in less than 5% solution	Common component of RIPA lysis buffer Suitable for isolating membrane-associated proteins and liposome preparation. Disrupts protein–lipid interactions
Digitonin	Nonionic detergent pH: data not available CMC: <0.5 mM, at 20–25°C. Micelle molecular weight: 70,000 g (average) Soluble in water at ~5% (w/v); must be heated to 95–98°C first, then cooled to room temperature. Soluble in ethanol at 10 mg/mL	Suitable for analyzing membrane-bound proteins and solubilizing lipids
Octyl-beta-glucoside	Nonionic detergent pH: data not available CMC: 23–25 mM (0.6716–0.7300%, w/v); Micelle Molecular Weight: 8,000 g Soluble in water	Suitable for studying membrane-associated proteins Readily integrated with mass spectrometry studies (i.e.., does not interfere as much as other detergents with ionization in MALDI MS experiments)
Nonidet P-40	Nonionic detergent pH 5.0–8.0 (5% aqueous solution) CMC: 0.059 mM (20–25 C) Soluble in water	Milder alternative to Triton X-100; depending on the concentration it may not penetrate nuclear membranes

Table 2
Examples of lysis buffers used for immunoaffinity purification of protein complexes

Bait (GFP-tagged)	Description and localization	Species and sample type	Optimized lysis buffer	References
Apl1	beta-adaptin of AP-2 complex; cellular membrane	*S. cerevisiae* cells	20 mM K-HEPES, pH 7.4, 110 mM KOAc, 2 mM $MgCl_2$, 0.1% Tween, 0.4% Triton, 200 mM NaCl, 1/100 (v/v) protease inhibitor mixture (20 mg/mL PMSF + 0.4 mg/mL pepstatin A), and 1/200 (v/v) protease inhibitor cocktail	(2)
Nup37	Member of nuclear pore complex; subunit of Nup107-160 subcomplex	*Homo sapiens* cells	20 mM K-HEPES, pH 7.4, 110 mM KOAc, 2 mM $MgCl_2$, 0.1% Tween, 0.5% Triton, 200 mM NaCl, 1/100 (v/v) protease inhibitor mixture (20 mg/mL PMSF + 0.4 mg/mL pepstatin A), and 1/200 (v/v) protease inhibitor cocktail	(2)
HDAC5	Histone deacetylase 5; nucleus and cytoplasm	*Homo sapiens* cells	20 mM HEPES-KOH, pH 7.4, 0.1 M potassium acetate, 2 mM $MgCl_2$, 0.1% Tween-20, 1 μM $ZnCl_2$, 1 μM $CaCl_2$, 0.5% Triton X-100, 250 mM NaCl, 4 μg/mL DNase, 1/100 (v/v) protease, and phosphatase inhibitor cocktails	(7)
PAP I	poly(A) polymerase I; cytoplasm and inner membrane	*Escherichia coli* cells	20 mM HEPES, pH 7.4, 0.11 M KOAc, 2 mM $MgCl_2$, 0.1% Tween-20 (v/v), 1 μM $ZnCl_2$, 1 μM $CaCl_2$, 1% Triton X-100, 0.5% Deoxycholate, 150 mM NaCl, 1/100 protease inhibitor cocktail, 1/200 phenylmethylsulphonyl fluoride	(6)
H3	Histone 3 isoforms; nucleus	*Mus musculus* ES cells	20 mM K-HEPES, pH 7.4, 110 mM K-acetate, 0.1% Tween 20, 0.5% Triton, 300 mM NaCl, and 1/100 (v/v) protease inhibitor cocktail	(5)

(continued)

Table 2 (continued)

Bait (GFP-tagged)	Description and localization	Species and sample type	Optimized lysis buffer	References
nsP3	Sindbis nonstructural protein 3; cytoplasm	Sindbis-infected *Rattus Norvegicus* cells	20 mM K-HEPES, pH 7.4, 110 mM KOAc, 2 mm $MgCl_2$, 0.1% Tween 20, 1% Triton, 0.5% deoxycholate, 500 mM NaCl, 25 units/mL DNase, 1/100 (v/v) protease inhibitor mixture	(3)
PSD (via VGluRδ2)	postsynaptic densities; cerebellar excitatory synapses	*Mus musculus* tissue	10 mM HEPES, pH 7.4, 2 mM $CaCl_2$, 132 mM NaCl, 3 mM KCl, 2 mM $MgSO_4$, 1.2 mM NaH_2PO_4, 0.5% Triton X-100, 1/100 (v/v/) protease inhibitor cocktail	(8)

6. Assess the levels of bait protein in the supernatant and pellet fractions (5–10% aliquots) using western blotting and antibodies against either the affinity tag (if present) or the endogenous protein.
7. To proceed with a larger scale immunoaffinity purification for proteomics analyses, select the lysis buffer condition that provides the highest proportion of bait protein in the soluble fraction, at the lowest necessary stringency. This will allow for a balance between an efficient extraction of the protein of interest and maintenance of interacting partners (see Note 3).

3.3. Immunoaffinity Purification of Protein Complexes

3.3.1. Conjugation of Magnetic Beads

This protocol has been optimized for conjugation of Dynabeads M-270 Epoxy; however, it can also be applied for conjugation of other types of magnetic beads with larger or smaller diameters. In such cases, it is important to adjust the amount of antibody used for conjugation, depending on the bead capacity of binding. This protocol can be utilized for conjugating beads with high-affinity purified in-house developed antibodies as well as commercially available ones, provided their storage buffer doesn't hinder covalent conjugation to Epoxy.

It is best to start this protocol in the late afternoon and perform all washing steps (steps 7–9) in the morning of the following day. Unless otherwise stated all steps should be performed at room temperature. Do not allow the beads to dry out (i.e., do not keep the beads without a washing solution in between the steps).

1. Weigh out the necessary amount of magnetic Dynabeads in a round-bottom tube.
   (a) Round-bottom tubes are preferred to avoid the trapping of beads in conical-shape tubes during the conjugation.
   (b) The necessary amount of beads is dependent on the purpose of the experiment and the abundance of the protein that will be immunoaffinity purified. An approximate guidance: 1–2 mg beads are appropriate for small-scale optimization experiments (as described in Subheading 3.2), 5–7 mg beads are usually sufficient for performing single immunoaffinity purifications, and 10–20 mg beads are suitable for highly abundant proteins.
2. Add 1 mL Sodium Phosphate buffer pH 7.4 to the beads; mix by vortexing for 30 s, followed by 15 min on a tube shaker (vigorous setting).
3. Place the tube on a magnetic rack. After all the beads settle towards the magnetic side, discard the buffer.
4. Remove the tube from the rack. Add 1 mL Sodium Phosphate buffer pH 7.4. Mix by vortexing for 30 s and remove the buffer in the same manner as above.
5. Remove the tube from the rack. Add, in this order, the necessary amount of antibodies, Sodium Phosphate buffer, and Ammonium Sulfate solution.
   (a) The optimal total volume during the beads conjugation (that includes the antibody, Sodium Phosphate buffer, and Ammonium Sulfate solution) is ~20 μL/mg beads
   (b) As a guideline of amounts of antibodies or IgG that we routinely use: 8–10 μg Ab/mg beads for commercially available antibodies and IgG, and 3–5 μg Ab/mg beads for purified high-affinity antibodies (e.g., in-house developed anti-GFP antibodies).
   (c) The 3 M Ammonium Sulfate solution is added last and will be one-third of total volume to give a final concentration of 1 M.
   (d) For example, to conjugate 18 mg beads, use a total volume of 360 μL. For this, add 54 μg antibody to beads (if using 3 μg Ab/mg beads), then add 0.1 M Sodium Phosphate Buffer (volume 0.1 M Sodium Phosphate Buffer = 360 μL − volume of antibody used − 120 μL 3 M Ammonium Sulfate), then add 120 μL of 3 M Ammonium Sulfate.
6. Secure the tube with parafilm and rotate bead slurry overnight on rotator at 30°C.

7. The next morning, place the tube with bead slurry on a magnetic rack. Remove and reserve the supernatant to assess the efficiency of bead conjugation by SDS-PAGE (see Note 4).
8. Wash the beads sequentially with the following buffers. For each wash, gently resuspend the beads in 1 mL of the buffer and then place the tube on the magnet and remove buffer. "FAST" indicates that buffer should not be in contact with beads for longer than it takes to resuspend them:
   (a) 1 mL of Sodium Phosphate buffer.
   (b) 1 mL 100 mM Glycine-HCl, pH 2.5 (FAST).
   (c) 1 mL 10 mM Tris–HCl pH 8.8.
   (d) 1 mL 100 mM Triethylamine solution (FAST).
   (e) 4 × 1 mL DPBS.
   (f) 1 mL DPBS containing 0.5% Triton X-100. Leave the tube on a Tomy shaker (gentle setting) for 15 min.
   (g) 1 mL DPBS.
9. Beads can be used immediately or stored at 4°C in DPBS containing 0.02% NaN_3. For beads stored for future use, measure the final volume of the bead slurry (the bead size will contribute to the final volume) and make a note of the volume required for 1 mg beads as this will permit known aliquots of beads to be removed for multiple immunoaffinity purifications. Beads should be used within 2 weeks of conjugation. After 1 month of storage, their efficiency for isolation decreases by approximately 40%.

3.3.2. Immunoaffinity Purification

Basic Elution of Immunoisolates

It is important to prepare all necessary reagents beforehand. Carry out all procedures on ice unless otherwise noted. At several steps during the protocol aliquots of samples (indicated with RESERVE) are taken to assess bait protein extraction and isolation efficiency (see Subheading 3.3.3).

Day 1

1. Prepare appropriate volume of optimized lysis buffer as determined in Subheading 3.2.2. Precool to 4°C. Add protease inhibitors just before use. Prepare 10 mL of wash buffer per sample (used in steps 6, 12–15), which is usually identical in composition to the optimized lysis buffer, except protease and phosphatase inhibitors cocktails are not included.
2. Incubate the frozen tissue powder on ice for 1–2 min, but do not thaw. Proceed immediately to step 3.
3. Resuspend the frozen tissue powder in appropriate volume of lysis buffer by first adding a small amount of lysis buffer and swirling homogenate to solubilize pellet. Continue to add lysis

buffer and gently mixing by hand until tissue powder has been completely solubilized (see Note 5).

4. Run Polytron 10 s in ultrapure dH_2O to wash. Ensure that the tissue homogenate occupies ≤1/3 of the conical tube volume. Subject tissue lysates to Polytron homogenization for 2 × 15 s (speed = 22.5k), resting the sample on ice for a few minutes between homogenizations. If processing additional samples, rinse and run Polytron in ultrapure dH_2O to wash out excess lysate. When finished, perform a final rinse with methanol.
5. Centrifuge the lysate at 8,000 × *g* at 4°C for 10 min.
6. During centrifugation, place tube containing antibody-conjugated magnetic beads on a magnetic rack for 30–60 s. Discard the storage buffer and wash with 3 × 1 mL wash buffer by gently pipetting up and down to resuspend the beads. Do not vortex antibody-conjugated beads. Suspend beads in 100–200 μL of wash buffer.
7. Carefully pour the clarified lysates (supernatant) into new 50 mL conical tubes (see Note 6). RESERVE (i) the cell pellet and (ii) 40 μL of the input supernatant (see Subheading 3.3.3 for analysis).
8. Gently flick tube of antibody-conjugated beads to mix beads in solution. Pipette the appropriate amount of beads into the clarified lysates.
9. Rotate the lysates with beads on a rotator at 4°C for 1 h. Do not use longer incubation times as this promotes the accumulation of nonspecific binding and loss of weak interacting partners.
10. During incubation, prepare base elution buffer and 1× LDS elution buffer (see Note 7).
11. Attach a bar magnet to the lysates/bead suspension tube using a rubber band. Incubate on ice for 5 min. RESERVE the flow-through (unbound) fraction by pouring the supernatant into a clean conical tube (see Subheading 3.3.3 for analysis).
12. Resuspend the beads in 1 mL of wash buffer and transfer the bead slurry to a round-bottom eppendorf tube.
13. Place on a magnetic rack for 30 s to pellet the beads and discard wash buffer. *Perform this procedure between all subsequent wash steps.*
14. Wash the beads 3 × 1 mL wash buffer. On the third wash, transfer the bead slurry to a clean round-bottom eppendorf tube, then pellet beads and discard wash buffer.
15. Wash the beads 2 × 1 mL with wash buffer.
16. Add 1 mL DPBS to beads and transfer slurry to clean round-bottom eppendorf tube. Repeat wash once with 1 mL of DPBS

to remove residual detergent. Quantitatively remove DPBS wash.

17. Add 750 μL of base elution buffer. Incubate at RT for 20 min while shaking.
18. Place the tube on the magnetic rack and transfer the eluate to an Axygen microcentrifuge tube. Freeze primary eluate in liquid nitrogen and evaporate to dryness overnight by vacuum centrifugation.
19. Perform a second elution from the beads by suspending the beads in 40 μL of 1× LDS elution buffer containing 50 mM DTT, incubating at 70°C for 10 min, and then at RT for 10 min while shaking. Place the tube on a magnetic rack and transfer the eluate to a clean microcentrifuge tube. RESERVE 10% (4 μL) of the secondary eluate in a clean microcentrifuge tube (see Subheading 3.3.3 for analysis). Freeze remaining 90% of secondary eluate in liquid nitrogen and store at −20°C.
20. Proceed to Subheading 3.3.3 to prepare reserved samples for Western blot analysis. Continue with step 21 the following day.

Day 2

21. Remove dried eluate from the Speedvac.
    (a) If performing SDS-PAGE-in-gel digestion (see Note 8), suspend dried eluate in 40 μL of 1× NuPAGE sample buffer containing 1× reducing agent and heat at 70°C for 10 min. RESERVE 10% (4 μL) of the eluate in a clean microcentrifuge tube. Add 4 μL of 1 M iodoacetamide to remaining 90% of primary eluate and incubate 30 min at RT protected from light. Freeze in liquid nitrogen and store at −20°C or proceed immediately to proteomic analysis (see Chapter 22, Subheading 2.1).
    (b) If in-solution digestion is to be performed (see Note 8), suspend dried eluate in 40 μL of 1× LDS elution buffer containing 50 mM DTT and heat at 70°C for 10 min. RESERVE 10% (4 μL) of the eluate in a clean microcentrifuge tube. Freeze remaining 90% of primary eluate in liquid nitrogen and store at −20°C or proceed immediately to proteomic analysis (see Chapter 22, Subheading 2.2).

Alternate Procedure (Detergent Elution of Immunoisolates; See Note 7)

Day 1

1. Perform steps 1–16 as described earlier, except preparation of base elution buffer can be omitted (step 10).
2. Add 40 μL of 1× LDS elution buffer to beads. Incubate 10 min at 70°C, then 10 min at RT while shaking.

3. Pellet beads on magnetic rack and transfer primary eluate to an Axygen microcentrifuge tube.
4. Repeat steps 2 and 3, transferring secondary eluate to an Axygen microcentrifuge tube.
5. Add 2.0 μL of 1 M DTT to primary and secondary eluates. Heat at 70°C for 10 min. RESERVE 10% (4 μL) of eluates in clean microcentrifuge tubes.
6. If performing in-gel digestion (see Note 8), add 4 μL of 1 M iodoacetamide to remaining 90% of eluates and incubate at RT for 30 min protected from light. If performing in-solution digestion (see Note 8), proceed directly to step 7.
7. Freeze remaining 90% of eluates in liquid nitrogen and store at −20°C or proceed immediately to proteomic analysis (see Chapter 22, Subheadings 2.1 or 2.2).

3.3.3. Assessment of Immunoaffinity Purification

Sample Preparation

Reserved sample aliquots from Subheading 3.3.2 are prepared for SDS-PAGE and Western blot analysis. The recommended amount of each aliquot analyzed is provided as a starting point, and may need further optimization for differences in input material and bait protein abundance.

1. Cell pellet
   (a) Wash the cell pellet with 1.0 mL dH_2O. Homogenize washed cell pellet in 1.0 mL of 2% SDS, transfer to microcentrifuge tube, and heat at 70°C for 10 min. Centrifuge at maximum speed for 5 min at RT (see Note 9).
   (b) Remove aliquot of SDS-soluble pellet fraction that corresponds to an identical percent of the input supernatant. For example, assuming a 40 μL aliquot of input supernatant was reserved from a total lysis volume of 10 mL, a 4 μL aliquot of the pellet sample should be removed.
   (c) Dilute aliquot of SDS-soluble pellet fraction to a final volume of 60 μL containing 1× NuPAGE LDS Sample Buffer/1× Reducing Agent.
2. Input supernatant
   (a) Dilute 40 μL of reserved input supernatant to a final volume of 60 μL containing 1× NuPAGE LDS Sample Buffer/1× Reducing Agent.
3. Flow-through (see Note 10)
   (a) Transfer 10% of flow-through to clean tube. Slowly add 4 volumes of −20°C acetone. Vortex briefly. Incubate at −20°C for at least 1 h.
   (b) Centrifuge at 3,000 × *g* at 4°C for 10 min. Pour off supernatant.
   (c) Briefly wash pellet with 4 volumes of 80% acetone/20% dH_2O and discard.

(d) Air dry pellet for 5 min, then partially solubilize in 40 μL of 1× NuPAGE LDS Sampler Buffer/1× Reducing Agent by gentle agitation.

4. Secondary eluate

   (a) Dilute 4 μL (10%) of the secondary eluate into a final volume of 40 μL 1× NuPAGE LDS Sample Buffer/1× Reducing Agent.

5. Primary Eluate (prepared the following day, see Note 11)

   (a) Dilute 4 μL (10%) of the primary eluate into a final volume of 40 μL 1× NuPAGE LDS Sample Buffer/1× Reducing Agent.

6. Heat all samples at 70°C for 10 min. Freeze the samples at −20°C until ready to proceed with SDS-PAGE and Western Blot Analysis.

SDS-PAGE and Western Blot Analysis

In this protocol the efficiency of immunoaffinity isolation of the bait protein is assessed by comparing five samples reserved at progressing stages of the immunoaffinity purification procedure: (1) cell pellet, (2) input supernatant, (3) flow-through, (4) secondary eluate, and (5) primary eluate (see Subheading 3.3.2). While the amount of sample prepared (see Subheading "Sample Preparation") is sufficient for duplicates analyses, the protocol below details a single experiment. A representative western blot is shown below (Fig. 2a) indicating efficient isolation (no protein remaining in the

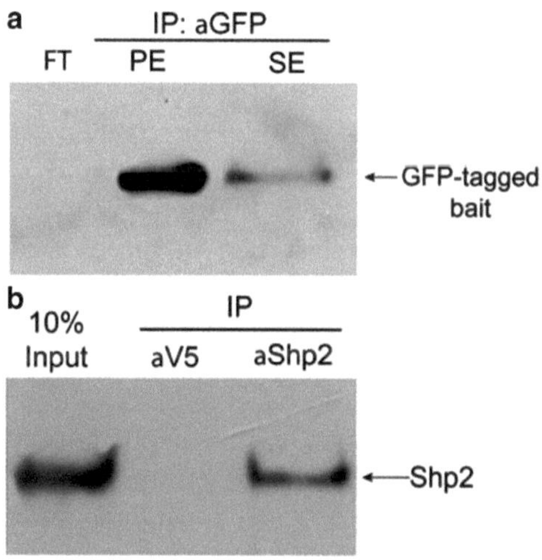

Fig. 2. Assessment of isolation efficiency and specificity of immunoaffinity purification. (**a**) A GFP-tagged bait protein was isolated (see Subheading 3.3.2) and eluted with the alternate detergent-based procedure. Ten percent of the following fractions are analyzed to assess the efficiency of isolation of a GFP-tagged bait protein: *FT* flow-through; *PE* primary eluate; *SE* secondary eluate. The majority of bait protein was found in the primary eluate. (**b**) Immunoaffinity purification of endogenous Shp2 protein from 100, stage 40 *Xenopus tropicalis* whole embryos. As a control for the specificity of the Shp2 antibody, a second immunopurification was performed in parallel using an anti-V5 antibody.

flow-through) and recovery of the intended GFP-tagged bait protein (a majority of the protein was in the primary eluate).

1. Set up the Xcell SureLock Mini-Cell electrophoresis system:
   (a) Remove the white strip and the upper comb from the NuPAGE 4–12% Bis-Tris precast SDS-PAGE gel and rinse wells with ultrapure dH_2O. Place the gel in the apparatus, using a buffer dam for the opposing side, then lock the assembly in place.
   (b) Fill the inner chamber with 200 mL and the outer chamber with 500 mL of 1× Running Buffer.
2. Thaw fractions if previously frozen and load into wells as follows: Lane 2, 10 μL Molecular Weight Standards; Lane 3, 30 μL of input supernatant; Lane 4, 30 μL of cell pellet; Lane 5, 20 μL of primary eluate; Lane 6, 20 μL of secondary eluate; Lane 8, 20 μL of flow-through; Empty lanes, 20 μL of 1× LDS Sample Buffer.
3. Electrophorese for 5 min at 100 V, then 45–50 min at 200 V, or until the dye front has migrated all the way down the gel.
4. While the gel is running, cut two pieces of Whatman filter paper and one PVDF membrane to the appropriate gel size. Always handle the membrane with tweezers.
5. Prewet PVDF in methanol and soak along with 2 transfer sponges, 2 filter papers, in precooled 1× Transfer Buffer at least 15 min at 4°C.
6. Open gel cassette to expose gel and discard the wells. Working with wet gloves, transfer gel (by the thick ridge at the bottom) into a plastic tray containing 1× Transfer buffer. Remove bottom ridge.
7. Assemble a WB "sandwich" in the tray with presoaked items. Layer them in a transfer cassette as follows, starting from the clear side of the cassette: sponge, filter paper, PVDF membrane, SDS-PAGE gel, filter paper (gently roll out any bubbles using a 15 mL conical tube), sponge. Close the sandwich and place into transfer apparatus with black side of cassette (gel side) facing black (anode) side transfer core. Place ice tray and stir bar into apparatus. Pour ice-cold 1× Transfer Buffer into the apparatus until it covers fully the cassette.
8. Transfer at 100 V for 1.5 h at 4°C while stirring.
9. Open transfer cassette and discard the filter papers and gel. Verify the prestained Molecular Weight Standards transferred to the membrane.
10. Wash PVDF membrane with 1× TBST, discard, then add Blocking Buffer and incubate for 1 h at RT on a rocking platform. Do not let the membrane dry.

11. Discard Blocking Buffer and add the primary antibody diluted in Blocking Buffer. Incubate for 1–2 h at RT or overnight at 4°C.
12. Wash the membrane 3×20 mL in 1× TBST for 5 min while rocking at RT.
13. Add the secondary horseradish peroxidase-conjugated antibody diluted in Blocking Buffer. Incubate for 1 h at RT.
14. Wash the membrane 4×20 mL in 1× TBST for 5 min while rocking at RT.
15. Mix ECL chemiluminescence substrates in 1:1 ratio (a total volume of 1 mL is usually sufficient to cover the membrane). Using tweezers, place the membrane on a dry plastic surface and apply the ECL solution, incubating for 1 min. Blot off the excess substrate and place the membrane between a sheet protector or Saran-wrap and tape into a film cassette.
16. Carefully place a piece of autoradiography film on top of the membrane and close the cassette avoiding any film shifts as this will result in smeared bands. An initial exposure of 30 s will indicate whether subsequent exposures of longer or shorter duration are required.

3.3.4. Appropriate Controls

To assess the specificity of immunoaffinity isolation it is critical to have a control sample analyzed in parallel, beginning from tissue lysis through the isolation of the target protein. The controls described below account for nonspecific binding of proteins to magnetic beads, antibody, and affinity tag (when present).

GFP Tagged

Often the bait protein is expressed as a GFP fusion protein. Here, parallel isolations of the GFP-tagged protein and GFP alone are highly desirable. If the GFP-tagged protein is stably overexpressed by mRNA injection into embryos, it is necessary to generate a GFP-only tissue/animal in the same manner as the GFP fusion protein, i.e. mRNA injection of a GFP-only expression construct. Moreover, all treatment conditions and experimental variables, such as lysis buffer composition, should be identical between the two isolations. Thus, both the tag alone and the fusion protein can be immunoaffinity purified using identical preparations of antibody-conjugated magnetic beads. If lysis buffer conditions or other isolation variables are altered, such as incubation time of sample with antibody, an additional control should be performed.

FLAG Tagged

If the bait protein is tagged with FLAG, then a more appropriate control (compared to the GFP tag) is the respective cell/tissue/animal but under wild-type conditions. For both FLAG-tagged and wild-type conditions, magnetic beads conjugated with an antibody against the FLAG tag should be utilized. Although this doesn't control for proteins that bind nonspecifically to FLAG itself, nonspecificity due to the antibody and beads can be determined.

Endogenous, Nontagged Protein

In many experiments it is essential to immunoaffinity purify the bait protein at an endogenous level of expression (under its native promoter). Here, the control sample will use the identical cell/tissue/animal as for the experimental condition. In contrast to an affinity tag, the beads for the negative control are conjugated with either nonspecific IgG or an IgG that lacks reactivity towards the endogenous bait protein. The control protein isolation will reveal interactions that bind nonspecifically to the antibody and magnetic beads. A representative western blot from an immunoisolation of endogenous Shp2 from *X. tropicalis* embryos is shown in Fig. 2b.

4. Notes

1. The number of embryos needed for each immunopurification is dependent on a number of criteria including protein abundance and solubility as well as the affinity of the antibody used for immunoisolation. For this reason, the optimal number of embryos and resulting amount of tissue must be optimized individually for each protein studied. Generally, 20–50 embryos per immunoisolation is a good starting point for small-scale experiments and may be scaled up for larger immunoaffinity purifications for proteomics studies.
2. If starting directly with a small amount of tissue, the sample could be ground using round-bottom eppendorf tubes or used directly for incubation with the various lysis buffers. While performing the cryogenic grinding is preferred to mimic the conditions that will be utilized for protein isolation, direct resuspension in lysis buffer is an alternative that can adequately guide selection of an optimal buffer composition.
3. The balance between extraction efficiency and maintenance of interacting partners can be further optimized by increasing the stringency of the buffer used to wash the magnetic beads relative to the buffer used for tissue homogenization.
4. To test the efficiency of bead conjugation, prepare the following samples for SDS-PAGE analysis: (1) Dilute ≥1 μg neat antibody with 2.5 μL 10× Reducing Agent and 6.25 μL 4× NuPAGE LDS Sample Buffer. Bring the sample to 25 μL total volume with dH_2O. (2) Calculate the amount of bead supernatant that is equivalent to the amount of neat antibody in the previous sample, assuming that no antibody successfully conjugated

to the beads. Dilute this amount with 2.5 μL 10× Reducing Agent and 6.25 μL 4× NuPAGE LDS Sample Buffer. Bring sample to 25 μL with ultrapure H_2O. Heat samples for 10 min at 70°C. Centrifuge the samples at 20,000×*g* for 3 min at room temperature and load onto an SDS-PAGE gel. Stain the gel with SimplyBlue SafeStain and look for a reduction in the amount of antibody in the lane containing the bead supernatant. Refer to Cristea et al. (2) for the expected amount of unbound antibody resulting from differing amounts of antibody used in the conjugation.

5. After suspending tissue powder in lysis buffer, the solution may be slightly turbid, but should be devoid of tissue "clumps." Do not proceed to Polytron (step 4) until a homogenous suspension is observed. If necessary, additional rotation for 10–20 min at 4°C can be performed.
6. If insoluble particles are present in supernatant after centrifugation, a pipette can be used to selectively transfer supernatant to clean 50 mL conical tube.
7. The base elution buffer is preferred as significantly less background protein contamination is observed under these elution conditions. However, if low recovery of bait protein is observed, e.g., when a high affinity antibody is used, an "Alternate Procedure" can be followed that uses a harsher detergent-based elution buffer (see above).
8. The selection of an in-gel or in-solution digestion approach depends largely on the properties and nature of the proteins within the samples, e.g. p*I*, molecular weight, hydrophobicity, complexity, dynamic range, and total yield. In general, for high complexity and large yield and/or dynamic range of protein abundances, an in-gel approach is often desired. For further discussion, see Chapter 22, Subheading 1.
9. For viscous samples, brief sonication can be used to aid solubilization of cell pellet.
10. As the flow-through fraction often has higher protein concentration, the percent of material analyzed may need to be adjusted to prevent overloading of the SDS gel. For NuPAGE gels of 1.0 mm thickness, between 100 and 150 μg is recommended.
11. If base elution buffer is used, processing of the primary eluate and its reserved fraction is not performed until the day after the immunoisolation was started (see Subheading 3.3.2, step 20).

References

1. Nieuwkoop P, Faber J (1994) Normal table of Xenopus laevis. Garland Publishing, New York
2. Cristea IM et al (2005) Fluorescent proteins as proteomic probes. Mol Cell Proteomics 4(12):1933–1941
3. Cristea IM et al (2006) Tracking and elucidating alphavirus-host protein interactions. J Biol Chem 281(40):30269–30278
4. Moorman NJ et al (2010) A targeted spatial-temporal proteomics approach implicates multiple cellular trafficking pathways in human cytomegalovirus virion maturation. Mol Cell Proteomics 9(5):851–860
5. Goldberg AD et al (2010) Distinct factors control histone variant H3.3 localization at specific genomic regions. Cell 140(5): 678–691
6. Carabetta VJ, Silhavy TJ, Cristea IM (2010) The response regulator SprE (RssB) is required for maintaining poly(A) polymerase I-degradosome association during stationary phase. J Bacteriol 192(14):3713–3721
7. Greco TM et al (2011) Nuclear import of histone deacetylase 5 by requisite nuclear localization signal phosphorylation. Mol Cell Proteomics 10(2):M110.004317
8. Selimi F et al (2009) Proteomic studies of a single CNS synapse type: the parallel fiber/purkinje cell synapse. PLoS Biol 7(4):e83

Chapter 22

Complementary Proteomic Analysis of Protein Complexes

Todd M. Greco, Yana Miteva, Frank L. Conlon, and Ileana M. Cristea

Abstract

Proteomic characterization of protein complexes leverages the versatile platform of liquid chromatography-tandem mass spectrometry to elucidate molecular and cellular signaling processes underlying the dynamic regulation of macromolecular assemblies. Here, we describe a complementary proteomic approach optimized for immunoisolated protein complexes. As the relative complexity, abundance, and physiochemical properties of proteins can vary significantly between samples, we have provided (1) complementary sample preparation workflows, (2) detailed steps for HPLC and mass spectrometric method development, and (3) a bioinformatic workflow that provides confident peptide/protein identification paired with unbiased functional gene ontology analysis. This protocol can also be extended for characterization of larger complexity samples from whole cell or tissue *Xenopus* proteomes.

Key words: *Xenopus*, Proteomics, Protein complex, Protein interactions, In-gel digestion, In-solution digestion, Mass spectrometry, HPLC, Bioinformatics

1. Introduction

Technological advances in sample preparation and instrumentation for application of mass spectrometry-based proteomics to biological samples has permitted an increasing depth of proteome coverage. For example, biochemical and analytical separation strategies, such as multidimensional chromatography (1, 2) and affinity enrichment (3, 4), have seen significant advances that include specific optimization for downstream mass spectrometry analysis. These methods are complemented by continued development of nanoscale liquid chromatography, affording improved separation efficiency (5), as well as mass spectrometry technology that provides improved mass accuracy,

Stefan Hoppler and Peter D. Vize (eds.), *Xenopus Protocols: Post-Genomic Approaches, Second Edition*,
Methods in Molecular Biology, vol. 917, DOI 10.1007/978-1-61779-992-1_22,

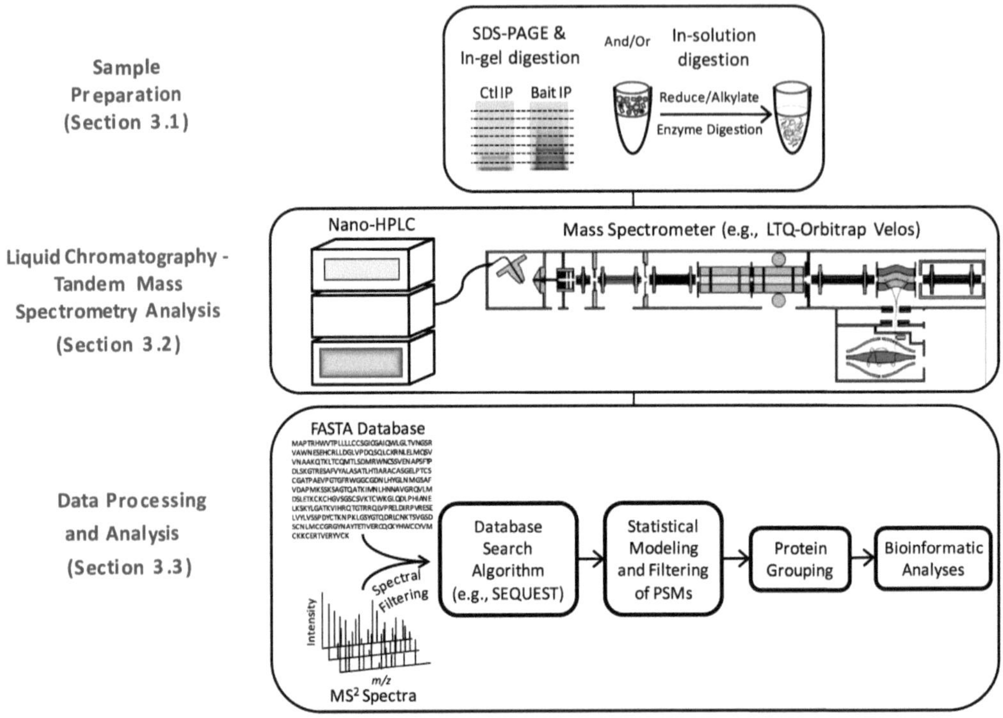

Fig. 1. Workflow summary for proteomic analysis of immunoisolated protein complexes; diagram of LTQ-Orbitrap Velos from Thermo Fisher Scientific.

resolution, and ion transmission and fragmentation (6, 7). Collectively, these achievements have enabled proteomic analysis to answer diverse biological questions, ranging from large scale, discovery-based experiments (8) to targeted, hypothesis-driven studies (9–11).

Leveraging several of these technologies, this chapter describes a mass spectrometry-based proteomics workflow (Fig. 1), tailored towards the analysis of protein complexes and assemblies obtained from affinity purification strategies ((12); see Chapter 21). Complementary sample preparation strategies are described (Subheading 3.1); a gel-based approach, which is readily used for optimization experiments and advantageous for complex samples (e.g., >1,000 proteins) of high dynamic range, and a solution-based approach that employs a single dimension of separation, providing greater reproducibility and optimal for lower complexity and sample amounts (13). Peptide analysis is performed by reverse phase nanoscale liquid chromatography coupled online to tandem mass spectrometry (Subheading 3.2). Data processing and analysis comprise the greatest overall proportion (based on time) of the workflow (Subheading 3.3). This multi-step process involves filtering of MS/MS spectra, matching of experimental spectra to theoretical spectra generated from in silico digestion of a protein sequence database. Next, database search software provides putative Peptide

Spectrum Matches (PSMs). Then, post-search validation algorithms calculate *p*-values for all PSMs (i.e., the likelihood of an incorrect match by chance) as well as *q*-values, which allow filtering of PSMs to achieve a desired global false discovery rate (FDR). Peptides passing these criteria are then assembled into protein groups, requiring a minimum number of ***unique*** peptides (e.g., 2) to define a protein group. Finally, bioinformatic and data reduction strategies are presented to aid in transitioning from hypothesis-generating to hypothesis-driven studies. Overall, these strategies are applicable not only to affinity-purified samples, but to a broad range of cellular and organism proteomes.

2. Materials, Equipment, and Instrumentation

2.1. Sample Preparation for Mass Spectrometry

2.1.1. SDS-PAGE and In-Gel Protein Digestion

Electrophoresis and protein staining

1. Primary eluate from immunoaffinity purification (see Chapter 21, Subheading 3.3.2).
2. NuPAGE 4–12% Bis-Tris pre-cast SDS-PAGE gel, 10 well (Invitrogen).
3. Xcell SureLock Mini-Cell electrophoresis system (Invitrogen).
4. 20× NuPAGE MOPS SDS Running Buffer (Invitrogen).
5. 1× Running Buffer: Dilute 20× NuPAGE MOPS SDS Running Buffer to 1× in 700 mL of ultrapure water.
6. Precision Plus Protein Dual Color Molecular Weight Standards (BioRad).
7. 4× NuPAGE LDS Sample Buffer (Invitrogen).
8. 10× NuPAGE Sample Reducing Reagent (Invitrogen).
9. 1× LDS Sampler Buffer: Mix 50 μL of 4× NuPAGE LDS Sampler buffer, 20 μL of 10× NuPAGE Sample Reducing Reagent, and 130 μL of ultrapure water.
10. SimplyBlue SafeStain (Invitrogen).

In-gel digestion and peptide extraction

Store stock solutions in glass containers that have been thoroughly rinsed with ultrapure water. Avoid using glassware that has been washed with detergents.

1. SimplyBlue SafeStain (Invitrogen).
2. Ultrapure dH_2O (e.g., from a Milli-Q Purification System).
3. Rocking platform.
4. Sheet protector and Scanner (Officemax).
5. Solution basins (Fisher).

6. Axygen Maxymum Recovery microcentrifuge tubes, 1.5 mL (VWR).
7. Axygen 96-well plates with sealing mat (VWR).
8. Multichannel pipet (Fisher).
9. Non-autoclaved, low binding pipet tips, 200 μL (VWR).
10. Ceramic Plate, Forceps, Razor Blade, Windex.
11. Mickle Gel Slicer (Brinkmann).
12. Lens Paper (Fisher).
13. Water, HPLC grade (Fisher).
14. Acetonitrile (ACN), HPLC grade (Fisher).
15. Formic acid stock (FA), LC-MS grade, 99+% (Pierce).
16. 1% FA solution: Mix 1.0 mL of FA stock in 99.0 mL ultrapure water. Store at RT.
17. 0.1 M ammonium bicarbonate (ABC): Dissolve 0.80 g of ABC solid in 0.1 L of HPLC grade water. Store at RT and use within 1 month, or sterile filter for long-term storage.
18. Destain solution: Mix 2 mL of ACN with 2 mL of 0.1 M ABC. Prepare fresh before use in a solution basin.
19. Rehydration solution: Mix 2.5 mL of ultrapure water with 2.5 mL of 0.1 M ABC. Prepare fresh before use in solution basin.
20. 0.5 μg/μL Trypsin stock, sequencing grade (Promega). Store at −80°C, limit to <5 freeze/thaw cycles.
21. 12.5 ng/μL Trypsin solution: Mix 8.5 μL of trypsin stock with 331.5 μL of rehydration solution. Prepare fresh immediately before use (step 16).
22. Vortex device with an adaptor for 96-well plates (Fisher).
23. Glass autosampler vials, MS Certified with 200 μL fused inserts and pre-slit caps (National Scientific).
24. 50% ACN/0.5% FA solution. Mix 0.5 mL of ACN and 0.5 mL of FA solution (1%) in microfuge tube. Prepare fresh on Day 3.
25. Savant SpeedVac Concentrator (Thermo Scientific).

2.1.2. In-Solution Digestion (Alternative Workflow)

Unless otherwise stated, store solutions at RT in glass containers that have been thoroughly rinsed with ultrapure water.

1. Primary eluate from immunoaffinity purification (see Chapter 21, Subheading 3.3.2).
2. HPLC grade water (Fisher).
3. Dithiothreitol (DTT), solid (Invitrogen). Store at −20°C.
4. Tris-base, solid (Fisher).
5. Urea, solid (Sigma).

6. Trifluoroacetic acid (TFA), LC-MS grade (Fisher). Always use a glass syringe to pipet concentrated acid stocks.
7. Iodoacetamide (IAA), solid (Sigma). Store at 4°C.
8. 1 M DTT: Dissolve 0.77 g DTT in 5 mL of HPLC grade water. Aliquot and store at −80°C.
9. 0.2 M Tris–HCl, pH 8.0: Dissolve in 1.21 g Tris-base in 40 mL of HPLC grade water, pH to 8.0, and dilute to 50 mL. Filter sterilize.
10. 0.05 M ABC: Dissolve 0.40 g of ABC solid in 0.1 L of HPLC grade water. Sterile filter, store at RT and use within 3 months.
11. 1 M IAA (Sigma): Dissolve 0.185 g of IAA in 1 mL HPLC grade water. Dispense into 50 × 20 μL aliquots and store at −20°C.
12. 50% ACN in HPLC grade water.
13. 0.5 μg/μL Trypsin stock, sequencing grade (Promega). Store at −80°C, keep freeze/thaw cycles <5.
14. 10% TFA. Mix 1 mL of 100% TFA with 9 mL of HPLC grade water. Store at 4°C.
15. Non-autoclaved, low binding pipet tips, 200 μL (VWR).
16. Amicon-500, 10 kDa ultrafiltration devices (Millipore).
17. Refrigerated microcentrifuge capable of 14,000 × *g* (maintain at 20°C).
18. Eppendorf tube shaker (e.g., Tomy shaker).
19. Savant SpeedVac Concentrator (Thermo Scientific).
20. 8 M Urea buffer (UB): Dissolve 0.48 g urea in 160 μL HPLC water and 0.5 mL 0.2 M Tris–HCl, pH, 8.0. Rock gently until dissolved. Yields 1 mL of solution, sufficient for one sample. Keep solution at RT. Do not heat urea-containing solutions above 37°C.
21. 100 mM IAA solution: Prepare 200 μL of solution/sample by mixing 180 μL of UB with 20 μL of 1 M IAA.
22. 5 ng/μL Trypsin solution: Prepare 100 μL/sample by mixing 10 μL of 0.5 μg/μL trypsin stock with 90 μL of 0.05 M ABC. Prepare fresh immediately before use (step 8).
23. Glass autosampler vials, MS Certified with 200 μL fused inserts and pre-slit caps (National Scientific).

2.2. Liquid Chromatography-Tandem Mass Spectrometry Analysis

Unless otherwise stated, all solutions should use LC-MS grade solvents.

2.2.1. NanoLC Configuration and Peptide Separation

1. Optima LC-MS grade solvents: water, ACN, and TFA (Fisher).
2. Formic acid (FA), LC-MS grade, 99+% (Pierce).
3. Graduated glass cylinders, 100 and 500 mL.

4. Glass syringe with needle, 0.5 mL (Hamilton).
5. Nanoflow HPLC system, e.g., Dionex Ultimate, Waters nano Acuity, or Agilent 1200 series (see Note 1).
6. Mobile phase A (MPA): 0.1% FA/99.9% water. Dilute 0.5 mL FA with water to final volume of 0.5 L in a glass graduated cylinder. Transfer to amber bottle and store for up to 6 months (see Note 2).
7. Mobile phase B (MPB): 0.1% FA/97% ACN/2.9% water. Mix 0.5 mL FA and 14.5 mL water. Dilute with ACN to final volume of 0.5 L in a glass graduated cylinder. Transfer to amber bottle and store for up to 6 months (see Note 2).
8. Mobile phase C (MPC): 0.5% TFA/1% ACN/98.5% water. Mix 1.5 mL TFA and 3.0 mL ACN. Dilute with water to final volume of 0.3 L in a glass graduated cylinder. Prepare fresh and transfer to HPLC solvent bottle. Replace monthly.
9. Self-Pack IntegraFrit Columns, 360 μm OD × 100 μm ID (New Objective).
10. Micro Tool Kit (New Objective).
11. Trap column cartridge assembly (Upchurch).
12. Polyimide-coated fused silica capillary tubing (Upchurch).
13. Trap column, 360 μm × 100 μm ID × 2.5 cm. Construct by packing reverse phase resin (Magic C18 AQ, 3 μm, Michrom Bioresources) into Self-Pack IntegraFrit columns, cleaving to 2.5 cm length, and inserting into Trap column cartridge assembly (see Note 3).
14. Analytical column, 75 μm ID × 25 cm, e.g., Acclaim PepMap RSLC (Dionex) (see Note 3).

2.2.2. Mass Spectrometry Configuration and Peptide Analysis

1. LTQ-Orbitrap Velos hybrid mass spectrometer (Thermofisher Scientific) (see Note 4).
2. Nanospray ESI source (Thermofisher Scientific).
3. SilicaTip Emitter, Tubing (OD × ID) 360 μm × 50 μm; Tip (ID) 10 μm (New Objective).

2.3. Data Processing and Analysis

1. Multi-core/multi-CPU 64-bit PC workstation with at least 12 GB of RAM and 2 TB of storage (see Note 5).
2. Software for generating peaklists and scoring of PSMs, e.g., Proteome Discoverer 1.3 (Thermofisher Scientific), Mascot 2.3 (Matrix Science), Scaffold 3.0 (Proteome Software), or COMPASS (14) (see Note 6).
3. Spreadsheet software (e.g., Microsoft Excel).

3. Methods

3.1. Sample Preparation for Mass Spectrometry

3.1.1. SDS-PAGE and In-Gel Protein Digestion

This protocol is performed over the course of 3 days. To minimize keratin and other environmental contaminants, it is recommended to wear a lab coat and hair protection, to avoid close contact and limit environmental exposure of the pre-case gel during sample loading and protein staining.

Electrophoresis and protein staining (Day 1)

1. If necessary, thaw primary eluate sample(s) at 70°C.
2. Set up the Xcell SureLock Mini-Cell electrophoresis system:
   - Remove the white strip and the comb from the NuPAGE 4–12% Bis-Tris pre-cast SDS-PAGE gel and rinse with ultrapure water. Place the gel in the apparatus, using a buffer dam for the opposing side, then lock the assembly in place.
   - Fill the inner chamber with 200 mL of 1× Running Buffer.
   - Fill the outer chamber with 500 mL of 1× Running Buffer.
3. Load samples into wells as follows: Lane 2, 10 μL Molecular Weight Standards. Lane 4, 90% of the eluate (~40 μL) (see Note 7).
4. Load 20 μL 1× LDS Sample Buffer into all empty wells.
5. Electrophorese for 5 min at 100 V, then 15–25 min at 200 V, or until the dye front has migrated approximately one third of the way down the gel (see Note 8).
6. Open gel cassette to expose gel and discard the wells. Working with wet gloves, transfer gel (by the thick ridge at the bottom) into a plastic tray containing ultrapure water. Remove bottom ridge.
7. Wash the gel 3×5 min with ultrapure H_2O while rocking.
8. Dispense SimplyBlue SafeStain to cover the gel completely. Incubate for 1–3 h until the protein bands become clearly visible. Pour off the solution and replace it with ultrapure water. Destain 1–2 h at room temperature or overnight at 4°C with gentle rocking (see Note 9).

In-gel digestion (Day 2)

Working solution volumes are calculated for 16 samples (based on 90 μL of solution/well except where noted). For higher throughput, a 96-well plate with sealing mat is recommended (see Note 10).

All solutions can be added/removed from the wells using a multichannel pipettor. For each sample set, use a different set of multichannel tips, though one set of tips can be used for all steps performed on Day 2.

1. Wash the destained gel several times with ultrapure water to remove excess staining reagent.
2. Place the gel in a sheet protector. Keep the gel wet with a little ultrapure water. Scan the gel, save the image, and print it out as it will serve as a reference during processing of gel slices (step 7).
3. Wash the ceramic plate, forceps, blade of the gel slicer, and a razor blade first with Windex then with ultrapure water.
4. Place the gel on the plate. Cut out the entire vertical lane that contains the sample. Trim lane to remove lower portion just at the dye front and the top 2 mm near the well.
5. Place a small piece of lens paper on the cutting stage of the Mickle Gel Slicer, wet the lens paper with ultrapure water, and flatten it out. Place the excised gel lane on the lens paper.
6. Cut the gel lane with the Mickle Gel Slicer to 1 mm slices (approximately 20–30 slices). Push down any gel pieces that get stuck to the gel slicer blade with a pair of tweezers.
7. Lift the lens paper and flip the cut gel lane back onto the ceramic plate. Carefully remove the lens paper. Keep cut gel slices slightly wet for ease of sample handling.
8. Group together 2–3 gel pieces, starting from the bottom of the gel, and cut each 1 mm slice into three equal pieces (see Note 11).
9. Transfer gel pieces using forceps or the flat side of razor blade into one well. Do not exceed 3×1 mm slices/well.
10. Add 90 μL of destain solution to each well, seal plate, and agitate on a vortexer (medium setting) at room temperature (RT) for 10 min. Aspirate and discard. For all aspiration steps, make sure the gel pieces are not stuck to pipet tips when discarding waste solution as sample loss could occur.
11. Repeat step 10 with fresh destain solution. While destaining, prepare and aliquot rehydration solution and ACN (5 mL each) in *separate* solution basins.
12. Aspirate the destain solution, add 90 μL ACN to the gel pieces, seal plate with mat, and invert manually several times. Let it stand at RT until the gel pieces are white (<5 min).
13. Remove ACN and add 90 μL of rehydration solution, seal the plate, and manually invert several times. Incubate at RT for <5 min, or until the gel pieces have swelled and are translucent. Aspirate rehydration solution.
14. Repeat dehydration/rehydration (steps 12/13) once.
15. Add 90 μL of ACN to dehydrate as previously described (step 12). Aspirate ACN and let residual ACN evaporate.
16. While ACN is evaporating, prepare trypsin solution (12.5 ng/μL). Mix gently and add 20 μL of trypsin solution to the dried gel pieces. Let the gel pieces swell (<5 min).

17. Add ~30 μL of rehydration solution to the gel pieces (use enough volume to ensure all gel pieces are fully covered). Seal the plate and incubate overnight at 37°C.

Peptide extraction (Day 3)

1. In the morning, add 30 μL 1% FA to each sample and incubate for 4 h at RT on bench. *CRITICAL STEP*: Transfer the extracted peptides to pre-labeled autosampler vials. Make sure gel pieces are not transferred into autosampler vials. Store vials at 4°C.
2. Add 30 μL of 50% ACN/0.5% FA solution to wash the gel pieces. Incubate for 2 h at RT on bench.
3. Transfer and combine the respective washes with their corresponding FA extracts from step 1.
4. Speedvac the samples to ~10–12 μL, avoiding concentration to dryness (see Note 12). If necessary, adjust volumes with 1% FA. Proceed to Subheading 3.2. Store at 4°C in autosampler vials if the sample will be analyzed for mass spectrometry analysis within the next 2 days, otherwise store at −20 or −80°C for long-term storage.

3.1.2. In-Solution Digestion (Alternative Workflow)

All procedures are performed at either RT (20°C) or higher (where indicated) to avoid precipitation of detergent or urea from solutions. For this purpose, preset the temperature of the microcentrifuge to 20°C. This protocol is performed over the course of 2 days.

In-solution digestion (Day 1)

1. If necessary, thaw eluate samples previously prepared by immunoaffinity purification (see Chapter 21, Subheading 3.3.2) by heating the sample at 70°C for 10 min.
2. Mix each eluate sample (max. 150 μg total protein in <100 μL) with 400 μL of UB in an ultrafiltration device. Centrifuge at 14,000 ×*g* for 30 min.
3. Discard flow-through. Add 400 μL of UB to the top of the filter unit and centrifuge at 14,000 ×*g* for 20 min.
4. Discard the flow-through. Add 200 μL of IAA solution and mix on an eppendorf tube shaker, medium setting, for 1 min.
5. Incubate for 15 min at RT protected from light. Centrifuge the filter units at 14,000 ×*g* for 15 min. Discard flow-through.
6. Add 200 μL of 0.05 M ABC to the top of the filter and centrifuge at 14,000 ×*g* for 15 min.
7. Repeat step 6 an additional two times discarding flow-through after each centrifugation. Ensure that the retained volume is ~30 μL. During the centrifugation steps rinse a fresh collection tube with 200 μL 50% ACN and discard ACN.
8. Transfer filter units to the pre-rinsed collection tubes and add 100 μL of 5 ng/μL trypsin (see Note 13). Mix on a tube shaker at medium speed for 1 min.

9. Seal caps with parafilm and incubate in a water bath at 37°C for at least 4 h to overnight.

In-solution digestion (Day 2)

1. Centrifuge the filter units at 14,000 × *g* for 15 min.
2. Add 50 μL of HPLC water to top of filter unit and centrifuge at 14,000 × *g* for 10 min.
3. Repeat step 2.
4. *Retain* flow-through and discard filter unit. Transfer flow-through to an autosampler vial containing 22 μL of 10% TFA.
5. Concentrate samples by vacuum centrifugation until total volume is between 10 and 30 μL. Final volume will be dependent upon the expected peptide yield (see Note 14).

3.2. Liquid Chromatography-Tandem Mass Spectrometry Analysis

3.2.1. NanoLC Configuration and Peptide Separation

1. Recommended solvent configuration for dual-pump LC system: Connect nanopump (NP) A and B solvent lines to 100 mL of MPA and MPB, respectively. Replace MPA and MPB monthly with solvent from stock solution. Connect loading pump (LP) solvent line to 300 mL of MPC. The LP flow path carries the sample to the trap column, while the NP delivers solvents to either (1) the analytical column alone or (2) both the trap and analytical column. These two conditions can be achieved using a six-port, two-position valve. Before sample injection, the trap column is equilibrated in 100% MPC for 1 min at 5 μL/min. The analytical column mobile phase should be held at 96% MPA/4% MPB by the NP at a flow rate of 250 nL/min.
2. Using a 20 μL sample loop and 25 μL syringe, injection volumes between 5 and 10 μL should be attainable. Optimally, the total amount of peptides analyzed on-column should be ≤4 μg (≈2 μg provides optimal chromatography). On-column sample load should be achieved by keeping *injection* volumes constant, while adjusting the total *sample* volume.
3. Desalting: Upon injection, the sample is washed onto the trap column and desalted in 100% MPC for 5 min at 5 μL/min. All solvent is directed to waste. The analytical column should remain under the same conditions as in step 2.
4. Elution: Using a two-position valve, the trap column is placed in line with the analytical column. Desalted peptides bound to the trap column are eluted and resolved over the analytical column at 250 nL/min using an increasing percent MPB delivered by the NP. For samples from in-gel digestion, a 90 min, discontinuous linear elution gradient is performed as follows: 4–20% MPB over 50 min, followed by 20–40% MPB over 40 min. For in-solution digestion samples, a 180 min continuous linear elution gradient is performed as follows: 4–40% MPB over 180 min.

5. Washing and Re-equilibration: Columns are washed for 5 min at 60% MPB, followed by 10 min at 97% MPB, and then re-equilibrated at 4% MPB for 15 min. The LC system is now ready to perform the next analysis. The total run time is calculated by adding the gradient time + 35 min (see Note 15).
6. Quality Controls: A standard, known sample should be injected at regular intervals to monitor LC and MS performance. Tryptic digests of a known protein, such as enolase, or a complex peptide mixture from whole cell lysate digestion can be used to monitor chromatographic reproducibility and overall depth of analysis. Also, between analyses of different biological samples, a "blank" or "wash" run should be conducted to minimize carryover of abundant analytes. Optionally, a high organic (60% ACN/0.5% TFA) solution can be injected to aid in removal of abundant analytes from the trap/analytical column.

3.2.2. Mass Spectrometry Configuration and Peptide Analysis

For global proteomic analysis of complex biological samples, data-dependent acquisition using collision-induced dissociation (CID) on a hybrid mass spectrometer, such as the LTQ-Orbitrap Velos, is the preferred method. The following steps summarize recommended instrument parameters for a hybrid LTQ-Orbitrap Velos mass spectrometer. While acquisition methods are often instrument-specific, the overall goal of the method should be to optimize the duty cycle or cycle time of the method, often defined as the time between repeated full scan MS acquisition events or the time to accomplish a single round of peptide fragmentation events. For most nanoLC chromatographic separations, an optimal cycle time should average ≤4 s.

1. MS^1 (Precursor Ion) Acquisition: Precursor ions are detected in the Orbitrap analyzer at a resolution of 60,000 at $m/z = 400$. It is recommended that the "Preview mode for FTMS master scan" is disabled, so that data dependent selection of precursors is performed using the MS^1 analytical scan collected at maximum resolution (see Note 16).
2. MS^2 (Fragment Ion) Acquisition: In data-dependent experiments, precursor ions detected in the MS^1 spectra are selected for fragmentation primarily as a function of intensity. A data-dependent CID Top 20 method is created, which selects the top 20 most intense precursor ions for acquisition of MS^2 (tandem or MS/MS) CID spectra. To optimize MS^2 acquisition, a maximum injection time of 100 ms, AGC target value of 5,000, and one microscan is recommended. To increase the dynamic range in abundance over which fragmentation spectra can be acquired, the dynamic exclusion function should be enabled (see Note 17). Using these settings, a single MS^1 scan at 60,000 resolution and 20 tandem (MS^2) scans, corresponding to a single acquisition cycle, can be completed in about 4 s.

3. Alternating MS^1–MS^2 (Top 20) cycles are repeated continuously throughout the LC gradient.

3.3. Data Processing and Analysis

To date, there are several well-developed software suites that process and analyze tandem mass spectra from large-scale, bottom-up proteomic experiments and provide a list of high confidence protein identifications. An overview of this workflow is presented, and for each stage of analysis a brief description is provided below. Although the spectral filters and algorithms may differ between software, the conceptual workflow is similar. For clarity, the individual steps described below use Proteome Discoverer and Scaffold as representative software.

Peptide spectrum matching by database searching

1. Database Creation: The UniProt knowledgebase of protein sequences in FASTA format are recommended as two levels of redundancy are available: the Swiss-Prot database, which is manually curated and reviewed, and the TrEMBL database which is automatically annotated and not reviewed (http://www.uniprot.org/downloads). For samples derived from well-annotated organisms, such as *Homo sapiens*, the Swiss-Prot sequences alone are adequate. For the other organisms, such as *Xenopus*, both the Swiss-Prot and TrEMBL sequences must be utilized. It is recommended that common contaminants, such as keratins, cell culture serum components, immunoglobulins, and digestion enzymes, be included in the database. For convenience, a custom database containing protein sequences from *H. sapiens*, *Mus musculus*, *Xenopus laevis*, and *Xenopus tropicalis* appended with common contaminants has been made available at http://www.xenbase.org.
2. Spectral Filtering: MS^2 spectra should be filtered before submitting to a database search engine. While some software has built-in spectral filters (e.g., to remove noise), others allow specification of individual spectrum filtering. At minimum, spectra should be filtered based on minimum number of peaks ($N=30$) and intensity ($N=200$).
3. Sequence-to-Spectrum Matching: Putative PSMs are generated by submission of filtered spectra to the database search algorithm (e.g., SEQUEST or Mascot), which assigns scores based on the similarity between the experimental and theoretical tandem mass spectra. The search algorithm should be configured to specify the following parameters: precursor tolerance, 10 ppm; fragment tolerance, 0.5 Da; static modification, cysteine carbamidomethylation. Methionine oxidation and serine, threonine, and tyrosine phosphorylation should be specified as dynamic modifications.
4. Statistical Modeling of PSMs: Most search engine raw scores do not provide a direct statistical assessment of confidence, i.e., the

probability of an incorrect match. To accurately estimate error rates for large-scale proteomic datasets, it is essential to utilize a post-processing probabilistic algorithm, such as Percolator (15) or PeptideProphet (16) to re-rank PSM scores (see Note 18). SEQUEST search results (*.msf files from Proteome Discoverer) are analyzed by Scaffold as follows: (a) Search result files from each sample are loaded into distinct biological samples. For in-gel digestion, the MuDPIT option in Scaffold is used to combine search result files corresponding to multiple gel fractions from the biological sample. (b) Two additional features of Scaffold that should be routinely enabled are (i) an X! Tandem re-search with the subset database option and (ii) high mass accuracy scoring. The former conducts a second search against a subset database containing only those proteins identified in the initial SEQUEST search. (c) Additional dynamic modifications, such as deamidation of asparagine and glutamine, should be included for X! Tandem re-searches. Optionally, other biological or chemical modifications of interest not included in the primary SEQUEST search can be added, such as acetylation or ubiquitination of lysine. (d) After input spectra are scored by X! Tandem, both SEQUEST and X! Tandem PSMs are analyzed by PeptideProphet to calculate estimated local FDRs (probability of an incorrect match of a single PSM or protein). The global FDR (error rate within a group of PSMs or proteins) is estimated from the statistical modeling of PSM score distributions (see Note 19).

Protein level filtering and bioinformatics analysis

1. Scaffold protein identifications should be filtered, removing keratins and contaminant proteins (Accession prefix "CON_").
2. Protein identifications are exported from Scaffold to a xls (tab-separated) format by the command: "Export," "To Excel," "Current View...." The desired export values (unique peptides, assigned spectra, etc.) can be controlled under "Display Options." It is recommended that separate protein identification spreadsheets containing unique peptides and unweighted spectrum count are exported.
3. Assessment of Bait-Specific Protein Interactions: Using a spreadsheet program, open Excel file containing unweighted spectrum count values. Calculate fold enrichment in spectrum counts: Divide spectral counts for each protein in the bait IP by the corresponding spectral counts in the control IP (e.g., GFP). Proteins with ≥5 spectral counts in one condition and either unique to bait IP or ≥2.0–2.5-fold change in spectral counts are the best candidates for specific protein interactions (see Note 20).
4. Bioinformatic Analyses: Identification of targets that are of biological interest can be pursued by two approaches

(a) Protein-Specific: Accession numbers for proteins of interest can be queried through protein knowledge bases such as UniProt (http://www.uniprot.org), Genecard (http://www.genecards.org/), or the Comprehensive Resource of Mammalian protein complexes (CORUM) (http://mips.helmholtz-muenchen.de/genre/proj/corum). This approach should be used with the caveat that it may lead to biased interpretation of the data.

(b) Global: Accession numbers for enriched proteins are submitted to functional enrichment software, which often converts accessions to gene-centric identifiers (e.g., gene symbols) for gene ontology and functional annotation mapping. The functional relationship between candidate interactions can be examined using STRING (http://www.string-db.org/). To examine over- or under-representation of functional annotations in a set of candidate interactions vs. a reference genome, the FatiGO tool within the Babelomics software suite can be used (http://babelomics.bioinfo.cipf.es/). Alternatively, a highly versatile open-source software platform called Cytoscape (http://www.cytoscape.org) can be used with functional enrichment plugins, such as BiNGO (http://www.psb.ugent.be/cbd/papers/BiNGO/Home.html) to perform both functional enrichment analyses and data visualization within a single software environment.

4. Notes

1. While many HPLC configurations can be used, the system should have the capacity for nanoliter per minute flow rates (ideally ≤500 nL/min). Also, the system should have dedicated nano and capillary pump capabilities that allow for shorter overall run times.
2. Mobile phases should be replenished from the stock solution monthly.
3. Many chemistries for column packing material are available and should be selected based on relative retention of desired analytes. For example, Magic C18 AQ resin was selected for the trap column for its increased retention of hydrophilic peptides, such as shorter length phosphopeptides. The Acclaim PepMap material was selected for the analytical column as it is available in sub-2 μm particle size. When selecting a column, instrument pressure limits should be noted as smaller particle sizes and longer columns increase operating pressures.
4. While several mass spectrometer configurations are adequate for these experiments, a hybrid instrument such as the LTQ-Orbitrap Velos offers (1) resolution up to 100,000, (2) <2 ppm

mass accuracy (external calibration) in the FT Orbitrap detector, and (3) attomol sensitivity in the linear ion trap detector.

5. While a computing cluster is preferred, the high initial and maintenance costs often make this option impractical. As a second option, a multi-CPU or multi-core workstation can be purchased for about $5,000. If multi-threaded applications will be used (e.g., Mascot Server), then a dual CPU system should be considered. Otherwise, maximizing the overall processing speed of a single multi-core CPU should be the goal. For a more details on hardware specifications see http://www.matrixscience.com/help/pc_specs.html.
6. Currently there are many excellent options for open-source or commercial software suites that include database search engines. The most critical aspect when selecting a software suite or database search tool is the capability to accurately estimate probability values (e.g., *q*-values or posterior error probabilities) of individual PSMs. All software cited above satisfies this requirement.
7. If loading additional eluates, skip two lanes between the samples. Reserve the remaining eluate (10%) for western blot analysis.
8. For more complex samples or greater protein load, resolving proteins for entire gel length may be beneficial.
9. If necessary, the stained gel can be stored in ultrapure water at 4°C, wrapped tightly, up to 2–3 weeks until proceeding to in-gel digestion, though some sample loss may occur.
10. Axygen microfuge tubes can be used as an alternative to a 96-well plate.
11. Do not mince gel slices as this increases likelihood of pipeting gel pieces into autosampler vials during extraction.
12. If samples are concentrated to dryness, add 30 μL of 50% ACN/0.5% FA solution, vortex briefly, and SpeedVac to reduce the volume as above.
13. Assuming an enzyme:protein ratio of 1:50–1:100, then the working solution of trypsin is sufficient for digestion of 25–50 μg of protein. If more protein is digested then increase concentration accordingly.
14. If peptide yield is known or can be estimated based on input material, concentrate sample to a final volume containing a peptide concentration of 0.5 μg/μL. Peptide concentration can be estimated by measuring A_{280} on a Nanodrop and dividing by 1.1 (extinction coefficient estimated based on the tryptophan frequency in mammalian proteins from a 0.1% protein solution).
15. Gradient and total run times provide a starting point for method development and should be optimized for specific LC hardware configurations, chromatographic material, and sample complexity.

16. For some instrument configurations, disabling the "preview scan" function may cause cycle times to routinely exceed 4 s. If this occurs, then decreasing resolution to 30,000 is recommended.
17. Recommended dynamic exclusion settings for a 90 min gradient are as follows: a repeat count of 1, repeat duration of 30, exclusion list size of 500, exclusion duration 70, and exclusion mass width 10 ppm. The exclusion duration should be adjusted based on chromatographic peak widths.
18. Percolator has been implemented in Mascot 2.3 and Proteome Discoverer 1.3, while PeptideProphet is available using either Scaffold (http://www.proteomesoftware.com) or the Trans-Proteomic Pipeline open-source software (http://tools.proteomecenter.org/wiki/index.php?title=Software:TPP).
19. PeptideProphet and ProteinProphet probability filters should be selected to reduce FDR to less than 1% at both the peptide and protein level. While confidence thresholds can vary between datasets, a general starting point is 99.0% protein probability, 95% peptide probability, and a minimum of 2 peptides/protein. For a more detailed discussion of error rate estimation see ref. (17).
20. Spectral counting analysis is a computationally simple method to evaluate which proteins may have large relative abundance differences between control and experimental samples. However, in the absence of biological replicates or independent validation assays, the statistical power of these analyses is limited and the results should be interpreted conservatively.

References

1. Washburn MP, Wolters D, Yates JR III (2001) Large-scale analysis of the yeast proteome by multidimensional protein identification technology. Nat Biotechnol 19(3):242–247
2. Wang H, Chang-Wong T, Tang HY, Speicher DW (2010) Comparison of extensive protein fractionation and repetitive LC-MS/MS analyses on depth of analysis for complex proteomes. J Proteome Res 9(2): 1032–1040
3. Gygi SP, Rist B, Gerber SA, Turecek F, Gelb MH, Aebersold R (1999) Quantitative analysis of complex protein mixtures using isotope-coded affinity tags. Nat Biotechnol 17(10):994–999
4. Fang X, Zhang WW (2008) Affinity separation and enrichment methods in proteomic analysis. J Proteomics 71(3):284–303
5. Kocher T, Swart R, Mechtler K (2010) Ultra-high-pressure RPLC hyphenated to an LTQ-Orbitrap Velos reveals a linear relation between peak capacity and number of identified peptides. Anal Chem 83(7):2699–2704
6. Hossain M, Kaleta DT, Robinson EW, Liu T, Zhao R, Page JS et al (2011) Enhanced sensitivity for selected reaction monitoring mass spectrometry-based targeted proteomics using a dual stage electrodynamic ion funnel interface. Mol Cell Proteomics 10(2):M000062–MCP000201
7. Olsen JV, Schwartz JC, Griep-Raming J, Nielsen ML, Damoc E, Denisov E et al (2009) A dual pressure linear ion trap Orbitrap instrument with very high sequencing speed. Mol Cell Proteomics 8(12):2759–2769
8. Olsen JV, Vermeulen M, Santamaria A, Kumar C, Miller ML, Jensen LJ et al (2010) Quantitative phosphoproteomics reveals widespread full phosphorylation site occupancy during mitosis. Sci Signal 3(104):ra3
9. Moorman NJ, Sharon-Friling R, Shenk T, Cristea IM (2010) A targeted spatial-temporal proteomics approach implicates multiple cellular trafficking pathways in human cytomegalovirus virion maturation. Mol Cell Proteomics 9(5): 851–860

10. Picotti P, Bodenmiller B, Mueller LN, Domon B, Aebersold R (2009) Full dynamic range proteome analysis of S. cerevisiae by targeted proteomics. Cell 138(4):795–806
11. Selimi F, Cristea IM, Heller E, Chait BT, Heintz N (2009) Proteomic studies of a single CNS synapse type: the parallel fiber/Purkinje cell synapse. PLoS Biol 7(4):e83
12. Greco TM, Yu F, Guise AJ, Cristea IM (2011) Nuclear import of histone deacetylase 5 by requisite nuclear localization signal phosphorylation. Mol Cell Proteomics 10(2): M110.004317
13. Kramer T, Greco TM, Enquist LW, Cristea IM (2011) Proteomic characterization of pseudorabies virus extracellular virions. J Virol 85(13): 6427–6441
14. Wenger CD, Phanstiel DH, Lee MV, Bailey DJ, Coon JJ (2011) COMPASS: a suite of pre- and post-search proteomics software tools for OMSSA. Proteomics 11(6):1064–1074
15. Kall L, Canterbury JD, Weston J, Noble WS, MacCoss MJ (2007) Semi-supervised learning for peptide identification from shotgun proteomics datasets. Nat Methods 4(11):923–925
16. Keller A, Nesvizhskii AI, Kolker E, Aebersold R (2002) Empirical statistical model to estimate the accuracy of peptide identifications made by MS/MS and database search. Anal Chem 74(20):5383–5392
17. Nesvizhskii AI (2010) A survey of computational methods and error rate estimation procedures for peptide and protein identification in shotgun proteomics. J Proteomics 73(11):2092–2123

Part V

Imaging *Xenopus* Development

Chapter 23

Antibody Development and Use in Chromogenic and Fluorescent Immunostaining

Eamon Dubaissi, Niki Panagiotaki, Nancy Papalopulu, and Peter D. Vize

Abstract

Antibody-based detection of protein distribution patterns both in wholemount and on sections revolutionized *Xenopus* research and ushered in the visual-based era of *Xenopus* data presentation. The ability to view the distribution of a gene product throughout an embryo makes it possible to rapidly map normal expression profiles and profiles that have been altered by an experimental intervention. The main limiting element in *Xenopus* immunostaining techniques has always been the availability of antibodies that work well on fixed whole embryos, a problem that persists. However, new antibodies are constantly being generated and improvements in detection systems allow antibodies that were once below the limits of detection to be utilized in multichannel immunofluorescence using tyramide amplification.

Key words: Immunohistochemistry, Immunofluorescence, Monoclonal antibody, Polyclonal antibody

1. Introduction

The concept of immunolocalization within *Xenopus* embryonic tissue is very simple. An antibody that recognizes an antigen within the embryonic tissue is applied to the sample, given time to bind, and then excess antibody washed away. The distribution of the bound antibody is then detected using either an enzyme covalently bound to the antibody or using a secondary antibody–enzyme complex that binds to the primary antibody. The linked enzyme is capable of catalytically converting a soluble colorless substrate into a colored insoluble product and the distribution of the colored precipitate provides a representation of the distribution of the antigen. In fluorescent techniques the principle is similar, but the primary or

Stefan Hoppler and Peter D. Vize (eds.), *Xenopus Protocols: Post-Genomic Approaches, Second Edition*,
Methods in Molecular Biology, vol. 917, DOI 10.1007/978-1-61779-992-1_23,

secondary antibody is covalently linked to a fluorochrome that can then be detected using fluorescence microscopy.

The original application of these techniques was to characterize the distribution of a protein during normal embryonic development (1, 2). It was also applied to characterizing how the distribution of a protein changed in response to manipulations of embryogenesis and microinjection of mRNAs (3). Most recently, antibodies that detect cellular processes like cell division, apoptosis, and responses to signal transduction have been used to characterize embryonic responses to experimental interventions (4–6). They have been used to detect the distribution of a lineage tracer or a transgene product when that product is not fluorescent, is expressed at too low a level to be detected, or when processing with alcohols/bleach/fixatives kills fluorescence in the target (7, 8). Protocols applicable to all of these applications are provided in this chapter.

2. Materials

2.1. Fixation

By far the most commonly used fixative is formaldehyde based (MEMFA) as this reagent works for most antibodies and is also compatible with other protocols, for example in situ hybridization and histology. Paraformaldehyde (PFA) based fixation offers better preservation of fine cell structure, such as the actin and microtubule cytoskeletons, and cell biologists tend to use it instead of MEMFA. However, some antibodies are very picky about antigen presentation and do not work on aldehyde fixed samples, for example the atp1a1 antibody alpha-6F (9). Others only work (or are enhanced) upon heat or acid denaturation of their protein targets—one example of this is the anti-BrdU antibody discussed in Chapter 26. Recipes for both MEMFA and Dent's fixatives are provided. For more alternatives, see ref. (2).

1. 10× MEMFA salts: 0.5 M MOPS pH 7.4, 20 mM EGTA, 10 mM $MgSO_4$. This solution is autoclaved then kept at room temperature. Store away from light.
2. MEMFA fix: For 5 mL of MEMFA, mix 4.0 mL of sterile water with one-tenth volume (0.5 mL) of 10× MEMFA salts and one-tenth volume of fresh (opened less than 1 month) 37% (w/v) formaldehyde. This concentration of formaldehyde is optimal for 2 h fixation at room temperature, but users can increase the formaldehyde concentration for shorter fixes (e.g., twice this concentration to fix in 1 h) or decrease the concentration to fix for longer periods or lower temperatures (e.g., one half this amount to fix overnight at 4°C). Make immediately before use and discard (appropriately) excess or used solution.

3. PFA fix: PFA is best made fresh from powder. To prepare a 10× PFA solution: add water or buffer to PFA powder to correct volume for making a 37% stock (e.g., 10 mL) and heat in a closed tube, in a water bath at 75°C for 1 h. Then, add one drop of 10 N NaOH and heat for a further hour. PFA should be completely in solution by this point. If not, add one more drop of NaOH and heat until it is in solution. Store at room temperature and use within a week.
4. Dent's fix: 80% methanol, 20% DMSO. Store at -20°C.

2.2. Solutions for Antibody Generation, Binding, and Detection

1. Column buffer: 50 mM MOPS pH 7.5, 200 mM NaCl, 1 mM EDTA.
2. Tris buffered saline (TBS): 150 mM sodium chloride plus 10 mM Tris pH 7.5.
3. TBS plus Tween 20 (TBST): TBS plus 0.05% (v/v) Tween 20.
4. TBS plus EDTA (TBSE): TBST plus 20 mM.
5. Blocking solution: TBST plus 2 mg/mL BSA and 10% heat-treated goat serum.
6. Alkaline phosphatase buffer: 20 mM Tris–HCl pH 9.5, 100 mM NaCl, 50 mM $MgCl_2$. Levamisole is added just prior to use.
7. Phosphate buffered saline (PBS): 150 mM NaCl, 3 mM KCl, 10 mM Na_2HPO_4, 1.76 mM KH_2PO_4, pH 7.4.
8. PBS plus Triton-X-100 (PBST): PBS plus 0.1% (v/v) Triton-X-100.
9. PBS plus Tween/BSA/imidazole (PBTI): PBS plus 10 mM imidazole, 0.2% BSA and 0.05% Tween 20. Store aliquots at -20°C.
10. Methanol bleach. Mix 70 mL of 100% methanol with 30 mL 30% (v/v) H_2O_2.

Store in dark at 4°C indefinitely.

11. Formamide bleach; 5% formamide, 0.5× SSC, 1% H_2O_2.
12. Fish gelatin; Sigma G7765.
13. Poly-l-lysine; Sigma P4707.
14. Mowiol 4-88; Polysciences.
15. BCIP/NBT; Roche 11681451001.
16. PAP pen; Abcam 2601.

2.3. Antibodies

Both monoclonal and polyclonal antibodies work with these protocols.

Monoclonals are available either as a cell culture supernatant or as ascitic fluid. The concentration of antibody is usually low in culture supernatant, and thus the reagent is used with very little dilution, sometimes it needs to be used neat. Following use, the antibody solution can be recovered and reused many times—we have recycled monoclonals up to ten times. Ascites contains very concentrated antibodies. The dilution utilized needs to be determined empirically, but is often in the realm of 1:5,000.

Over 20 monoclonals that work against *Xenopus* proteins are available from the Developmental Studies Hybridoma Bank (DSHB, http://dshb.biology.uiowa.edu/) and many more will soon be available from the European *Xenopus* Resource Centre (EXRC, http://www.port.ac.uk/research/exrc/). These, along with some commercial reagents that work in *Xenopus*, are listed in Table 1. Some manufacturers will provide small trial samples of

Table 1
Monoclonal antibodies useful in *Xenopus* research

Monoclonal	Target	Source
12/101	Somites	DHSB
12G10	Alpha-tubulin	DHSB
1H5	Cytokeratin type II	DHSB
39.4D5	Islet-1 homeobox	DHSB
3A10	Neurofilament-associated antigen	DHSB
3G8	Pronephros	EXRC
4A4	p63 (basal layer of epidermis)	Abcam, Millepore, Santa Cruz
4A6	Pronephros	EXRC
4D9	Engrailed/invected	DSHB
5A3	Somites and heart	EXRC
6.21E+07	Slug/snai2	DSHB
8C8	Integrin beta 1	DSHB
AA4.3	Alpha tubulin	DSHB
Alpha 6F	Na/K ATPase alpha 1/atp1a1	DSHB- note! use Dent's fix
C464.4	Na/K ATPase alpha 1/atp1a1	Millipore cat. 16-243, MEMFA fix
E Cadherin	E cadherin	DSHB- note! use Dent's fix
Active JNK/mapk8	Dual phosphorylated JNK	Promega V7931
erk 1/2 mapk1	Phospho-mapk1	Sigma m96292
Alpha tubulin	Microtubules	Sigma T6199
Gamma tubulin	Basal bodies, microtubule minus ends	Abcam ab27076
zn-12	Neural	DSHB- note! use Dent's fix

Table 2
Polyclonal antibodies useful in *Xenopus* research

Target	Fix, dilution	Source
Alpha-catenin ctnna1	MEMFA 60' 1:300	Sigma C2081
Caspase 3 cleaved casp3	MEMFA 30' 1:300	Cell Signaling Technology #9661
Centrin cetn1 (basal bodies)	MEMFA 60' 1:200	Abcam ab11257
Phospho (ser 10) histone H3 (metaphase)	MEMFA 120' 1:500	MillIpore 06-570
Phospho erk1/2 mapk1	MEMFA 60' 1:400	Cell Sig Technology #9101
PKC mu/prkd1	MEMFA 30' then Dent's. 1:300	Santa Cruz SC-639
Phospho-SMAD 1/5/8	MEMFA 60' 1:1,000	Cell Signaling Technology #9511
Troponin (raised in goat)	Dent's	Fitzgerald 70-xg79
ZO-1, tight junctions	MEMFA 60' 1:100	Invitrogen 61-7300

their products before you invest in what can be very expensive reagents—especially when they do not work on *Xenopus*. Others will sometimes exchange an antibody that does not work, so for one purchase you can work through antibodies until finding one that meets your needs.

Some polyclonals also work well on *Xenopus*, with antibodies directed against bacterial fusion proteins generally having a higher success rate than those directed against synthetic peptides. A list of useful polyclonals is provided in Table 2. We have successfully generated polyclonals using both approaches. Although many researchers are put off by the time and effort involved in generating antibodies, particularly in the case of bacterial fusions, there are many reputable companies that provide significant technical help. The research benefits of having a working antibody to the protein of interest usually far outweigh the effort put into producing it. The authors have generated six polyclonals that worked well; foxg1, foxi1e, cal2, sox3, myt1, Alsin all in rabbits, and four that did not; pax8 and p27xic in rabbits, sox3 in chickens, and wt1 in guinea pigs.

Many antibodies work well against denatured proteins on Western blots, but show no immunoreactivity against whole fixed embryos. This is particularly true of antibodies generated against peptides. Should this occur it is recommended that a bacterial fusion be used as the antigen. Alternative fixation protocols (Dent's, TCA) and epitope uncovering methods (heat, low pH) should also be tested.

3. Methods

3.1. Generation of Polyclonal Antibodies

In general, polyclonal antibodies can be synthesized either by production of a fusion protein in bacteria or by synthesis of a short peptide sequence that is coupled to a carrier protein, followed by injection of the fusion protein or peptide into an animal in order to raise an antigenic response. Many companies offer services to generate antibodies against your required target. In our experience, generation of polyclonal antibodies in rabbits has been most successful. However, it is worth spending some time thinking about how the antibody will be used. If it will be primarily used together with an existing successful antibody generated in rabbits, it is worth raising the new one in a different species (e.g., chickens or guinea pigs). Double antibody staining with antibodies raised in the same species is possible (e.g., using the Zenon kit, Invitrogen); however, it requires a lot of optimization.

When monoclonals work they work well and have extremely low background staining (Table 1). However, their generation is complex and beyond the scope of this chapter. Once again, many companies provide monoclonal generation and testing services to make the production of these reagents less time consuming.

3.1.1. Antibodies Against Synthetic Peptides

The production of polyclonal antibodies against peptides is the least laborious, as companies do most of the work. A region of the protein of interest is selected, based on lack of homology with other proteins of the same class, and then the sequence is used to predict peptides with the greatest antigenicity, taking into account hydrophobicity, hydrophilicity, and steric hindrance. Companies offer extensive help at this step. The peptides are chemically synthesized and often coupled to a carrier protein such as BSA or KLH, and then injected into the chosen host animal (e.g., rabbit). Serum is harvested from the animal and can be tested by immunofluorescence and/or western blot. To affinity purify the antibody, the peptide is covalently attached to a column and the serum is passed down the column so that antibodies recognizing the target bind to the peptide. The antibody can then be eluted at low or high pH to obtain the final purified product. Companies also offer these affinity purification services (e.g., Eurogentec and Cambridge Research Biochemicals). An example of a peptide antibody that we have generated is the antibody to foxi1e (Fig. 1).

3.1.2. Antibodies Against Bacterial Fusion Proteins

Generating an antibody against a larger portion of a protein or the whole protein is a more involved process. This is exemplified by our generation of the antibody against carbonic anhydrase 12 (Fig. 1). In this approach, a fusion protein is generated prior to

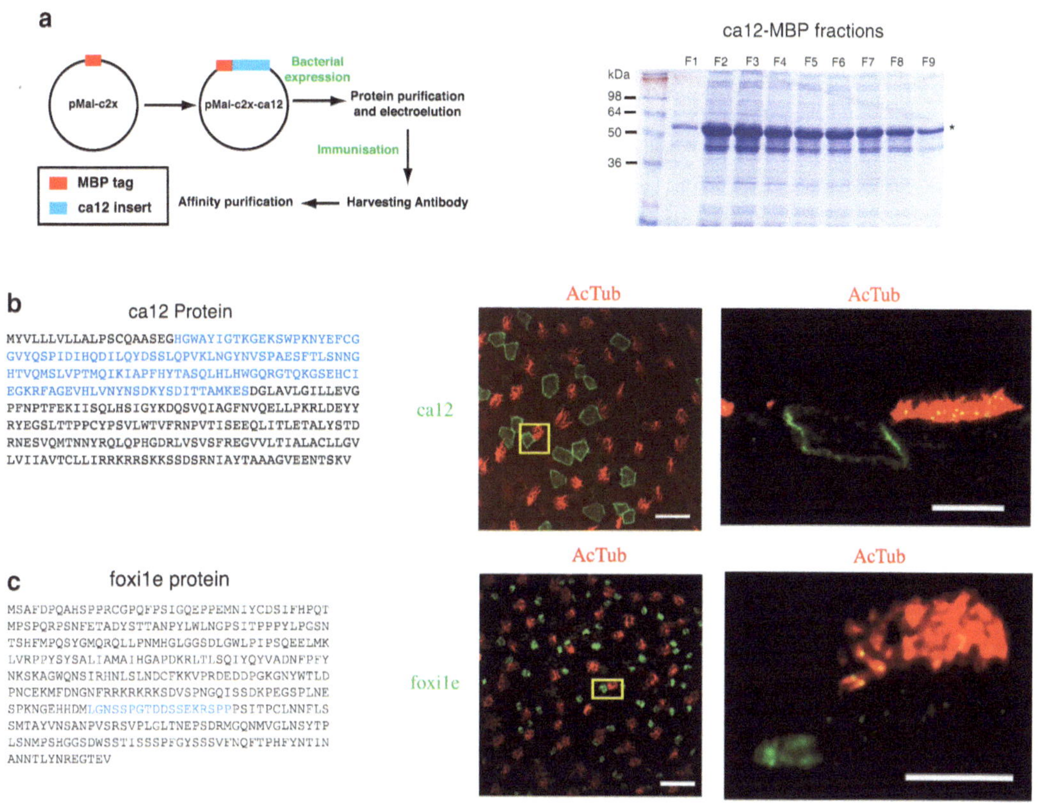

Fig. 1. Generation of polyclonal antibodies. (**a**) Schematic for generation of an antibody against a fusion protein, exemplified by MBPca12. Fractions (F1-F9) of the purified MBPca12 fusion protein are shown on the gel and *highlighted* (*asterisk*). (**b**) ca12 polyclonal antibody. The portion of the ca12 protein used to generate the fusion protein is shown in *blue*. Immunostaining for the ca12 antibody (*green*) is shown in wholemount (*center*) and sections (*right*) of *Xenopus tropicalis* embryos. Ca12 stains the ionocytes in the epidermis. The commercially available monoclonal antibody to acetylated α-tubulin (*red*) marks the ciliated cells. (**c**) foxi1e polyclonal antibody was generated against a peptide sequence (*blue*) within the foxi1e protein. Immunofluorescence for foxi1e (*green*) shows nuclear staining in wholemounts and sections of *X. tropicalis* embryos. Foxi1e also marks ionocytes in the epidermis. Ciliated cells are highlighted with the acetylated α-tubulin antibody (*red*). Scale bars, 50 μm for wholemount antibody staining and 10 μm for sections.

immunization and purification. Production of the antibody against a fusion protein can be separated into three stages:

1. Cloning, bacterial expression, and purification of the fusion protein.
2. Immunization of animals with the fusion protein.
3. Affinity purification of the antibody.

Companies or in-house institutional facilities usually perform the immunization step, but the other steps can easily be performed in the laboratory and these are described below. Again, the portion of the protein chosen should be predicted to have good antigenicity and be significantly different from other related proteins. Either the full-length protein, or most commonly, a portion of the protein fused to a tag such as glutathione-*S*-transferase (GST) or maltose-binding protein (MBP) is used. Tagging the protein facilitates its purification by enabling specific binding to an immobilized ligand. For GST-tagged proteins, glutathione beads can be used to bind the tagged protein, whilst amylose beads can be used to bind MBP-tagged proteins. Amylose is a polymer of glucose residues and since maltose is a glucose disaccharide, MBP can bind to it. We prefer using an MBP tag because it enhances the solubility of the fusion partner protein and enables greater yield of your fusion protein (10, 11).

Generating a Fusion Protein

Specifically designed primers are used to clone the whole gene or region of interest by PCR amplification into a vector in-frame with the tag. The vector we have used is pMal-c2x, which is available from New England Biolabs (NEB). It is recommended that the open reading frame of your protein be inserted in such a way that it encodes the downstream C-terminal end of the fusion protein, if Affigel is to be used for affinity purification, with MBP located at the N-terminal end. Restriction sites should be chosen so that the gene of interest is cloned into the appropriate end of MBP. It is a good idea to sequence the construct to check for accuracy of the fusion prior to bacterial expression.

A common method of expressing fusion proteins is to transform the fusion construct into bacteria and induce expression of the protein product. This was done for the MBP-ca12 fusion protein (Fig. 1). For details and troubleshooting, see company protocols that accompany the vectors (e.g., pMal-c2x, NEB). The protocol takes 4 days and is summarized below.

Day 1

Transform fusion construct into competent bacterial cells (e.g., DH5). As a control, the construct containing just the tag (e.g., MBP) should also be transformed into competent cells. Spread the transformed cells onto LB plates seeded with a selective antibiotic (e.g., ampicillin for pMal-c2x). The antibiotic used depends on the resistance conferred by the plasmid used to make the fusion construct. Leave the plate overnight at 37°C to form colonies containing transformed bacterial cells.

Day 2

Pick a colony from both the control plate (containing tag only) and the fusion construct plate and resuspend each in 2 mL LB media (containing selective antibiotic) for several hours. Use this culture to seed a 10 mL overnight culture.

Day 3

1. Use the 10 mL overnight culture to inoculate 200 mL LB (+ antibiotic) and after 1 h growth increase volume to 1 L. Add glucose to a concentration of 0.2%. (Glucose is added specifically for genes cloned into the pMal vector because it

represses the maltose genes on the chromosome of the *E. coli* host, one of which is amylase, which can degrade the amylose on the affinity resin.)

2. Grow culture to 2×10^8 cells/mL ($A_{600} = 0.5$) so that the cells are in logarithmic growth phase. Add IPTG (Sigma) to a final concentration of 0.3 mM and incubate at 37°C for 3 h. (The temperature and time of incubation depends on several factors such as stability of protein or host strain, and it can be varied in order to find optimum conditions for expression.) IPTG induces expression of the fusion protein.
3. Centrifuge the culture at $4{,}000 \times g$ for 20 min at 4°C to separate the bacterial cells from the media. After centrifugation, discard the supernatant and resuspend the cells in 50–100 mL column buffer. Centrifuge again ($4{,}000 \times g$ for 20 min at 4°C), discard supernatant, and resuspend the cells in 50 mL column buffer containing protease inhibitors (1× Complete tablet, Roche, 100 μM PMSF, Sigma and 100 μg/mL lysozyme). Freeze sample overnight at −20°C in 2×50 mL tubes containing 25 mL of cell suspension in each tube.

Day 4

1. Slowly defrost cells on ice.
2. Keep cells on ice and sonicate four times for 15–60 s at an amplitude of 40%. Allow 1-min recovery time between sonication steps.
3. Centrifuge at $5{,}000 \times g$ for 1 h at 4°C. Harvest the supernatant. This is the crude extract.
4. Affinity purification of fusion protein: To purify the fusion protein, prepare the substrate, which will bind to the tag. In the case of MBP, this is amylose resin. To prepare the amylose, which comes stored in 20% ethanol (NEB), wash three times in column buffer to equilibrate the substrate. Centrifuge at 500 rpm and pour off the buffer. Repeat washes and centrifugation steps three times. Use approximately 8 mL amylose to get a final volume of 4 mL after centrifugation steps.
5. After washing the amylose resin, mix with the crude extract (25 mL max) and roll at 4°C for 2 h. This allows the MBP-tagged fusion protein to bind to the amylose.
6. Load a column (Bio-Rad, 20 mL, Econo-Pac chromatography columns, cat: 731-1550EDU) with the amylose-crude extract mix and collect the flow-through. Re-load the flow-through to ensure as much of the fusion protein has bound to the amylose.
7. Wash with column buffer (30 mL).
8. To elute fusion protein, add 10 mM maltose (in column buffer) to column and take 10×1 mL fractions.

9. Take a small aliquot of each fraction and test on a small SDS/PAGE gel using coomassie blue staining to detect the proteins. The fusion protein should be evident as a strong band. Combine fractions positive for the fusion protein.

Immunization

A company or institutional facility usually carries out the immunization procedure. How the fusion protein is sent for immunization depends upon the company's protocol. The most successful method used in our laboratory involves running half of the volume of each positive fraction on a large gel, excising the fusion protein from the gel, homogenizing to a finer gel, and combining with the other half of the fraction volume. This allows both denatured protein (from the gel) and native protein (from the column fractions) to be injected into the animal. It is possible to cleave the tag prior to immunization but we have found that this process is often inefficient and it does not offer any advantages, as any antibodies raised against the tag can be removed from the postimmunization serum by affinity chromatography.

The immunization project usually takes 1–3 months. In summary, a species of animal is chosen and injected with your protein of interest. The animal is usually given 2–3 doses of the protein and regular harvest bleeds are taken for testing. In this case, testing by antibody staining can show quite early on whether the antibody is effective. At the end of the project, the final harvest bleed is taken and this should contain antibodies specific to your protein. The antibody can then be affinity purified.

Affinity Purification of Polyclonal Antibodies

Affinity purification greatly reduces background staining, which is often a serious problem with polyclonal antibodies. This process links specific targets of the antibody, usually a bacterial fusion protein or a synthetic peptide, to beads that are then used to purify serum via column chromatography (Biorad 20 mL columns). The purification process also removes antibodies directed against the tag portion of the fusion protein.

Coupling of target proteins or peptides to a matrix

We use affi-gel from Biorad as the column matrix, but many others are available and work well. Affi-gel comes in two varieties, Affi-gel 10 (Biorad cat# 153-6046) and Affi-gel 15 (Biorad cat#153-6052). The different products prefer proteins with different *p*Is. You can either use your favorite software to tell you the *p*I of your protein and use the appropriate product, or be lazy and use a 1:1 mixture of both products. Affi-gel couples to primary amine groups. The common buffer, Tris, will therefore couple to the column very effectively and occupy all active sites, rendering the matrix inert. If the slightest amount of Tris is in your protein the coupling will fail (see Note 1).

Detailed instructions on performing the coupling of your protein in 0.1 M MOPS to the matrix come with the product.

We use 0.5 mL (packed volume) of matrix, which can bind up to 20 mg of protein, and couple for 4 h at 4°C. You test the efficiency of coupling by comparing the free protein concentration before and after the coupling reaction. It normally approaches 80%.

Following coupling you should block any unreacted binding sites on the matrix. This is done by removing your protein and replacing it with 1 M ethanolamine pH 8.0, mix at 4°C for an hour. Rinse with TBS and store at 4°C in TBS plus 0.2% sodium azide.

Two different columns should be made for purification of polyclonal antibodies directed against fusion proteins. The first consists of the tag portion of the fusion protein, typically MBP. Sera is first passed through this column so all antibodies directed against MBP are removed. The depleted serum is then applied to a second column coupled to the MBP fusion protein to isolate antibodies to the protein of interest.

If your antibody is generated against a peptide, then a single column coupled to the peptide is sufficient. There are many ways to couple peptides. The best choice depends on the sequence of your peptide, and which end you want linked to the matrix. If you want to link the amino end you can use a matrix similar to Affi-gel, but you usually use anhydrous coupling in DMSO rather than the protocol described for fusion proteins. Details are supplied with the Affi-gel products. If you would like to link via a cysteine residue you can make a very stable thio-ether bond to Epoxy-Sepharose-6B (Pharmacia). This is a handy way to link the C-terminus to the resin, just add a cys to the end when it is made. This matrix can also link to amine and hydroxyl groups, so is a very flexible reagent. Do the coupling to 1.0 mL of preswollen Epoxy-Sepharose-6B in 0.1 M $NaHCO_3$ pH 9.0 in a small reaction volume with constant mixing at room temperature for 16 h. Wash with 50 mL of 0.1 M $NaHCO_3$ pH 9.0, then block unreacted groups on the matrix by incubating the 1.0 mL of resin in 5.0 mL of 0.1 M 2-mercapto-ethanol for 2 h with gentle agitation. An alternative method of blocking is to use 1 M ethanolamine pH 8.0 for 4 h at 4°C.

Different antibodies are eluted off affinity columns under different conditions. Most will come off in the glycine elution protocol outlined here. You can test if they came off or not by applying the eluate fractions to a small piece of nitrocellulose that has a spot of your protein on it and developing it by normal protocols. Glycine is the best option, as not only does it work with most antibodies it also causes less damage to the column itself. If the antibody is not eluted with glycine, try the ethanolamine base protocol, which is also not too harsh on the column. If nothing comes off with either of these, try the high magnesium elution, while this will probably work it may also destroy your column. Protocols for ethanolamine and magnesium elution are available on the Xenbase wiki at www.xenbase.org.

Preparation of serum

1. Thaw serum at room temperature.
2. Place in 56°C water bath for 30 min to inactivate complement.
3. Cool on ice.
4. Spin in Eppendorf for 2 min to remove scum, transfer clear serum to clean tube, and store on ice.
5. Dilute heat treated and chilled serum 1:10 with TBS, apply to the Affigel MBP column.
6. Collect the eluate—it will contain the antibodies of interest.
7. Pass the eluate over the column a second time. Collect the eluate from this column and store on ice.

Preparation of fusion protein column for first use
Except where noted, perform this entire procedure at 4°C.

The column should be prewashed with the solution that will be used in the purification to ensure that any materials that will be eluted/removed by them do not do so during the purification. If you have tested your antibody and found it comes off well in glycine, use glycine.

Use a column bed volume of 0.5–1.0 mL.

1. Wash with 5 mL of TBS, allow to drain.
2. Wash with 20 mM Tris pH 7.5, 500 mM NaCl, drain.
3. Wash with 5 mL of glycine elution buffer, allow to drain.
4. Wash with 5 mL of 50 mM Tris pH 8.8, allow to drain.
5. Wash with 2 × 5 mL TBS. Ready to use.
6. Prepare ten Eppendorf tubes each containing 42 μL of 1 M Tris Base and 50 μL 5 mg/mL BSA. Label these tubes with the antisera number and Gly#1 to Gly#10, these will be used for glycine elution.
7. Apply the prepared serum (stored on ice earlier) to the Affigel fusion protein column and allow to drain through.
8. Collect the serum and pass over the column again. Collect and pass over the column a third time. The antibodies of interest will be bound to the column—but collect serum and store at 4°C—just in case.
9. Wash column with 2 × 5 mL of TBS.
10. Wash column with 2 × 5 mL 20 mM Tris pH 7.5, 500 mM NaCl.
11. Wash column with 10 × 0.5 mL of glycine, as you add each 500 μL to the column, collect 500 μL aliquots from the bottom of the column into the prelabeled tubes containing Tris base and BSA. Add the first 0.5 mL, allow it to run through and collect it, then add the next 500 μL, etc.

12. Wash column with 5 mL of glycine.
13. Wash column with 5 mL of 50 mM Tris pH 8.8.
14. Wash column with 10 mL TBS.
15. Store all samples at 4°C temporarily. Add azide to 0.1% for long-term storage in fridge.
16. Check antibody on wholemounts or nitrocellulose spot blots. Try 1/20 and 1/100 dilutions. Label with details and freeze positive aliquots.

Controls

Once a new antibody has been generated, tests should be performed to ascertain that the antibody recognizes the protein of interest and that it is specific for it. The antibody production company typically runs ELISA assays on the crude antiserum, but these tests are not a good predictor of success in the embryo. We start by injecting mRNA for myc or HA tagged versions of the protein and testing whether the antibody recognizes these in Western blots. The Western blot is stripped and hybridized with anti-myc or anti-HA antibody. If the result is positive, we proceed to test detection of the endogenous protein using developmental stages and/or tissues that we know are positive or negative for expression of the gene from RNA expression studies, as positive and negative controls, respectively. It is not uncommon for antibodies to work well on Western blots but not in immunostaining, but we have encountered exceptions to this rule; therefore, we always check new antibodies by immunostaining, with a variety of fixative conditions, irrespective of the outcome of Western blots. We perform these tests with the serum prior to affinity purification to decide whether the serum ought to be affinity purified. The ultimate test to show specificity of the antibody is to test it in a genetic null background. In *Xenopus*, we test our antibodies in MO-injected embryos. If the antibody is specific, the band or staining should not be present in extracts or tissues from MO-injected embryos. Unfortunately, this is only interpretable if one knows that the efficiency of the MO knockdown is good. The preimmune serum is also a good negative control, as is preabsorbing the postimmune serum with the antigen used in the immunization.

3.2. Application of Antibodies

The immunohistochemistry method consists of preparation of the sample (fixation of the embryos and optional sectioning), antibody staining (with optional amplification of fluorescence signal), bleaching, and clearing.

3.2.1. Fixation and Permeabilization

Place embryos in a 5 mL vial and remove excess culture medium. Fill the vial with fresh MEMFA except for a small air space. Place on a rocking or rolling platform for 2 h at room temperature. Following fixation, replace the MEMFA with 100% methanol and store at –20°C for up to 2 years (see Note 2). The methanol enhances the permeability of fixed embryos and is an important step in wholemount protocols. Incubation in methanol for at least 24 h is essential.

To fix with Dent's fixative, place embryos in Dent's and store overnight at –20°C. Replace with 100% methanol the following day and store at –20°C.

3.2.2. Sectioning

Embryos can be sectioned prior to antibody staining or after antibody staining. For epidermal targets it is easy to perform wholemount antibody staining and then section the embryos, but for internal targets it may be necessary to section the embryos first, particularly on later stage embryos. Embryos are embedded in fish gelatin prior to sectioning. For preneurula and neurula stage embryos, embed embryos in 15% fish gelatin (+15% sucrose, w/v). For postneurula stage embryos, embed embryos in 25% fish gelatin (+15% sucrose, w/v). Leave embryos to rock in fish gelatin for a few hours or overnight. Place embedded embryos in plastic moulds and mark position of embryo on mould. Freeze samples on dry ice for 20–30 min to solidify the fish gelatin. Remove fish gelatin from block and place on cryostat for sectioning. Trim block so that the region containing the embryo is exposed to the blade. Sections of 10–12 μm are usually sufficient. Transfer sections onto polylysine-coated glass slides (by heat transfer of the cold fish gelatin sections onto the warmer slide). Freeze slides for 1 h or overnight at –80°C, and then allow to air dry before removing the fish gelatin by washing for 2 min in acetone. If sectioning prior to antibody staining, the slides can be washed in PBS before proceeding with the antibody staining protocol. If sectioning after antibody staining, remove slides from acetone, wash in PBS, and air-dry, before mounting in Mowiol and sealing with a glass coverslip.

3.2.3. Wholemount Antibody Staining

The specific sequence of steps required are rehydrate, block, incubate with primary antibody, wash, block, incubate with secondary antibody, wash, detect bound antibody.

1. Rehydrate. Although dehydration can be performed quickly, rehydration needs to be done at a slower pace as rapid hydration will cause swelling and loss of integrity.
2. Replace the 100% methanol with 98% methanol, then 95%, then 75%, then 50%, then 25%, then replace with TBS. With each change mix the sample and allow it to sit for a few minutes before moved down the methanol series.
3. Nonspecific antibody binding is blocked by incubation with blocking solution. Remove the TBS and replace with 300 μL of blocking solution. Incubate at room temperature with rocking for 30 min.

4. Add the primary antibody to the blocking solution already in the tube. Typical dilutions for tissue culture supernatants is 1:5 and for ascites 1:2,000. Incubate with rocking at 4°C overnight.
5. Remove most of the solution, leaving the embryos just covered with liquid (the blocking solution/primary antibody can be stored at −20°C and reused many times if the antibody is in short supply).
6. Fill the vial with TBST except for a small space, cap, and place on a rocker. The small bubble generated by the space will agitate the embryos as they rock.
7. Change the solution every 30 min for 3 h.
8. Remove the TBST and replace with 300 μL of blocking solution. Incubate at room temperature with rocking for 15 min (see Subheading 3.2.5).
9. Add the secondary antibody directly to the blocking solution. Most secondary antibodies such as goat or donkey anti-mouse-alkaline phosphatase and FITC conjugated anti-mouse secondary antibodies are supplied at around 500 times working concentrations, so approximately 0.5 μL of secondary is normally used per sample. Incubate with rocking at 4°C overnight.
10. Remove most of the solution, leaving the embryos just covered with liquid.
11. Fill the vial with TBST except for a small space, cap, and place on a rocker. The small bubble generated by the space will agitate the embryos as they rock.
12. Change the solution every 30 min for 3 h. At this point, experiments using fluorescent secondary antibodies can be viewed immediately. If histochemistry is required the embryos should be placed in a buffer appropriate for the conjugated enzyme, for example alkaline phosphatase.
13. Fill the vial with 1× alkaline phosphatase buffer, add 50 μL of 100 mM levamisole, rock for 5 min.
14. Remove all but 500 μL of solution. Add 10 μL of BCIP/NBT (9.4 mg/mL; 18.75 mg/mL, respectively, in dimethyl formamide), mix and incubate at room temperature while monitoring color development.
15. Stop the color reaction by rinsing away the substrate solution with TBST, then store in TBSE. The EDTA is an effective preservative and embryos can be stored in this solution for long periods at 4°C.

3.2.4. Antibody Staining on Sections

This method is based on the same principles as wholemount antibody staining, but has shorter incubation times and requires different apparatus.

1. Air-dry slides (containing sections) for 1 h in the fumehood.
2. To remove fish gelatin from sections, submerge slides in acetone for 2 min.
3. Wash slides for 20 min in PBST.
4. Surround perimeter of each slide with a PAP pen to create a hydrophobic barrier and retain the staining reagents in a localized area over the tissue sections.
5. Transfer slides to a humidified chamber (box lined with a wet tissue) and block with 150–200 μL PBST and 5% heat-treated lamb serum (HTLS) for 1 h at room temperature. Apply the blocking reagent to each slide so that it is retained on the tissue sections within the hydrophobic barrier created by the PAP pen.
6. Remove blocking reagent. Dilute primary antibody in PBST + 5% HTLS and apply 150 μL of the antibody to the sections. Incubate overnight in chamber at 4°C.
7. Wash slides 3 × 5 min in PBST to remove excess primary antibody. Block with PBST + 5% HTLS for 1 h.
8. Add 150 μL secondary antibody onto each slide and incubate for 1 h at room temperature.
9. Wash for 5 min in PBST.
10. If staining DNA, dilute DAPI (1:2,000) and apply 150–300 μL to each slide of sections.
11. Incubate for 5 min at room temperature.
12. Wash 3 × 10 min in PBS-Triton.
13. Dry slides, mount in Mowiol, and cover with coverslip.
14. Leave overnight at 4°C.
15. Seal with nail varnish, dry, and image.

3.2.5. Tyramide Fluorescent Amplification

If fluorescent antibodies do not generate a strong enough signal to be visualized or documented, the strength of the signal can be greatly amplified by using tyramide-based reactions. This is done using the protocol described in Subheading 3.2.3 up to step 8. At step 9, continue as follows:

1. Add a HRP (also known as POD) conjugated secondary antibody, for example donkey anti-mouse-HRP from Jackson Immuno. As with other secondary antibodies, these are often supplied at approximately 500× working concentration.
2. Incubate overnight at 4°C
3. Remove most of the solution, leaving the embryos just covered with liquid.
4. Fill the vial with TBST except for a small space, cap, and place on a rocker. The small bubble generated by the space will agitate the embryos as they rock.

5. Change the solution every 30 min for 3 h.
6. Change TBST solution to PBTI, mix for 10 min at room temperature.
7. Discard PBTI, replace with PBTI plus BSA, mix at room temperature for 30 min.
8. Discard solution, replace with 100 μL PBTI plus 1% BSA. Add 1 μL of fluorescent tyramide. Mix and incubate for 30 min at room temperature.
9. Add H_2O_2 to 0.001% (v/v). Incubate at room temperature for 60 min.
10. Wash with TBST at room temperature for 30 min.
11. Change TBST, wash overnight (or longer) at 4°C with rocking. If background fluorescence is strong, wash for additional 24 h at 4°C.

Tyramide reaction products are very resistant to photobleaching and very stable in both TBSE and methanol.

3.2.6. Bleaching

Both the maternal pigment that makes the epidermis tan/brown and the pigmented melanocytes that develop starting at stage 33 can be effectively bleached using H_2O_2.

Two protocols are commonly used, methanol bleach or formamide bleach (see below). We use methanol/bleach.

1. Rinse embryos in 70% methanol.
2. Remove methanol, replace with 4 mL of 70% methanol/10% H_2O_2.
3. Place under a source of light overnight at room temperature. We use a light box containing fluorescent bulbs.
4. Rinse away bleach with 70% methanol. Embryos can be dehydrated for clearing or rehydrated for viewing in aqueous solution.

The alternative protocol is performed in aqueous solution. Transfer embryos directly from TBST or TBSE to formamide bleach. Place on light box. When bleached rinse with TBSE.

Bleaching can be performed prior to incubation with antibodies, but this does reduce the stability of some antibody targets. For example, 3G8 works well on previously bleached embryos, but if bleached embryos are stored at −20°C for more than a few days, 3G8 no longer binds.

3.2.7. Clearing

The yolk filled opaque tissues of early embryos can be rendered transparent by clearing. Traditionally this was performed using Murray's clear, but can also be done using methyl salicylate. The later compound is not as pungent and works extremely well. Neither clearing solution is miscible with water, so embryos must be thoroughly dehydrated prior to clearing. If there is any residual water in the embryo the clearing solution will not penetrate these regions and a portion of the embryo will remain opaque. Multiple washes with 100% methanol will ensure complete dehydration. Once dehydrated, remove the majority of the methanol and replace with 2 mL of methyl salicylate. Embryos will slowly settle into the clearing solution and rapidly become transparent. Caution should be used in handling the clearing solution both to avoid contact with the skin and to avoid any contact with plastic, as it will rapidly dissolve most plastics. For this reason clearing should be performed only in glass containers and glass pipettes, though good micropipette tips are also safe to use.

3.2.8. Performing Immunolocalization on Embryos Previously Stained by In Situ Hybridization

The hybridization of an RNA probe during in situ hybridization is performed in 50% formamide, a denaturant, at 60°C. It is not surprising then that many monoclonal antibodies, which are very particular about the 3D shape of their antigens, do not work on embryos that have previously been processed by this protocol, e.g., 3G8. However, some monoclonal staining actually benefits from this treatment. Three examples are 12/101 (somites), the atp1a1 monoclonal C464.4 (pronephric epithelia), and 4D9 (anti-engrailed). Applying either a chromogenic or fluorescent immunolocalization counterstain using these, and possibly other monoclonals, can be a very useful technique. After in situ hybridization, embryos are simply rehydrated and processed for antibody staining as outlined in Subheading 3.2.3. It is not possible to process embryos for in situ hybridization after they have already been stained with antibodies as mRNA will have been degraded.

4. Notes

1. If you use dialysis to remove Tris beware, as it will reduce but not completely remove the buffer. We have found that the only effective way to have no Tris is to not add it in the first place. If you are purifying fusion proteins on some sort of affinity matrix, such as maltose, or glutathione, simply substitute a good buffer for the Tris in your affinity protocol. For example, we usually use the NEB pMal vectors, which generate maltose-binding protein fusions. The cell lysis, wash, column loading washing, and elution are then all performed using buffers made with MOPS or HEPES rather than Tris. After removing the maltose and salt via dialysis against 0.1 M MOPS the protein is ready to couple. The pH of the final MOPS dialysis will depend on whether you use Affigel 10 or 15 (see BioRad instructions).

2. While most antibody targets are stable for long periods in methanol, this is not universal. For example, the target bound by pronephric antibody 3G8 only lasts weeks in methanol and only relatively fresh embryos will stain with this reagent. If staining for an antibody that should work fails, testing fresher embryos or nonbleached embryos is a sensible first problem solving step.

Acknowledgements

This work was supported by the Wellcome Trust to NP and NIH R01 HD45776 and P41 HD064556 to PDV. The authors would like to thank Xiaolan Zhou for his role in developing tyramide protocols.

References

1. Dent JA, Polson AG, Klymkowsky MW (1989) A whole-mount immunocytochemical analysis of the expression of the intermediate filament protein vimentin in Xenopus. Development 105:61–74
2. Klymkowsky MW, Hanken J (1991) Whole-mount staining of Xenopus and other vertebrates. Methods Cell Biol 36:419–441
3. Hemmati-Brivanlou A, Melton DA (1992) A truncated activin receptor inhibits mesoderm induction and formation of axial structures in Xenopus embryos. Nature 359:609–614
4. Saka Y, Smith JC (2001) Spatial and temporal patterns of cell division during early Xenopus embryogenesis. Dev Biol 229:307–318
5. Faure S, Lee MA, Keller T, ten Dijke P, Whitman M (2000) Endogenous patterns of TGFbeta superfamily signaling during early Xenopus development. Development 127: 2917–2931
6. Dubaissi E, Papalopulu N (2011) Embryonic frog epidermis: a model for the study of cell-cell interactions in the development of mucociliary disease. Dis Model Mech 4:179–192
7. Vize PD, Melton DA, Hemmati-Brivanlou A, Harland RM (1991) Assays for gene function in developing Xenopus embryos. Methods Cell Biol 36:367–387
8. Kroll KL, Amaya E (1996) Transgenic Xenopus embryos from sperm nuclear transplantations reveal FGF signaling requirements during gastrulation. Development 122:3173–3183
9. Drummond IA, Majumdar A, Hentschel H, Elger M, Solnica-Krezel L, Schier AF et al (1998) Early development of the zebrafish pronephros and analysis of mutations affecting pronephric function. Development 125:4655–4667
10. Kapust RB, Waugh DS (1999) Escherichia coli maltose-binding protein is uncommonly effective at promoting the solubility of polypeptides to which it is fused. Protein Sci 8:1668–1674
11. Fox JD, Waugh DS (2003) Maltose-binding protein as a solubility enhancer. Methods Mol Biol 205:99–117

Chapter 24

Multicolor Fluorescent In Situ mRNA Hybridization (FISH) on Whole Mounts and Sections

Robert Lea, Boyan Bonev, Eamon Dubaissi, Peter D. Vize, and Nancy Papalopulu

Abstract

In situ hybridization involves the hybridization of an antisense RNA probe to an mRNA transcript and it is a powerful method for the characterization of gene expression in tissues, organs, or whole organisms. Performed as a whole mount (WISH), it allows the detection of mRNA transcripts in three dimensions, while combined with sectioning, either before or after hybridization, it provides gene expression information with cellular resolution. FISH relies on the fluorescence detection of probes and is the method of choice for the simultaneous detection of transcripts with similar or overlapping expression patterns, as each can be clearly distinguished by the selection of fluorophore. Here, we describe a protocol for performing multicolor FISH in *Xenopus* embryos in whole mounts and sections that can be further combined with antibody staining.

Key words: *Xenopus*, In situ hybridization, FISH

1. Introduction

The inclusion of whole-mount RNA in situ hybridization (WISH) to the repertoire of standard techniques in *Xenopus* has been a major boost to our ability to perform three-dimensional analysis of gene expression. Nonradioactive WISH was first described in *Drosophila* embryos (1) and it was quickly adapted for use in *Xenopus* (2) and other vertebrate embryos. The central premise of any RNA in situ hybridization (ISH) protocol is the hybridization of the target mRNA to an in vitro synthesized antisense RNA probe, labeled with some form of tag, for which a reliable and sensitive detection system exists. The detection system typically consists of an antibody specific to the tag; coupled to an enzyme that catalyzes a chromogenic reaction (in chromogenic ISH) or deposition of fluorophore labels (in FISH).

Stefan Hoppler and Peter D. Vize (eds.), *Xenopus Protocols: Post-Genomic Approaches, Second Edition*, Methods in Molecular Biology, vol. 917, DOI 10.1007/978-1-61779-992-1_24, © Springer Science+Business Media, LLC 2012

Multicolor FISH can be combined with fluorescent immuno histochemistry and is advantageous in allowing the simultaneous detection of multiple expression patterns. In FISH, signal detection involves a fluorophore-labeled tyramide, which is deposited next to immobilized horseradish peroxidase (HRP) enzyme. FISH for *Xenopus* embryos has been described previously (3–6) based on modifications of the protocol of standard chromogenic ISH (7). However, it was initially limited to one or two fluorescent probes and required in-house synthesis of tyramides (5). Recently, the sensitivity of FISH has improved greatly, presumably due to increased quality and concentration of tyramides in the commercial kits. It is now possible to perform multiple FISH on whole mounts or sections of *Xenopus* embryos, with good sensitivity and reproducibility, based entirely on commercially available tyramides. In the method we describe here, each antisense RNA probe is labeled with one of three haptens (digoxigenin, DIG; dinitrophenol, DNP; or fluorescein, FITC) and detected in sequence, using anti-Dig, anti-DNP, and anti-FITC antibodies coupled to HRP (also known as peroxidase, POD), with a peroxidase inactivating step between each antibody. Detection involves a fluorophore-labeled tyramide amplification reagent, coupled to cy3, cy5, or FITC and visualized by standard or confocal fluorescent microscopy (Fig. 1). The protocol works equally well in whole mounted embryos, which can then be sectioned posthybridization (Fig. 1), or directly on cryo-sections (Fig. 1). Direct detection of fluorescently labeled probes, without amplification of the signal, has been difficult in *Xenopus* embryos, presumably due to size and yolk content of the embryo. FISH can be carried out on all *Xenopus* NF stages between 10.5 and 45 but later stages are less reliable due to impaired penetration of reagents into the specimen.

2. Materials

2.1. Fixation Materials

1. Wheaton glass Scintillation vials (20 mL VWR).
2. 0.1× Marc's Modified Ringer (MMR): 10 mM NaCl, 0.2 mM KCl, 0.1 mM $MgCl_2$, 0.2 mM $CaCl_2$, 0.5 mM HEPES, pH 7.5. Prepare a 10× stock solution, adjust to pH 7.5 with NaOH. Sterilize the 10× solution by autoclaving.
3. MEMFA fixative: 0.1 M MOPS pH 7.4, 2 mM EGTA, 1 mM $MgSO_4$, 3.7% Formaldehyde. MEMFA should be made fresh from stock solutions of 1 M MOPS pH 7.4, 0.5 M EGTA pH 8, and 1 M $MgSO_4$. Fresh formaldehyde should always be used. Stock solutions of MOPS should be kept away from light.

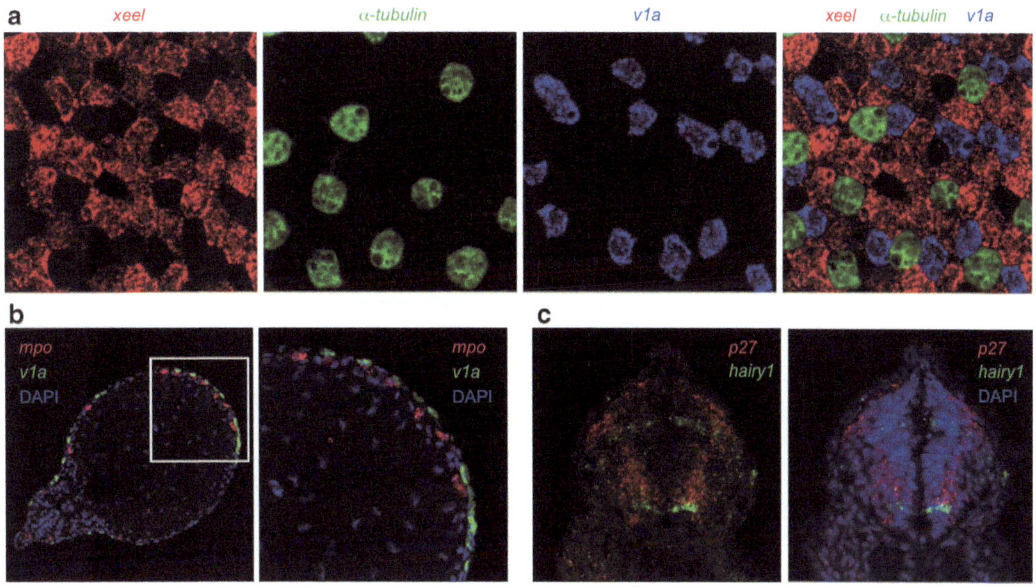

Fig. 1. (**a**) Triple whole mount FISH of stage 32 *X. tropicalis* epidermis with lectin (*xeel*) in *red*, α-tubulin in *green*, and V-ATPase (v1a, (8)) in *blue*. *Xeel* was DIG tagged and detected using the Cy3-tyramide substrate, α-tubulin was FITC tagged and detected using FITC-tyramide, v1a was DNP tagged and detected using Cy5-tyramide. The image is a lateral view taken at ×40 magnification using an Olympus FluoView FV1000 confocal microscope, with FluoView software. The channels are shown individually and as a merged image. (**b**) Transverse Coronal cryo-section (12 μm thick) of a double whole mount FISH using Myeloperoxidase (mpo, (9)) as a probe, shown in *red* and v1a, in *green* (stage 28 *X. tropicalis*). A mpo probe was DIG tagged and detected using the Cy3-tyramide substrate, a v1a probe was FITC tagged and detected using FITC-tyramide. DAPI staining has been carried out after sectioning. Images were taken on Nikon Eclipse 80i microscope using the ×10 objective. (**c**) Transverse Coronal cryosection (12 μm thick) of a double whole mount FISH (stage 30 *X. tropicalis*). A p27Xic1 probe (10) was tagged with DIG and detected using Cy3-tyramide (*red*), and a hairy1 probe (6) was tagged with FITC and detected using FITC-tyramide. DNA was stained using DAPI (*blue*). Images were taken on Nikon Eclipse 80i microscope using the ×20 objective.

4. 100% Methanol.
5. Bouin's posthybridization fixative: 10% formaldehyde, 5% glacial acetic acid in H_2O.

2.2. Probe Transcription Materials

1. 3 M sodium acetate (Sigma).
2. Ethanol.
3. DEPC water (Invitrogen).
4. RNA polymerase T3, T7 or SP6 (Roche).
5. 10× transcription buffer supplied with polymerase (Roche).
6. RNase inhibitor (Roche).
7. Ribonucleoside triphosphate set lithium salts (Roche).
8. Turbo DNsae (Ambion).
9. Micro Bio-spin columns (Biorad Cat. N: 732-6221).

2.3. WISH Reagents and Materials

1. Hybridization baskets 9 mm, 55 μm mesh (Intavis Cat. N: 16.33025).
2. Rack 40 wells (Intavis Cat. N: 33008).
3. 7 mL plastic vials (Greiner).
4. 100% Methanol.
5. 1× PBS 0.1% Tween 20 (Sigma).
6. Proteinase K 20 mg/mL (Ambion).
7. 0.1 M triethanolamine (pH 7.5).
8. Acetic anhydride.
9. 4% Formaldehyde (Baker) in 1× PBS 0.1% Tween 20.
10. Hybridization Buffer: 50% formamide (Ambion), 5× SSC, 1 mg/mL Torula RNA (Sigma), 100 μg/mL heparin (Sigma), 1× Denhardt's solution (Sigma), 0.1% Tween-20, 0.1% CHAPS (Sigma).
11. 2× SSC: 300 mM NaCl, 30 mM sodium citrate. Prepare a 20× stock solution, adjust to pH 7.0 with NaOH. Sterilize the 20× solution by autoclaving.
12. 0.2× SSC.
13. RNase cocktail (Ambion). The solution contains 500 U/mL RNase A and 20,000 U/mL RNase T1.
14. 1× MAB: 100 mM maleic acid, 150 mM NaCl. Prepare a 2× stock solution, adjust to pH 7.5 with NaOH. Sterilize the 2× solution by autoclaving.
15. 30% H_2O_2 (Sigma).
16. 10% Blocking Medium (BM) (Roche).
17. 10% Heat treated lamb serum (HTLS) (Sigma).
18. 1× PBS 0.1% Triton X-100 (Sigma).
19. 1× BB1: 1× PBS with 1% BSA (Sigma) and 0.1% Triton X-100 (Sigma).

2.4. Additional Reagents and Materials for ISH on Sections

1. SuperFrost Plus slides (VWR—Cat. N: 631-0448).
2. Cover Glass coverslips (VWR—Cat. N: 631-0448).
3. Acetone (Fisher Scientific).
4. Washing Buffer—50% formamide, 2× SSC, 0.05% Tween 20.
5. TBSTw—100 mM TRIS 7.5, 150 mM NaCl, 0.05% Tween-20.
6. Liquid Blocker PAP pen (Daido Sangyo, Japan).
7. Blocking Solution: 0.5% BM + 1% HTLS in TBSTw.
8. Secondary antibodies (Molecular Probes).
9. Vectashield Mounting Medium with DAPI (Vector Labs—Cat. N: H-1200).

2.5. Reagents

2.5.1. Probes

DIG-11-UTP (Roche Cat. N: 11209256910).

FITC-12-UTP (Roche Cat. N: 11427857910).

DNP-11-UTP (Perkin Elmer Cat. N: NEL555001EA).

2.5.2. Antibodies

Anti-DIG-HRP (Anti-DIG-POD) (Roche Cat. N: 11207733910).

Anti-DNP-HRP (Anti-DNP-POD) (Perkin Elmer Cat. N: NEL747A001KT).

Anti-FITC-HRP (Anti-FITC-POD) (Roche Cat. N. 11426346910).

Nuclear stain, Sytox green (Invitrogen Cat. N: S7020) or DAPI (Vector Labs—Cat. N: H-1200).

Alexa Fluor 405 goat anti-mouse (Invitrogen Cat. N: A-31553).

2.5.3. Tyramide Amplification Reagents

TSA Plus Cyanine 3 kit (Perkin Elmer Cat. N: NEL744001KT).

TSA Plus Cyanine 5 kit (Perkin Elmer Cat. N: NEL745001KT).

TSA Plus Fluorescein kit (Perkin Elmer Cat. N. NEL741001KT).

2.5.4. Mowiol Mounting Media

Mix 6 mL of glycerol, 2.4 g of Mowiol 4-88 (Calbiochem), and 6 mL of distilled deionized water. Mix at room temperature for 2 h. Add 12 mL of 200 mM Tris–HCl pH 8.5 and incubate at 50°C with occasional shaking until the Mowiol has completely dissolved. Filter on 0.45 μm membrane filter and store aliquots at −20°C.

3. Method for Multiple FISH on Whole Mounts

3.1. Fixation

1. Embryos are collected at the desired developmental stage and fixed with fresh MEMFA for 1 h at room temperature in 20 mL glass vials.
2. Samples are then washed three times 10 min in 100% methanol and can be stored for several months at −20°C.

3.2. Probe Transcription

1. Linearize plasmid DNA by digesting in appropriate buffer, according to standard molecular biology protocols.
2. Resuspend DNA pellet in DEPC water at a final concentration of 1 μg/μL.
3. Set up the transcription reaction.
   - 5 μL Roche 10× transcription buffer
   - 10 μL 2.5 mM NTP-DIG, NPT-DNP, or NTP-FITC (see Note 1)
   - 28 μL DEPC treated water
   - 2.5 μL 1 μg/μL linearized template DNA

- 0.5 μL 40 U/μL RNase inhibitor
- 4 μL RNA polymerase (T3, T7 or SP6)

  Incubate for 2 h at 37°C.

4. Add 1 μL of 1 mg/mL DNase and incubate for 10 min at 37°C. Add DEPC water to up to 100 μL.
5. Purify on a Micro Bio-spin column, as follows:
   - Mix column by inverting it to remove any bubbles, remove end and cap and place over collection tube to drain.
   - Discard flow through and return to collection tube.
   - Centrifuge at 4,000 rpm in an Eppendorf microfuge for 2 min.
   - Place column in clean microfuge tube.
   - Load the probe on the column.
   - Centrifuge for 4 min at 4,000 rpm.
6. Add 100 μL of DEPC water, 20 μL of 3 M sodium acetate pH 5.3, and 500 μL of ethanol. Vortex vigorously and incubate at −80°C for at least 30 min.
7. Wash with 70% ethanol.
8. Air dry pellet and resuspend in 50 μL of DEPC water.
9. Check 1 μL on a freshly made agarose gel (see Note 2).
10. Store the stock probe at −80°C.

3.3. In Situ Day 1

In all steps, samples should be rocked or shaken at moderate speed at room temperature, unless otherwise noted.

1. Place a 40-well Intavis rack in a bath containing 100 mL of 100% methanol (see Note 3). The fixed embryos are then distributed as appropriate into separate mesh baskets sitting in the rack, using a trimmed and fire polished glass Pasteur pipette. For best results, do not place more than 10–15 embryos in the same mesh basket.
2. Re-hydrate the samples in 100 mL of the relevant solution for 5 min.
   - 75% Methanol, 25% H_2O
   - 50% Methanol, 50% H_2O
   - 25% Methanol, 75% 1× PBS 0.1% Tween 20.
   - Three times 5 min washes in 1× PBS 0.1% Tween 20.
3. Use forceps to transfer each basket (with the embryos inside) from the 100 mL baths to an individual 7 mL plastic tube (see Note 4).
4. Incubate the embryos for 10 min in approximately 1 mL of 5 μg/mL proteinase K per tube (see Note 5).

5. Rinse the embryos twice for 5 min in 5 mL 0.1 M triethanolamine pH 7.8.
6. Do not remove the last 5 mL triethanolamine wash. Instead add 12.5 μL of acetic anhydride to each basket and rock for 5 min then add another 12.5 μL of acetic anhydride. Rock for 5 min (see Note 6).
7. Transfer the baskets back to the rack and wash two times 5 min in 100 mL of 1× PBS 0.1% Tween 20.
8. If proteinase K was used, re-fix for 20 min with 4% formaldehyde in 1× PBS 0.1% Tween 20 (see Note 3).
9. Wash five times 5 min in 1× PBS 0.1% Tween 20.
10. Transfer each basket to 7 mL individual plastic tubes containing 500 μL of hybridization buffer and prehybridize for 4–6 h at 60°C in a shaking incubator.
11. Replace the hybridization buffer with 0.5 mL of probe solution (hybridization solution with 1 μg/mL of a mixture of labeled RNA probes; see Note 7). Hybridize overnight at 60°C.

3.4. In Situ Day 2

1. Remove the probes and store at −20°C. Replace the probes with 1 mL of hybridization buffer and shake at 60°C for 10 min (see Note 8).
2. Transfer back to the rack and wash three times for 20 min at 60°C in 2× SSC.
3. Wash for 20 min at 37°C in 2× SSC.
4. Transfer the embryos to individual glass 5 mL Wheaton vials for RNase treatment at 37°C. Dilute 10 μL of RNase cocktail solution per mL of 2× SSC. Use 0.5 mL of the final solution per sample (see Note 9).
5. Transfer back to the rack and wash for 10 min in 2× SSC at room temperature to remove excess RNase cocktail.
6. Wash twice for 20 min at 60°C in 0.2× SSC.
7. Wash twice in 1× MAB for 5 min at room temperature.
8. Wash for 30 min in 3% H_2O_2 diluted in 1× MAB (see Note 10).
9. Wash twice in 1× MAB for 5 min.
10. Transfer the baskets to 7 mL individual plastic tubes and block for 1 h in 0.5 mL 1× MAB + 2% BM + 2% HTLS.
11. Incubate for 4 h with 0.5 mL of antibody diluted in 1× MAB + 2% BM + 2% HTLS. We usually detect the Dig labeled probe first, so use anti-DIG-POD (1:1,000) in this step (see Note 11).
12. Transfer the samples back to the rack and wash three times in 1× MAB, 5 min for each wash. Store overnight in 1× MAB on a rocker.

3.5. In Situ Day 3

1. Wash five times in 1× MAB at room temperature, 20 min per wash.
2. Remove the embryos from the baskets using a glass pipette and transfer them to 5 mL glass Wheaton vials. Incubate for 10 min in 300 μL of amplification buffer.
3. Aspirate the amplification buffer and incubate for 20–30 min in 300 μL of Cy3-tyramide solution (see Note 12) (1:100 dilution in amplification buffer). Ensure that all the embryos are fully submerged in the solution. We prefer to use Cy3 first, due to its stability (see Note 13).
4. Transfer the samples back to the baskets and wash the samples three times in 1× PBS 0.1% Tween 20, 10 min for each wash. Leave overnight in 1× PBS 0.1% Tween 20.

3.6. In situ Day 4

1. Wash in 1× PBS 0.1% Tween 20 for 20 min.
2. Wash for 30 min in 3% H_2O_2 diluted in 1× PBS 0.1% Tween 20 (see Note 14).
3. Wash three times 5 min in 1× PBS 0.1% Tween 20.
4. Transfer the baskets to 7 mL individual plastic tubes and block for 1 h in 0.5 mL 1× MAB + 2% BM + 2% HTLS.
5. Incubate for 4 h with 0.5 mL of antibody diluted in 1× MAB + 2% BM + 2% HTLS. We use anti-DNP POD (1:500) as the second antibody.
6. Transfer the samples back to the rack and wash three times in 1× MAB, 5 min for each wash. Wash overnight in 1× MAB on a rocker.

3.7. In Situ Day 5

Same as day 3; the only difference is the use of Cy5-tyramide as a substrate in the enzymatic reaction. See Note 15 if doing a double in situ.

3.8. In Situ Day 6

Same as day 4 but use anti-FITC POD (1:500) as the final antibody.

3.9. In Situ Day 7

Same as day 3 but use FITC-tyramide as a substrate for the enzymatic reaction. Since FITC is light sensitive, all the subsequent steps need to be performed in tin foil wrapped vials or baskets (see Note 16).

3.10. In Situ Day 8

1. Remove the embryos from the baskets using a glass pipette and transfer them to 5 mL glass Wheaton vials.
2. Aspirate the 1× PBS 0.1% Tween 20 and wash two times in 1× BBT at room temperature, 1 h per wash.
3. Block for 1 h in 1 mL 1× BBT + 2% HTLS.
4. Incubate overnight at 4°C with 0.5 mL of primary antibody diluted appropriately in 1× BBT + 2% HTLS.

3.11. In Situ Day 9

1. Wash four times 1 h in 1× BBT.
2. Block for 1 h in 1 mL 1× BBT + 2% HTLS.
3. Incubate overnight at 4°C with 0.5 mL of secondary antibody diluted in 1× BBT + 2% HTLS. We use Alexa Fluor 405 anti-mouse (see Note 21).

3.12. In Situ Day 10

1. Wash for 1 h in 1× BBT.
2. Wash three times 1 h in 1× PBS 0.1% Tween 20 and leave overnight at 4°C.
3. Store in 1× PBS 0.1% Tween 20 ready for imaging.

3.13. In Situ Day 11

1. Wash three times 5 min in 1× PBS (see Note 17).
2. Wash for 1 h in Bouin's to postfix.
3. Remove Bouin's by washing in 1× PBS five times, 5 min per wash.
4. Store in 1× PBS and away from light (see Notes 18–21).

4. Method for Fluorescent In Situ Hybridization on Cryosections

Whole mount in situ and in situ on sections give similar results; however, in situ on sections may be preferable to detect weakly expressed genes present in the neural tube. Here, we describe a protocol for performing FISH combined with immunofluorescent antibody staining on sections (see Note 22. For details on fixing and sectioning *Xenopus* embryos, please see Chapter 26). We generally use 12 μm sections and they can be stored at −80°C for up to 2 months. It is essential that the cryosections are stored at −80°C immediately after sectioning to minimize the exposure to RNAses. Other important parameters are the fixation of the embryos with fresh MEMFA and timing the duration of the incubation with the tyramide substrate. Generally no proteinase K treatment is necessary but if there is no signal, then it can be included before the acetylation step (1–2 μg/mL PK for 5 min). RNAse-free solutions and conditions should be used until the hybridization step.

4.1. In Situ Day 1

1. Leave the slides to dry for 1 h in the fume hood.
2. Place the slides in a rack and place it in a container filled with acetone for 2 min to remove the residual fish gelatine.
3. Transfer the rack to a container filled with 1× PBS + 0.1% Triton for 2 min.
4. Pipette 250 μL 0.1 M TEA on top of the slide for 5 min
5. Replace with 0.25% acetic anhydride in 0.1 M TEA pH 7.8 and incubate for 10 min.

6. Place the slides in a rack and wash two times 5 min in a container filled with 1× PBS 0.1% Tween 20. All washes are performed in a glass beaker.
7. Remove the slides from the rack and pipette 400–500 μL 4% formaldehyde in 1× PBS 0.1% Tween 20 for 15 min.
8. Place slides in a homemade hybridization chamber (slide box with a piece of wet tissue on the bottom).
9. Apply 500–750 μL of prewarmed hybridization buffer and incubate for 1–2 h in an oven, set to 60°C.
10. Cut small stripes from coverslips with a diamond cutter (two for each slide).
11. Remove the hybridization buffer from the slide by tipping it to the side.
12. Place the strips on the two sides of the slide; add 50 μL of 1 μg/mL probe in hybridization buffer per slide. Using forceps carefully put a coverslip on top so that there are no air bubbles (crucial).
13. Incubate overnight at 60°C.

4.2. In Situ Day 2

1. Carefully dip the slides one by one in 2× SSC until the coverslips gently falls off (important—do not force the coverslip as it might damage the sections).
2. Incubate for 10 min in 2× SSC at room temperature.
3. Pipette 500 μL of wash Buffer on top of the slide and incubate for 30 min at 60°C.
4. Wash slides twice in 2× SSC for 15 min at 60°C.
5. Wash slides twice in 0.2× SSC for 15 min at 60°C.
6. Wash once for 10 min in 1× TBSTw at room temperature.
7. Wash for 20 min in 3% H_2O_2 diluted in 1× TBSTw.
8. Wash twice for 10 min each in 1× TBSTw.
9. Carefully wipe the edges of the slide with a tissue, draw a line at the edges with a PAP pen, and pipette 150 μL Blocking Solution. Incubate for 1 h at room temperature.
10. Replace with 150 μL of 1:1,000 anti-DIG-POD diluted in Blocking Solution (if you want to combine FISH with immunostaining—add the primary antibody at the appropriate dilution).
11. Incubate overnight at 4°C.

4.3. In Situ Day 3

1. Wash the slides three times for 10 min in 1× TBSTw.
2. Carefully wipe the edges of the slide with a tissue, pipette 100 μL Cy3 or FITC conjugated tyramide (1:100 in Amplification Buffer) on top of the slide and incubate for 10–15 min (see Note 23).

3. Wash three times 10 min with 1× TBSTw.
4. If you are also detecting protein reblock for 30 min with 5% HTLS in TBSTw.
5. Incubate with secondary antibody 1:500 in 5% HTLS (we use antibodies conjugated with Alexafluor 488 when combined with Cy3 tyramide or Alexafluor 568 when combined with FITC tyramide) for 1 h.
6. Wash three times 10 min with TBSTw.
7. Mount with Vectashield mounting medium containing DAPI (see Note 24).

5. Notes

1. To prepare 2.5 mM NTP-DIG, NTP-FITC, and NTP-DNP mixes.
   - 7.1 μL 100 mM CTP
   - 7.1 μL 100 mM GTP
   - 7.1 μL 100 mM ATP
   - 4.6 μL 100 mM UTP
   - 25 μL 10 mM DIG-11-UTP or 10 mM FITC-12-UTP or 10 mM DNP-11-UTP
   - 234 μL DEPC treated water

   The mix can be dispensed into 10 μL aliquots and stored at −20°C. 10 mM RNA labeling mixes for DIG (Roche Cat. N: 11277073910) and FITC (Roche Cat. N: 11685619910) can also be purchased from Roche.
2. The gel should be prepared using DEPC treated 0.5× TAE and RNase free gel apparatus. An RNA loading buffer (like the Ambion Gel loading buffer provided with the MEGAscript kit) should be used. The probe quality and concentration can be roughly estimated from the gel. For a more accurate method of estimating probe concentration, please refer to Sive et al. (7). A good transcription reaction should yield about 10 μg of probe dissolved in 100 μL of DEPC water. 100 μg/mL and is equivalent to 100× stock. This stock is diluted 1/100 in hybridization buffer to the working concentration of 1 μg/mL.
3. We have found that Gilson tip box lids work well as baths. 100 mL is the optimum volume of liquid recommended per wash using this method. While this protocol uses Intavis baskets and racks intended for use with an in situ machine it is also possible to use homemade baskets and racks described in Sive et al. (7). While using baskets is not essential, it is recommended

when performing large-scale fluorescent in situ screens. Using individual vials for all the steps can increase the risk of damaging embryos during aspiration.

4. Care must be taken to record the order of the samples when they are removed from the rack.
5. Proteinase K is used to permeabilize the embryo allowing better access for the probe. However, extensive proteinase K incubation can damage the tissue and is not recommended when studying genes expressed in the epidermis as it may digest away the superficial layer. We have found some variability between different batches of proteinase K so the time of incubation may need to be adjusted, if excessive damage of the embryos is observed.
6. Acetic anhydride treatment is required to acetylate and neutralize free amines, which reduces electrostatic interactions between basic proteins and the probe.
7. All three probes (DIG, FITC, and DNP-labeled) are added to the samples together.
8. Probes can be reused a maximum of three times and should be stored at –20°C.
9. In this protocol, the samples are changed from the baskets to individual glass 5 mL Wheaton vials for this step, avoiding the contamination of baskets with RNase. In some cases, RNAse treatment can be omitted without affecting the specificity of the signal.
10. This step is essential for quenching the endogenous peroxidase and therefore for preventing nonspecific background.
11. We develop the DIG-probe first as we have found it to be the most reliable of the three haptens in terms of signal-to-noise ratio. We therefore recommend using DIG for probes that are expressed at the lowest abundance. FITC gives the most background and is sensitive to light; therefore, we suggest detecting this last.
12. Incubation length with the substrate is dependent upon the expression levels of the gene. For highly expressed genes incubation with the substrate can be decreased to 10 min and for genes expressed at lower levels incubation length can be increased to 45 min.
13. Cy3 and Cy5 are used first from the selection of substrates, as they are more stable to light in compared to FITC. In general, do not use an FITC-tyramide substrate before using an anti-FITC antibody, as the antibody will recognize the cross-linked substrate.
14. This step is very important as it inactivates the peroxidase activity of the first antibody (in our case anti-DIG-POD). The time of incubation may need adjustment to the individual probe.

15. If a double, rather than triple in situ is required stop here, wash, and postfix, as in day 8.
16. Once the FITC solution has been added all subsequent washes in the protocol must be carried out in foil because FITC is light sensitive.
17. In order to obtain the strongest signal it is recommended that the embryos are imaged before fixation.
18. Once the embryos have been fixed they can be stored for up to 2 weeks. However, the signal strength will decrease over time. The FITC is light sensitive and fades when exposed to UV light, so it needs to be imaged first.
19. When the gene of interest is not expressed in the epidermis and/or is not very abundantly expressed, the signal from FISH might not be evident without clearing the embryo or cryosectioning it (Fig. 1; for sectioning see Chapter 30). If you wish to carry out DAPI staining after sectioning you can use either VECTASHIELD mounting medium with DAPI (Vector Laboratories) or alternatively, incubate the slides for 10 min with 5 μg/mL DAPI (4′,6 diamidino-2-phenylindole) in 1× PBS. Wash with 1× PBS two times for 5 min and mount using Mowiol.
20. In order to clear the embryos after FISH follow these steps:
    - In scintillation vials wash the embryos three times 10 min in methanol.
    - Clear the embryos in benzoate solution in a watch glass or 9 well glass plate. Benzoate solution is a 1:2 mix of benzyl alcohol: benzyl benzoate.
    - The embryos will clear within 5–30 min. The samples are kept in the benzoate solution. Image the samples in the desired manner.
    - Return the embryos to methanol and wash three times 10 min.
    - Gradually rehydrate the embryos in 1× PBS.

    The benzoate solution used for clearing is an irritant and a solvent. Therefore, any steps that use this must be performed in a glass container. Remember that the scintillation vials have plastic lids so the embryos are usually transferred to a watch glass. A pair of forceps should be designated for use only with this application, as it is difficult to remove contaminating benzoate solution from them. Contaminated material can be cleaned with methanol. We have found the fluorophores described in this chapter to be stable in clearing solution.
21. It is possible to perform immunofluorescence on whole mounts after triple FISH, in which case, do not fix the embryos after

day 7 and proceed to the protocol described in Chapter 30. We have found that the combination of fluorophores described here (cy3, cy5, and FITC) for FISH works well combined with Alexa Fluor 405 anti-mouse (Invitrogen) (A-31553) as the secondary antibody for IHC.

22. To perform multiple FISH on sections, adapt the protocol along the lines of multiple FISH on whole mounts.
23. If the FISH reaction is to be coupled with an antibody staining, we generally perform the tyramide enzymatic reaction first, before incubating with a secondary antibody, but we have found that for some proteins it was necessary to do the secondary antibody first. If doing multiple FISH coupled with antibody staining, one could add the primary antibody at the same time as the second antibody of the FISH or complete the FISH protocol before adding the primary antibody.
24. If a nuclear counterstain is required, DAPI works well.

Acknowledgments

We thank Enrique Amaya for donation of the mpo probe and Samantha Winkler for carrying out the mpo/v1a FISH. The work was funded by a Wellcome Trust Senior Research Fellowship to NP. BB and ED were funded by Wellcome Trust 4 year PhD studentships.

References

1. Tautz D, Pfeifle C (1989) A non-radioactive *in situ* hybridization method for the localization of specific RNAs in *Drosophila* embryos reveals translational control of the segmentation gene hunchback. Chromosoma 98:81–85
2. Brivanlou AH, Harland RM (1989) Expression of an engrailed-related protein is induced in the anterior neural ectoderm of early *Xenopus* embryos. Development 106:611–617
3. Davidson LA, Keller RE (1999) Neural tube closure in *Xenopus laevis* involves medial migration, directed protrusive activity, cell intercalation and convergent extension. Development 126:4547–4556
4. Zhou X, Vize PD (2004) Proximo-distal specialization of epithelial transport processes within the *Xenopus* pronephric kidney tubules. Dev Biol 271:322–338
5. Vize PD, McCoy KE, Zhou X (2009) Multichannel wholemount fluorescent and fluorescent/chromogenic *in situ* hybridization in *Xenopus* embryos. Nat Protoc 4:975–983
6. Bonev B, Pisco A, Papalopulu N (2011) MicroRNA-9 reveals regional diversity of neural progenitors along the anterior-posterior axis. Dev Cell 20:19–32
7. Sive HL, Grainger RM, Harland RM (2000) Early development of *Xenopus laevis*. Cold Spring Harbor Laboratory Press, Cold Spring Harbor
8. Dubaissi E, Papalopulu N (2011) Embryonic frog epidermis: a model for the study of cell-cell interactions in the development of mucociliary disease. Dis Model Mech 4:179–192
9. Costa RM, Soto X, Chen Y, Zorn AM, Amaya E (2008) Spib is required for primitive myeloid development in Xenopus. Blood 112: 2287–2296
10. Carruthers S, Mason J, Papalopulu N (2003) Depletion of the cell-cycle inhibitor p27(Xic1) impairs neuronal differentiation and increases the number of ElrC(+) progenitor cells in Xenopus tropicalis. Mech Dev 120:607–616

Chapter 25

Methods to Analyze microRNA Expression and Function During *Xenopus* Development

Boyan Bonev and Nancy Papalopulu

Abstract

microRNAs are a class of small noncoding RNAs that regulate gene expression at a posttranscriptional level. microRNAs are transcribed as primary transcripts, characterized by specific hairpin secondary structure that undergo stepwise processing to yield mature microRNAs of approximately 22 nt length. The function of the majority of vertebrate microRNAs has not yet been established and *Xenopus* offers a powerful system to test their biological function. Working with microRNAs is based on well-established protocols for the detection of mRNAs and manipulation of gene expression; however, the small size of mature microRNAs and their unique biogenesis require modifications to the existing protocols. Here, we present methods that can be used to detect, overexpress, and inhibit microRNAs in *Xenopus tropicalis*.

Key words: *Xenopus*, microRNA, miR, In situ hybridization, RT-PCR

1. Introduction

microRNAs are a class of small noncoding RNAs that regulate gene expression at a posttranscriptional level. They have been implicated in many key developmental processes such as control of stem cell proliferation and differentiation during vertebrate development (1–3). There are estimated to be at least 700 miRNAs (and there might be as many as 1,000) in the human genome, comprising 1–4% of all expressed human genes and this makes miRNAs one of the largest classes of gene regulators (4, 5).

microRNAs are transcribed as primary transcripts, called *pri-miRs*, characterized by specific hairpin secondary structure that is recognized by the RNase III enzyme, Drosha (6), and the double-stranded RNA-binding protein Pasha (also known as DGCR8), and processed into ~70-nucleotide precursor miRNAs, called *pre-miRs* (7). Drosha cuts the RNA duplex with a staggered cut typical

Stefan Hoppler and Peter D. Vize (eds.), *Xenopus Protocols: Post-Genomic Approaches, Second Edition*,
Methods in Molecular Biology, vol. 917, DOI 10.1007/978-1-61779-992-1_25, © Springer Science+Business Media, LLC 2012

for RNAse III endonucleases, producing a 5′ phosphate and an approximately two nucleotide 3′ overhang (8). This can lead to destabilization of the pre-miRNA ends that could be important for the subsequent processing and selection of guiding strand. The precursor miRNAs are then exported into the cytoplasm by the RAN GTP-dependent transporter Exportin 5 (9) and undergo an additional processing step which involves the RNAse III enzyme, Dicer, that is thought to recognize preferentially the 5′ phosphate and 3′ overhang of the base of the stem loop (10). Subsequently, it cuts both strands leading to the excision of the terminal base pairs and pre-miRNA loop and the liberation of a dsRNA of approximately 22 nucleotides in length, the miRNA:miRNA* duplex.

For most of the known miRNAs one strand is then preferentially incorporated in the functional RISC complex and guides the binding to its target mRNA, while the other strand, called miRNA*, is present at much lower copy numbers and is usually targeted for degradation (11). However, some of the miRNA* retain functional importance and are also incorporated into the RISC silencing complex, albeit at lower frequency (miRBase database).

microRNAs are particularly attractive regulators for their potential to coordinate the response of many target genes by fine tuning or acting as a developmental switch to promote certain cell fate. Even though microRNA mechanism of action is fairly well established (12, 13), the biological function of the majority of the vertebrate microRNAs has not yet been elucidated. *X. tropicalis* is a great model to study microRNA function during development because of its external development and the ability to manipulate gene expression using morpholinos and microinjections. However, the small size of microRNAs and their unique biogenesis require modifications to the existing protocols. Here, we present methods that can be used to detect microRNAs in *X. tropicalis* and functionally characterize them through overexpression and inhibition.

The lack of genomic information in *Xenopus laevis* complicates the identification and characterization of microRNAs in this model organism, which is evident by the relatively low number of microRNAs validated in *X. laevis* (22 total primary microRNA transcripts, according to miRBase) compared to *X. tropicalis* (188 pri-miRs, according to miRBase at the time of writing).

2. Reagents

2.1. In Situ Hybridization (ISH)

2.1.1. LNA-Based ISH for Mature microRNA

1. Wheaton glass scintillation vials (20 mL VWR).
2. MEMFA fixative: 0.1 M MOPS pH 7.4, 2 mM EGTA, 1 mM $MgSO_4$, 3.7% formaldehyde. MEMFA should be made fresh from stock solutions of 1 M MOPS PH 7.4, 0.5 M EGTA pH

8, and 1 M $MgSO_4$. Fresh formaldehyde should always be used. Stock solutions of MOPS should be kept away from light.

3. PTw (PBS + 0.1% Tween-20).
4. 1× TBSTw: 100 mM Tris pH 7.5, 150 mM NaCl, 0.05% Tween-20 (see Note 1).
5. Methylimidazole buffer: 0.13 M 1-methylimidazole (Sigma), 300 mM NaCl, adjust to pH 8.0 with 12 N HCl. Prepare fresh.
6. EDC fixative: 0.16 M EDC (1-ethyl-3-(3-dimethylamino-propyl)—carbodiimide; Sigma) in methylimidazole buffer and re-adjust to pH 8.0 by adding approximately 100 μL 12 N HCl. Prepare fresh.
7. 5′ DIG-labeled LNA probe (Exiqon) (see Note 2). Probe is shipped at stock concentration of 25 μM in 10 μL. On arrival dilute the probe in 10 mL hybridization buffer for a final concentration of 25 nM, mix well, aliquot, and store at −20°C.
8. Proteinase K 20 mg/mL (Ambion).
9. TEA buffer—0.1 M triethanolamine pH 8.
10. Acetic anhydride.
11. 4% Formaldehyde (Baker).
12. Hybridization buffer: 50% formamide (Ambion), 5× SSC, 1 mg/mL Torula RNA (Sigma), 100 μg/mL heparin (Sigma), 1× Denhardt's solution (Sigma), 0.1% Tween-20, 0.1% CHAPS (Sigma).
13. 2× SSC: 300 mM NaCl, 30 mM sodium citrate. Prepare a 20× stock solution, adjust to pH 7.0 with NaOH. Sterilize the 20× solution by autoclaving.
14. 1× MAB: 100 mM maleic acid, 150 mM NaCl. Prepare a 2× stock solution, adjust to pH 7.5 with NaOH. Sterilize the 2× solution by autoclaving.
15. Blocking Solution—2% blocking medium (Roche) + 2% heat treated lamb serum (Sigma).
16. 5× TBST solution (700 mM NaCl, 13.4 mM KCl, 125 mM Tris pH 7.5 + 2% Tween-20).

2.1.2. Additional Reagent for ISH for Primary microRNA Transcripts

1. Reagents for fluorescent in situ hybridization (FISH), see Chapter 24.

2.2. RT-PCR-Based Method for Detecting Mature microRNAs

2.2.1. RNA Extraction

1. TRIzol Reagent (Invitrogen—Cat. 15596-018).
2. Chloroform (Fisher Scientific).
3. Isopropanol (Fisher Scientific).

2.2.2. TaqMan-Based qRT-PCR for Mature microRNAs

1. TaqMan microRNA Reverse Transcription Kit (ABI, Cat. 4366596).
2. Taqman microRNA Assay specific to the microRNA of interest (ABI).
3. TaqMan ready PCR machine (such as StepOne plus from ABI).

2.3. Overexpression of microRNAs

2.3.1. Generating Primary microRNA Transcripts for Microinjection in X. tropicalis Embryos

1. mMessage mMachine kit (Ambion).
2. Standard PCR reagents.
3. PCRII-TOPO and pCS2 or equivalent vector.

2.3.2. Injection of miRNA Synthetic Precursors

Pre-miR miRNA precursor mimics (Applied Biosystems). Upon arrival resuspend in H_2O to the desired concentration (we use 200 μM) and store at –20°C.

2.4. Inhibition of microRNAs Using Morpholinos and Target Protector Oligos

1. Lyophilized morpholino or target protector oligonucleotide (Gene Tools).
2. Nuclease-free H_2O (Ambion).
3. Microinjector capable of delivering nanoliter volumes into *Xenopus* embryos.

3. Methods for Detecting microRNAs

3.1. In Situ Hybridization

Although protocols for detection of mRNAs by ISH are well established (e.g., see Chapter 26), the small size of the mature microRNAs presents several technical challenges in labeling and detection. Two strategies are used to overcome this problem:

1. Since microRNAs are generated from longer (generally >1 kb long) primary precursor RNAs, one can use nascent transcript ISH to detect the microRNA precursors, while they are transcribed in the nucleus (14).
2. Chemically modified digoxigenin labeled probes with high melting temperatures have been successfully used to detect (15, 16).

Both methods have advantages and disadvantages that should be considered when planning experiments to detect microRNAs:

1. The advantage of detecting nascent microRNA transcripts is that it relies on minimal modifications of the well-established ISH and FISH protocols; however, the major challenge in detecting nascent microRNA primary transcripts is their short life and hence low abundance. Most microRNA precursors are

subject to rapid processing through Drosha/Pasha and Dicer, which results in their cleavage and loading of the mature strand into the RISC complex. Because of this, not all primary transcripts accumulate to sufficient levels to allow detection by ISH. In addition, it is at least theoretically possible that the nascent transcript is not processed to the mature, biologically active microRNA.

2. These problems can be overcome by detecting the mature microRNA, however, since mature microRNAs are very short, conventional methods of using antisense-labeled RNA probe for ISH do not produce satisfactory results. We have successfully used LNA (Locked-Nucleic Acid) modified probes to detect mature transcripts for miR-9 in *X. tropicalis.* LNA is a chemical modification of a RNA oligo, whereby the furanose ring of the ribose sugar is locked in a specific confirmation by a methylene bridge, resulting in an increased hybridization affinity towards complementary DNA and RNA sequences (15). In the miRCURY LNA-based probes (usually 20–25 nt long) designed by Exiqon A/S (Denmark) approximately every third nucleotide is replaced by a LNA analog (16), which dramatically increases the melting temperature and is essential for the high stringency required in the ISH protocol. However, one limitation with the use of LNA probes is their low signal strength. Therefore, additional modifications to the standard in situ protocol, that we describe below, are necessary to ensure strong specific signal and low background.

In most cases an LNA-based in situ is preferable, since it gives information about the expression of the biologically relevant mature microRNA. However in some cases, especially where there are multiple primary transcripts for the same mature microRNA or there is an interest into the regulation of microRNA transcription, the nascent pri-miRNA in situ protocol should be employed.

3.1.1. LNA-Based Detection of Mature microRNAs

Essential modifications to the basic ISH protocol described in Chapter 26 are necessary, because LNA probes are short and small RNAs tend to diffuse in the hybridization buffer during the prehybridization step, such as longer initial fixation and additional fixation using EDC fixative (17). Here we describe the complete protocol for chromogenic detection of mature microRNAs.

Day 1

1. Embryos are fixed in MEMFA overnight at 4°C to ensure complete fixation of small RNAs.
2. In scintillation vials rehydrate fixed embryos by 5 min washes in;
   - MeOH
   - 75% MeOH, 25% H_2O

- 50% MeOH, 50% H_2O
- 25% MeOH, 75% PTw
- Then 3 × 5 min washes in 100% PTw

3. Incubate embryos for 10 min at room temperature in approximately 5 mL diluted proteinase K per vial.
   laevis—5 μg/mL proteinase K in 5 mL PTw
   tropicalis—1.5 μg/mL proteinase K in 5 mL PTw
4. Rinse embryos 2 × 5 min in 5 mL TEA buffer.
5. Do not remove the last 5 mL TEA wash. Add 12.5 μL acetic anhydride to this. Rock for 5 min then add another 12.5 μL acetic anhydride. Rock for 5 min.
6. Wash 2 × 5 min in 5 mL PTw.
7. Refix for 20 min with 4% formaldehyde in PTw.
8. Wash 3 × 5 min 5 mL TBS.
9. Incubate 2 × 10 min with 5 mL methylimidazole buffer.
10. Incubate for 1 h with 5 mL EDC fixative (incubation can be extended up to 2 h for older embryos).
11. Wash 2 × 5 min with 5 mL TBS + 0.2% (w/v) glycine (see Note 3).
12. Wash 4 × 5 min with 5 mL TBS.
13. Aliquot 500 μL hybridization buffer into 2 mL screw cap plastic vials that have been autoclaved. Transfer embryos into the buffer carrying over as little TBS as possible.
14. Once the embryos have settled through the dense buffer, remove the buffer and replace with another 500 μL of hybridization buffer.
15. Prehybridize for 2–3 h at the melting temperature of the probe (T_m) −20°C (see Note 4).
16. Replace hybridization buffer with 0.5 mL of probe solution (25 nM probe in hybridization buffer). Hybridize overnight at T_m −20°C (Hyb Temp).

Day 2

Washing

Washes are performed in 2 mL screw cap plastic tubes. Solutions are removed by aspiration.

1. Remove probe and keep at −20°C (see Note 5).
2. Replace probe with hybridization buffer, at Hyb temp for 10 min.
3. Wash 3 × 20 min at Hyb temp +2°C in 2× SSC (see Note 6).

4. Wash 2 × 20 min at Hyb temp +2°C in 0.2× SSC.
5. Wash 2 × 10 min with MAB.

Antibody incubation

Keep the samples in the 2 mL screw cap plastic tubes. Rock tubes horizontally in blue plastic trays as for vials.

6. Block for 1 h in blocking solution.
7. Incubate for 4 h at RT or overnight at 4°C with anti-DIG AP antibody in blocking solution (1:2,000).
8. Transfer the embryos to a 50 mL falcon tube containing 1× MAB. Wash eight times for 20 min each in 1× MAB. One of the washes must be overnight at 4°C. For each wash the falcon tubes are laid horizontally in a blue tray and rocked, including the overnight wash.

Day 3

Chromogenic reaction

The following steps should be rocked as previously.

1. Transfer the embryos into Wheaton glass vials.
2. Wash 2 × 5 min at room temperature with 2 mL Alkaline Phosphatase (AP) buffer.
3. Add the chromogenic substrate diluted in AP Buffer, 2 mL per vial.
4. Develop ~2 h at 30°C and then for up to 2 days at RT (see Note 7).
5. Once color has developed, wash 3 × 20 min with MeOH.
6. Wash 2 × 20 min with PBS.
7. Fix samples with MEMFA.
8. If desired rehydrate the samples in PBS and bleach for 2–3 h (or until pigment is no longer visible) in 10% H_2O_2 in PBS under light.

3.1.2. In Situ Hybridization to Detect Primary microRNA Transcripts

Generating Antisense Probe

1. The hairpin precursor of the microRNA of interest is identified using miRBase (www.mirbase.org).
2. The precursor + 500 nt flanking sequence on either sides is PCR amplified from genomic DNA. The direction of the transcribed RNA (±) is verified *using* miRBase (see Note 8).
3. PCR product of the correct size (~1 kb) is inserted into pCRII-TOPO vector according to the manufacturer's protocol (Invitrogen) and the clone is verified by sequencing (see Note 9). Antisense DIG or FITC labeled (see Note 10) RNA probes are generated using standard methodology (see Chapter 24).

In Situ Hybridization

Chromogenic ISH (and FISH if desired) using antisense probes to primary microRNA precursors is performed as described earlier (for chromogenic in situ) or in Chapter 24 (for FISH), with only minor modifications:

- After step 7 in the LNA in situ protocol do the following:
  - 4 × 5 min washes with PBSTw.
  - Aliquot 500 μL hybridization buffer into 2 mL screw cap plastic vials that have been autoclaved. Transfer embryos into the buffer carrying over as little TBS as possible.
  - Once the embryos have settled through the dense buffer, remove the buffer and replace with another 500 μL of hybridization buffer.
  - Prehybridize for 4 h at 60°C.
  - Replace hybridization buffer with 0.5 mL of probe solution (1 μg/mL probe in hybridization buffer). Hybridize overnight at 60°C.
  - Replace probe with hybridization buffer at 60°C for 10 min.
  - Wash 3 × 20 min at 60°C in 2× SSC.
  - Wash 2 × 30 min at 60°C in 0.2× SSC.
  - Continue the in situ protocol from step 5 (day 2)
- Repeated cycles of substrate / TBST washing (see Note 7) are essential as most primary microRNA transcripts are short-lived and low abundant.
- When FISH is used to detect primary microRNA transcripts, the substrate incubation time should be increased to 45 min instead of the standard 30 min.
- microRNA primary transcripts can also be detected directly on cryosections using FISH (as described in Chapter 24) and this can be combined with immunohistochemistry to detect a protein in the same section (18).

3.2. RT-PCR-Based Method for Detecting Mature microRNAs

In addition to the ISH method, it is possible to use real-time PCR (RT-PCR) to detect mature microRNA transcripts. The advantage of RT-PCR is that it is a more quantitative method of detection and it can be used to verify results obtained by the ISH. In addition, it requires small amounts of total RNA and can be used to detect even microRNAs with very low abundance. We will present a protocol for purifying total RNA (including the small RNA fraction), generating cDNA, and using it to measure microRNA abundance using real-time PCR.

3.2.1. RNA Extraction

1. Flash freeze 5–10 *X. tropicalis* embryos in 1.5 mL tubes on dry ice for 1 h (take care to avoid any leftover liquid in the tube).

2. Add 500 μL TRIzol reagent and homogenize using pipette tip and tissue grinder (see Note 11).
3. Vortex well for 20 s.
4. Leave at RT for 5 min.
5. Add another 500 μL TRIzol reagent and vortex briefly.
6. Add 200 μL chloroform and mix by inverting the tube 2–3 times (do not vortex).
7. Spin down in a precooled centrifuge (4°C) for 15 min at 13,200 rpm.
8. Take the upper aqueous phase and transfer it to a new tube.
9. Add equal amount of chloroform (600 μL) and vortex well.
10. Spin down for 5 min and transfer the upper phase to a new tube.
11. Add equal amount of isopropanol and vortex briefly.
12. Precipitate the RNA overnight in –20°C (we have found that the extended precipitation step at low temperatures gives us the best small RNA yield).
13. Centrifuge at max speed for 25 min in a precooled centrifuge (4°C).
14. Wash the pellet with 70% EtOH.
15. Centrifuge for 5 min at RT at max speed and resuspend the pellet in 20 μL of nuclease-free H_2O.
16. Measure RNA yield on a Nanodrop machine.
17. Since subsequent procedures discriminate between genomic and cDNA there is no need for DNAse treatment of the samples.

3.2.2. qRT-PCR for Mature microRNA Transcripts Using TaqMan-Based Chemistry

Applied Biosystems (ABI) have recently developed a method that allows the detection of microRNAs using TaqMan-based assays (a preoptimized set of PCR primers and probe), which significantly reduces the amount of RNA required and also improves the reproducibility of results. The first step of the assay involves a specific amplification of the miRNA of interest through a stem-loop reverse-transcription primer. The stem-loop primer recognizes only the mature microRNA to generate first cDNA strand and this increases the subsequent specificity of the qPCR reaction.

For real-time PCR we start with 1.5 μL of cDNA per reaction and we use three technical replicates per point. The small nuclear RNAs U2 or U6 can be used as housekeeping genes (19). The U2 and U6 TaqMan-based probes can be obtained from Applied Biosystems. The reverse transcription and real-time detection is done according to the manufacturer's protocol (see Note 12).

3.3. Functional Analysis of microRNAs in X. tropicalis

3.3.1. Overexpression of microRNAs

One potential way to study the function of microRNAs during *Xenopus* development is to examine the physiological consequences of their overexpression. There are two strategies for introducing exogenous microRNAs into *Xenopus* embryos through microinjections, with different advantages and disadvantages:

1. The first method (modified from Martello et al. (18)) relies on amplifying and cloning the region hosting the microRNA transcript. It is not necessary to clone the entire primary transcript, especially since in most cases the actual size of the entire pri-miR transcript is unknown. This is followed by transcription of RNA in vitro and injection into the embryo. This method has the advantage of being able to express the RNA transcript behind a promoter of choice and also to tag it.
2. The second method involves direct injection of synthetic microRNA precursor mimics. This is advantageous in that it is easier, as the mimic is made commercially, and it does not rely on processing in the nucleus to produce an active microRNA. miRNA precursor molecules (Ambion) are small, chemically modified double-stranded RNA molecules designed to mimic endogenous mature miRNAs. These ready-to-use miRNA mimics, which are similar to, but not identical to siRNAs, and can be introduced into *Xenopus* embryos by microinjection. The advantage of using synthetic miRNA precursor mimics is that they do not require additional processing by Drosha/Pasha in the nucleus and thus can be injected at lower concentrations, reducing the chance of nonspecific effects. Furthermore, they can be optimized at source for the preferred microRNA strand utilization by the RISC complex. This modification increases the specificity of the target repression. We have microinjected precursors mimics in X. tropicalis and checked the processing efficiency by RT-PCR for the mature miR. The guide strand was processed much more efficiently than the star strand, as claimed (Bonev and Papalopulu, unpublished data).

Generating Primary microRNA Transcripts for Microinjection in *X. tropicalis* Embryos

1. The hairpin precursor of the microRNA of interest is identified using miRBase (example—xtr-miR-9a-1–scaffold_427: 10032-14119(+)).
2. PCR amplify the precursor + 150 nt flanking sequence on both sides and subclone in pCRII-TOPO vector (see Note 13).
3. Transfer the DNA fragment into pCS2+ vector in the correct 5′–3′ orientation (see Note 14).
4. In vitro transcribe the RNA as per manufacturer's instructions (Ambion).
5. Before injection, incubate the pri-miRNA at 65°C for 5 min and slowly cool to RT to facilitate the formation of secondary structures (hairpins) (see Note 15).

6. Inject 2–5 ng RNA in one- or two-cell stage embryo (see Note 16).
7. Verify the efficiency of the processing and the abundance of the mature microRNA at the desired stage by qRT-PCR as described in Subheading 3.2.

Injection of miRNA Synthetic Precursors

1. Select the miRNA precursor for the desired microRNA—Applied Biosystems (see Note 17).
2. Reconstitute the oligo at 200 μM concentration in ddH_2O.
3. Incubate the precursor at 65°C for 5 min and slowly cool to RT to facilitate the formation of secondary structures (hairpins).
4. Inject 5nL (1pmol) at one- or two-cell stage.
5. Verify the efficiency of the processing of the mature microRNA by Dicer by qRT-PCR, as described in Subheading 3.2 of this chapter.

3.3.2. Inhibition of microRNAs Using Morpholinos

Even though overexpression of the desired microRNA can give insight about its potential function, quite often it leads to a variety of nonspecific effects during early development. This is particularly true for microRNAs that are expressed later during embryogenesis and/or in specific tissues and when overexpressed can modulate the expression of targets that are normally not under their regulation. Therefore, loss-of-function experiments should be performed in combination with gain-of-function studies to confirm the function of the endogenous microRNA without potential side effects.

Morpholino oligonucleotides are stable, uncharged, and water-soluble molecules that bind to complementary sequences of RNA and are traditionally used to inhibit mRNA processing, protein translation, and protein \binding at those sites. In addition, morpholinos form stable base pairs with complementary nucleic acid sequences but do not undergo protein-mediated degradation; nor do they mediate RNA cleavage by RNAse H or the RISC complex.

Morpholinos Blocking miRNA Processing and Incorporation into the RISC Complex

We and others have used morpholinos to block microRNA function by interfering with their processing, in addition to preventing the incorporation of the mature microRNA into the RISC complex. 25-base oligonucleotides designed to be complementary to the mature miRNA but extending one or two bases over the flanking sequence of the guide strand, past the Drosha and Dicer cleavage sites are generally used to block miRNA function. The morpholinos should be designed to carry a fluorescein tag so that their delivery and distribution in the embryo can be followed.

Morpholinos That Act by Interfering with miRNA–mRNA Binding

Instead of using morpholinos to block miRNA maturation and function, they can be used to selectively prevent the interaction between a miRNA and one of its targets, while leaving the regulation of other targets of the same miRNA unaffected. This could be especially useful to investigate the contribution of a particular miRNA-target interaction to the physiological function of the miRNA of interest. The method was originally developed in zebrafish (20, 21) but we have successfully used it in *X. tropicalis*. The target protector morpholino is designed to overlap with the seed-complementary region in the target 3′UTR but extend 10–15 nt to the 3′ to confer specificity. Care is to be taken to confirm not only that the target protector can alleviate miRNA-induced repression, but that it does not affect the regulation of other targets. One possible way to do that is to examine the expression of a luciferase-based reporter fused to the target 3′UTR in the presence or absence of a target protector combined with miRNA precursors in cell lines.

Injecting morpholinos in X. tropicalis embryos

Procedure

- Add the appropriate amount of nuclease-free H_2O to the vial to make a 20 μg/μL stock solution.
- Vortex the vial, wait 5 min, and incubate it for 5 min at 65°C.
- Aliquot the stock solutions into several tubes and store at −20°C. Aliquots could be used for several (6–7 months) without any noticeable change in efficiency.
- Before microinjection dilute the stock solution to the desired concentration (5–10 μg/μL) with nuclease-free water and incubate for 5 min at 65°C—spin down and keep at 37°C to prevent needle clogging.
- Inject the appropriate volume (we generally use 20 ng total—i.e., inject 2 nL of 10 μg/μL) into one- or two-cell stage embryo using standard techniques.
- Let the embryos develop until the desired stage and then verify miRNA knockdown using LNA-based in situ and/or qRT-PCR as described in Subheadings 3.1 and 3.2 of this chapter.

4. Notes

1. TBS instead of PBS is essential to minimize the amount of phosphate in the buffer, which interferes with the subsequent EDC-based fixation.
2. Single 5′ DIG tag is usually sufficient for abundant microRNAs; however, for some microRNAs with low abundance, double-labeled LNA probe (5′ and 3′ DIG) might increase the signal. If the relative abundance of the microRNA of interest is unknown, we recommend to start with double-labeled LNA probe.
3. Glycine is used to inactivate the EDC.
4. This temperature has been determined as optimal for signal-to-background ratio (15).

5. Probe can be reused several times. Best results are obtained with probes, which have been used once but not more than three times.
6. The slightly higher temperature than the one used for hybridization helps decrease the background.
7. If background develops, this can be minimized by removing the substrate solution and washing the embryos in 5× TBST until background levels are reduced. We usually do one short (~10 min) wash and then a longer one for ~2–3 h. Cycles of substrate incubation and washing can be repeated until background staining is reduced and chromogenic signal is clear. Embryos can be left overnight in washing solution and fresh substrate can be added the next day. We have found that 2–3 days of substrate incubation / washing is necessary for weakly expressed microRNAs.
8. We aim for about 1 kb of amplified sequence to generate a sufficiently long probe for a good signal.
9. The pCRII-TOPO vector contains two promoters (SP6 and T7) flanking the inserted PCR product and thus allows the generation of antisense probes regardless of the orientation of cloning.
10. The detection of FITC labeled probes is generally weaker than DIG-labeled probes and this can be a problem for some weakly expressed transient microRNA transcripts.
11. We have found that using TRIzol reagents (Invitrogen) for RNA extraction compared to a column-based methods such as RNAeasy (QIAGEN) work better in preserving the small RNA fraction, which is critical for the success of detecting microRNAs by RT-PCR.
12. The Applied Biosystems protocol for the reverse transcription of microRNAs I the TaqMan-based assay for qPCR can be accessed at: http://www3.appliedbiosystems.com/cms/groups/mcb_support/documents/generaldocuments/cms_042167.pdf
13. 150 nt in both directions is the minimum flanking sequence, which is required for efficient processing. If a longer sequence is already available (e.g., from the plasmid to generate the in situ probe for the primary microRNA transcript) it can generally be used to the same effect.
14. It is essential that the in vitro transcribed RNA is properly capped and polyadenylated. Standard vectors used for overexpression in *Xenopus*, such as the pCS2 vector, contain SV40 polyadenylation site. The Ambion kit contains a 5′ cap analog, which is incorporated during transcription.
15. This step is required to facilitate the formation of secondary structures (hairpins) that are essential for the proper processing of the microRNA transcripts (17).

16. Since the primary precursor has to undergo additional processing in the nucleus a relatively high amount of injected RNA is required.
17. The search tool for microRNA precursor mimics is accessible at the following address: https://bioinfo.appliedbiosystems.com/genome-database/mirna-mimics-inhibitors.html. Applied Biosystems have recently developed an even newer generation of precursor mimics (mirVana miRNA mimics), which should increase the specificity by further reducing the amount of microRNA star strand. While we have not evaluated them, we advise new researchers to compare the two precursor mimics before selecting one.

Acknowledgments

This work was funded by a Wellcome Trust Senior Research Fellowship to NP and a Wellcome Trust 4-yr studentship to BB.

References

1. Gangaraju V, Lin H (2009) MicroRNAs: key regulators of stem cells. Nat Rev Mol Cell Biol 10:116–125
2. Kosik KS (2006) The neuronal microRNA system. Nat Rev Neurosci 7:911–920
3. Stefani G, Slack FJ (2008) Small non-coding RNAs in animal development. Nat Rev Mol Cell Biol 9:219–230
4. Bentwich I, Avniel A, Karov Y, Aharonov R, Gilad S, Barad O, Barzilai A, Einat P, Einav U, Meiri E, Sharon E, Spector Y, Bentwich Z (2005) Identification of hundreds of conserved and nonconserved human microRNAs. Nat Genet 37:766–770
5. Berezikov E, Guryev V, van de Belt J, Wienholds E, Plasterk RH, Cuppen E (2005) Phylogenetic shadowing and computational identification of human microRNA genes. Cell 120:21–24
6. Lee Y, Ahn C, Han J, Choi H, Kim J, Yim J, Lee J, Provost P, Radmark O, Kim S, Kim VN (2003) The nuclear RNase III Drosha initiates microRNA processing. Nature 425:415–419
7. Lee Y, Jeon K, Lee JT, Kim S, Kim VN (2002) MicroRNA maturation: stepwise processing and subcellular localization. EMBO J 21: 4663–4670
8. Basyuk E, Suavet F, Doglio A, Bordonne R, Bertrand E (2003) Human let-7 stem-loop precursors harbor features of RNase III cleavage products. Nucleic Acids Res 31:6593–6597
9. Lund E, Guttinger S, Calado A, Dahlberg JE, Kutay U (2004) Nuclear export of microRNA precursors. Science 303:95–98
10. Hutvagner G, McLachlan J, Pasquinelli AE, Balint E, Tuschl T, Zamore PD (2001) A cellular function for the RNA-interference enzyme Dicer in the maturation of the let-7 small temporal RNA. Science 293:834–838
11. Lau NC, Lim LP, Weinstein EG, Bartel DP (2001) An abundant class of tiny RNAs with probable regulatory roles in Caenorhabditis elegans. Science 294:858–862
12. Pillai RS, Bhattacharyya SN, Filipowicz W (2007) Repression of protein synthesis by miRNAs: how many mechanisms? Trends Cell Biol 17:118–126
13. Guo H, Ingolia NT, Weissman JS, Bartel DP (2010) Mammalian microRNAs predominantly act to decrease target mRNA levels. Nature 466:835–840
14. Walker JC, Harland RM (2008). Expression of microRNAs during embryonic development of Xenopus tropicalis. Gene expression patterns : GEP 8(6):452–456. doi:10.1016/j.gep.2008.03.002
15. Kloosterman WP, Wienholds E, de Bruijn E, Kauppinen S, Plasterk RH (2006) In situ detection of miRNAs in animal embryos using LNA-modified oligonucleotide probes. Nat Methods 3:27–29

16. Obernosterer G, Martinez J, Alenius M (2007) Locked nucleic acid-based in situ detection of microRNAs in mouse tissue sections. Nat Protoc 2:1508–1514
17. Pena JTG, Sohn-Lee C, Rouhanifard, SH, Ludwig J, Hafner M, Mihailovic A, Lim C, et al. (2009) miRNA in situ hybridization in formaldehyde and EDC–fixed tissues. Nature Methods 6(2):139. doi:10.1038/nmeth.1294
18. Bonev B, Pisco A, Papalopulu N (2011) MicroRNA-9 reveals regional diversity of neural progenitors along the anterior-posterior axis. Dev Cell 20:19–32
19. Martello G, Zacchigna L, Inui M, Montagner M, Adorno M, Mamidi A, Morsut L, Soligo S, Tran U, Dupont S, Cordenonsi M, Wessely O, Piccolo S (2007) MicroRNA control of Nodal signalling. Nature 449:183–188
20. Choi W-Y, Giraldez AJ, Schier AF (2007) Target protectors reveal dampening and balancing of Nodal agonist and antagonist by miR-430. Science 318:271–274
21. Staton AA, Giraldez AJ (2011) Use of target protector morpholinos to analyze the physiological roles of specific miRNA-mRNA pairs in vivo. Nat Protoc 6:2035–2049

Chapter 26

A Bromodeoxyuridine (BrdU) Based Protocol for Characterizing Proliferating Progenitors in *Xenopus* Embryos

Hélène Auger*, Raphaël Thuret*, Warif El Yakoubi, and Nancy Papalopulu

Abstract

BrdU is a thymidine analog that is incorporated into DNA during the S-phase of the cell cycle. BrdU incorporation can be used to quantify the number of cells that are in S-phase in the time period that BrdU is available. Thus, BrdU incorporation is an essential method in the quantitative analysis of cell proliferation, during normal embryonic development or after experimental manipulation. It is a reliable and versatile method that can be easily combined with immunohistochemistry and in situ hybridization to relate cell proliferation with gene expression. BrdU incorporation has been used in all model organisms; here, we describe a protocol adapted for use in *Xenopus* embryos.

Key words: *Xenopus*, Neural progenitor, Neurons, Cell cycle, EdU, BrdU

1. Introduction

Control of cell division is key for correct embryonic growth, patterning, and organogenesis. As we understand more about the integration of cell signaling and transcription factor function with cell cycle control, it has become increasingly important to be able to monitor the proliferative properties of a tissue, and how these may change upon experimental manipulation of gene expression.

Here, we describe the characterization of proliferating progenitors based on BrdU incorporation, adapted for use in *Xenopus*

*The authors Hélène Auger and Raphaël Thuret contributed equally

Stefan Hoppler and Peter D. Vize (eds.), *Xenopus Protocols: Post-Genomic Approaches, Second Edition*, Methods in Molecular Biology, vol. 917, DOI 10.1007/978-1-61779-992-1_26, © Springer Science+Business Media, LLC 2012

embryos. BrdU is incorporated into DNA during the S-phase of the cell cycle. Consequently, it is widely used as a marker of S-phase, although it is technically a marker for DNA synthesis, which could also occur during other processes, such as DNA repair (1). The use of BrdU immunohistochemistry for the study of cell proliferation in a developing embryo was described in a seminal paper in the late 1980s (2). Since then, BrdU incorporation has most often been used to identify sites of proliferation and determine the number of dividing cells in a tissue. Changes in the number of BrdU positive cells have often been taken as an indication of increase or decrease in proliferation. An increase in BrdU positive cells in a tissue may indicate that cells have failed to become postmitotic or that there has been a change in cell cycle kinetics that causes a change in the number of cells that are in S-phase for the time that BrdU is available. However, the depth of information that one can derive from the absolute number of S-phase cells is limited.

More useful is the Labeling Index of a tissue, which is the proportion of cells labeled in the total proliferating population at any given time. Since the fraction of cells in S-phase over the total proliferating cells is proportional to the fraction of S-phase length over the total cell cycle length, the Labeling Index can be taken as a measure of the cell cycle length. For example, a decrease in the Labeling Index denotes an increase in cell cycle length (e.g., ref. (3)). This is based on the assumption that the length of the S-phase does not change much, which agrees with findings reporting increase in the G-phases of the cell cycle during neurogenesis (4–6). Nonetheless, differences in S-phase length in various populations of neural progenitors have been recently reported (7). There are more accurate ways to estimate the length of the S-phase and of the total cell cycle, including cumulative BrdU labeling (2, 5), dual pulse labeling with BrdU, and other thymidine analogs such as IdU, CldU, or EdU ((8–11), see Note 1) or FACS analysis (6).

Theoretically, cumulative and dual pulse labeling are equivalent and presuppose that all cells studied are proliferating, are uniformly distributed in the cell cycle, and the population grows at a steady state (2, 5, 11). The aim of those methods is to estimate the number of cells in S-phase at various time points in order to extrapolate the length of the S-phase and of the total cell cycle.

Cumulative BrdU incorporation requires saturation of the cycling population to be reached, which can be challenging (see ref. (2, 5) for discussion). Dual S-phase pulse labeling only requires two injections of different thymidine analogs in a short time range that will be later on distinguished by immunochemistry (8, 9, 11). However, discriminating thymidine analogs such as IdU, CldU, and BrdU is tedious and can require some important set up, since all those molecules are structurally close to each other and present a potentially comparable affinity for different anti-BrdU antibodies ((8), see Note 2).

The recent development of EdU molecule eases the use of the dual phase labeling method. EdU is a new kind of thymidine analog, which can be directly coupled with a fluorophore by a copper catalyzed reaction (12). This presents two main advantages, namely that it does not require DNA denaturation and that it is not an antibody-based detection system. This technology can then be easily associated with BrdU detection in order to perform dual pulse labeling (10), especially since particular clones of anti-BrdU antibody, commercially available, have been shown to not cross-react with EdU (see Note 2).

To address whether changes in the proliferating cell population are accompanied by changes in cell differentiation, it is useful to combine S-phase labeling with immunohistochemistry or in situ hybridization for cell type-specific markers. Such combination allows BrdU/EdU labeling also to be used in characterizing the fate of progeny of cells that have been dividing at the time of labeling and to determine the birth date of a cell type of interest.

Here, we provide a detailed account of how these different protocols can be combined. The reader is referred to other publications on the theoretical aspects of using dual pulse labeling or cumulative S-phase labeling to calculate cell cycle length (2, 5, 11). We focus on neural tissue, but as far as we know, these protocols can be applied to any tissue in the embryo. The method presented here starts with BrdU delivery (see Note 3), it includes an optional in situ hybridization step and then, it is split into two alternative paths: route A involves immunostaining on sections, while route B involves whole mount immunostaining, followed by sectioning (see Fig. 1). Route A gives better quality results in terms of cellular definition; however, route B is preferable if a whole mount image of the embryo or tissue is desired, prior to sectioning. The in situ hybridization step should be performed prior to immunohistochemistry, which may include one or two antibodies. One antibody is for BrdU and the other can be either a cell cycle antibody (e.g., PH3, (6, 13)) or a cell type-specific antibody (e.g., Sox3 for neural progenitors or Myt1 for postmitotic neurons (13, 14)).

2. Materials

2.1. BrdU or EdU Injection in Xenopus laevis Embryos (See Note 3)

1. Injection buffer: 1% Ficoll in 0.1× MMR (0.1× MMR is made as a 10× stock solution and diluted 100 times with distilled water. 10× MMR stock solution: 1 M NaCl, 20 mM KCl, 10 mM $MgCl_2$, 20 mM $CaCl_2$, 50 mM HEPES, pH 7.5, sterilize by autoclaving).
2. Injection needles: GC100TF-10 glass capillaries (Harvard Apparatus Cat. No. 30-0038) are pulled with a Flaming/

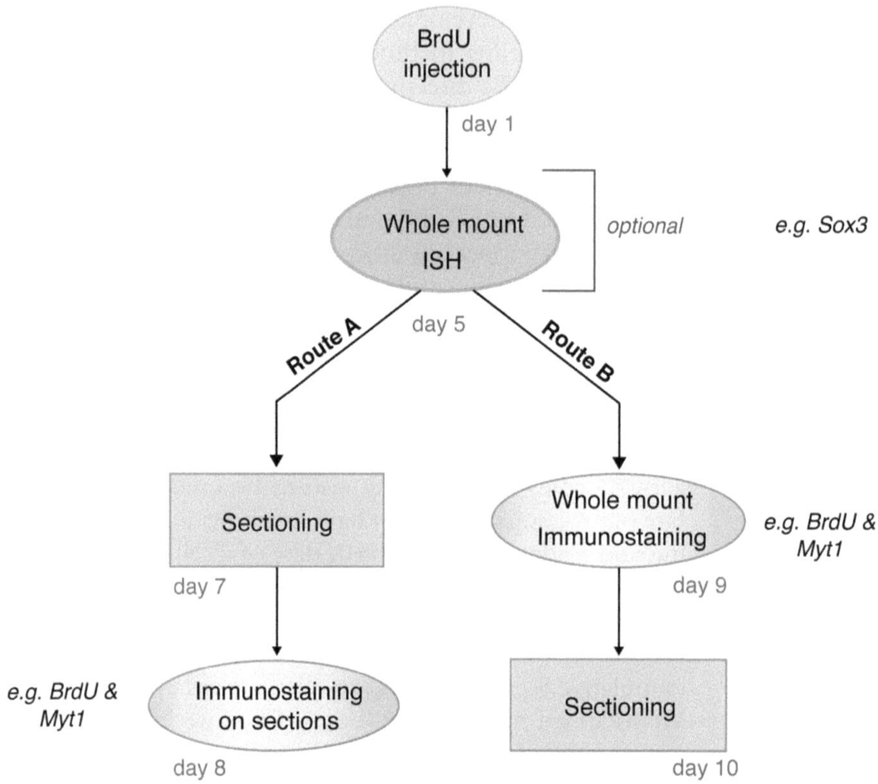

Fig. 1. Overview of the method. Following BrdU injection, two different routes can be taken. Route A involves immunostaining on sections while Route B involves whole mount immunostaining followed by sectioning. An optional step for in situ hybridization can be added following BrdU injection.

Brown micropipet puller Model P-87 (Sutter Instruments Co.). Conditions used are temp = ramp test, pull = 50, vel = 100, and time = 5.

3. Microloader tips (Eppendorf).
4. Air-pulse-based nano-injector type Pico-Spritzer.
5. 10 mM BrdU solution (Roche).
6. 1 mM EdU solution (Life Technology, part of Click-iT® EdU Imaging kit, dilute EdU according to manufacturer's recommendations).
7. 60 mm Petri dish coated with nylon mesh (Spectrum, Spectra Mesh Cat. No. 146410).

2.2. Fixing Samples

1. Wheaton glass scintillation vials (20 mL, VWR).
2. Freshly prepared MEMFA: 0.1 M MOPS pH 7.4, 2 mM EGTA, 1 mM $MgSO_4$, 3.7% formaldehyde.

2.3. Whole Mount In Situ Hybridization

1. 100 μg/mL Digoxigenin (DIG) RNA probe.
2. PTw: 1× PBS, 0.1% Tween-20 (Sigma).

3. 100% Methanol, 50% methanol in distilled deionized water, 25% methanol in PTw.
4. 0.1 M Triethanolamine (TEA) adjusted to pH 7.8.
5. Autoclaved 2 mL screw cap tubes.
6. Acetic anhydride.
7. 4% Formaldehyde in PTw.
8. Hybridization buffer: 50% formamide, 5× SSC, 1 mg/mL Torula RNA, 100 μg/mL heparin, 1× Denhardt's, 0.1% Tween-20, 0.1% CHAPS.
9. 60°C water bath.
10. Probe hybridization solution: 100 μg/mL probe in hybridization buffer.
11. 20× SSC stock: 3 M NaCl, 0.3 M Sodium Citrate, pH 7.0.
12. 2×, 0.2× SSC.
13. RNAse cocktail (Ambion) diluted in 2× SSC (dilute 10 μL of RNase cocktail per mL of 2× SSC).
14. Heat treated lamb serum (HTLS): Incubate lamb serum at 60°C for 1 h, aliquot in 50 mL Falcon tubes, and store at −20°C.
15. 2× Maleic acid buffer (MAB) stock: 200 mM Maleic acid, 300 mM NaCl, pH 7.5.
16. 1× MAB.
17. 1× MAB, 2% BMB.
18. 1× MAB 2% BMB 20% HTLS.
19. Sheep anti-DIG antibody coupled to alkaline phosphate (Roche) diluted 1:2,000 in 1× MAB, 2% BMB, 20% HTLS.
20. Alkaline phosphate (AP) buffer: 100 mM Tris pH 9.5, 50 mM $MgCl_2$, 100 mM NaCl, 0.1% Tween-20.
21. NBT/BCIP chromogenic substrate diluted in AP buffer. Add 4.5 μL of 75 mg/mL of NBT (Roche) and 3.5 μL of 50 mg/mL of BCIP (Roche) per mL of AP buffer.
22. Wheaton glass vials (4 mL, VWR).
23. 100% Methanol, 50% methanol in 1× PBS.

2.4. Route A: Immunostaining on Sections

2.4.1. Sectioning

1. 15% Fish gelatin (Sigma), 15% sucrose dissolved in distilled deionized water for embryos up to NF stage 30. Use within a week of preparation.
2. 25% Fish gelatin (Sigma), 15% sucrose dissolved in distilled deionized water for embryos from NF stage 30 onwards. Use within a week of preparation.
3. Cryomolds (Fischer).

4. Cryostat (Leica).
5. OCT compound (Lamb).
6. Superfrost plus slides (VWR).

2.4.2. Combining BrdU/EdU Detection and Immunostaining on Sections

1. 2 N HCl in fresh 1× PBS.
2. 1× PBS, 0.1% Triton X-100.
3. 1× PBS, 0.1% Triton X-100, 5% HTLS.
4. 1× PBS, 0.5% Triton X-100, 5% HTLS.
5. Click iT® Reaction cocktail (Life Technologies, part of Click-iT® EdU imaging kit, prepare reagents and cocktail according to manufacturer's guidelines).
6. Primary and secondary antibodies (such as anti-BrdU, anti-Acetylated Tubulin, anti-Myt1, and anti-Sox3, see Note 3).
7. Mowiol mounting media: mix 6 mL of glycerol, 2.4 g of Mowiol 4-88 (Calbiochem), and 6 mL of distilled de-ionized water. Mix at room temperature for 2 h. Add 12 mL of 200 mM Tris–HCl pH 8.5 and incubate at 50°C with occasional shaking until the Mowiol has completely dissolved. Filter on 0.45-μm membrane filter and store aliquots at −20°C.
8. DAPI diluted at 10 mg/mL in distilled de-ionized water (Sigma). Make 50 μL aliquots and keep at −20°C.
9. Slides tank type Coplin jar.
10. Glass coverslips (22 × 50, thickness N°1, VWR).
11. Nail varnish.

2.5. Route B: Whole Mount Immunostaining Followed by Sections

2.5.1. Combining BrdU Detection and Whole Mount Immunostaining

1. 100% Methanol, 50% methanol in 1× PBS.
2. 2 N HCl in fresh 1× PBS.
3. BBT: 1% BSA (Sigma), 0.1% Triton X-100 (Sigma) in 1× PBS.
4. HTLS (see Subheading 2.3, item 14).
5. BBT + 5% HTLS.
6. Primary and secondary antibodies (such as anti-BrdU, anti-Acetylated Tubulin, anti-Myt1, and Sox3, see Note 3).
7. PTw (see Subheading 2.3, item 2).
8. 2 mL screw cap tubes.
9. Wheaton glass scintillation vials (20 mL, VWR).

2.5.2. Sectioning

See Subheading 2.4.1.

3. Methods

3.1. BrdU or EdU Injection in X. laevis Embryos

1. Fill up an injection needle with 5 μL of BrdU or EdU solution using a microloader tip. Install the needle on the holder, break the needle tip, and set up the injector in order to obtain a 4.2 nL drop using an ocular micrometer for measurement. Place embryos in injection buffer on a dish coated with nylon mesh or use forceps to immobilize them and inject them as required depending on their stage of development (see Note 4).
2. Let embryos recover in injection buffer for the required amount of time (see Note 5).

3.2. Fixing Samples

1. Once the embryos have reached the developmental stage of interest, remove the vitelline membrane (if still present) with fine forceps and fix embryos in MEMFA for 1 h at room temperature (see Note 6).
2. Wash three times in 100% methanol and store at least overnight in 100% methanol at −20°C.

3.3. Whole Mount In Situ Hybridization

This protocol is a standard in situ protocol using chromogenic reaction except that the Proteinase K treatment is omitted in order to keep any antigen intact for further immunostaining (for Fluorescent In Situ Hybridization, FISH, please see Chapter 24 in the same issue). This method works both in *X. laevis* and *Xenopus tropicalis* (see Note 7).

NB: This protocol can be run in 3 days, in which case, step 11 should be reached on day 1 and step 21 (fourth 1× MAB wash) on day 2.

1. Gradually rehydrate fixed embryos by 5 min washes from 100% methanol to 50% methanol in ddH_2O and then 25% methanol in PTw.
2. Wash three times in PTw.
3. Wash embryos twice for 5 min in 5 mL 0.1 M TEA pH 7.8.
4. Add 12.5 μL acetic anhydride to the second TEA wash and rock for 5 min. Add another 12.5 μL acetic anhydride and rock for 5 more minutes.
5. Wash twice 5 min in 5 mL PTw.
6. Refix for 20 min with 4% formaldehyde in PTw.
7. Wash five times in PTw for 5 min.
8. Aliquot 500 μL hybridization buffer into 2 mL screw cap tubes. Transfer embryos into the buffer carrying over as little PTw as possible.

9. Once the embryos have settled through the dense buffer, remove the buffer and replace with another 500 μL of hybridization buffer.
10. Prehybridize for 4–6 h at 60°C.
11. Replace hybridization buffer with 500 μL of probe solution (hybridization solution with probe at 1 μg/mL). Hybridize overnight at 60°C.
12. Remove probe and keep at −20°C. Probe can be reused several times. Replace probe with hybridization buffer and put at 60°C for 10 min.
13. Wash three times for 20 min at 60°C in prewarmed 2× SSC.
14. Treat with RNase cocktail for 30 min at 37°C.
15. Wash for 10 min in 2× SSC at room temperature.
16. Wash twice for 30 min in 0.2× SSC at 60°C.
17. Wash twice in 1× MAB, 10–15 min per wash at room temperature.
18. Preblock for between 15 min and 1 h in 0.5 mL 1× MAB + 2% BMB. Dilute 2 mL 10% BMB stock 1 in 5 with 8 mL 1× MAB.
19. Block for 1 h in 0.5 mL 1× MAB + 2% BMB + 20% HTLS. Add 2 mL 10% BMB and 2 mL HTLS to 6 mL 1× MAB.
20. Incubate with 0.5 mL of anti-DIG antibody diluted in 1× MAB + 2% BMB + 20% HTLS. Incubate for 4 h at room temperature or overnight at 4°C.
21. Transfer the embryos to a 50 mL Falcon tube containing 1× MAB. Wash eight times for 20 min each in 1× MAB. One of the washes must be done overnight at 4°C. For each wash, Falcon tubes are laid horizontally in a blue tray and rocked, including overnight wash.
22. Transfer the embryos into Wheaton glass vials. Wash twice for 5 min at room temperature with 2 mL of AP buffer.
23. Replace the AP buffer with 2 mL of AP buffer with the chromogenic substrate diluted in. Rock in the dark at room temperature. In order to accelerate the color reaction, samples can be incubated the same way at 30°C.
24. Once the color has developed sufficiently, wash the samples in 50% methanol in 1× PBS for 5 min then wash twice in 100% methanol for a total of 2 h maximum. This step will allow the inactivation of the alkaline phosphatase and the removal of excess color reaction. Embryos kept for more than 2 h in methanol will turn black and be useless for analysis.
25. Rehydrate gradually the sample in 1× PBS. From this point, you should proceed to route A or route B protocol in the next 24 h. You can store the samples in 1× PBS at 4°C overnight if needed but not for longer.

3.4. Route A: Immunostaining on Sections (See Note 8)

3.4.1. Sectioning

1. Freshly prepare some fish gelatin (15% for embryos up to NF30, 25% for later stages, see Note 9).
2. If needed (continuing from Subheading 3.2, step 2), gradually rehydrate the fixed samples in 1× PBS to finally put them in fish gelatin. Let them embed at least overnight at room temperature on a rocker.
3. Mount them in embedding moulds of the appropriate size. First put a bed of fish gelatin and let set on dry ice. Then place each sample and orient it with forceps under the dissecting microscope quickly enough to keep the fish gelatin bed frozen. Mark the mould in order to locate the sample, top up with fish gelatin, and let freeze on dry ice.
4. Once set, mark the position of the sample on the block on the dry ice cold block with a pencil and let the block cool to cryostat temperature for at least 10 min. Mount it on the cryoblock holder with OCT compound and let it set. Set up your block on the cryostat head, let it cool down to object temperature for 5 min and start sectioning. First, trim between 30 and 50 μm until the sample is reached. Section at 12 μm and gather sections on superfrost slides (see Note 10).
5. Keep slides in −80°C for at least 1 h before processing for immunostaining.

3.4.2. Combining BrdU/EdU Detection and Immunostaining on Sections (See Fig. 2 and Notes 11 and 12)

1. Take slides out of the −80° and let them dry under a fume hood on a tray for 1 h.
2. Dip slides in acetone for 2 min and rehydrate them in 1× PBS 0.1% Triton X-100 in a slide tank type Coplin jar then draw a boundary with a PAP pen.
3. Block 30 min in 1× PBS, 0.5% TritonX-100 5% HTLS, incubate 30 min in the dark in a humidity chamber with 100 μL per slide of Click-iT® Reaction cocktail. Subsequently, wash three times for 5 min with 1× PBS 0.1% Triton X-100.
4. In order to perform BrdU staining, incubate the slides in 2 N HCl for 30 min at 37°C. Then wash three times for 10 min in 1× PBS 0.1% Triton X-100 (see Note 13).
5. Block with 1× PBS 0.1% Triton X-100 5% HTLS for 30 min in a humidity chamber.
6. Incubate with primary antibodies (against BrdU and your antigen of interest, see Note 14) for 2 h at room temperature or overnight at 4°C in a humidity chamber.
7. Wash three times for 10 min in 1× PBS, 0.1% Triton X-100 then block with 1× PBS, 0.1% Triton X-100, 5% HTLS for 30 min in a humidity chamber.

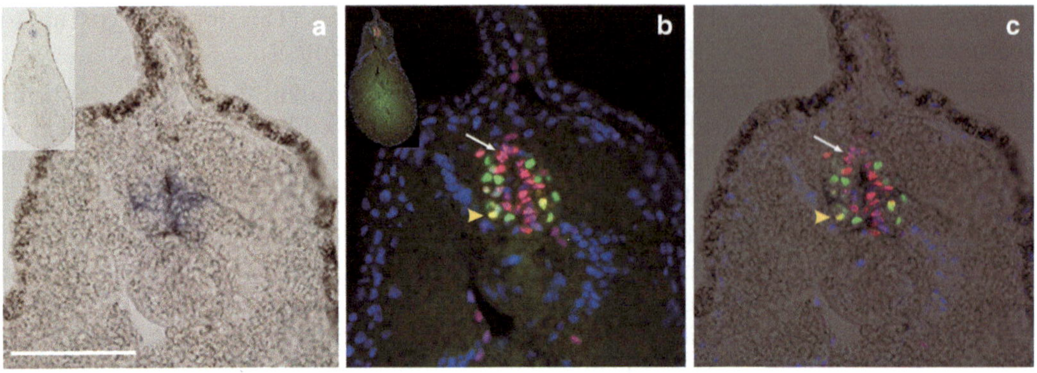

Fig. 2. Characterization of proliferating neural progenitors and newborn neurons in NF stage 14 embryo. Embryos were injected with BrdU at NF stage 14 and grown at 18°C for 24 h allowing them to reach NF stage 28. They were then fixed in MEMFA and processed following Route A protocol. After in situ hybridization for Sox3, the sample was thin sectioned (12 μm) and processed for immunostaining against BrdU and Myt1. (**a**) Bright field image of a section of the sample showing Sox3 signal in the spinal chord. *Inlay* shows the entire section through the embryo. (**b**) Fluorescent image of the same section showing BrdU (*red*), Myt1 (*green*), and DAPI (*blue*). Newly born neurons are double positive for BrdU and Myt1 (*yellow*, *arrowhead*) and proliferating progenitors are only positive for BrdU (*red*, *arrow*). (**c**) Merge image combining bright field and fluorescent images. Scale bar: 100 μm.

8. Incubate with fluorescent-labeled secondary antibodies for an hour at room temperature. Usually, secondary fluorescent antibodies will be used from 1:300 to 1:500 (see Note 15).
9. Wash three times 10 min in 1× PBS, 0.1% Triton X-100.
10. Incubate with DAPI diluted at 5 μg/mL for 10 min and wash three times for 10 min in 1× PBS.
11. Take slides out of the washing tank and partially dry them before mounting them in Mowiol. Seal them with nail varnish.

3.5. Route B: Whole Mount Immunostaining Followed by Sections

3.5.1. Combining BrdU Detection and Whole Mount Immunostaining

1. If needed quickly rehydrate the samples in 50% methanol, 50% BBT for 2 min (continuing from Subheading 3.2, step 2).
2. Wash the samples in scintillation vials in BBT for 1 h twice.
3. Treat the samples with 2 N HCl for 1 h at room temperature, and then wash three times in 1× PBS for 5 min.
4. Transfer the samples into 2 mL screw capped tubes and block for 1 h in BBT + 5% HTLS.
5. Incubate overnight at 4°C in the same tubes in 1 mL of primary antibodies diluted in BBT + 5% HTLS.
6. Transfer the embryos back into scintillation vials and wash in BBT for 1 h four times.
7. Transfer the embryos back into 2 mL screw capped tubes and block for 1 h in BBT + 5% HTLS.

8. From now on, embryos have to be protected from light. Incubate overnight at 4°C in the same tubes in 1 mL of secondary antibodies dilution in BBT + 5%HTLS.
9. Transfer embryos back into scintillation vials and wash for 1 h in BBT.
10. Wash embryos in PBT three times 1 h and then overnight at 4°C.
11. The embryos can be imaged after immunostaining using either a stereomicroscope or a microscope get higher magnification images (confocal or not). Embryos can be cleared in Murray's solution (2 volumes of benzyl benzoate, 1 volume of benzyl alcohol) to allow visualization of deep structures. To do so, gradually dehydrate embryos in methanol before transferring them in Murray's solution.

3.5.2. Sectioning

Follow the sectioning protocol presented in Route A (Subheading 3.4.1) to process stained embryos.

4. Notes

1. BrdU, CldU, and IdU can be used at the same concentration range (10 mM stock for BrdU and IdU, 100 mM stock for CldU, see Note 3 for administration) and are all identified by anti-BrdU antibodies that show variable affinities for these different analogs (8, 9, 11). EdU is another thymidine analog enabling a direct coupling with a fluorophore postincorporation. We are injecting it at a concentration of 1 mM, the same way we do for BrdU (see Note 3). The staining protocol is carried out according to manufacturer's guideline (ClicK-iT® EdU Imaging kits) and is routinely performed on sections. In the context of dual pulse labeling experiments, it is performed between rehydration and HCl treatment (Subheading 3.4.2, step 3). EdU can easily replace BrdU but, in our hands, BrdU proved itself to be more reliable. Moreover, we have not attempted to combine in situ hybridization with EdU injection and staining.
2. Different efficient primary anti-BrdU antibodies are available. They are raised in rabbit (from Abbiotec), sheep (from Abcam), rat (from AbD Serotec), or mouse (from Roche or Phoenix), allowing the combination of BrdU staining with any other primary antibody directed against an antigen of interest without risking cross-reactivity. CldU (Invitrogen) can be used instead

of BrdU, possibly with better sensitivity of detection (in this case, the primary antibody used is a rat anti-BrdU from AbD Serotec diluted at 1:2,000). In case of dual pulse labeling using EdU and BrdU, anti-BrdU antibody clone MoBU-1 (Invitrogen, B35128) can be used without cross-reacting with EdU molecule.

3. BrdU can be administered in two ways: by soaking embryos in BrdU solution or by injecting embryos with BrdU. The soaking is routinely used by fish' researchers (8, 15). However, on a side-by-side comparison of soaking and injection in *Xenopus* early embryos (NF stage 10, soaking in 100 μM BrdU in 0.1× MMR vs. a single injection of 4.2 nL of BrdU 10 mM in the blastocoel), we have found that the later method leads to a more homogeneous distribution of BrdU signal, particularly in deep tissues. Moreover, injection has the advantage of using less quantity of thymidine analogs, especially on late stages where animals need to be kept in big volumes of rearing media.
4. For early embryos (up to NF stage 12), one injection in the blastocoel is sufficient for labeling the entire embryo, whereas two medial injections in the neural region are preferable in embryos from NF stage 12 to NF stage 25. The number of injections is increased to 3, for embryos up to NF stage 35 and 4, from NF stage 35 to NF stage 40. From this stage, tadpoles secrete mucus and the epidermis is tougher, which makes injection more difficult. However, injecting in the gills from NF stage 40 enables homogenous BrdU distribution. Ensure that you comply with appropriate animal experimentation regulations in your country.
5. Cells that are in S-phase during the injection start to incorporate BrdU immediately. In our hands, BrdU incorporation can be detected by immunostaining after 10–20 min of injection. The length of incubation after injection determines the number of proliferating cells that will be labeled by BrdU. This depends on the length of the cell cycle, which may vary in a tissue and stage specific manner, and the length of time for which BrdU is available. It may be important to consider how long BrdU is available for labeling after a single injection ("pulse"). One way to do this is to administrate a saturating amount of BrdU to a set of animals and fix some for BrdU analysis in regular intervals, for example, every 30 min. The percentage of BrdU positive cells should increase linearly, reaching a plateau after a certain period of time, which is equal to the time that BrdU is available for labeling after a single injection in this system (also called bioavailability, retention, or clearance time). In practice, the curve may not reach a "flat" plateau as BrdU positive cells continue to divide. In this case,

the inflection point of the curve represents the BrdU retention time. Using such a protocol, we found that the clearance time of BrdU in early gastrula embryo (NF stage 9.5) is about 90 min and that 1.5 h incubation labels most cells in the ectoderm (around 83%, H. Auger. unpublished observations). In other systems, it has been estimated that BrdU was available for 2 h in adult rodent brain and 4 h in adult fish brain (16, 17). BrdU molecules incorporated into the DNA are thought to be diluted to undetectable levels after about three cell divisions (18).

6. Tadpoles from NF stage 50 are fixed overnight at 4°C. After three washes in PBS, the CNS or any other tissue of interest can be dissected in PBS in order to facilitate further processing by reducing the size of the sample.
7. The Proteinase K treatment takes place between Subheading 3.3, steps 2 and 3. Usually, we treat embryos for 10 min at room temperature with Proteinase K diluted in PTw (10 ng/L for *X. laevis*, 1.5 ng/L for *X. tropicalis*). A mild proteinase K treatment could be needed in some cases to allow a good penetration of the probe. If so, tests should be performed in order to optimize the results of combined in situ hybridization and immunochemistry.
8. Immunostaining on sections gives better results than sectioning whole mount immunostained embryos, especially if the aim is to count stained cells. This might be partly due to a more efficient penetration of the antibodies on sections compared to whole mount embryos.
9. Fish gelatin is used instead of the classically used OCT embedding because OCT does not penetrate tissues (19). Hence, fish gelatin lead to better results in terms of conservation of tissue integrity.
10. Thickness of the sections can be increased to 25 μm, but it might affect how well sections stick on slides. The overall results obtained on thick sections (25 μm) is usually less clear than on thin sections (12 μm) due to an overlap of signal, giving light blurriness to stained samples.
11. In the example shown in Fig. 2 (immunostaining on sections), the embryos were thin sectioned at 12 μm. Thicker sections (from 20 to 30 μm) may be preferred if the samples have already been processed for in situ hybridization in order to be able to clearly identify the signal (see Notes 8 and 10).
12. Omit Subheading 3.4.2, step 3 if no EdU has been injected.
13. HCl is essential for BrdU staining. Any good antibody, that does not require any tricky pretreatment, has good chances of working after this treatment.

14. The concentration of primary antibody used depends on the antibody itself. For instance, the primary anti-BrdU antibody commonly used in our lab (Roche anti-BrdU Cat. No. 11 170 376 001) is used at 1:200. We have also made primary antibodies for Mytl, used at 1:1,000 and Sox3, used at 1:2,000.
15. Secondary fluorescently labeled antibodies used in our lab mainly come from Molecular Probes and are generally diluted at 1:400.

Acknowledgments

We are grateful to Dr. Federico Calegari, Muriel Perron, and Morgan Locker for many useful discussions and protocols. This work was funded by a Wellcome Trust Senior Research Fellowship to NP. HA and RT are Research Associates, funded by the Wellcome Trust.

References

1. Taupin P (2007) BrdU immunohistochemistry for studying adult neurogenesis: paradigms, pitfalls, limitations, and validation. Brain Res Rev 53:198–214
2. Nowakowski RS, Lewin SB, Miller MW (1989) Bromodeoxyuridine immunohistochemical determination of the lengths of the cell cycle and the DNA-synthetic phase for an anatomically defined population. J Neurocytol 18:311–318
3. Chapouton P, Adolf B, Leucht C, Tannhauser B, Ryu S, Driever W, Bally-Cuif L (2006) her5 expression reveals a pool of neural stem cells in the adult zebrafish midbrain. Development 133:4293–4303
4. Calegari F, Haubensak W, Haffner C, Huttner WB (2005) Selective lengthening of the cell cycle in the neurogenic subpopulation of neural progenitor cells during mouse brain development. J Neurosci 25:6533–6538
5. Salomoni P, Calegari F (2010) Cell cycle control of mammalian neural stem cells: putting a speed limit on G1. Trends Cell Biol 20:233–243
6. Locker M, Agathocleous M, Amato MA, Parain K, Harris WA, Perron M (2006) Hedgehog signaling and the retina: insights into the mechanisms controlling the proliferative properties of neural precursors. Genes Dev 20:3036–3048
7. Arai Y, Pulvers JN, Haffner C, Schilling B, Nusslein I, Calegari F, Huttner WB (2011) Neural stem and progenitor cells shorten S-phase on commitment to neuron production. Nat Commun 2:154
8. Alunni A, Hermel JM, Heuze A, Bourrat F, Jamen F, Joly JS (2010) Evidence for neural stem cells in the medaka optic tectum proliferation zones. Dev Neurobiol 70:693–713
9. Conboy MJ, Karasov AO, Rando TA (2007) High incidence of non-random template strand segregation and asymmetric fate determination in dividing stem cells and their progeny. PLoS Biol 5:e102
10. Gomez-Nicola D, Valle-Argos B, Pallas-Bazarra N, Nieto-Sampedro M (2011) Interleukin-15 regulates proliferation and self-renewal of adult neural stem cells. Mol Biol Cell 22:1960–1970
11. Martynoga B, Morrison H, Price DJ, Mason JO (2005) Foxg1 is required for specification of ventral telencephalon and region-specific regulation of dorsal telencephalic precursor proliferation and apoptosis. Dev Biol 283:113–127
12. Salic A, Mitchison TJ (2008) A chemical method for fast and sensitive detection of DNA synthesis in vivo. Proc Natl Acad Sci U S A 105:2415–2420

13. Sabherwal N, Tsutsui A, Hodge S, Wei J, Chalmers AD, Papalopulu N (2009) The apicobasal polarity kinase aPKC functions as a nuclear determinant and regulates cell proliferation and fate during Xenopus primary neurogenesis. Development 136:2767–2777
14. Zhang C, Basta T, Jensen ED, Klymkowsky MW (2003) The beta-catenin/VegT-regulated early zygotic gene Xnr5 is a direct target of SOX3 regulation. Development 130:5609–5624
15. Rothenaigner I, Krecsmarik M, Hayes JA, Bahn B, Lepier A, Fortin G, Gotz M, Jagasia R, Bally-Cuif L (2011) Clonal analysis by distinct viral vectors identifies bona fide neural stem cells in the adult zebrafish telencephalon and characterizes their division properties and fate. Development 138:1459–1469
16. Zupanc GK, Ott R (1999) Cell proliferation after lesions in the cerebellum of adult teleost fish: time course, origin, and type of new cells produced. Exp Neurol 160:78–87
17. Hayes NL, Nowakowski RS (2000) Exploiting the dynamics of S-phase tracers in developing brain: interkinetic nuclear migration for cells entering versus leaving the S-phase. Dev Neurosci 22:44–55
18. Kiel MJ, He S, Ashkenazi R, Gentry SN, Teta M, Kushner JA, Jackson TL, Morrison SJ (2007) Haematopoietic stem cells do not asymmetrically segregate chromosomes or retain BrdU. Nature 449:238–242
19. Fagotto F, Brown CM (2008) Detection of nuclear beta-catenin in Xenopus embryos. Methods Mol Biol 469:363–380

Chapter 27

Microscopy Tools for Quantifying Developmental Dynamics in *Xenopus* Embryos

Sagar D. Joshi*, Hye Young Kim*, and Lance A. Davidson

Abstract

Early *Xenopus* embryos, and embryonic tissues isolated from them, are excellent model systems to study morphogenesis. Cells migrate, change shape, and differentiate to form new tissues as embryos mature and recapitulate those same processes in tissue isolates. Both large-scale and small-scale cell and tissue movements can be visualized with a range of microscopy techniques. Furthermore, protein dynamics, fine-scale cell movements, and changes in cell morphology can be observed simultaneously as multicellular structures are sculpted. We provide an overview of complementary methods for visualizing macroscopic tissue movements, cell shape changes, and subcellular protein dynamics. Time-lapse imaging followed by quantitative image analysis aims to provide answers to some of the long-standing questions in developmental biology: How do tissues form? How do cells acquire specific shapes? How do proteins localize to specific positions? To address these questions we suggest strategies (1) to visualize whole embryos and tissue isolates using stereoscopes and epifluorescence imaging techniques, and (2) to visualize cell shapes and protein expression using high-resolution live imaging using confocal microscopy. These imaging approaches along with simple image analysis tools provide us with ways to understand the complex biology underlying morphogenesis.

Key words: *Xenopus laevis*, Stereo dissecting microscope, Epifluorescence, Confocal microscopy, Epithelia, Mesenchyme, Morphogenesis, Cell and tissue mechanics, Kinematics

1. Introduction

Early embryonic multicellular tissues provide us with a remarkable model system to understand tissue- and cell-mechanics during morphogenesis. Understanding how particular cells or tissues obtain their shape or differentiated state requires us to collect time-lapse movies using multiple imaging techniques and then analyze them to gather quantitative information. We have developed several independent preparatory and imaging protocols that can be

*These authors contribued equally to this work

Stefan Hoppler and Peter D. Vize (eds.), *Xenopus Protocols: Post-Genomic Approaches, Second Edition*,
Methods in Molecular Biology, vol. 917, DOI 10.1007/978-1-61779-992-1_27,

used in complementary ways. They are outlined in this chapter: (1) visualizing whole embryos and tissue isolates (stereoscope), (2) lineage mapping and for correlation with introduced protein expression (epifluorescence), and (3) visualization of fluorescent probes for cellular morphology and protein dynamics (e.g., GFP-fusion proteins; confocal microscopy). We begin by discussing the preparatory steps and raw materials (embryo culture, handling, and care) and then discuss each imaging approach in detail.

2. Materials

2.1. Materials for imaging whole embryos and tissue isolates under stereoscope

1. Stereoscope with trinocular, video-, or camera-port.
2. Computer controllable CCD camera.
3. Computer with software to control camera, acquire images, and time-lapse sequences (we highly recommend ImageJ, a freely available software package for Windows, OSX, iOS, and Linux operating systems) (Rasband, W.S., ImageJ, U. S. National Institutes of Health, Bethesda, Maryland, USA, http://rsb.info.nih.gov/ij/, 1997–2011).
4. Modeling clay (Van Aken International, CA; black clay is preferred as it is less fluorescent and less toxic than other colors).
5. Glass Pasteur pipettes. One pipette with the heat-rounded tip to make depressions in clay.
6. Plastic or hair "spears" to position and orient embryos.
7. Nitex mesh of a variety of sizes (Amazon Supply, Seattle WA) to hold multiple embryos.

2.2. Additional Materials for imaging with epifluorescence-equipped stereomicroscope

1. Fluorescence-equipped stereomicroscope, nonfluorescent black background (e.g., the white/black plate used by stereoscopes).
2. Color CCD camera.

2.3. Additional Materials and Equipment for confocal microscopy

1. Confocal imaging microscope. Our group has access to a laser-scanning confocal (Leica TCS SP5, Leica Microsystems, Bannockburn, IL), however, other laser-scanning as well as spinning disk confocal systems are suitable (available from Zeiss, Nikon, Olympus, and Yokogawa).
2. Immersion oil (Cargille Laboratories, Cedar Grove) is required for oil-immersion lenses (40× and 63×) for high-resolution imaging.
3. Small cover-slip fragments (2 by 8 mm) are prepared in advance using a diamond pencil. We routinely store these fragments in a Petri dish along with a convenient supply of silicone grease dabbed on the undersurface of the lid of the dish.

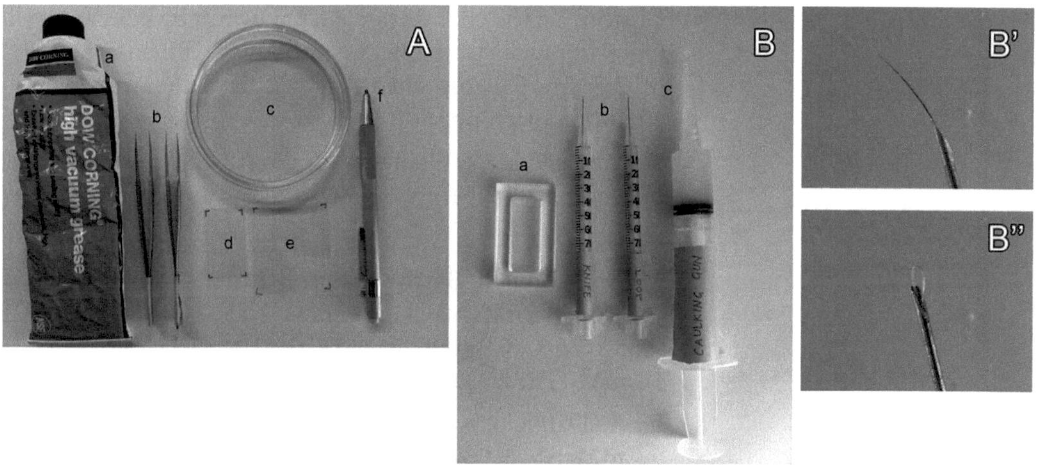

Fig. 1. Handy materials for preparing *Xenopus* embryos and tissues for imaging. (**A**) Commercially available tools. Silicone grease (a), forceps (b), Petri dish (c), large cover-slip (d), small cover-slip (e), and diamond pencil (f). (**B**) Custom-made tools. Custom acrylic culture chamber (a), hair tools (b) and silicone grease "caulking-gun" (c). (**B′**) Close-up of the hair knife, and (**B″**) hair loop.

4. Prepare lint-free tissue lined Petri dishes to store culture chambers and to keep their lower glass surface clean.
5. Sparkle glass cleaner.

2.4. Materials for embryonic culture/ microsurgery

1. Commercially available tools (Fig. 1A):
   (a) Silicone grease (high vacuum, Dow Chemical).
   (b) Forceps (Dumont #5 stainless steel; Fine Science Tools, Foster City, CA).
   (c) 60 or 100 mm Petri dishes.
   (d) Cover glass (small), 24 by 40 mm (12-544C, #1.5; Fisher Scientific).
   (e) Cover glass (large), 45 by 50 mm (12-544-F, #1.5; Fisher Scientific, Hampton, NH) diamond pencil (retractable; Ted Pella Inc., Redding, CA).
   (f) Stage micrometer (1 mm; Fine Science Tools, Foster City, CA).
   (g) AmpliCap Transcription kit; (Epicentre Biotechnologies, Madison WI).
   (h) Microinjection system (PLI-100; Harvard Apparatus, Holliston, MA).
2. Custom-made tools (Fig. 1B):
   (a) Hair tools (hair knife and hair loop): Create the tools using 1 mL syringes. Insert eyebrow hair (for hair knife, (Fig. 1B′)) or baby hair (for hair loop, (Fig. 1B″)) in the syringe needle and fasten the hair in place using melted wax.

(b) Custom acrylic chambers: Use 2.5 by 5.0 by 0.5 cm acrylic block to produce an inner chamber with dimensions of 1.2 by 2.8 cm. Custom chambers can be milled in a suitably equipped machine shop.

(c) Silicone grease "caulking-gun": Prepare a grease dispenser by loading silicone grease into a 5 or 10 mL syringe.

3. Solutions

(a) MBS (Modified Barth's Saline): Embryo culture media (MBS; 88 mM NaCl, 2.4 mM $NaHCO_3$, 1 mM KCl, 0.33 mM $CaCl_2$, 0.41 mM $(CaNO_3)_2$, 0.82 mM $MgSO_4$, 10 mM HEPES (H3375, Sigma-Aldrich, St. Louis MO)).

(b) DFA (Danilchik's for Amy): Explant culture media (DFA; 53 mM $NaCl_2$, 5 mM Na_2CO_3, 4.5 mM potassium gluconate, 32 mM sodium gluconate, 1 mM $CaCl_2$, 1 mM $MgSO_4$).

- Add 0.1 % bovine serum albumin with DFA (BSA; A7906, Sigma-Aldrich).
- Add antibiotics and antimycotics (0.8 % in media; A5955, Sigma-Aldrich) fresh to inhibit bacterial growth.

(c) Calcium- and Magnesium-Free DFA: Prepare the same as DFA above except omitting $CaCl_2$, $MgSO_4$, BSA, and antibiotics.

(d) 1× MBS with 3 % Ficoll (Sigma, St. Louis, MO) for microinjection.

3. Methods

Before we begin the discussion of various imaging techniques, it is important to understand that there are several preparatory steps that should be completed in advance. Standard practices are followed to obtain embryos (1). Injection strategies and microsurgeries are discussed in detail previously (2) and the materials for the process are listed in the Subheading 2.4. Preparatory steps are listed below:

Preparation
Steps for preparing live samples for imaging include preparation of mRNA, morpholinos, or lineage tracers, obtaining and culturing embryos, and microinjection and have been described previously (1). In the following sections, we describe the various imaging approaches for cells, isolated tissues, and whole embryos.

3.1. Imaging whole embryos and tissue isolates using a stereomicroscope

The first and most effective way to explore a specific morphogenetic process is to monitor movements in the intact *Xenopus* embryo or within a microsurgically isolated tissue fragment. Early developmental processes in *Xenopus* occur rapidly and can be observed either directly or in tissue fragments. The *Xenopus* embryo develops rapidly, for instance, gastrulation takes about 5 h to complete at room temperature. Advanced microsurgical techniques allow the researcher to isolate cells and tissues from surrounding tissues so that the role of nonautonomous signaling or mechanics can be explored. Such isolated cells or explants often recapitulate developmental programs of differentiation and local morphogenetic movements. Multipurpose culture chambers can provide experimental access to both gross morphogenetic movements and subcellular processes. In addition, pigment granules and yolk platelets in the *Xenopus* embryo provide good contrast when using side-illumination on a simple stereo dissecting microscope so that one can follow tissue and cell movements without the additional complication of fluorescent-labeling techniques. In this section, we will introduce practical methods to image and analyze the morphogenesis of developing embryos and isolated tissues, and discuss tips to improve the image quality.

Protocol:

1. Prepare the embryos or microsurgically isolated tissues at desired developmental stage.
2. Transfer the embryos or tissue isolates using a disposable plastic Pasteur pipettes into a small Petri dish filled with culture media (DFA for isolates, 1/3× MBS for whole embryos).
3. Position embryos or tissues with hair tools. For whole embryo imaging, there are several approaches to hold the embryos in place during development (Fig. 2A–C; see Note 1).
4. Gently fill or top-off the culture media in the dish with plastic pipette until the small Petri dish is slightly overfilled. Be careful to avoid disturbing the positioned embryos.
5. Seal or cover the Petri dish with large glass cover-slip by gently pressing the cover-slip from the top to avoid introducing air bubbles inside the culture chamber (see Note 2).
6. Adjust factors such as zoom, focus, brightness, contrast, and color balance in live image mode, and set optimal acquisition parameters in the software (see Note 3).
7. Collect single images at long time intervals or time-lapse movies (see Note 4).

Analysis

It is often said "a photo is worth a thousand words," however, you will often need to reduce a photo or a time-lapse series to a small set of representative numbers. Quantitative image analysis is key to comparing the effects of treatments or defects associated with genetic lesions. We present here examples of several

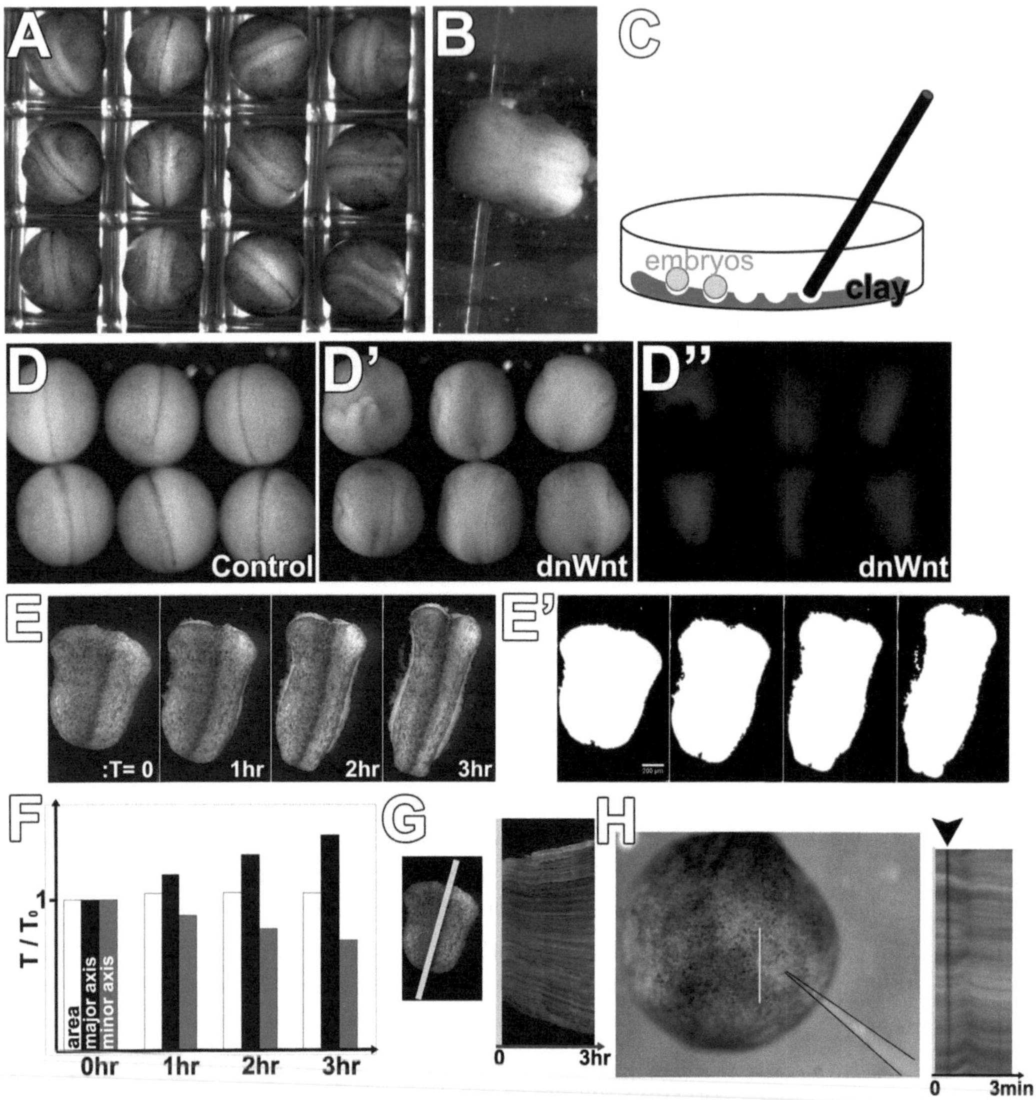

Fig. 2. Approaches for imaging intact whole *Xenopus* embryos and large-scale changes in tissue explants. Whole embryos can be held with (**A**) nylon mesh, (**B**) a thin plastic piercing, and (**C**) dented clay. (**D**–**D″**) Sorting embryos with an epifluorescence-equipped stereomicroscope. Embryos had been injected with mRNA encoding a dominant negative Wnt11 mixed with Rhodamine-dextran, a fluorescence lineage tracer. ((**D**) Control embryos (**D′**) dorsal targeted dnWnt (*red* in (**D″**)) expressing embryos) (**E**) time-lapse sequences from stereoscopic time-lapse movie of developing dorsal isolate tissue. (**E′**) Threshold-based auto-segmented dorsal isolates from (**E**). (**F**) Quantified results of elongation of the isolated dorsal axis over time. The ROIs (region of interest) segmented from (**E′**) are added to ROI Manager and measured for morphological properties. (**G**) Kymograph drawn from line across the major axis of the elongating dorsal isolate. (**H**) Kymograph used for analyzing induced contraction. Line for kymograph and *triangular outline from bottom right corner* shows a needle tip for adding solution (*arrow head* and line in *right panel* demonstrate the instance when the solution was added).

morphogenetic movements observable with the stereomicroscope and quantitative methods to assess their progress.

1. Measuring rates of dorsal isolate elongation or animal cap explant spreading (Fig. 2E, E′, F, G): Time-lapse sequences of these explants, whose widths range from 200 to 5,000 μm,

change shape over 1–5 h of elapsed time. Methods can be developed to automatically segment the tissue from the background. However, it is often simpler to manually identify the boundary of the explant or tissue.

(a) Automated segmentation. Threshold the background, detect the edge of the explant, and auto-measure a region of interest (ROI) that contains the tissue but excludes the surroundings. Once the boundaries of a structure can be defined by a threshold intensity, the structure can be segmented from the background using the AnalyzeParticles function in ImageJ. AnalyzeParticles can produce a list of ROIs using the ROI-Manager tool. This tool allows the user to track, maintain, save, and reopen a set of ROIs associated with a particular image sequence. The ROI-Manager provides many functions for nondestructive manipulation of ROIs and can be used to batch measure morphological properties of the ROI or the pixels contained within them.

(b) Manual measurement: Use the mouse to draw an ROI manually. Hand-selected ROIs can be stored within the ImageJ ROI-Manager and then used to measure properties of the tissue such as width, perimeter, or area.

2. Blastopore closure

(a) Kymograph produced from a time-lapse sequence collected as bottle cells form, tissues involute, and the blastopore closes. A kymograph is a 2D image which visualizes the time-rate of change of a one-dimensional line-profile in a time-lapse image sequence. One axis of a kymograph represents the intensity along the line (e.g., the profile) and the other axis represents that intensity at different times. Timing of individual events, such as the time of blastopore closure, the formation of dorsal or ventral most bottle cells, and the rate of tissue involution can be measured from the kymograph. A kymograph can be prepared within ImageJ by first drawing the desired line-segment across the image and using the "reslice" command from the STACKS menu.

(b) Manual tracking of individual landmarks such as cell clusters or patches of high or low pigmentation can provide rates of convergence or extension. Deformation of rectangular ROIs can provide semi-quantitative measures of convergence, contraction, or dilation.

(c) Automated segmentation of the blastopore morphology is challenging; however, deformation maps of tissue shape changes can be estimated by following movements of discrete pigment marks or cells using texture mapping, image warping, or registration algorithms (e.g., using the UnwarpJ plug-in (3, 4) for ImageJ).

3. Induced contraction (Fig. 2H):
   (a) Kymograph produced from a time-lapse sequence collected as the tissue undergoes a contraction. The animal cap is a naturally pigmented epithelial tissue. Irregular patterns of pigment can be used as an indicator of tissue movements.
   (b) Manual tracking of individual landmarks such as triple-cell junctions.
   (c) Automated tracking using image warping to estimate tissue deformation.

3.2. Imaging with a epifluorescence-equipped stereomicroscope

Genetically encoded fluorescent proteins such as green fluorescent protein (GFP) and many new variants of fluorescent proteins have accelerated the use of fluorescent microscopy in developmental biology. A stereomicroscope equipped with bright-field or side-illumination uses endogenous contrast to visualize tissues and cells on the embryo; alternatively, fluorescence illumination lets the researcher visualize specific tissues, cells, or cellular structures that have been labeled with exogenously introduced fluorophores. In addition, an epifluorescence-equipped stereomicroscope allows the observation of a large field-of-view with an extended working distance between the sample and the objective, allowing the researcher to manipulate embryos or tissues with hand tools such as forceps and hair tools or with micromechanical devices. An epifluorescence-equipped stereomicroscope is useful to verify the expression of fluorescent proteins and to locate labeled cells within the embryo. When tissue-specific, targeted expression is required for protein over-expression study or further high-resolution imaging, hundreds of embryos can be easily inspected and identified quickly (Fig. 2 D, D′, D″). Optimally expressing embryos or tissues can then be prepared for other imaging experiments. Lastly, we often mix lineage-tracking fluorophores (e.g., fluorescent-conjugated dextran or membrane targeted GFP/RFP) with RNA-antisense morpholinos or experimental mRNA encoding proteins that block or induce a signaling pathway; co-injection of a lineage label allows rapid identification of mutant cells and tissues. In this section, we will introduce the materials, methods, and provide examples of epifluorescence-equipped stereomicroscopy.

Protocol:

1. Inject fluorophore-conjugated dextran (see Note 5) or desired mRNA at one cell stage. (See preparation section above for details on microinjection.)
2. Incubate injected embryos to the desired developmental stage.
3. Turn on the fluorescent light source (i.e., mercury vapor lamp) attached to the fluorescent microscope.

4. Choose the appropriate filter block for your fluorophore excitation wavelength, and place the sample over the black background. Many black materials are auto-fluorescent; we have found that magnetized rubber on the back of refrigerator magnets provides excellent black background.
5. Select suitable fluorescent-labeled embryos using a hair loop (see Note 6).

3.3. Visualizing cellular morphology with GFP-fusion proteins with confocal microscopy

Confocal microscopy is used primarily to examine very small structures (in microns) where high-resolution images and time-lapse movies can enhance our understanding of biological events that occur in morphogenesis. Confocal microscopy allows one to eliminate out-of-focus light and improve image quality (6). This is done by detection of fluorescence light that is in or very close to the focal plane. There are several advantages of using this imaging technique as a stand-alone method or in combination with other techniques. It enables us to collect high-resolution images of biological samples that improve our understanding of biological tissue architecture and protein locations. Collecting a set of images at fixed time intervals allows us to understand, on a subcellular level, the nature of the proteins and their motion pathway to their specific locations. Collecting long-term and short-term time-lapse movies enables us to monitor tissue and cell behaviors. Also, by collecting z-stacks of images, it is possible to reconstruct the three-dimensional tissue architecture by combining the set of images. Thus, confocal microscopy has wide applications in morphogenesis where there is a need to image cellular and subcellular activities to better understand the role of different cell types and various proteins in development.

Materials/Equipment

In addition to the materials mentioned in the "2. Materials" section above, the following items are required for an experiment involving imaging using confocal microscopy:

Preparation of chambers for long-term culture

1. Chambers for long-term culture: Custom acrylic chambers are prepared as mentioned earlier. Large cover glass is used as the bottom of the acrylic chamber. The chamber is sealed to the cover glass using silicone grease. The cover glass serves two purposes for confocal imaging: (a) it serves as a transparent bottom that allows imaging and (b) it serves as a thin boundary between the microscope objective and the tissue housed in the chamber and does not hinder imaging.
2. Prepare the substrate in the chamber according to your application and tissue.
   (a) Epithelial tissue isolated from the embryo needs anti-adhesive surfaces between the apical surface and the

cover-slip glass surface. To reduce unwanted adhesion, coat the glass surface inside the chamber with 1 % BSA in 1/3× MBS (BSA-MBS) overnight at 4 °C. Chambers are rinsed several times in DFA before use.

(b) Deep or mesenchymal cells isolated from the embryo may require either a nonadhesive (see above) or an adhesive surface. To provide an adhesive surface, we coat the glass surface with 20 μg/mL human plasma fibronectin diluted in 1 mL of 1/3× MBS overnight at 4 °C. If the experiment involves cell motility, then fibronectin serves as a matrix for cells to attach and crawl. Chambers are rinsed several times in DFA before use.

3. Tissues of interest are microsurgically isolated and transferred into the chamber (7). Tissue isolates are fixed in place under cover-slip bridges, i.e., glass fragments held in place with small dabs of silicone grease on either end. Avoid excessive pressure on the fragments as they are fragile. With practice, one can make an educated guess about the amount of pressure required to keep them in place without smashing the tissues. It is good practice to position the tissues very near to each other in straight lines or along a grid, so that during imaging, there is less translation of the objective and make it easier to note the location of isolates in case there is a need to image multiple samples over time. With practice, 16–20 tissues can be housed in a single chamber.
4. Once the tissues are mounted on the chambers, fill the chambers until nearly overflowing with DFA, apply silicone grease to the rim of the chamber, and seal the chamber by pressing a 24 by 40 mm cover glass into place (see Note 7).

Protocol

1. Microinject fertilized embryos with desired mRNAs at 1-cell to 2-cell development stage and culture embryos till gastrulation.
2. Depending upon the experiment, excise specific tissues and prepare them as described before in the "Chambers for long-term culture" section above.
3. Power up the confocal imaging system.
4. Move the appropriate objective lens into place. For high-resolution imaging, oil-immersion objectives such as 40× and 63× are recommended. Apply a drop of immersion oil on the objective and then slowly place the chamber on the microscope stage. Move the focus to bring the objective toward the stage so that the oil on the objective wets the chamber bottom without creating any air bubbles in the oil.
5. Position small weights on either sides of the chamber to hold it in place.

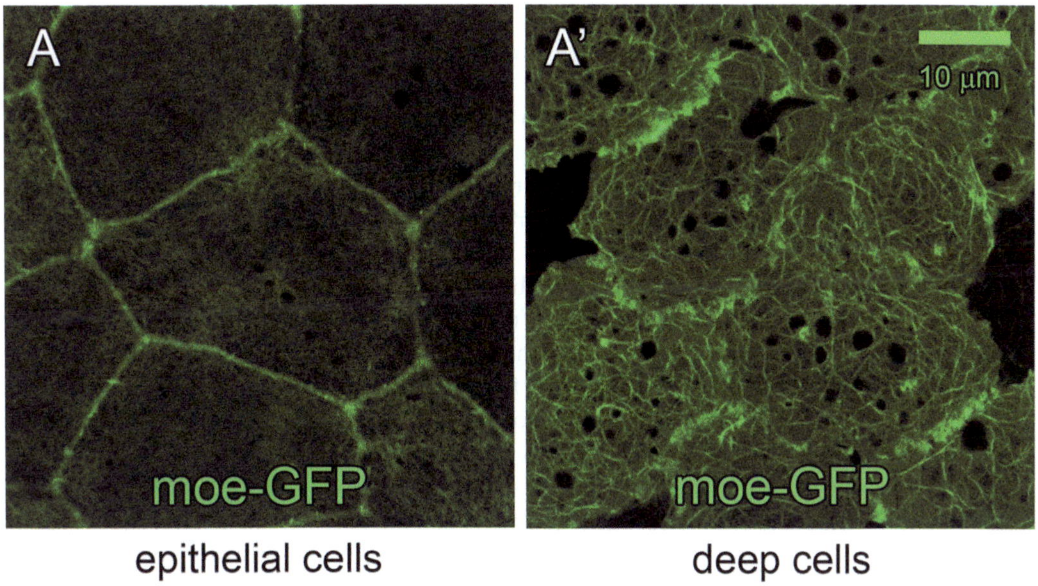

Fig. 3. *Xenopus* embryonic cells observed using confocal microscope. (**A**) X-Y view of a moesin-GFP labeled F-actin apical cell cortex in epithelial cells. (**A′**) X-Y view of moe-GFP labeled deep cells.

6. Using bright-field illumination, adjust the focus cautiously. Once tissues are in focus, one is ready to collect fluorescence images.
7. Select a scan format. 1,024 by 1,024 pixels provide good image quality.
8. Adjust the spectral emission range and select the excitation filters. Adjust pinhole between 1 and 1.5 Airy units. The pinhole value is decided based on how wide or thick the structure is that needs to be resolved and on the level of the protein expression (see Notes 8 and 9).
9. Single images at a single z-step can be captured using X-Z mode or X-Y mode (Fig. 3). Epithelial (Fig. 3A) and deep cells (Fig. 3A′) can be imaged as earlier stated in (b) in "Chambers for long-term culture."
10. The image quality can be improved by (1) Accumulation (Line or Frame) or by (2) Averaging (Line or Frame).
11. Time-lapse movies are the best way to describe precise cell behaviors in cell and developmental biology. For capturing different types of confocal time-lapse movies, the four different collection modes are: (1) XYT, for collecting a series of images over time keeping a constant z-plane, (2) XYZ, for collecting a series of images as a single z-stack, (3) XZT for collecting a series of transverse images over time, or (4) XYZT, a mode combining the characteristics of XYZ and XYT to collect z-stacks over time (see Note 10).

Image Analysis

Images of developing whole embryos or tissue isolates and time-lapse movies of dynamically changing cell shapes provide us with information on the likely events that happen in real time. Single-image panels from various time points can illustrate the important events but they need to be supported by quantitative approaches. There are several tools including MATLAB's Image Processing Toolbox (Mathworks, Natick, MA), PAX-it Image Analysis Software (Midwest Information Systems, Inc., Villa Park, IL), and IDL (ITT Visual Information Solutions, Boulder, CO) to analyze images or time-lapse movies. In our lab, we utilize an open source tool, ImageJ, for image analysis.

For the purpose of this image analysis section, we will limit our discussion to analysis of the high-resolution images obtained from confocal microscopy. Image data obtained from a Leica confocal systems system can be easily converted to .TIF files. The original files from the confocal microscope are initially saved and later archived in .LIF format. To convert .LIF into .TIF images for publication or post-processing we use the freely available Bio-Formats plug-in for ImageJ. Bio-Formats (Kevin Eliceiri, LOCI, University of Wisconsin, Madison, WI) is a standalone Java library for reading image file formats including most proprietary formats from major imaging companies. We open the .LIF experiment file (includes image and information files collected during a particular experiment) using Bio-Formats and then save these individual files separately as .TIF stack files.

In the following section, we offer one example of quantitative analysis of confocal image data. Previously, we have presented approaches to induce contractility in an epithelial cell sheet (8). One approach discussed was to induce contraction by ablating a single cell in the tightly connected cell sheet (Fig. 3A). We observed that the rest of the cells in the field-of-view would contract when a cell is ablated. To demonstrate the contraction, we utilize complementary approaches to represent the experimental data.

1. While measuring image properties, it helps if one increases the zoom in the ROI (Fig. 4A′) to improve visualization. A short stepwise procedure is as follows:
   (a) Open a .TIF file in ImageJ.
   (b) Select the any type of selection tool (which could be rectangular, elliptical, or polygon: these tools can be selected by clicking the tabs on the tools panel on ImageJ). Generally, it is a good practice to use rectangular tool to select the ROI.
   (c) Draw a rectangle surrounding the ROI.
   (d) Select IMAGE > DUPLICATE. This will provide you with a new image with only the ROI outlined by the rectangular selection (Fig. 4A′).

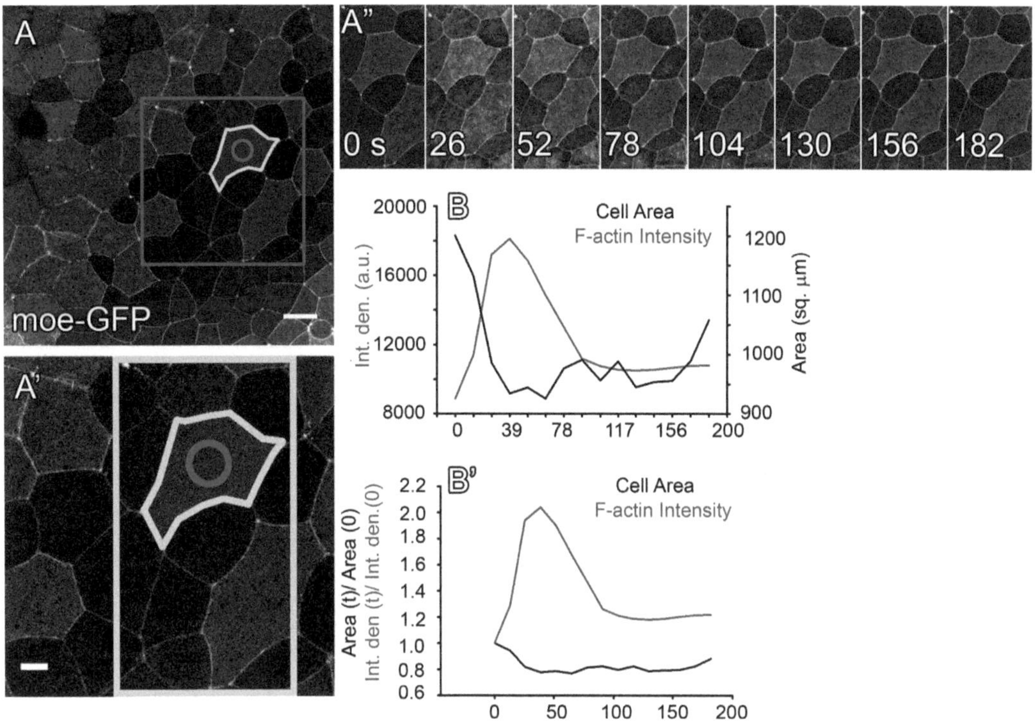

Fig. 4. Image analysis. (**A**) Original image of a moe-GFP labeled cell sheet. A high-resolution view (**A′**) of the *square box* inset in (**A**) shows three "regions-of-interest" or ROIs. (**A′**) The *polygonal outline* indicates the boundary of a single cell selected using the polygon tool and the *circle* indicates the area inside the cell where we intend to measure the protein intensity. The *vertical rectangular* inset is used to demonstrate a strip kymograph in (**A″**). Scale bars in (**A**) and (**A′**) indicate 10 μm. (**B**) F-actin intensity (*circle* ROI in (**A′**)) is plotted against cell area (polygon in (**A′**)) along with the relationship between their normalized values (**B′**).

2. To visualize the changes in cell properties, a simple yet effective method is to create a strip kymograph of the time-lapse images (Fig. 4A″). This helps us to pictorially represent the changes in convincing way for the reader. A stepwise procedure to create a kymograph is as follows:

   (a) Open a .TIF file in ImageJ.

   (b) Add "Substack Maker" (authors: Anthony Padua and Daniel Barboriak) plug-in from ImageJ website to your ImageJ. For this, download "Substack_Maker.class" file to the "plug-ins/stacks" within the ImageJ folder on your computer and restart your ImageJ. This will add a Substack Maker tab in PLUGINS > STACKS on your ImageJ menu.

   (c) Open a .TIF file (FILE > OPEN).

   (d) Click PLUGINS > STACKS > SUBSTACK MAKER. This will provide with a window to enter the text. Enter all the frame numbers (separated by commas) that should be part of the strip kymograph. We recommend a nonpartisan method, i.e., to select frames at regular intervals (Fig. 4A″).

Another common method is to select special frames that the author believes can convey relevant information about the experiment.

3. Intensity changes in confocal time-lapse data contain information on changing rates of protein abundance, localized assembly or aggregation, and protein activity levels. To acquire these data from image sequences requires a consistent methodology. Here, we illustrate a simple method where changes in F-actin assembly can be quantified from a series of images. In case of induced contraction experiment, quantitative data is obtained by collecting information about the changing cell areas and the F-actin intensities (labeled with moe-GFP) within the cell. This can be done using simple built-in plug-ins in ImageJ, the procedure for which is as follows:
   (a) Open the .TIF file in ImageJ.
   (b) Choose "Area" mode for measurements. (ANALYZE > SET MEASUREMENTS).
   (c) Open the ROI-Manager tool (ANALYZE > TOOLS > ROI MANAGER).
   (d) Select on the "polygon-selection" tool from the ImageJ menu.
   (e) Starting in the first image outline the cell or area of interest (yellow polygon is Fig. 4A) and add it to the ROI Manager by typing the "t" key or by clicking the button that reads "ADD (t)" in the ROI Manager.
   (f) Advance through the time-lapse by clicking ">" and outline a new area until the end of the time-window of interest.
   (g) Save your ROI set to a file that may be reopened for later analysis (see Note 11).
   (h) Measure areas of the ROI cell outlines by selecting "MEASURE" in ROI Manager. A new "RESULTS" window will now open with a list of AREA for each ROI.
   (i) Capture the dynamic changes in cell area in more cells by repeating steps (e)–(h).
   (j) Select "Integrated Density" in (ANALYZE > SET MEASUREMENTS) to record F-actin intensities.
   (k) Select the "Elliptical selection" tool from the ImageJ toolbar.
   (l) Draw a circle in the region of interest (red circle is Fig. 4A) add it to ROI Manager as before.
   (m) Advance through the time-lapse and repeat step (l) to record circular areas for F-actin intensity measurements.
   (n) Measure intensities within the ROI circles by selecting "MEASURE" in ROI Manager. A new "RESULTS"

window will now open. The results window contains Integrated Densities of all the recorded ROIs and can be saved as spreadsheet or text file.

(o) Repeat the steps (b)–(n) to record cell area and F-actin intensities for as many cells as needed.

(p) Graphical demonstration: Visual representation of the quantitative data is a critical aspect of research. Microsoft Excel (Microsoft Corporation, Seattle, WA) or Sigma Plot (Systat Software Inc., San Jose, CA) and SPSS (IBM Corporation, Armonk, NY) can be used to plot data graphically. Here we have shown an example of our study where cell areas decrease and F-actin intensities increase. The changes can be reported in real values (Fig. 4B) or as ratios (Fig. 4B′). We prefer to report both the aspects since real values are critical for the reader to make a good judgment of the experimental setup and ratios provide us with a method to compare values across experiments. The graphs constructed here are using Sigma Plot (Fig. 4B, B′). These are first steps in image analyses. More detailed analytical approaches have been discussed previously (9).

4. Notes

1. Since embryos often rotate and shift their body as they undergo large-scale tissue morphogenesis, keeping embryos in place is essential for acquiring long-term time-lapse sequences. To minimize the sliding of embryos during imaging, vitelline membrane can be carefully removed with forceps before positioning. Larger tissue explants can also shift position during extended time-lapse imaging. Explants can be placed within narrow slots or boxes formed from glass cover-slip fragments positioned with silicone grease.
2. Covering the Petri dish with a cover-slip provides higher quality optics than imaging through either an air–water interface or imaging through plastic. In addition, a tightly sealed culture chamber prevents evaporation of media that can cause dramatic shifts of imaging focus for long-term live imaging.
3. Use a high-contrast background surface (e.g., black paper under a light-colored explant) to improve subsequent image analysis efforts. Increase the camera gain and minimize the intensity of the light source or add an IR or UV filter to the light source to reduce the heat and other hazardous effects of long-term light exposure during time-lapse movie collection. Often, the resolution or field-of-view of the stereoscope can be

improved by either decreasing or increasing the magnification on the video-coupler between the camera and port.

4. Use a manual stage to collect images at long intervals (e.g., every hour) or use a computer-controlled stage to automatically collect at short intervals. Automatically collected images are aligned within the same frame-of-reference during collection whereas manually collected images can be aligned after collection.
5. Dextran conjugates commonly used for lineage tracing may need to be treated with DEPC to remove RNase activity if present.
6. The fluorescent expression levels of mRNAs are often low at early stages of development, but will increase as embryos develop and accumulate more fluorescent proteins. To improve the fluorescent signal detection from each sample, adjust the fluorescent illumination via lamp intensity. Avoid intense long-term fluorescent illumination which can bleach the fluorophore and cause phototoxic effects in cells. Caveat: over-expression of fluorescently tagged reporter constructs can perturb embryonic development. Often the best embryos will be the lowest expressing samples!
7. The goal is to seal the chamber with as few air bubbles as possible since these may cause the chamber to flex and images to drift over long-term imaging sessions. Keep a lint-free tissue handy to catch overflowing culture media.
8. While conducting experiments with confocal microscopy, use lowest possible laser power that provides good-quality images. This depends on the protein expression levels as well as the type of protein that is being expressed. Low laser power ensures the least possible photo-bleaching, reduces photo-toxicity, and cell damage.
9. Photomultiplier gain values must be kept to a minimum to reduce noise. The values depend on the protein expression and background tolerance levels.
10. The researcher should follow a meaningful folder- and file-naming convention that allows easy archiving and the ability to track image information without the need to open the file. Modern operating systems allow long filenames but deep folder structures may exceed to character limit of pathnames in many applications.
11. ROI Manager can serve as a library for segmented structures within image sets. ROI data from an entire study can be archived in ROI Manager files and used for future calculations, validation of image analysis protocols, or assessment of segmentation quality.

Acknowledgements

We would like to thank past and present members of the Davidson, Keller, and DeSimone Labs for their care and patience in helping to develop these methods through trial and mostly error. In addition, we would like to thank Lin Zhang for her assistance. This work was made possible through grants from the NIH (R01 HD044750, R21 ES019259) and from a CAREER award from the NSF (IOS-0845775).

References

1. Sive HL, Grainger RM, Harland RM (eds) (2000) Early development of *Xenopus laevis*: a laboratory manual. Cold Spring Harbor Laboratory Press, Cold Spring Harbor, New York, p 338
2. Kim HY, Davidson LA (2010) Methods to investigate molecular mechanisms and cellular mechanics responsible for morphogenesis in *Xenopus laevis* embryos. In: Wong R, Sharpe J (eds) Imaging in developmental biology: a laboratory manual. Cold Spring Harbor Laboratory Press, New York
3. Arganda-Carreras I et al (2006) Consistent and elastic registration of histological sections using vector-spline regularization. Computer Vision Approaches to Medical Image Analysis, ser. Lecture Notes in Computer Science, vol 4241. Springer, Berlin, p 85–95
4. Sorzano CO, Thevenaz P, Unser M (2005) Elastic registration of biological images using vector-spline regularization. IEEE Trans Biomed Eng 52(4):652–663
5. Lane MC, Davidson L, Sheets MD (2004) BMP antagonism by Spemann's organizer regulates rostral-caudal fate of mesoderm. Dev Biol 275(2):356–374
6. Tsien RY, Waggoner AF (1995) Fluorophores for confocal microscopy. In: Pawley JB (ed) Handbook of biological confocal microscopy, 2nd edn. Plenum Press, New York, pp 267–280
7. Joshi SD, Davidson LA (2010) Live-cell imaging and quantitative analysis of embryonic epithelial cells in *Xenopus laevis*. J Vis Exp 23(39):pii: 1949
8. Joshi SD, von Dassow M, Davidson LA (2010) Experimental control of excitable embryonic tissues: three stimuli induce rapid epithelial contraction. Exp Cell Res 316(1):103–114
9. Kim HY, Davidson LA (2011) Punctuated actin contractions during convergent extension and their permissive regulation by the non-canonical Wnt-signaling pathway. J Cell Sci 124(4):635–646

Part VI

Novel Approaches in *Xenopus*

Chapter 28

Mathematical Modeling of Gene Regulatory Networks in *Xenopus* Development

Yasushi Saka

Abstract

Mathematical modeling has become increasingly indispensable for scientists who study the dynamics of gene regulatory networks (GRN) that underlie cell differentiation and pattern formation in animal development including *Xenopus* embryogenesis. Here I outline a step-by-step procedure for constructing a mathematical model of GRN based on ordinary differential equations (ODE), using the network of Activin and its downstream target genes *Xenopus Brachyury* (*Xbra*) and *Goosecoid* (*Gsc*) as an example. I also briefly explain methods to analyse the dynamics described by an ODE model.

Key words: Mathematical modeling, Ordinary differential equation, Gene regulatory network, Nonlinear dynamics, *Xenopus*, *Activin*, *Brachyury*, *Xbra*, *Goosecoid*, *Gsc*

1. Introduction

Mathematical modeling has been a valuable tool in biology for many years. Yet, it is only recently that application of mathematics has become increasingly popular among biologists who want to understand interactions among molecules inside the cell. This is because networks one might come across often have a structure whose dynamics is not easily understood by simple reasoning alone, although recent advancement in technology has provided a detailed picture of such interaction networks. This sort of nonlinear networks of interacting genes, proteins, or any other biological molecules could show rich variety of counterintuitive behavior. This is true even for a small network of two interacting factors (e.g., Turing's reaction-diffusion model) (1, 2).

Here I describe a method to construct a model of gene regulatory network (GRN) based on ODEs. Deterministic models using ODE serves as a basis for analyzing nonlinear dynamics of GRN

Stefan Hoppler and Peter D. Vize (eds.), *Xenopus Protocols: Post-Genomic Approaches, Second Edition*,
Methods in Molecular Biology, vol. 917, DOI 10.1007/978-1-61779-992-1_28, © Springer Science+Business Media, LLC 2012

(see Note 1). This chapter is intended for experimentalists, including *Xenopus* embryologists, who wish to model and analyze the dynamics of a GRN of interest for the first time, but have no experience in mathematical modeling or do not know where to start (see Note 2). Taking the GRN of *Activin* gene and its downstream target genes *Xenopus Brachyury* (*Xbra*) and *Goosecoid* (*Gsc*) as an example (3), we will see a step-by-step protocol to write down ODEs corresponding to the GRN and construct a deterministic model from scratch (see Note 3). The resulting model is an approximation of the system. However, it illuminates the essential feature of the GRNs dynamics, which explains how Activin, a morphogen, induces Xbra and Gsc expression in a concentration-dependent manner and also why a threshold concentration is generated in its gradient. I will only describe the most essential steps for constructing and analyzing a simple ODE-based model, including numerical simulations. Although the example I use here has only a few variables (i.e., Xbra and Gsc), a similar approach of model construction may be applied to larger systems and analyze the GRN of your interest mathematically.

2. Materials

A pencil and a piece of paper is all what you need before going on to the analysis of the dynamics of your constructed model. For plotting, analysis, and appreciation of the model dynamics, a general mathematics software package, such as Mathematica (Wolfram Research), MATLAB (MathWorks), or Maple (Maplesoft), is recommended. These software packages are widely used in many fields including physics and engineering. Alternatively, a free utility software called Grapher (Apple Inc.) is available on MacOSX platform, although its functions are limited (see Note 4). For numerical simulations using these softwares, a standard PC would be sufficient to run them smoothly, although it depends on the size of your model (see Note 5).

3. Methods

3.1. Network Diagram

Choose key constituents of the network to model and define interactions (positive or negative) among them (see Note 6). In our example, they are Activin, Xbra, and Gsc, which form the GRN shown in Fig. 1 (see Note 7).

3.2. Ordinary Differential Equations

ODEs describe how fast each component's concentration (or the number of molecules per unit volume) changes as a result of

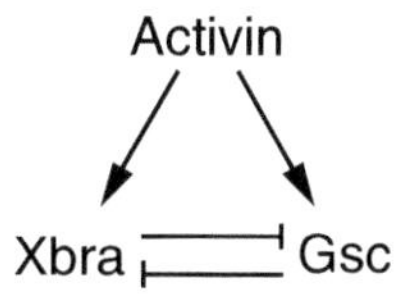

Fig. 1. The network of Activin, Xbra, and Gsc. In the mathematical model. Activin, Xbra, and Gsc are represented by the variables *w*, *u*, and *v*, respectively.

interactions with other components in the system. A basic ODE describing the dynamics of the component "*x*" is

$$\frac{dx}{dt} = (\text{ production term}) - (\text{ degradation term}).$$

The left-hand side of the equation is the first derivative of *x*, that is, the velocity of *x* (see Note 8). The production term is a function of other variables that represent the other components interacting with *x*. This is written down as a function of standard Michaelis–Menten or Hill kinetics with cooperativity (Fig. 2) (see Note 9). The degradation term is usually a simple linear function of *x*, that is, *x* multiplied by a certain parameter that defines the rate of degradation (see Note 10). Let us now consider the ODE for Xbra, which can be written as follows:

$$\frac{du}{dt} = f(v,w) - \mu_1 u, \tag{1}$$

where *u*, *v*, and *w* stand for Xbra, Gsc, and Activin concentrations, respectively, and μ_1 is a degradation rate constant. The production term is a function of *v* and *w*, $f(v, w)$, because Xbra expression (*u*) is controlled by Gsc (*v*) and Activin (*w*) in our example. We now define this function *f*. To do this, we need to make some assumptions based on biological observations (see Note 11). First, the expressions of Xbra (and Gsc) depends on Activin, i.e., if no Activin, no Xbra expression. Second, Gsc exerts a sigmoidal effect of repression on Xbra expression, which is represented by Hill coefficient *n*. Third, Activin concentration does not change over time and Hill coefficient is not considered for the induction of Xbra or Gsc by Activin for the sake of simplicity of the model. The response of Xbra and Gsc to Activin therefore follows Michaelis–Menten kinetics (Fig. 2a, $n=1$). These assumptions lead to the function *f* as follows:

$$\begin{aligned} f(v,w) &= k_1\left(\frac{1}{1+v^n}\right)\left(\frac{w}{1+w}\right) \\ &= \frac{k_1 w}{(1+v^n)(1+w)} \end{aligned} \tag{2}$$

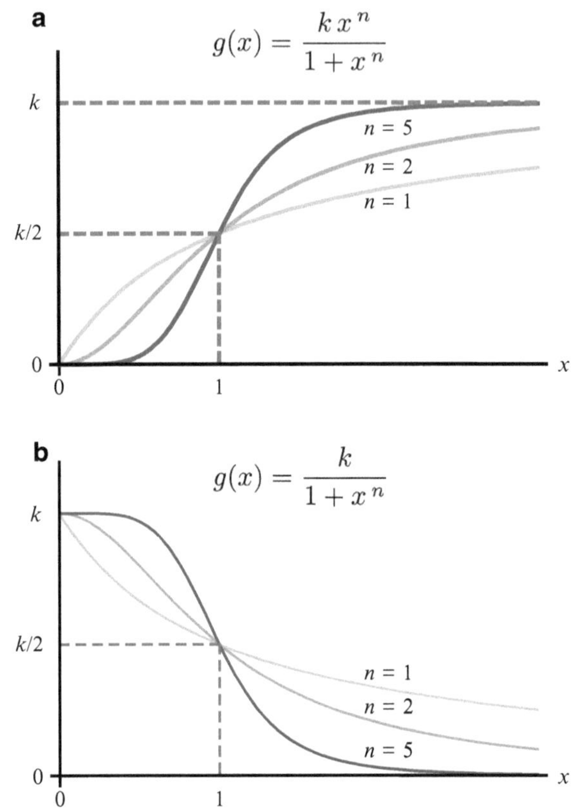

Fig. 2. Hill kinetics and Michaelis–Menten Kinetics. Hill kinetics, described by $g(x)$, is plotted for (**a**) activation and (**b**) inhibition with the cooperativity $n=1,2$ and 5. Michaelis–Menten kinetics is a special case when $n=1$ in Hill kinetics. Note that $g(1)=\frac{k}{2}$.

where k_1 is a production rate parameter (see Notes 12 and 13). Combining Eqs. 1 and 2 leads to the complete ODE for Xbra (u):

$$\frac{du}{dt} = \frac{k_1 w}{(1+v^n)(1+w)} - \mu_1 u. \tag{3}$$

Similarly, the ODE for Gsc (v) is

$$\frac{dv}{dt} = \frac{k_2 w}{(1+u^n)(1+w)} - \mu_2 v, \tag{4}$$

where k_2 is a production rate parameter and μ_2 is a degradation rate parameter. Finally, Activin concentration (w) does not change, therefore,

$$\frac{dw}{dt} = 0,$$

which is equivalent to

$$w = C_1,$$

where C_1 is a constant. Equations 3 and 4 therefore describe the dynamics of the GRN in Fig. 1 (see Note 14).

3.3. Steady State and Phase Portrait

Now we are ready to analyze the dynamics of the system of these ODEs. In general, it is hopeless to solve nonlinear ODE system like ours analytically (see Note 15). We therefore resort to the so-called phase portrait, which is a qualitative method, and numerical simulation, which is a quantitative approach and we shall see it later. The phase portrait is a basic but powerful method to illustrate and understand the system's dynamics intuitively. The dynamics can be understood as a trajectory of a point (phase point) on the plane with u and v coordinates (phase plane) (see Fig. 3). The velocity of a phase point is composed of the one in u direction and the other in v direction, which is represented by the velocity vector (u', v') (see Note 8).

The first thing to do is to examine what happens to the system as $t \to \infty$ and whether it ends up in a steady state, starting from an arbitrary state. At steady states, $u' = v' = 0$ by definition. Replacing the left-hand side of Eqs. 3 and 4 with 0 leads to

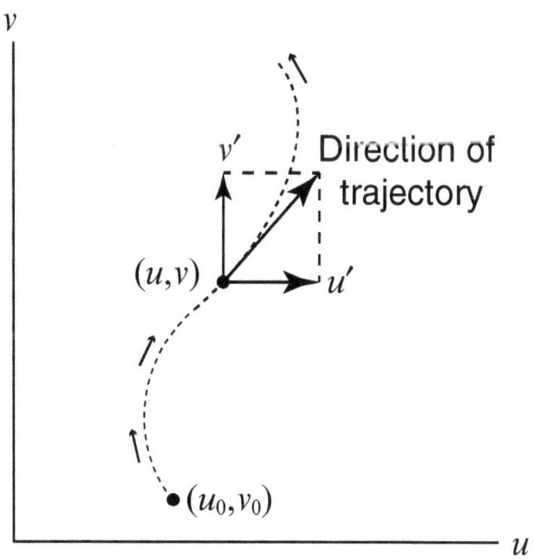

Fig. 3. Trajectory of a phase point on the phase plane. Schematic trajectory of a phase point (u, v) is shown on the phase plane, which is generally called a phase space for higher dimensions. The vector (u', v') indicates the direction and the velocity of the phase point and is tangent to the trajectory (*dotted line*). (u_0, v_0) indicates the initial condition.

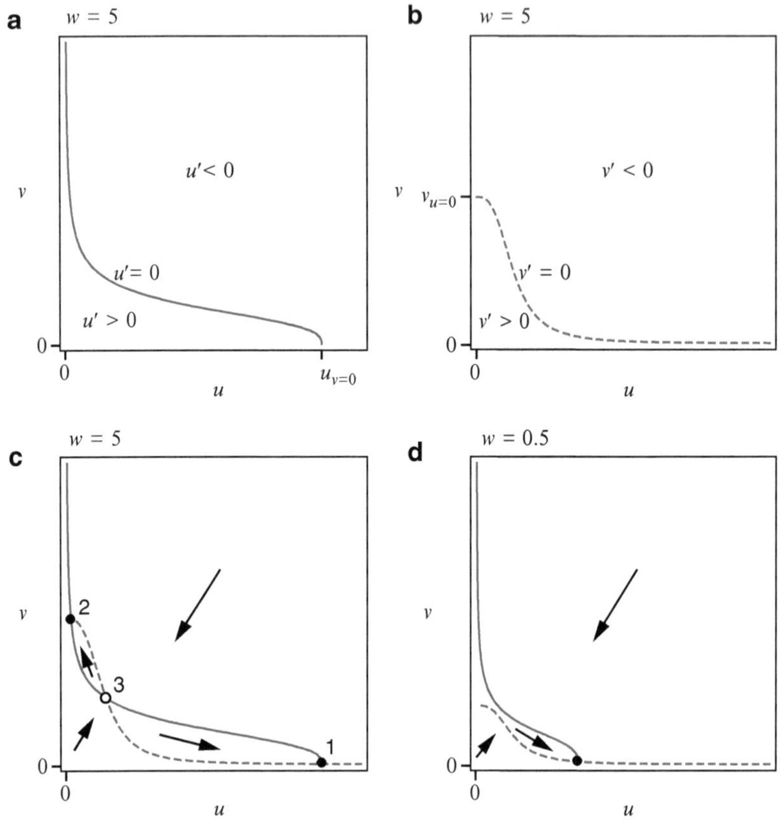

Fig. 4. Steady state and phase portrait. (**a**) Nullcline for $u'=0$. It divides the phase plane in two regions: one with $u'<0$ and the other one with $u'>0$. The direction of any trajectories on this nullcline is perpendicular to u axis. The nullcline crosses with u axis at $(u_{v=0}, 0)$. (**b**) Nullcline for $v'=0$. This nullcline crosses with v axis at $(0, v_{u=0})$. (**c**) Phase portrait for $w=5$. The nullclines divide the phase plane into six regions, including narrow strips on the *lower-right* and *upper left corner* between axes and nullclines. Stable fixed points are indicated by *black dots* and unstable fixed point by an *open circle*. (**d**) Phase portrait for $w=0.5$. In this example, the nullclines cross each other only once and the system has a single stable fixed point. *Arrows* indicate general flow direction of phase points in each region. Parameter values used for these plots are $k_1=0.5$, $k_2=0.7$, $\mu_1=0.06$, $\mu_2=0.15$, and $n=3$.

$$0 = \frac{k_1 w}{(1+v^n)(1+w)} - \mu_1 u \tag{5}$$

and

$$0 = \frac{k_2 w}{(1+u^n)(1+w)} - \mu_2 v. \tag{6}$$

Steady states must satisfy both of these equations, but let us consider Eq. 5 alone for the moment. Equation 5 corresponds to a curve on the phase plane, which is called a nullcline in mathematics (Fig. 4a) (see Note 16). To draw this curve, we assign a certain set of parameter values and the number for w, i.e., with a fixed value of Activin concentration (see Note 17). It divides up the phase plane into two regions: in the upper-right area, u decreases ($u' < 0$); in the lower-left area, u increases ($u' > 0$). Similarly, you can plot the Eq. 6, which is another nullcline on the phase plane (see Fig. 4b). The phase portrait is obtained by combining these plots (Fig. 4c). In this particular example, the phase plane is divided into six areas. The arrows indicate the rough direction of (u', v') in each area and illustrate how phase points move in the phase plane (also see Figs. 7 and 8). The crossing points of these curves correspond to the steady states. In Fig. 4c, there are three such points, which are called fixed points. Two of these fixed points (indicated by 1 and 2 in the figure) are locally stable, because the arrows surrounding these points all direct toward them. On the contrary, the middle one (indicated by 3) is unstable because some arrows are pointing away from it (see Note 18).

You may realize that the shape of nullclines differs depending on the parameter values and on the value of w. Consequently, there are not always three fixed points. Note that, however, with a given set of parameters (k_1, k_2, μ_1, μ_2, and n) and a fixed w, the nullclines do *not* change over time but only (u, v), the phase point, moves. With the same parameter values as Fig. 4c but a different value of w, we obtain a phase portrait shown in Fig. 4d. In this example, there is only a single stable fixed point (see Note 19). From these phase portraits, we can predict that steady states are different according to the value of w. This provides an important clue about the concentration-dependent induction of Xbra and Gsc by Activin.

3.4. Numerical Simulations

Phase portrait is a powerful means to analyze the global dynamics of a system of ODE qualitatively. However, it is not sufficient to predict exactly what happens to the system starting from a particular state at $t = 0$, which is called the initial condition. We therefore need numerical simulations to complement the phase portrait approach.

For our model, the initial condition is the concentration of Xbra (u) and Gsc (v) (and Activin, w, which remains unchanged over time) at $t = 0$, that is, $(u_0, v_0) = (0, 0)$. The resulting data of a numerical simulation is a time series of (u, v) (see Note 20). The first thing you may want to do is to run simulations with various set of parameter values you have chosen rather randomly (Fig. 5; also see Figs. 9 and 10). Examining the dynamics by visual inspection of plots, you may notice that, depending on the parameter values, the system exhibits very different behavior. This is how the author found the most crucial aspect of the GRN's dynamics (3). After plotting with various parameter values, you may also notice that the system's dynamics falls into just a few categories of qualitative pattern. Each of these pattern corresponds to a particular type of phase portrait or "landscape" (Figs. 7 and 8) (see Note 21).

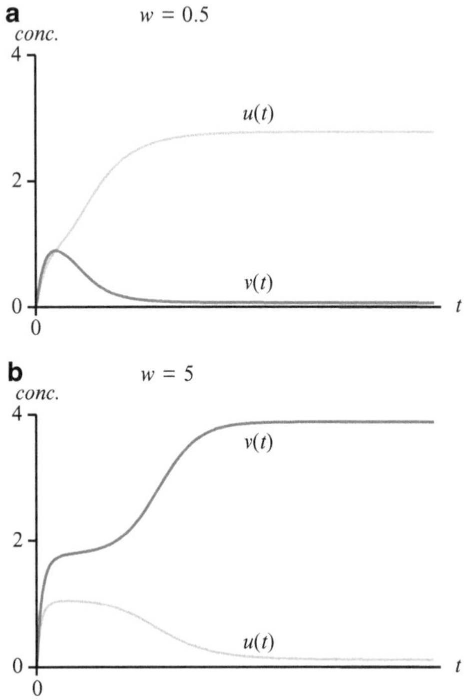

Fig. 5. Plots of numerical simulations. Plots of numerical simulations with the same parameter values as Fig. 4 are shown. *w* value is indicated for each panel. Note that Xbra (*u*) is on and Gsc (*v*) is off in (**a**) and vice versa in (**b**) at steady states. Also note that concentration is in arbitrary unit, and in the same scale in (**a**) and (**b**).

The next step is to test parameter values more systematically to understand how each parameter affects the dynamics of the system, and which condition the model behaves as we expect (see Note 22). For our present example, we expect the model reproduces the concentration-dependent induction of Xbra and Gsc by Activin. It is often very difficult to know the real parameter values, because of lack of information or experimental limitations. You could guess and estimate the parameter values. For instance, if we suppose that the half-lives of Xbra and Gsc are similar, μ_1 and μ_2 should be in the same range. Rather, it is more useful to examine in which range of the parameters realizes the desired dynamics and whether that range is biologically relevant. For our model, relatively broad range of parameter values reproduces the concentration-dependent action of Activin and a sharp transition of gene expression pattern in the gradient of Activin. In fact, examination of many parameter sets is not a trivial task, because the simulations and the resulting data may become very large. You may thus need a certain expertise, and therefore a cross-disciplinary collaboration becomes useful.

The above example just scratches the surface of the art of mathematical modeling, but I hope it has demonstrated a huge benefit of the approach and may serve as an entry point for (*Xenopus*) embryologists who wish to apply mathematical modeling for solving their own questions.

4. Notes

1. It is also the starting point for extending the model to include molecular diffusion and study pattern formation. To incorporate molecular diffusion in the system, partial differential equations (PDE) will be required. Also real GRNs have noisy dynamics, whose effect can be analyzed using stochastic simulation methods. Modeling using these methods is out of scope in this chapter.
2. Mathematics is a sort of foreign language required for cross-disciplinary collaborations, for instance, with physicists. Like any other ordinary foreign language, it requires a serious investment in terms of time and money to learn. Being comfortable with the language of mathematics, albeit not necessarily proficient, cross-disciplinary collaboration becomes much more productive. A series of video lectures of undergraduate level are available for calculus, differential equations, and linear algebra (4). The author found these lectures most useful to learn or refresh his memory on these areas of mathematics, which are all necessary for the analysis of nonlinear dynamics. It is also useful to have standard textbooks for these subjects. A textbook by Strogatz on nonlinear dynamics is highly recommended (5). Edelstein-Keshet (6) is more oriented to applications of mathematics to biology. You do not need to know everything to start modeling. It helps to have your own problem (e.g., GRN dynamics) to solve and you may learn necessary mathematics along the way.
3. Deterministic model does not take into account random variation of reactions and predicts outcomes precisely. ODE model is deterministic and *determines* unique states of variables (e.g., protein concentration in the system) for a given set of initial conditions. On the other hand, stochastic model considers randomness of reactions and produces probability distribution of states.
4. A mathematics software package (e.g., Mathematica) itself requires another kind of (computer) language, which may not be so difficult to learn, if you have ever written any computer programming codes. Plotting and visualizing mathematical functions greatly enhance one's comprehension. The only way to learn it is to use it. Also some reference books would be useful, for instance, see (7) for Mathematica. Figures 6–10 show just several examples of Mathematica code and the screen shot

```
Module[{k1 = 0.5, μ1 = 0.06, n = 3, w = 5},
 ContourPlot[0 == (k1 w)/((v^n + 1) (w + 1)) - μ1 u, {u, 0, 8}, {v, 0, 8}, MaxRecursion → 3]
]
```

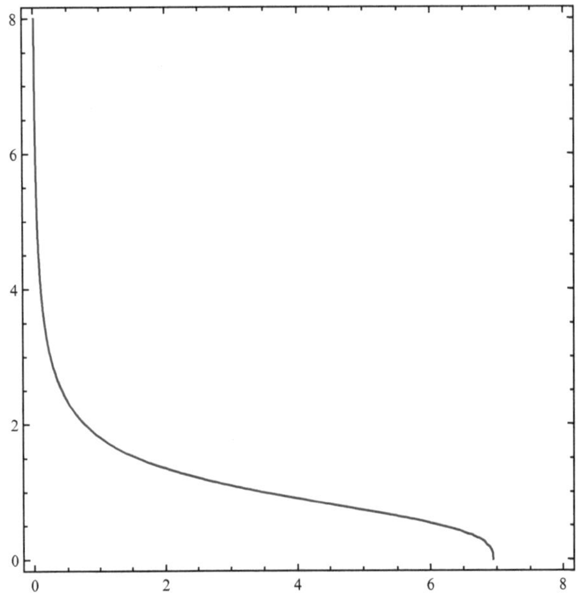

Fig. 6. Mathematica code for plotting nullclines and its output. Nullcline for Eq. 5 is shown as an example. It can be plotted as a contour curve that satisfy the equation, using a Mathematica function `ContourPlot`. The function `Module` makes the assigned parameter values local, so that they are not interfering with any operations outside this `Module`.

of its output. For details, see the figure legends. These codes should work on your PC by just copy & paste and executing them by shift+return (or enter) keys, and you should get the same plots in the figures (some functions may not work with an old version of the software). Perhaps your institution has a network license for either one of these software packages. If you need to buy one of these, academia or student discounts may be available. The author uses Mathematica. Besides these general mathematics software packages or Grapher on MacOSX, there might be freeware or shareware available for plotting ODEs, which may allow you to visualize network dynamics sometime even without writing down a single ODE. However, this sort of free softwares generally have only limited functions and more bugs than commercial ones. They also often have hidden algorithm that cannot be controlled. By the time you become familiar with these shortcomings, you would find a general mathematics software package handier.

5. For example, the author uses a laptop (MacOSX, 2.13 GHz Intel Core 2 Duo, 2 GB memory, and NVIDIA GeForce

```
Module[{du, dv, g1, g2, k1 = 0.5, μ1 = 0.06, k2 = 0.7, μ2 = 0.15, n = 3, w = 5},
 du = k1 w / ((v^n + 1) (w + 1)) - μ1 u;
 dv = k2 w / ((u^n + 1) (w + 1)) - μ2 v;
 g1 = StreamPlot[{du, dv}, {u, 0, 8}, {v, 0, 8}];
 g2 = ContourPlot[{0 == du, 0 == dv}, {u, 0, 8}, {v, 0, 8},
   ContourStyle → {Red, Green}];
 Show[{g1, g2}]
]
```

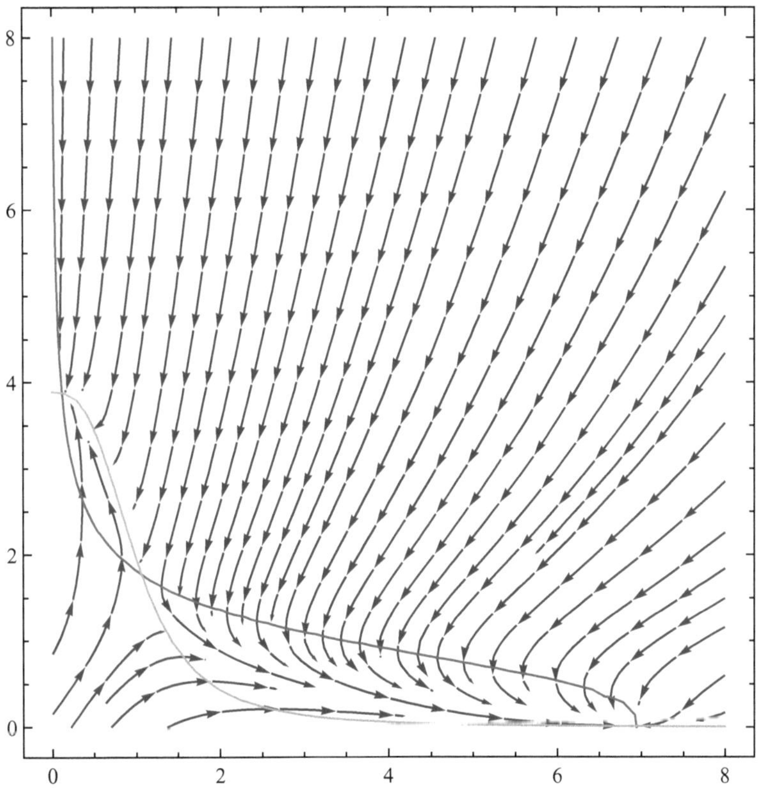

Fig. 7. Mathematica code for drawing a phase portrait. Phase portrait can be drawn as a compound plot using the Mathematica function `StreamPlot` and `ContourPlot` as shown in this example. *Arrows* correspond to the velocity vector (u', v'). Note that stream lines are perpendicular on the nullcline $u = 0$, and horizontal on the other nullcline $v' = 0$. Starting from a given point on the phase plane and following the arrows, you can determine the trajectory of a phase point and its destination, i.e., steady state.

9400M for graphics) and gets a plot of a simulation result in a fraction of a second for a model with several variables.

6. It is important to construct the first model with minimal number of components. The model can be refined later if you wish. Only include the most essential ones and omit linear intermediate steps, which can be treated as a single reaction in the model. For example, a cascade of gene activation can be

```
Module[{du, dv, k1 = 0.5, μ1 = 0.06, k2 = 0.7, μ2 = 0.15, n = 3, w = 5},
  du = (k1 w)/((v^n + 1) (w + 1)) - μ1 u;
  dv = (k2 w)/((u^n + 1) (w + 1)) - μ2 v;
  Plot3D[Norm[{du, dv}], {u, 0, 8}, {v, 0, 8}, AxesLabel → {"u", "v", None}]
]
```

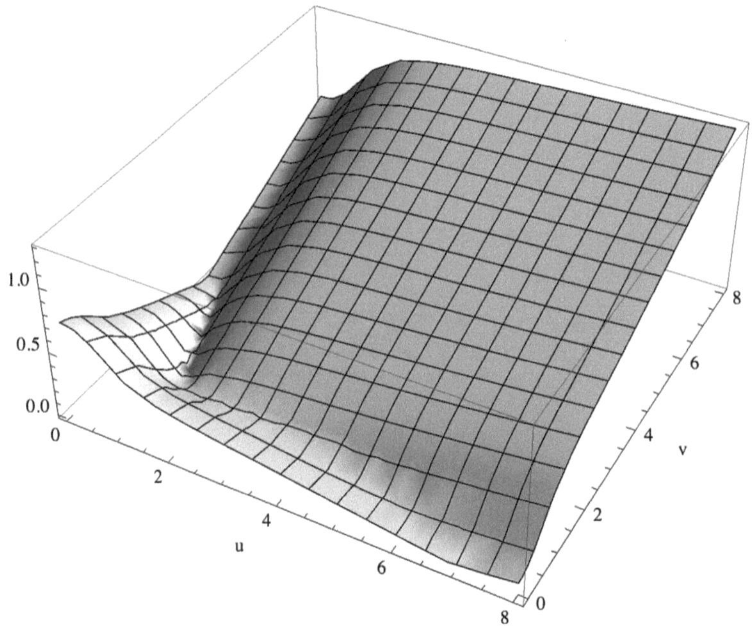

Fig. 8. Phase portrait as a landscape. Phase portrait can also be drawn as a 3D plot as shown in this example. It uses the function `Norm` which converts a vector, for example, (u', v') in this case, into its size. In this "landscape" view, the trajectory of a phase point is understood as the path of a ball rolling down the slope. This illustrates phase portrait in a more intuitive way. Nullclines correspond to troughs in the landscape.

represented as one-step reaction. Transcription and translation steps can also be regarded as a single reaction. Any model is an approximation of the real biological system and is not supposed to be an exact reproduction of the system with every detail. Complex model does not necessarily mean complex dynamics and may even obscure the critical aspect of the GRN. Simple network model often captures the essential feature of the GRN better.

7. For those who are not familiar with the biological background of this network, please first refer to (3), which also contains a detailed analysis of the model. For more general analysis of the cross-repression network and its switching behavior, see (8).

```
Module[{eqns, soln, k1 = 0.5, μ1 = 0.06, k2 = 0.7, μ2 = 0.15, n = 3, w = 5},
  eqns =
   {u'[t] == (k1 w)/((v[t]^n + 1) (w + 1)) - μ1 u[t],
    v'[t] == (k2 w)/((u[t]^n + 1) (w + 1)) - μ2 v[t],
    u[0] == 0,
    v[0] == 0};
  soln = NDSolve[eqns, {u, v}, {t, 0, 200}];
  Plot[Evaluate[{u[t], v[t]} /. soln], {t, 0, 200}, AxesLabel → {"time", None}]
]
```

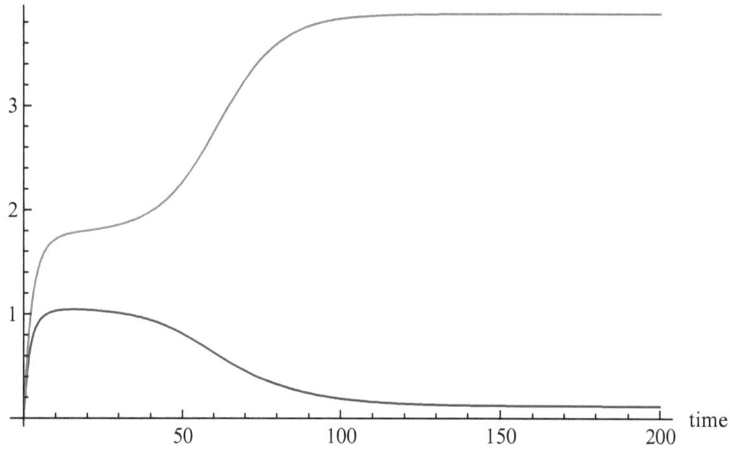

Fig. 9. Numerical integration of ODE and its plot by Mathematica. The function called `NDSolve` in Mathematica does numerical integration.

8. Another equivalent notation of the first derivative of x is x', or $\dot{x}$, which does not explicitly express that it is a derivative of independent variable, the time t for our current example.
9. Hill (or Michaelis–Menten) kinetics is a model of enzymatic reactions, which you can find in any biochemistry textbook. We adopt them to approximate the saturating effect of reactions between factors in the network, which are not restricted to enzymatic reactions. The plots in Fig. 2 illustrate Hill kinetics with different values of cooperativity represented by the Hill coefficient n. Michaelis–Menten kinetics can be regarded as a special case when $n=1$. As you see in Fig. 2, steepness of a sigmoidal curve depends on the value of cooperativity n, and the maximum value of these functions corresponds to the parameter value of k. Alternatively, you can use mass action kinetics to describe a reaction in the network, in which its reaction rate is simply proportional to each reactant's concentration.

```
Manipulate[
 Module[{eqns, soln},
  eqns =
   {u'[t] == k1 w / ((v[t]^n + 1) (w + 1)) - μ1 u[t],
    v'[t] == k2 w / ((u[t]^n + 1) (w + 1)) - μ2 v[t],
    u[0] == 0,
    v[0] == 0};
  soln = NDSolve[eqns, {u, v}, {t, 0, 200}];
  Plot[Evaluate[Tooltip[#[t] /. soln, #] & /@ {u, v}], {t, 0, 200},
   AxesLabel → {"time", "conc."}]
 ],
 {{k1, 0.5}, 0, 1, Appearance → "Labeled"},
 {{μ1, 0.06}, 0, 0.2, Appearance → "Labeled"},
 {{k2, 0.7}, 0, 1, Appearance → "Labeled"},
 {{μ2, 0.15}, 0, 0.2, Appearance → "Labeled"},
 {{n, 3}, 1, 4, Appearance → "Labeled"},
 Delimiter,
 {{w, 2}, 1, 10, Appearance → "Labeled"}
]
```

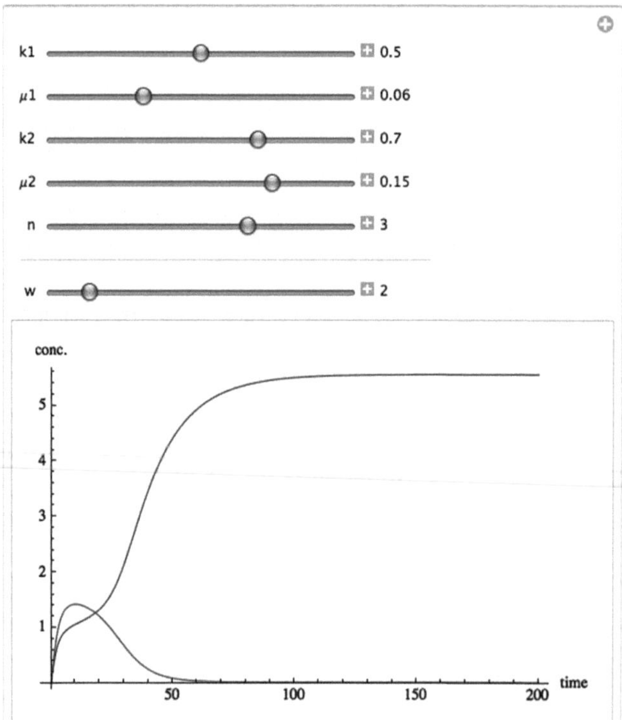

Fig. 10. Exploring parameter values in an interactive manner. In order to see how the dynamics of the system changes with various parameter values, it is very useful to have an interactive interface as shown in this example. This code uses the function `Manipulate`, which lets you literally manipulate parameter values and update a plot as you move each slider.

When the total amount of a certain reactant is conserved (e.g., gene copy number), saturating effects would be observed in mass action kinetics.

10. Unless considering an active regulation, degradation is formulated as a first-order kinetics, that is, the degradation rate of the molecule is proportional to its own concentration. This kinetics is described by the ODE:

$$\frac{dx}{dt} = -\mu x,$$

where μ is the degradation rate parameter. The general solution of this ODE is an exponential decay function:

$$x(t) = C e^{-\mu t},$$

where C is a constant. Note that the half-life of x is $\frac{\ln 2}{\mu}$ (i.e., $x(\frac{\ln 2}{\mu}) = \frac{C}{2}$), which is a useful formula to remember.

11. Any mathematical modeling requires a certain degree of simplification (see Note 6). ODE models actually reflect the population average of dynamics. This simplification or abbreviation in turn relies on various assumptions. For example, cell volume may be treated as being constant. Alternatively, one may assume that cell volume is halved with every cell division in early *Xenopus* embryos. For the first modeling attempt of your GRN of interest, it is important to make bold assumptions and not to dwell on too much details of the system. The point here is that one understands on which condition these assumptions are valid. If in doubt, examine how the dynamics of the model changes by taking an alternative assumption later. Unlike living embryos, mathematical model can easily be modified if necessary. Your model may provide some insightful predictions about the system's dynamics, which is often counterintuitive and unimaginable without modeling. This is why mathematical modeling is useful. Those predictions could be tested and verified experimentally.

12. One may wonder why unity is in the denominator of the function f (e.g., 1 in $1 + v^n$) instead of another parameter. This is because the equations are rescaled and dimensionless. Usually quantities such as concentration are measured in a conventional unit, for example, nM. In a dimensionless equation, quantities are measured in scalar without unit. In fact, it is often the case that quantities in biological experiments are measured in arbitrary units, for example, intensity of fluorescence. Therefore, absolute values and units of parameters are not so important as relative scale of these values to each other. Let us consider a

compound Hill function with dimensions for Xbra production:

$$g(V,W) = a\left(\frac{b}{b+V^{n}}\right)\left(\frac{W}{c+W}\right),$$

where V and W are the concentration (with a unit dimension such as nM) of Gsc and Activin, respectively, and a, b, and c are parameters. Each of these parameters have a dimension, for example, s^{-1}, $(nM)^n$, and nM, respectively. The function $g(V, W)$ is equivalent to $f(v, w)$ (Eq. 2) with the following relationship:

$$w = \frac{W}{c}, \qquad v = \frac{V}{b^{\frac{1}{n}}}, \qquad k_1 = \frac{a}{\hat{a}},$$

where $\hat{a}$ is the unit of a (concentration/time, for example, nM/s). This procedure of rescaling reduces the function $g(V, W)$ into a simpler form $f(v, w)$ with only two parameters k_1 and n instead of four, and yet the functions f and g define the identical dynamics with different scales only. This is very useful because one of the keys for a successful modeling is to reduce the number of parameters. For an introduction to the dimensionless formulation, please refer to (5, 6).

13. If Hill coefficient m is introduced for Activin, f becomes as following:

$$f(v,w) = \frac{k_1 w^{m}}{(1+v^{n})(1+w^{m})}.$$

14. The resulting model has three variables (u, v, and w) and five parameters (k_1, k_2, μ_1, μ_2, and n).
15. A system is said to be linear when variables in the ODEs (e.g., u, v, and w in our example system) appear to the first power only. Our system is nonlinear because its ODEs contain terms of u and v to the power of n. Nonlinear terms include not only powers but also products and other functions such as cos, sin, and ln. Most nonlinear systems do not have *analytical solution*, that is, their solution cannot be expressed by a combination of a limited number of certain elementary functions such as exponentials, logarithms, and so on. In mathematics, analytical solution is called *closed-form solution*.
16. This curve can be sketched manually. When trying to do this, it is often useful to calculate the values that satisfy the equation for special cases. For instance, when $v = 0$,

$$u = \frac{k_1 w}{\mu_1(1+w)}.$$

This is the point where the curve crosses the u axis ($u_{v=0}$, 0) (see Fig. 4a). Similarly, when $v \to \infty$, $u \to 0$. Similar approach can be applied to any other ODE models.

17. Values of the parameters k_1 and μ_1 for the Eq. 5 are perhaps different from k_2 and μ_2 for the Eq. 6, if one considers that u and v correspond to different genes (i.e., Xbra and Gsc).
18. Stability of fixed points can be examined analytically by the technique called linear stability analysis, which uses linear algebra. For the description of linear stability analysis, please refer to (5, 6).
19. This is not true in a strict sense. It holds true for our system because we are only interested in the real situation where u and v are zero or positive.
20. Numerical simulation is also called numerical integration, which is itself a vast subject in mathematics. We do not need to know how it works in detail for our present purpose, but a brief introduction is available in (5).
21. Use geometrical or visual means of analysis where possible, which is often beneficial.
22. Because Eqs. 3 and 4 describing the dynamics of our model are rescaled and dimensionless, relative values of parameters, rather than their absolute values, become relevant. For instance, as long as the relative ratio of the parameters k_1, k_2, μ_1, and μ_2 in our model remains the same, the dynamics is identical with only a rescaling of the time (you can confirm this by plotting a numerical simulation using the parameters 10 × and the same n and w values, and rescaling the time by 0.1 ×). Rescaling of the time is not an issue here, because the time in Eqs. 3 and 4 is also dimensionless and measured in an arbitrary unit.

Acknowledgements

I wish to thank Stefan Hoppler, Jonathan Pettitt, and Ekkehard Ullner for critically reading the manuscript. This work is supported by Scottish Universities Life Sciences Alliance (SULSA).

References

1. Kondo S, Miura T (2010) Reaction–diffusion model as a framework for understanding biological pattern formation. Science 329(5999): 1616–1620
2. Turing AM (1952) The chemical basis of morphogenesis. Philos Trans R Soc B 237(641): 37–72
3. Saka Y, Smith JC (2007) A mechanism for the sharp transition of morphogen gradient interpretation in xenopus. BMC Dev Biol 7:47
4. MIT OpenCourseWare – Mathematics. URL http://ocw.mit.edu/courses/mathematics/
5. Strogatz SH (1994) Nonlinear dynamics and chaos: with applications in physics, biology, chemistry, and engineering. Addison-Wesley, Reading
6. Edelstein-Keshet L (2005) Mathematical models in biology. SIAM, Philadelphia
7. Ruskeepää H (2009) Mathematica navigator: mathematics, statistics, and graphics, 3rd edn. Academic, Amsterdam
8. Cherry JL, Adler FR (2000) How to make a biological switch. J Theor Biol 203(2): 117–133

Chapter 29

Stem-Cell-Like Embryonic Explants to Study Cardiac Development

Boni A. Afouda

Abstract

Cells from the animal pole of *Xenopus* blastula embryo possess pluripotency that can be used to generate various tissues and even functional organs ex vivo. This finding has sparkled development of a variety of experimental protocols to study mechanisms that underlie formation of various organs and explore strategies for organ engineering for clinical applications. In this chapter, key methods are described for using *Xenopus* stem-cell-like embryonic explants as an assay system for studying organ development, with a focus on cardiogenesis. This assay allows investigation of cardiogenesis in isolation from neighboring tissues, minimizes interference with other developmental processes, and presents the further advantage of a heterologous system to study cardiogenesis in isolation of endogenous development of the heart. The cardiogenic assays can be exploited to investigate molecular mechanisms and cellular processes that underlie function of different molecules involved in cardiogenesis.

Key words: Xenopus, Stem cell, Organ development, Pluripotent, Heart, Cardiogenesis, Cardiogenic assay, Embryonic explants, Organogenesis

1. Introduction

Xenopus blastula-stage pluripotent cells represent a simple yet reliable experimental system that can be quickly and inexpensively exploited for a broad range of investigations into mechanisms of embryonic development and organ differentiation (1–5), but also for screening and identification of genes and chemical compounds that might be useful for drug discovery and therapeutic applications (6). Many of the genes identified by differential screening using such ex vivo-induced organs from *Xenopus* animal explants have been found to function in development of these organs in mammalian embryos.

In addition, various protocols and methods for in vitro differentiation of mammalian embryonic stem (ES) cells have recently been established based on original discoveries from studies in the

Stefan Hoppler and Peter D. Vize (eds.), *Xenopus Protocols: Post-Genomic Approaches, Second Edition*, Methods in Molecular Biology, vol. 917, DOI 10.1007/978-1-61779-992-1_29, © Springer Science+Business Media, LLC 2012

Xenopus animal cap assay system (7). However, mammalian stem cell culture tends to be carried out in serum-containing medium with various soluble factors such as vitamins, extracellular matrices, and growth factors. The exact nature of the defined factors required for specific organ or tissue formation has therefore become an issue for the establishment of protocols for in vitro organogenesis with ES cells. To avoid this problem, a serum-free medium for mammalian ES cells is being developed (8). *Xenopus* stem-cell-like embryonic explants present several advantages for tissue and organ engineering since they can be easily cultured in a simple saline solution and induced to differentiate into different cell lineages by either addition of defined active inducers into this saline culture medium or by simple injection of mRNA encoding defined gene products.

The animal cap explant protocol consists of excising the animal pole tissue from the blastula or early gastrula and then placing it in saline solution for culture. The explant is cultured in isolation until desired stages for molecular analysis of tissue and organ differentiation using various techniques. This approach has previously been successfully used to study early embryonic development, such as mesoderm or neural induction and patterning. Here I describe experimental techniques for the use of the *Xenopus* animal cap explant system and other *Xenopus* explants as assays for ex vivo differentiation of heart tissue, and how they can be used to investigate mechanisms involved in organogenesis.

2. Materials

2.1. Xenopus Embryos

Fertilized embryos according to standard procedures (9).

2.2. Dissection Tools

Dissection forceps (Dumont no. 5 or no. 55), tungsten needle, or gastromaster (XENOTEK Engineering, Belleville, IL) (see Note 1).

2.3. Growth Factor

Human recombinant activin A solution (originally from Dr. Y. Eto Central Research Laboratories, Ajinomoto Co. Inc., Japan) is used as cardiac inducer as described in (10).

2.4. mRNAs

For cardiac induction both activin beta-B (originally from Professor Hugh Woodland's laboratory, University of Warwick) and inducible HA-tagged gata4 protein fused to glucocorticoid receptor (GR) mRNAs are used (3, 5) (see Note 2).

2.5. Antibodies

1. Antibodies for western blot analyses: For detection of the tagged gata4 protein, anti-HA (3F 10) rat monoclonal antibody from Roche Applied Science (cat no 1 867 423) is used as first antibody and a polyclonal goat anti-rat IgG H&L HRP-conjugated from Calbiochem (cat no 401416) is used as secondary antibody.

Protein loading is assayed with expression of the housekeeping gene MAP kinase erk2 using rabbit polyclonal IgG erk2 (C-14) from Santa Cruz Biotechnology as first antibody and a polyclonal swine anti-rabbit IgG/HRP (code no P0217) from DakoCytomation as secondary antibody.

2. Further useful antibodies for immunostaining of cardiac tissue are listed in Table 3.

2.6. Reporter Genes

Luciferase or chloramphenicol acetyl transferase (CAT) can be used as reporter genes (11) (see Note 3).

2.7. Components for Embryo and Explant Culture

1. Antibiotics: Combination of kanamycin (10 μg/mL), gentamycin (10 μg/mL), and ampicillin (50 μg/mL) is used in the culture medium.
2. Marc's Modified Ringer's (MMR) final concentration in 1×: 0.1 M NaCl, 2 mM KCl, 1 mM $MgSO_4$, 2 mM $CaCl_2$, 5 mM HEPES (pH 7.8), 0.1 mM EDTA. Sterilize by autoclaving and store at room temperature.
3. Holtfreter's saline (HS, also known as HS + Ca^{2+}): 60 mM NaCl, 0.7 mM KCl, 0.9 mM $CaCl_2$, 4.6 mM HEPES, 0.1 g/L kanamycin sulfate, 0.1 % BSA (A-7888, Sigma, USA, pH7.6).
4. Calcium-free Holtfreter's saline (HS-Ca^{2+}): 60 mM NaCl, 0.7 mM KCl, 4.6 mM HEPES, 0.1 g/L kanamycin sulfate, 0.1 % BSA (A-7888, Sigma, USA, pH7.6).

2.8. Solutions for Further Analysis

1. The solutions required for nucleic acid extraction and analysis are as described in (1).
2. The solutions required for protein extraction are as described in (1).
3. For western blot analysis solutions see standard protocols as described on Invitrogen website (http://www.invitrogen.com).
4. Solutions required for whole mount in situ hybridization as well as in situ hybridization on sections are as described in (12).
5. Solutions required for RNA Protection Assay (RPA) are as described on promega website (http://www.promega.com).

3. Methods

3.1. Isolation and Culture of Intact Animal Cap Explants

1. Inject synthetic capped mRNA (see Note 2 and Table 1) for a potential regulator in appropriate volume into the animal pole of one-cell stage embryo (Fig. 1). Embryos are injected in MMR containing 3 % ficoll. Around 2 h after injection, gradually replace culture solution with MMR without ficoll. It is important to culture embryos in a solution without ficoll to allow the process of gastrulation to proceed later on.

Table 1
mRNAs that influence cardiac differentiation when injected into explants

mRNAs	Site of injection	Phenotype	References
Activin beta-B	Animal pole	Promote cardiogenesis	(4, 5)
GATA4	Animal pole	Promote cardiogenesis	(3, 4, 11)
Dkk1	Ventral marginal zone (VMZ)	Promote cardiogenesis	(3, 16)
β-catenin	Dorsal marginal zone, or co-inject with activin beta-B or gata4 (into animal pole) or dkk1 (into VMZ)	Prevent cardiogenesis	(3, 17)

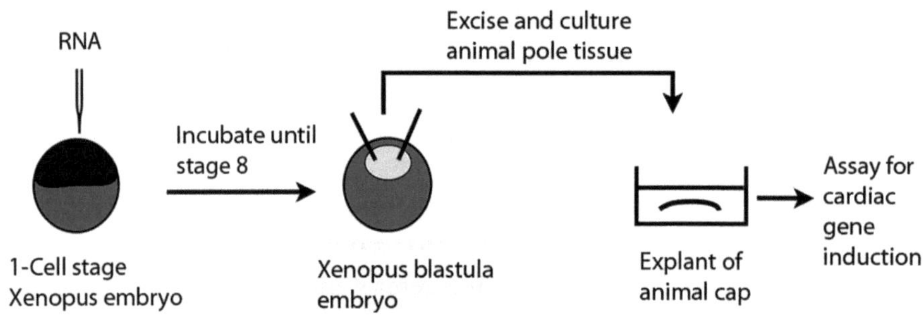

Fig. 1. The *Xenopus* animal cap explant strategy (figure modified after Afouda and Hoppler (4)).

2. Dissect animal caps (see Fig. 1) with tweezers, tungsten needles, or a gastromaster (see Note 1) around stage 7.5 in 0.6× MMR in Petri dishes coated with 1 % agarose. Explants size are usually around 0.7 mm. In the case of activin mRNA-injected explants, explants of this size strongly elongate from around stage 11 to 20+.
3. Remove all debris and put explants in fresh solution containing several antibiotics (see Subheading 2).
4. Incubate explants overnight at 23 °C (see Note 4). The next day, replace solution with fresh solution containing the above-mentioned antibiotics (see Subheading 2.7). Keep solution clean by replacing at least three times a day until desired stages for analyses (see Note 7).
5. When used as cardiogenic assay, by the end of the third day (st. 42), first beating tissue will appear with the number growing until the end of the fourth day at 23 °C. The longest period we have kept the beating explants were 7 days when they still show spontaneous rhythmic beating.

3.2. Isolation and Culture of Animal Cap Explant with Dissociation and Reaggregation Method

1. Remove explants from uninjected embryos as described earlier (see Subheading 3.1 steps 1–3).
2. Disperse explant cells by gentle agitation in calcium-free medium (or HS-Ca^{2+}) for 2 h.
3. Concomitantly treat cells with 10–100 ng/mL of human recombinant activin A (10) during this period.
4. After the dissociation step, cell reaggregation is achieved by culturing the explants in medium with calcium (or HS + Ca^{2+}) plus the following antibiotics: ampicillin, gentamycin, and kanamycin (see Subheading 2.7).

3.3. The Dorsal Marginal Zone (DMZ) and the Ventral Marginal Zone (VMZ) Explant Systems

After fertilization when the sibling embryos reach the four-cell stage, the dorsal blastomeres, from which the heart forming region later develops, are identified as the two less pigmented blastomeres (Fig. 2a) in contrast to the opposite blastomeres that are identified as the darker ventral blastomeres (Fig. 2b).

1. Inject mRNAs of interest (Table 1) in either the dorsal (for dorsal marginal zone (DMZ) explants) or the ventral (ventral marginal zone (VMZ) explants) blastomeres of the four-cell embryos (see Notes 5 and 6).
2. Place embryos into saline solution (0.6 × MMR) until gastrulation.

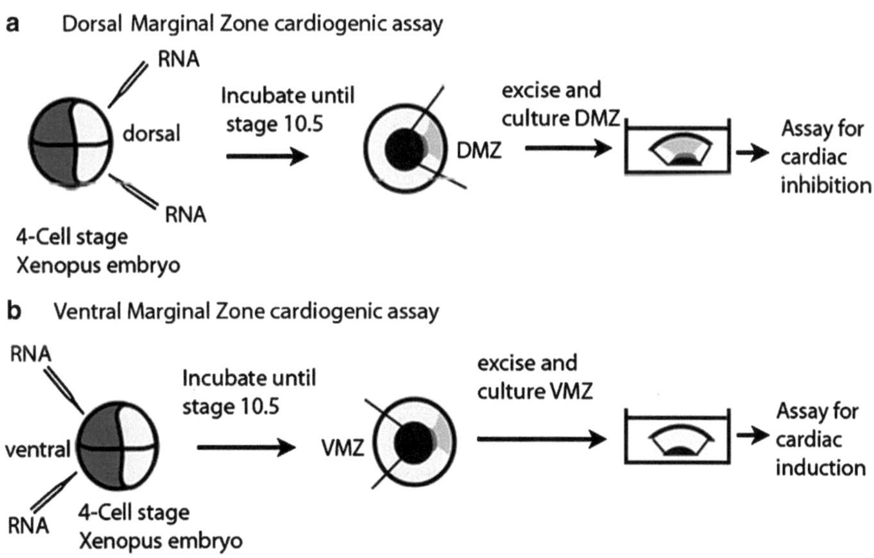

Fig. 2. The *Xenopus* dorsal marginal zone (**a**) and ventral marginal zone (**b**) explant assays (figure modified after Afouda and Hoppler (4)).

3. At stage 10.25, identify the dorsal side with the blastopore lip and carefully excise (as for animal cap explants see Note 1) this region (for DMZ explant) or the opposite region (for VMZ explant).
4. Culture excised marginal zone explants as mentioned earlier for animal cap explants (see Subheading 3.1 steps 3–4; see Note 7).

3.4. Analysis of Cardiac Development in Explants

1. Different stages of cardiac development can be assayed using stage appropriate markers (see Note 8 and Table 2). For example, to examine early cardiac determination use nkx2-5, gata4, gata6, tbx5, tbx20, and hand1, for in situ hybridization (http://www.xenbase.org and see Note 9), RT-PCR (3, 5) or RPA assays (11, 13) at stages 13–20. For terminal differentiation, use myh6 (myosin heavy chain 6 cardiac muscle α), myl2 (myosin light chain 2), actc1 (cardiac actin), tnni3 (cardiac troponin), and hand1 to detect cardiomyocytes differentiation from stages 24 to 32 (see Table 2).
2. Cardiac induction can also be monitored using reporter assays (see Note 3).
3. Cardiac proteins can also be detected using standard immunostaining protocols (14). Useful antibodies are listed in Table 3.

Table 2
Markers for assessing cardiac development by RT-PCR or in situ hybridization

Developmental stages	Genes	References
Early markers (from stages 13 onwards)	gata4[a]	(3)
	gata6[a]	(3)
	nkx2-5[a]	(3)
	wnt11[a]	(3)
	bmp2[a]	(18)
	bmp4[a]	(18)
	hand1	(19)
Intermediately early markers (from stage 16)	tbx5	(5, 20)
	tbx20	(21)
Markers from stage 20	actc1 (cardiac actin)	(22)
Markers from stage 24	myh6 (myosin heavy chain 6 cardiac muscle α) also known as Myosin Heavy Chain α	(5)
Cardiac differentiation markers (from stage28)	myl2 (myosin light chain 2)	(5)
	tnni3 (cardiac troponin)	(5)
	hand1	(19)

[a]Not strictly cardiac specific so should be used in conjunction with other cardiac differentiation markers

Table 3
Useful antibodies for immunostaining

Developmental stage	Protein	Source/references
From stage 30	tnni3 (cardiac troponin)	(23, 24)
From stage 28	myl2 (myosin light chain2) myh6 (myosin heavy chain 6 cardiac muscle α) also known as Myosin Heavy Chain α	http://www.abcam.com (23)

4. Notes

1. Depending on available equipment, tweezers, tungsten needles, or gastromaster could be used for the excision of the explants. The use of gastromaster will allow certain consistency regarding the size of explants. However, it will require some additional effort for cleaning the microsurgery tip and maintaining its shape identical throughout the experiment. For these reasons the use of tweezers or tungsten needles is considered as a cheaper alternative.
2. For beating animal cap from activin-injected explants, 50 fg of capped activin beta-B mRNA is injected in a volume of 10 nL into the animal pole of one-cell stage embryo. However, for beating animal cap from GATA4-injected explants, 1,000 pg of mRNA are injected.
3. Reporter assays could be used to investigate gene activities in *Xenopus* stem-cell-like embryonic explants. Reporter constructs and gene of interest are co-injected in explants that are cultured for desired period of time for assessing the reporter gene activity. Luciferase and CAT are among the reporter constructs used. Protein extracts are prepared as mentioned earlier (see Subheading 2.8, item 2) and the level of expression of the reporter gene is assessed with the extracts (11) using appropriate luminometer.
4. Lower temperatures (some lab use 13°) should be used for culturing embryo if surviving rate is low or if the infection rates are high.
5. Capped mRNA or morpholino antisense oligonucleotides (MO) for a potential regulator of cardiac development is injected into the dorsal (for DMZ explants) or ventral (for VMZ explants) blastomeres of four-cell stage *Xenopus* embryos (Fig. 2a, b) at required concentration in similar way

as mentioned for animal cap explants. At stage 10.25 the DMZ or VMZ (Fig. 2a, b) are excised as for animal cap explants.

6. When explanted and placed into culture, the uninjected DMZ forms heart tissue and therefore is used for positive control (as opposed to the VMZ which is used for negative control) in the experiments.
7. Just after removing the explants, these should be put into fresh and clean medium containing antibiotics. The explants should be kept in such media throughout the period of culture. Explants are left to develop until desired stages (e.g., st. 24 for analysis of pro-cardiogenic transcription factor gene expression, st. 32 for differentiation marker, and st. 42 for imaging beating heart muscle tissue).
8. The stages of detection may slightly vary depending on the methods used for assessing the gene expression due to the fact that RT-PCR technique is more sensitive than in situ hybridization.
9. The in situ hybridization techniques can be carried out in eppendorf tubes. This requires a careful attention to avoid the loss of explants due to their small size. To overcome this handicap a basket-based technique can be used (15).

Acknowledgement

This work was supported by the British Heart Foundation (PG/07/043).

References

1. Afouda BA, Ciau-Uitz A, Patient R (2005) GATA4, 5 and 6 mediate TGF beta maintenance of endodermal gene expression in Xenopus embryos. Development 132: 763–774
2. Warkman AS, Krieg PA (2007) Xenopus as a model system for vertebrate heart development. Semin Cell Dev Biol 18:46–53
3. Afouda BA, Martin J, Liu F, Ciau-Uitz A, Patient R, Hoppler S (2008) GATA transcription factors integrate Wnt signalling during heart development. Development 135:3185–3190
4. Afouda BA, Hoppler S (2009) Xenopus explants as an experimental model system for studying heart development. Trends Cardiovasc Med 19:220–226
5. Afouda BA, Hoppler S (2011) Different requirements for GATA factors in cardiogenesis are mediated by non-canonical Wnt signalling. Dev Dyn 240:649–662
6. Asashima M, Ito Y, Chan T, Michiue T, Nakanishi M, Suzuki K, Hitachi K, Okabayashi K, Kondow A, Ariizumi T (2009) In vitro organogenesis from undifferentiated cells in Xenopus. Dev Dyn 238:1309–1320
7. Asashima M, Michiue T, Kurisaki A (2008) Elucidation of the role of activin in organogenesis using a multiple organ induction system with amphibian and mouse undifferentiated cells in vitro. Dev Growth Differ 50(Suppl 1):S35–S45
8. Furue M, Okamoto T, Hayashi Y, Okochi H, Fujimoto M, Myoishi Y, Abe T, Ohnuma K, Sato GH, Asashima M, Sato JD (2005) Leukemia inhibitory factor as an anti-apoptotic mitogen for pluripotent mouse embryonic stem

cells in a serum-free medium without feeder cells. In Vitro Cell Dev Biol Anim 41:19–28
9. Hoppler S (2008) Studying Wnt signaling in Xenopus. Methods Mol Biol 469:319–332
10. Ariizumi T, Kinoshita M, Yokota C, Takano K, Fukuda K, Moriyama N, Malacinski GM, Asashima M (2003) Amphibian in vitro heart induction: a simple and reliable model for the study of vertebrate cardiac development. Int J Dev Biol 47:405–410
11. Latinkic BV, Kotecha S, Mohun TJ (2003) Induction of cardiomyocytes by GATA4 in Xenopus ectodermal explants. Development 130:3865–3876
12. Ciau-Uitz A, Walmsley M, Patient R (2000) Distinct origins of adult and embryonic blood in Xenopus. Cell 102:787–796
13. Gove C, Walmsley M, Nijjar S, Bertwistle D, Guille M, Partington G, Bomford A, Patient R (1997) Over-expression of GATA-6 in Xenopus embryos blocks differentiation of heart precursors. EMBO J 16:355–368
14. Broer S (2010) *Xenopus laevis* oocytes. Methods Mol Biol 637:295–310
15. Knecht AK, Good PJ, Dawid IB, Harland RM (1995) Dorsal-ventral patterning and differentiation of noggin-induced neural tissue in the absence of mesoderm. Development 121:1927–1935
16. Schneider VA, Mercola M (2001) Wnt antagonism initiates cardiogenesis in *Xenopus laevis*. Genes Dev 15:304–315
17. Martin J, Afouda BA, Hoppler S (2010) Wnt/beta-catenin signalling regulates cardiomyogenesis via GATA transcription factors. J Anat 216:92–107
18. Peterkin T, Gibson A, Patient R (2003) GATA-6 maintains BMP-4 and Nkx2 expression during cardiomyocyte precursor maturation. EMBO J 22:4260–4273
19. Sparrow DB, Kotecha S, Towers N, Mohun TJ (1998) Xenopus eHAND: a marker for the developing cardiovascular system of the embryo that is regulated by bone morphogenetic proteins. Mech Dev 71:151–163
20. Horb ME, Thomsen GH (1999) Tbx5 is essential for heart development. Development 126:1739–1751
21. Brown DD, Binder O, Pagratis M, Parr BA, Conlon FL (2003) Developmental expression of the *Xenopus laevis* Tbx20 orthologue. Dev Genes Evol 212:604–607
22. Pandur P, Lasche M, Eisenberg LM, Kuhl M (2002) Wnt-11 activation of a non-canonical Wnt signalling pathway is required for cardiogenesis. Nature 418:636–641
23. Goetz SC, Brown DD, Conlon FL (2006) TBX5 is required for embryonic cardiac cell cycle progression. Development 133:2575–2584
24. Lavery DL, Davenport IR, Turnbull YD, Wheeler GN, Hoppler S (2008) Wnt6 expression in epidermis and epithelial tissues during Xenopus organogenesis. Dev Dyn 237:768–779

Chapter 30

Studying Regeneration in *Xenopus*

Caroline W. Beck

Abstract

For most *Xenopus* embryos, life is very short. The majority of research labs working with this model organism study the processes of early vertebrate patterning and morphogenesis. And quite rightly too, since over the last two decades labs across the world have provided the fate maps, animal cap assays, expression patterns, and functional screens that put *Xenopus* firmly on the map as a developmental model organism. *Xenopus*, however, still has a lot more to offer. A new wave of interest in later developmental events has followed the development of transgenic technology, which has opened up opportunities for studying events that occur after stage 40. In this chapter, I will give a brief descriptive background of some of the different types of regeneration studied in *Xenopus*, and provide protocols and morphological scoring information with the aim of facilitating progress in understanding regeneration in this model system. Additionally, some particularly elegant recent examples are used to highlight the advantages of *Xenopus* as a model for regeneration and the future opportunities that this offers.

Key words: Regeneration, Tail, Limb, Lens, Lentectomy, Transdifferentiation

1. Introduction

All animals have evolved strategies to react to injury and disease, but only a minority can replace lost tissues, organs, or appendages. Among the model vertebrates, amphibians stand out as having the most impressive regenerative capabilities. The urodeles (newts and salamanders) are able to regenerate complex appendages such as the tail and limbs even as adults. On the other extreme, mammals are capable of only very limited regeneration, such as the extreme digit tips (1). *Xenopus* sits nicely in the middle, having the ability to produce either near-perfect regeneration of the tail of the tadpole and the early limb, or to fail to do so, depending on the developmental stage of the animal. While limb regeneration undergoes a gradual ontogenic decline, tail regeneration continues until the tail is absorbed at metamorphic climax, with the exception of a

Stefan Hoppler and Peter D. Vize (eds.), *Xenopus Protocols: Post-Genomic Approaches, Second Edition*,
Methods in Molecular Biology, vol. 917, DOI 10.1007/978-1-61779-992-1_30,

brief refractory period at stage 46/47 when regeneration completely fails. This as-yet unexplained phenomenon means that both gain and loss of function experiments can be designed using *Xenopus*.

1.1. Tail Regeneration

The larval *Xenopus* tail is composed of the same core tissues as the trunk, namely the spinal cord, notochord, and somites, and is capable of almost perfect regeneration (for review, see ref. (2)). Tail regeneration is particularly compelling because it involves the production of new and functional spinal cord in a model vertebrate. Whether partial amputation of the tail leads to successful or failed regeneration depends on the developmental stage at the time of amputation. Developmental staging in *Xenopus* is standardized and can be determined by reference to Nieuwkoop and Faber (3) or to the Xenbase website (see Note 1). Prior to stage 40, removal of part of the tail bud, from which the tail extends, always results in a loss of final tail structures (4). As shown in Fig. 1, the tadpole tail regenerates well between stages 40 and 44, but this ability is lost at stage 45–47, the time at which the animal begins to feed (5).

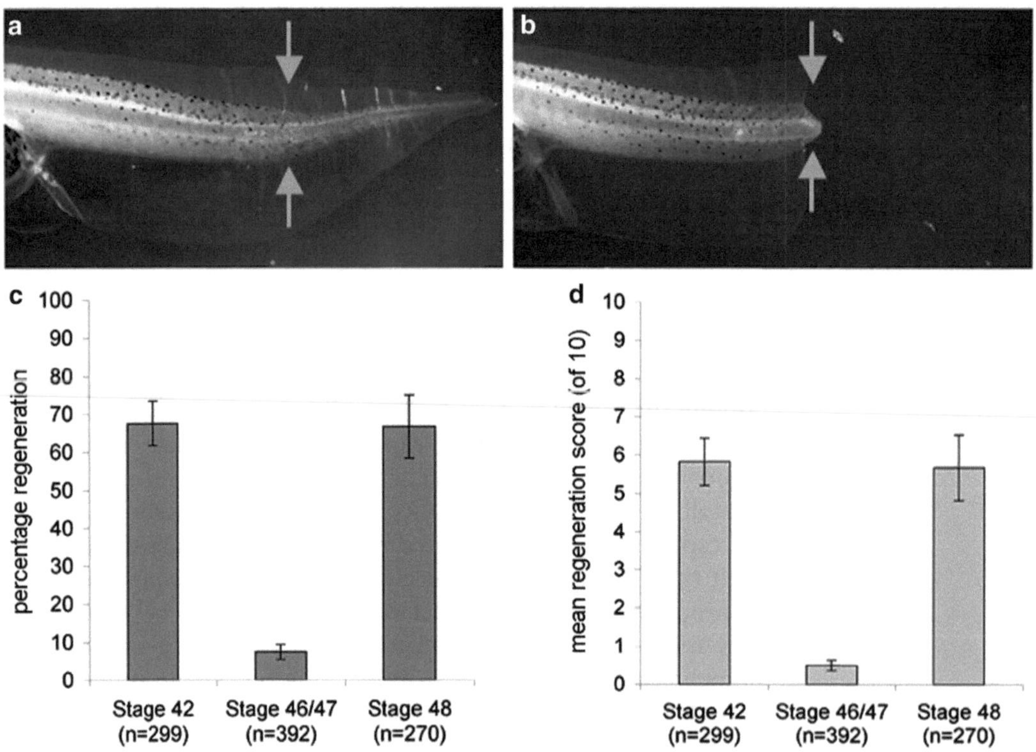

Fig. 1. Tail regeneration in *Xenopus laevis*. (**a**) Tail amputated at stage 42 (permissive) and (**b**) stage 46 (refractory), allowed to regenerate for 7 days. *Arrows* indicate the level of tail amputation. The small bulge at the end of the refractory tail in (**b**) is typical of a refractory tail and does not indicate regeneration (see Note 1). (**c**) Bar graph showing percentage of animals regenerating during the early permissive period, refractory period, and larval permissive period (post stage 48). (**d**) Bar graph showing mean regeneration score out of 10 for the same data set. Error bars indicate standard error.

When the tails are partially removed during this refractory period, there is no regeneration of any tail tissues, and the wound becomes covered in normal looking skin. Tadpoles continue to grow and can metamorphose normally provided they are left with enough tail to swim and feed effectively, but the tail never grows back. The refractory period is not the result of an ontogenic decline (Fig. 1c, d). If tails are partially amputated at around stage 48, when the hindlimb buds first appear as tiny rudiments, tail regeneration is successful and complete in around 70% of the cases. This ability is then retained until metamorphosis (6, 7).

Thanks to a series of elegant experiments following the fate of the tail tissues during the process of regeneration, we know how each individual tissue of the tail is renewed. Gargioli and Slack (8) used orthotopic grafts of transgenic (CMV-GFP) tissue to wild-type embryos at the neurula stage to label tail muscle, notochord, or spinal cord. The resulting grafted tadpoles were subsequently amputated through the labeled region of the tail and the identity of the graft-derived, GFP-positive cells was confirmed after the tails had been allowed to regenerate. The results of these experiments clearly demonstrated a lack of tissue switching (metaplasia) in tail regeneration: the new spinal cord cells were derived from the stump spinal cord cells, and the new notochord cells from the stump notochord cells. However, the muscle cells of the regenerate derive from the stem-like satellite cells rather than from the differentiated multinucleate cells of the stump somites tail (8, 9).

While the tail appears to regenerate perfectly in *Xenopus* tapoles, closer investigation reveals some deficits. The muscle tissue, derived from the satellite cell population, is not perfectly segmented (10), the neural crest derived dorsal root ganglia are almost entirely missing and spinal nerves do not re-establish a normal segmented pattern (11). Melanophores, which are also neural crest cell derivatives, regenerate from blastemal precursors presumably recruited from the fin tissue near the stump (12).

1.2. Tadpole Hindlimb Regeneration

Regeneration of the limb buds in *Xenopus laevis* was first described by Komala in 1957 (13). Unlike the tail, hindlimb regeneration undergoes a gradual ontogenic decline. This was first described by Dent (14), who observed that partially amputated limbs up to stage 52 were able to regenerate a normally patterned limb with five toes. Stage 53 hindlimbs regenerated less well, often failing to reform digit I, and by stage 55, many limbs failed to initiate regeneration of any structure at all (Table 1). Regenerative competence is determined by the age of the mesenchyme rather than the epidermis (15) and is not dependent on innervation (16). Unlike tail regeneration, which is very reproducible between animals, limb regeneration studies have reported a significant amount of variation in the regenerative ability of hindlimbs post stage 52/53 (see Note 2). The sequence of events of hindlimb regeneration has been well described and is summarized in Table 2.

Table 1
Ontogenic decline of hindlimb regeneration, after Dent (14)

Stage of amputation	Initial morphology	Number of digits/ft	*N*	Appearance at metamorphosis
51		4.8 (4.7)	25	
53	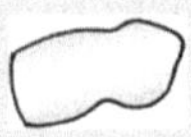	3.7 (3.7)	32	
55	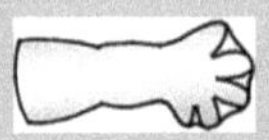	2.6 (2)	22	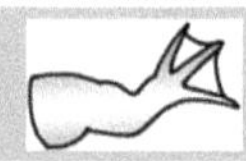
57	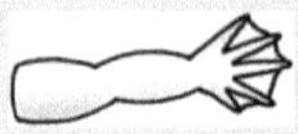	0.8	18	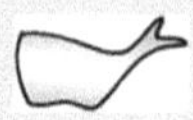
60		Hypomorphic spike	15	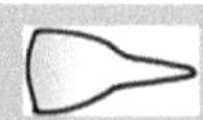

N represents the number of animals in each group, and representative phenotypes are shown. It should be noted that Dent operated through the distal femur and that most contemporary studies use more proximal amputation levels. Numbers in parentheses are from (20), with amputations at knee level

Table 2
Time course of regeneration following amputation of hindlimb at future knee level at stage 52

Days after amputation	Morphology of amputated limb	Developmental stage of control limb
1	Wound covered by migrating epithelial cells from stump	52
2–3	Apical epithelial cap (AEC) forms, with no basement membrane or dermis (40) basal epithelial cells become columnar (19), blastema forms (20)	Late 52
3–5	Cone shaped blastema, AEC starts to degenerate, hypertrophic cells seen	53
6–10	Re-differentiation	54–55
11–14	Completion of regeneration	55 +

Hindlimb regeneration studies have focused on improving regenerative outcomes at later stages, with mixed results. Topical application of Fgf10 via insertion of protein-soaked beads was reported to enhance hindlimb regeneration by Yokoyama et al. (17)

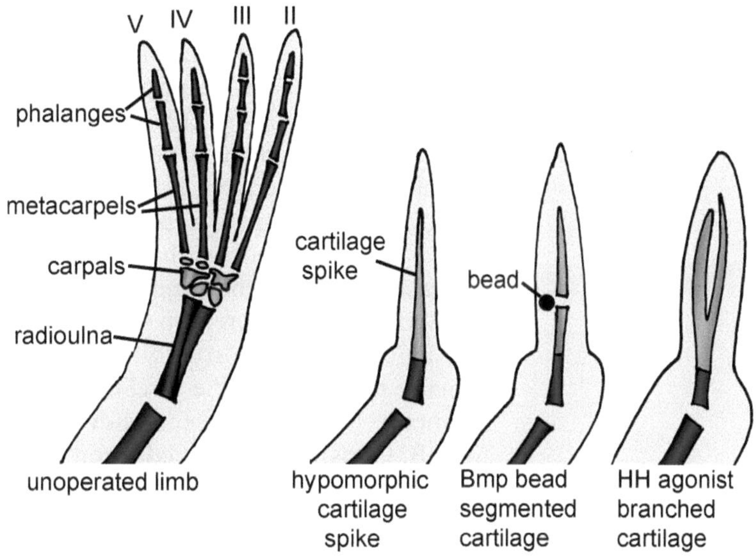

Fig. 2. Forelimb regeneration produces a pattern deficient spike with cartilage (*light gray*) but no bone (*dark gray*). Adding a Bmp-4 soaked bead induces segmented cartilage, and HH agonist treatment induces branching of the cartilage. Conventional digit numbering is shown in roman numerals, with digit II most anterior and digit V most posterior. Regenerates are drawn to show amputation midway through the radioulna.

but could not be repeated by Slack et al. (18). However, in Slack's study, transgenic upregulation of *fgf10* did result in improved regeneration scores, although this was not statistically significant. More recently, Barker and Beck (2009) overexpressed *msx1*, a mesenchymal gene associated with regenerative success, during hindlimb regeneration using heat shock inducible transgenics (see Note 3), but did not see significant improvement (19). While the attempts at enhancement have thus far yielded disappointing results, it has been possible to implicate the bone morphogenetic protein (BMP) signaling pathway in limb regeneration by inducing the inhibitor noggin, again using heat shock inducible transgenes, to eliminate regeneration in stage 52 limbs, which normally regenerate perfectly (20). When *noggin* is overexpressed, limb regeneration fails. Hedgehog has also been demonstrated to be essential for pattern formation in regenerates, since treatment with cyclopamine resulted in the formation of patternless spikes (21)(Fig. 2).

1.3. Froglet Forelimbs

Traditionally, post-metamorphic studies of limb regeneration entail a switch of focus from the hindlimb to the forelimb. Presumably, this is because the metamorphosed froglet, having lost its tail, cannot swim efficiently with a defective hindlimb. Forelimb amputations, however, are well tolerated. Forelimbs regenerate, much as Dent's stage 60 hindlimbs, as hypomorphic spikes. These spikes contain no ossified bone or muscle and normally contain a core of unsegmented cartilage, along with skin, nerves, and blood vessels. Although muscle satellite cells do not invade the blastema and contribute to the hypomorphic spike, transplanted satellite cells can differentiate normally into myofibers (22). Pattern is missing along the anterior posterior axis and there is no proximal to distal patterning either. Dorsal ventral patterning, however, is recovered in the spikes (23). The hypomorphic spike in post metamorphic animals does appear to be nerve dependent. In contrast to hindlimbs, denervated *Xenopus* froglet forelimbs cannot regenerate (24).

Because froglet forelimb spikes are pattern deficient, they provide an excellent opportunity to investigate the mechanisms that improve the ability of the forelimb blastema to create a patterned regenerate (Fig. 2). A recent study used a hedgehog agonist (Hh-Ag 1.8, Curis) to induce hedgehog signaling in the froglet blastema, producing improved regenerates (25). These spikes still have no muscle or ossified bone, but form multiple cartilaginous structures. Furthermore, implantation of bmp4 soaked beads into the forelimb regenerates induced joint-like structures in the cartilage of the spike (26). Studies such as these demonstrate the challenge of improving regenerative outcomes for vertebrates.

1.4. Lens Regeneration

X. laevis tadpoles are also capable of regenerating structures that would never be lost naturally in the wild, such as the lens of the eye (27). If the lens is surgically removed (lentectomy: Subheading 3.9), some of the overlying outer cornea cells transdifferentiate under the influence of an unidentified signal originating from the neural retina and residing in the vitreous fluid (28, 29). After 1–2 weeks, a new lens is formed from the cornea (Fig. 3a) (Table 3). As for hindlimb regeneration, there is an ontogenic decline, with around 70% of lenses regenerating after removal at stage 50, and almost none by stage 58 (27). However, this is not due to the lack of corneal cell competence to respond to the vitreous signal, but rather to increased rates of healing the neural crest derived inner cornea, which serves as a barrier to the vitreous signal (30). The continued competence can be demonstrated by implanting corneal epithelium directly into the vitreous chamber of the eye (corneal implant, Fig. 3b, Subheading 3.10). Interestingly, *Xenopus tropicalis* cannot regenerate lenses at any stage, due to an ability to heal the inner cornea much faster than *X. laevis* (31). Corneal implants demonstrate that this is not due to a lack of competence of the *X. tropicalis* corneal cells to respond to the vitreous signal.

Since lens regeneration can be clearly visualized using transgenic lines of *Xenopus* that have GFP or other fluorescent proteins expressed under the control of the gamma crystallin promoter (32), this provides an opportunity to apply the same transgenic and chemical genetic methods that have been used to study other regenerative processes in *Xenopus*. Two recent in vivo functional studies of lens regeneration have implicated roles for FGF and BMP signalling in *Xenopus* lens regeneration (33, 34).

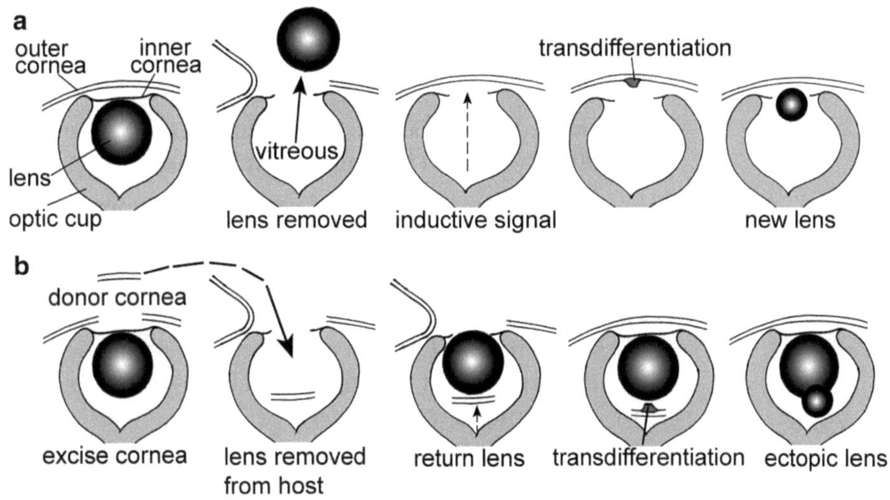

Fig. 3. Mechanisms of lens induction from the cornea in *Xenopus*. (**a**) Simple lentectomy (**b**) corneal implant.

Table 3
Freeman's descriptive stages of lens regeneration (27)

Freeman stage	Days after lentectomy	Morphological description
1	1–2	Epithelial cells of the inner layer of the outer cornea change from squamous to cuboidal
2	2–3	Small clusters of these cells form by aggregation
3	3–5	Aggregate has become a lens vesicle and is positioned over the open pupil
4	7–8	Lentoid delaminates and begins to migrate through pupil. Primary lens fibers formed
5	10–12	Secondary lens fibers formed, lens grows to normal size and is located in vitreous

2. Materials

2.1. Dissection Equipment

1. Forceps: Dumont no. 5, stainless steel jeweler's forceps.
2. Scalpel blades, e.g., Swann Morton non-sterile carbon steel protected surgical blades, ref 0106.
3. Vannas iridectomy scissors: 8 cm straight 5 mm blades (World Precision Instruments).

2.2. Solutions

1. Alcian blue stain: 10 mg alcian blue 8GX powder (Sigma #A5268) dissolved in 60 mL EtOH and 40 mL acetic acid. This keeps for a few weeks.
2. Alizarin red stain: dissolve 10 mg alizarin red S (Sigma #A5533) in 100 mL KOH. The stain is quite stable.
3. Frog water: tap water filtered through two commercial carbon based drinking water filters in series to remove chlorine (see Note 2).
4. Gentamycin: dissolve 50 mg gentamycin sulfate (Sigma #G3632) in Milli-Q water and store as frozen stocks. The normal dose to prevent infection in embryos or tadpoles is 50 μg/mL in the media (see Note 4).
5. 6% H_2O_2 in PBS. Dilute 10 mL 30% H_2O_2 1 in 5 with PBS.
6. 1% KOH: dissolve 1 g potassium hydroxide pellets in 100 mL distilled water.
7. MMR: Marc's Modified Ringers: Made as 10× stock and diluted with Milli-Q or frog water when required. 10× MMR is 1 M NaCl, 20 mM KCl, 10 mM $MgCl_2$, 20 mM $CaCl_2$, 50 mM HEPES, pH 7.5, and is sterilized by autoclaving.
8. MS222: ethyl 3-aminobenzoate methane sulphonate (tricaine) salt. Obtained from Sigma cat#A5040. Dissolve 1 g in 40 mL distilled water to make a stock solution that will keep for 2–3 weeks.
9. PBS: phosphate-buffered saline, Oxoid Dulbecco A. One tablet dissolved in 100 mL Milli-Q water.

2.3. Equipment

1. Aerator: aeration can be provided to recovering animals via a small pet shop pump such as the Aqua One 9500 connected via silicon tubing to a small air stone.
2. Nutator: small rotating benchtop platform such as the BD Clay Adams unit.

3. Methods

3.1. Pre-feeding (Embryonic) Tail Regeneration Assay

1. Partially amputate tails of embryos at stage 40–41 using a round-ended scalpel blade to cut at the level where the tail begins to taper. Tadpoles at this stage lie on their sides and can be cut without anesthetic.
2. Transfer to treatment Petri dishes in 0.1× MMR at a density not exceeding one tadpole per milliliter. Sibling and age-matched unoperated animal controls should be prepared to ensure treatments are not toxic. Gentamycin may be added to the media at 50 μg/mL if required.

3. Chemicals to be tested can be added immediately, or at specific time points after amputation. Stock solutions where water is not the solvent invoke the need for control vehicle treated animals.
4. Culture at 18–22°C. Treatment can be terminated by changing to fresh 0.1× MMR or continued over several days. If the latter is the case, then fresh solutions may need to be prepared daily or more frequently depending on the stability of the chemical compound.
5. Score regeneration after 7 days using either Subheading 3.4 or 3.5 (see Note 5).

3.2. Larval Tail Regeneration Assay

1. Tadpoles at stage 48 are identified by the appearance of gold iridophores surrounding the developing gut and the presence of hindlimb buds but not forelimb buds. Siblings from the same mating should be used where possible.
2. Tadpoles are captured using a small plastic tea strainer or soft fish net, and should be washed briefly to remove food debris. Anesthetize tadpoles in groups of no more than ten by transferring to 90 mm Petri dishes containing 30 mL 1/4,000 (w/v) MS222 (tricaine) in frog water until still. Do not leave in MS222 for extended times as prolonged anesthesia can be lethal. Ten minutes is an ideal limit.
3. Remove 30–50% of the post anal tail using a rounded scalpel blade and transfer tadpole to recovery dish containing frog water or 0.1× MMR. Use a gentle air bubbler to revive tadpoles for 1–2 min. The tail clip can be used for genotyping or sexing if required (CF method).
4. Continue until ten tadpoles are recovered in the dish. Add treatment as required. Tadpoles are cultured in their dishes, with approximate density of one tadpole per 3 mL frog water for 1 week at 20–22°C before scoring regeneration using Subheading 3.4 or 3.5 (refer to Fig. 1 and see Note 6).

3.3. Refractory Period Assay

One of the major advantages of *Xenopus* as a regenerative model is that the refractory period, when tail regeneration fails, can be used to test potential regenerative stimuli. Examples of how this has been achieved with transgenic (5) or chemical methods (35) are available.

1. Identify tadpoles in the refractory period (stage 45–47, see Note 7).
2. Follow steps 2–4 of Subheading 3.2.

3.4. Regeneration Index (RI) Method

RI (36) is calculated as follows:

1. Score embryos in each treatment as in Table 4.
2. Calculate the percentage of embryos in each category.

Table 4
Calculating the regeneration index (RI) for tails

Score	Regeneration	Description	Factor
++	Complete	Regenerated tail, indistinguishable from uncut controls	3
+	Robust	Has minor defects such as missing fin, curved axis	2
+/−	Poor	Hypomorphic regenerates	1
−	None	Appearance as Fig. 1b	0

3. Multiply percentages by the corresponding factor for each score and add together to obtain the RI. This ranges from 0 to 300, with 0 corresponding to no regeneration in any case and 300 to complete regeneration of all cases.

3.5. Mean Regeneration Score (RS)

Mean regeneration score method as described by Beck et al. (5)

1. Score animals at 7–10 days post amputation as for tail regeneration as total (Fig. 1a), partial (hypomorphic regenerates or missing tissues), or none (Fig. 1b). Ten points are awarded for total regeneration, five for partial, and zero for none.
2. Points are totalled for each treatment group and divided by the number in each group (*N*) to generate a mean score. Mean RS varies from 0 to 10 with 0 indicating complete failure to regenerate and 10 indicating perfect regeneration.

3.6. Hindlimb Regeneration

1. Anesthetize animals as Subheading 3.7 step 1, with up to five animals at a time.
2. Gently scoop each tadpole up using a large spatula and transfer to damp paper towel circle, placed on top of a laminated circle of black paper or card, inside a 90 mm Petri dish lid. Lay tadpoles on their left sides in a column.
3. Under a dissecting microscope, verify limb stage. Stage 52 is the most reliable stage to ensure perfect or near perfect regeneration.
4. Using Vannas iridectomy scissors, carefully remove the distal half of each tadpole's right hind limb. This corresponds to roughly the knee level.
5. Transfer to a small tank containing 1–2 L of 0.1× MMR or frog water with gentle aeration until recovered and swimming.
6. Return tadpoles to normal culture conditions until stage 58, when the forelimbs emerge and are clearly visible.

7. Score regeneration by counting digits. Score 0.5 for a spike, 1 for 1 toe, 2 for two toes, 3 for 3 toes, 4 for 4 toes, and 5 for 5 toes. Calculate the average quality score for each group or treatment by summing the scores and dividing by the number of animals (Σscore/*N*).
8. Alternatively, use skeletal preparations (Subheading 3.8) to get accurate identification of the regenerated elements (see Note 8).

3.7. Forelimb Regeneration Assay

1. Anesthetize froglets in batches of 3–4 in 30 mL 1/5,000 MS222 in frogwater in a 90 mm Petri dish.
2. Amputate one forelimb from each animal at mid-forearm level cutting through all tissues cleanly using a scalpel blade.
3. Recover animals in shallow frog water with aeration (bubbling) for 20–30 min. Recovery is much slower than for tadpoles and attention is required to avoid drowning.
4. Return animals to aquarium when recovered. Forelimbs take up to 6 weeks to regenerate.

3.8. Skeletal Staining of Xenopus Tadpoles >st.57

Adapted from Newman et al. (37).

1. Fix tadpoles or froglets in 4% formaldehyde in PBS in 20 mL glass scintillation vials. Samples can remain in this stage for months.
2. Wash fixed samples 3–4 times in PBS to remove the fixative.
3. Under a dissecting microscope, use small spring scissors and Dumont no. 5 forceps reserved for fixed material and cut open the abdominal cavity, removing the viscera as cleanly as possible.
4. Bleach samples to remove pigmentation by placing in 10 mL 6% H_2O_2 in PBS. Ensure the vials are capped tightly and place on a nutator with illumination from a desk lamp with a non-energy saving, 60 W or similar bulb which provides both heat and light to speed the bleaching reaction. Inspect tubes regularly and loosen lids carefully to release trapped gases building up. Bleaching can take several hours to overnight, and the 6% H_2O_2 may need replacing if it turns yellowish brown. If leaving overnight, refresh the 6% H_2O_2 and remove tubes from the nutator and stand them upright to avoid the risk of leaks.
5. Once the pigment has been removed, wash in PBS 3–4 times and inspect the samples for trapped air bubbles. Bleaching produces bubbles that may become trapped beneath the skin of the animals; these can be removed by placing uncapped vials under vacuum using a dessicator, or by manipulating manually with fine no. 5 forceps.
6. Transfer to 70% ethanol in PBS and either store tightly capped or proceed to step 7.
7. Cartilage is stained first. Leaving vials upright, replace 70% EtOH with fresh alcian blue stain for around 4 h, which stains

the cartilage bright blue. At this stage, much of the skeleton is still obscured by the surrounding tissues.

8. Wash samples in 70% EtOH in PBS to destain soft tissues. Place vials on the nutator. Replace the solution as it turns blue and continue until the solution stays clear. This can be done overnight.
9. Replace the ethanol with 1% KOH to clear the soft tissue and reveal the skeleton. This stage may take several days depending on the age and size of the sample. Replace 1% KOH daily until you can clearly see the skeleton.
10. Ossified bone is then stained pink/red by soaking the samples in alizarin red S for 1–2 h. Following bone staining, return the samples to 1% KOH and soak overnight. If background staining remains, replace the 1% KOH and repeat.
11. For storage and photography, wash in PBS 3–4 times and store in glycerin or a 1:1 mix of glycerol/ethanol.

3.9. Simple Lentectomy

1. Stage 50–51 tadpoles are anesthetized in MS222 (1/4,000 w/v) in 0.1× MMR, then placed on their left sides on damp paper towels for surgery.
2. Under a dissecting microscope, the outer cornea is first snipped with Vannas iridectomy scissors, cutting from the posterior side and then dorsally and ventrally around the eye before lifting up as a flap.
3. Rupture the inner cornea and gently remove the lens through the pupil using forceps (alternatively, a narrow bore pipette can be used to aspirate the lens).
4. Gently replace the flap of cornea and epidermis over the eye.
5. Discard animals with hemorrhage or damaged retina. Return good specimens gently to small 1–2 L tanks containing 0.1× MMR.
6. Allow animals to recover in 0.1× MMR overnight before returning to a marine biotech aquarium and fed as normal.

3.10. Corneal Implants ("Eye-Steck" Grafts)

After Filoni et al. (30): this assay can be used to determine the competence of corneal tissue to transdifferentiate into lens.

1. Remove lens from host animal as described for lentectomy (Subheading 3.9).
2. Using iridectomy scissors, remove a small piece of cornea from the donor animal and gently push through the open pupil of the host so that it ends up in the vitreous of the host eye.
3. Similarly, reinsert the original host lens through the pupil.
4. Close the host cornea gently over the eye and recover animals as for lentectomy (Subheading 3.9, step 6).

4. Notes

1. Line drawings and descriptions of the 66 normal stages of *X. laevis* development, to the completion of metamorphosis, can be found in Nieuwkoop and Faber (3). The line drawings can be viewed and downloaded from Xenbase, http://www.xenbase.org, but the copyright remains with these authors.
2. In our hands, stage 54–55 tadpoles amputated at knee level only regenerate around half of the time, and even then the average number of formed toes is 2, a little lower than that described by Dent (14, 19, 20). Nye and Cameron (38) suggest that variation can be minimized by selecting only animals that follow the normal growth curve and avoiding outliers. They also note better quality regenerates arising from animals amputated through a joint rather than through bone (39). Beck et al. (2) further suggest the need for sibling and age-matched controls in all experiments involving limb regeneration.
3. The *Xenopus hsp70* heat shock promotor has been extensively employed to create transgenic animals in which a developmentally significant gene can be expressed at specific time points during regeneration. Transgenic animals harboring such constructs can be induced by transferring them to frog water at 34°C or 20–30 min daily as required but will grow, develop, and regenerate normally at ambient temperatures, at or below 22°C.
4. Raising animals to adulthood that were initially exposed to gentamycin has not been successful in our hands, so for late stage experiments or transgenic lines, we avoid gentamycin, instead use gloves and sterile techniques when handling embryos.
5. It is important to include replicates for each treatment and to repeat the experiments with tadpoles from different matings, although tail regeneration does not seem subject to the same extent of variation as seen in limb regeneration.
6. In refractory tails, or tails in which regeneration has been prevented by treatment, a small bulge appears at the level of the axial tissues which is often mistaken for regeneration. In fact, this is due to vacuolation of notochord cells resulting in a distension of the overlying epidermis.
7. Staging animals to identify refractory stages is most easily done by observing pigment cell changes, since limb buds can be hard to spot. Stage 46 can be identified by the appearance of xanthophores (metallic silver pigment cells) on the retina of the eye and the abdomen and stage 47 when these xanthophores completely cover the abdomen, obscuring the underlying intestine. Once iridophores (golden) appear at stage 48, regeneration of the tail is restored.

8. With skeletal staining, it is possible to identify individual digits in hindlimbs. *Xenopus* hindlimbs have five digits labeled I to V from anterior to posterior. Claws are present on digits I to III. Digits I and II each have two phalanges, but digit I is shorter due to a much smaller proximal phalange. Digit III has three phalanges. Digit IV has four phalanges and no claw. Digit V has three phalanges and a longer metatarsal than the other digits. For a good example, refer to (21).

Acknowledgements

I would like to thank Elisha Wang and Amy Taylor for helpful comments on the draft versions of this chapter. Research in my lab is supported by a University of Otago Research Grant, Genetics Otago, and by the Department of Zoology.

References

1. Borgens RB (1982) Mice regrow the tips of their foretoes. Science 217:747–750
2. Beck CW, Izpisua Belmonte JC, Christen B (2009) Beyond early development: *Xenopus* as an emerging model for the study of regenerative mechanisms. Dev Dyn 238: 1226–1248
3. Nieuwkoop PD, Faber J (1967) A normal table of Xenopus laevis (daudin). Elsevier, North Holland, Amsterdam
4. Tucker AS, Slack JMW (1995) The *Xenopus laevis* tail-forming region. Development 121: 249–262
5. Beck CW, Christen B, Slack JM (2003) Molecular pathways needed for regeneration of spinal cord and muscle in a vertebrate. Dev Cell 5:429–439
6. Filoni S, Bosco L, Carlizzi C (1977) La rigenerazione del midello spinale della coda di larvae di *Xenopus laevis* operate negli tardivi. Rend Acc Naz Lincei (ser VIII) 63:440–446
7. Bosco L (1979) Expression of regenerative capacity of caudal spinal cord during the larval development of *Xenopus laevis.* Acta Embryol Exp 3:275–286
8. Gargioli C, Slack JMW (2004) Cell lineage tracing during *Xenopus* tail regeneration. Development 131:2669–2679
9. Chen Y, Lin G, Slack JM (2006) Control of muscle regeneration in the *Xenopus* tadpole tail by Pax7. Development 133:2303–2313
10. Slack JM, Lin G, Chen Y (2008) The *Xenopus* tadpole: a new model for regeneration research. Cell Mol Life Sci 65:54–63
11. Filoni S, Bosco L (1981) Comparative analysis of the regenerative capacity of caudal spinal cord in larvae of several anuran amphibian species. Acta Embryol Morphol Exp 2:199–226
12. Lin G, Chen Y, Slack JM (2007) Regeneration of neural crest derivatives in the *Xenopus* tadpole tail. BMC Dev Biol 7:56
13. Komala Z (1957) Comparative investigations of the course of ontogenesis and regeneration of the limbs in *Xenopus laevis* tadpoles in various stages of development. Folia Biol 5:1–51
14. Dent JN (1962) Limb regeneration in larvae and metamorphosing individuals of the South African clawed toad. J Morphol 110:61–78
15. Yokoyama H, Yonei Tamura S, Endo T, Izpisua Belmonte JC, Tamura K, Ide H (2000) Mesenchyme with fgf-10 expression is responsible for regenerative capacity in *Xenopus* limb buds. Dev Biol 219:18–29
16. Filoni S, Paglialunga L (1990) Effect of denervation on hindlimb regeneration in *Xenopus laevis* larvae. Differentiation 43:10–19
17. Yokoyama H, Ide H, Tamura K (2001) FGF-10 stimulates limb regeneration ability in *Xenopus laevis.* Dev Biol 233:72–79
18. Slack JM, Beck CW, Gargioli C, Christen B (2004) Cellular and molecular mechanisms of regeneration in *Xenopus.* Philos Trans R Soc Lond B Biol Sci 359:745–751
19. Barker DM, Beck CW (2009) Overexpression of the transcription factor Msx1 is insufficient to drive complete regeneration of refractory stage *Xenopus laevis* hindlimbs. Dev Dyn 238: 1366–1378

20. Beck CW, Christen B, Barker D, Slack JM (2006) Temporal requirement for bone morphogenetic proteins in regeneration of the tail and limb of *Xenopus* tadpoles. Mech Dev 123:674–688
21. Satoh A, Endo T, Abe M, Yakushiji N, Ohgo S, Tamura K, Ide H (2006) Characterization of *Xenopus* digits and regenerated limbs of the froglet. Dev Dyn 235:3316–3326
22. Satoh A, Ide H, Tamura K (2005) Muscle formation in regenerating *Xenopus* froglet limb. Dev Dyn 233:337–346
23. Tassava RA (2004) Forelimb spike regeneration in *Xenopus laevis*: testing for adaptiveness. J Exp Zool A Comp Exp Biol 301:150–159
24. Endo T, Tamura K, Ide H (2000) Analysis of gene expressions during *Xenopus* forelimb regeneration. Dev Biol 220:296–306
25. Yakushiji N, Yokoyama H, Tamura K (2009) Repatterning in amphibian limb regeneration: a model for study of genetic and epigenetic control of organ regeneration. Semin Cell Dev Biol 20:565–574
26. Satoh A, Suzuki M, Amano T, Tamura K, Ide H (2005) Joint development in *Xenopus laevis* and induction of segmentations in regenerating froglet limb (spike). Dev Dyn 233: 1444–1453
27. Freeman G (1963) Lens regeneration from the cornea in *Xenopus laevis*. J Exp Zool 154:39–65
28. Reeve JG, Wild AE (1978) Lens regeneration from cornea of larval *Xenopus laevis* in the presence of the lens. J Embryol Exp Morphol 48:205–214
29. Bosco L, Filoni S, Cioni C (1980) Lens formation from cornea in the presence of the old lens in larval *Xenopus laevis*. J Exp Zool 213:9–14
30. Filoni S, Bernardini S, Cannata SM, D'Alessio A (1997) Lens regeneration in larval *Xenopus laevis*: experimental analysis of the decline in the regenerative capacity during development. Dev Biol 187:13–24
31. Henry JJ, Elkins MB (2001) Cornea-lens transdifferentiation in the anuran, *Xenopus tropicalis*. Dev Genes Evol 211:377–387
32. Smolich BD, Tarkington SK, Saha MS, Stathakis DG, Grainger RM (1993) Characterization of *Xenopus laevis* gamma-crystallin-encoding genes. Gene 128:189–195
33. Day RC, Beck CW (2011) Transdifferentiation from cornea to lens in *Xenopus laevis* depends on BMP signalling and involves upregulation of Wnt signalling. BMC Dev Biol 6:11–54
34. Fukui L, Henry JJ (2011) FGF signaling is required for lens regeneration in *Xenopus laevis*. Biol Bull 221(1):137–145
35. Tseng AS, Beane WS, Lemire JM, Masi A, Levin M (2010) Induction of vertebrate regeneration by a transient sodium current. J Neurosci 30:13192–13200
36. Adams DS, Masi A, Levin M (2007) H + pump-dependent changes in membrane voltage are an early mechanism necessary and sufficient to induce *Xenopus* tail regeneration. Development 134:1323–1335
37. Newman SM Jr, Dumont JN (1983) Thiosemicarbazide-induced osteolathyrism in metamorphosing *Xenopus laevis*. J Exp Zool 225:411–421
38. Nye HL, Cameron JA (2005) Strategies to reduce variation in *Xenopus* regeneration studies. Dev Dyn 234:151–158
39. Wolfe AD, Nye HL, Cameron JA (2000) Extent of ossification at the amputation plane is correlated with the decline of blastema formation and regeneration in *Xenopus laevis* hindlimbs. Dev Dyn 218:681–697
40. Tschumi PA (1957) The growth of the hindlimb bud of *Xenopus Laevis* and its dependence upon the epidermis. J Anat 91:149–173

Chapter 31

On-Line Resources for *Xenopus*

Jeff Bowes

Abstract

Since the advent of computational methods in biology, the quantity of biological data has grown exponentially. These data support genomic, genetic, developmental, and other forms of biological experimentation. The number of on-line resources has kept pace with the growth in data. *Xenopus* has perhaps lagged some of the other model organisms in developing resources, but is now quickly catching up. There are now a number of well-established and developing resources for *Xenopus*. This chapter looks beyond the widely known public databases, Genbank and the EBI, and describes how the researcher can use a number of central sites such as Xenbase, UniProtKB, and major genome browsers to navigate to a variety of different resources.

Key words: Database, Website, Genome browser, Gene model, Xenbase, ECR

1. Introduction

Xenbase (1), the model organism database for *Xenopus laevis* and *tropicalis*, is the natural starting point for the researcher to explore on-line resources for *Xenopus*. It is the only *Xenopus*-specific resource to integrate different experimental data. Xenbase includes genomic, gene expression, tissues and development, literature, and community information. It incorporates a gene catalog, gene expression search, a GBrowse genome browser (2), Textpresso (3) literature search and mining tool, an anatomy browser, community directory, and a wiki. Importantly, Xenbase ties into many related resources.

Major institutional resources such as the NCBI (4) and EBI (5) are also very powerful integrated resources that focus upon multiple organisms, including *Xenopus*. The reader is expected to already have some experience with these resources. Therefore, they will not be covered in this chapter except for some specific tools such as the Ensembl genome browser.

The resources section of this book will introduce a number of on-line resources, some widely used and some interesting resources

Stefan Hoppler and Peter D. Vize (eds.), *Xenopus Protocols: Post-Genomic Approaches, Second Edition*,
Methods in Molecular Biology, vol. 917, DOI 10.1007/978-1-61779-992-1_31, © Springer Science+Business Media, LLC 2012

that are a little off the beaten path. Because of its importance to *Xenopus* research, gene expression is excluded here and covered in detail in Chapter 19.

The methods section describes how to research a gene of interest using Xenbase as a hub. From Xenbase, UniProtKB (6) will be used to link out to protein and other resources, and the genome browsers will be used to link to other genomics resources. As well, several effective literature mining tools will be demonstrated.

2. Resources

2.1. Xenbase (www.xenbase.org)

Xenbase accesses a wealth of integrated data for *Xenopus* research both internal to the site and linked to in external resources. Xenbase uses a combination of computational and manual curation, with a staff of two to three curators.

The body of the front page of Xenbase provides an excellent summary of the resources available in Xenbase. A sidebar provides announcements and links to other relevant sites. The Xenbase header (Fig. 1) is found on every page and provides the principal means of navigating through Xenbase. The menu items in the navigation bar mirror the menu sections of Xenbase described on the home page. These are as follows:

1. The **Blast Tool** provides access to a number of different *Xenopus*-specific databases. Results provide links to Xenbase gene and clone pages as well as to external resources.
2. The **gene catalog** is the heart of Xenbase. The gene pages provide detailed information on genes, gene expression evidence, literature associated with a gene (see Note 1), sequences and potential interactants (see Note 2), and the gene wiki. Each gene page provides information for *X. tropicalis* and *X. laevis*

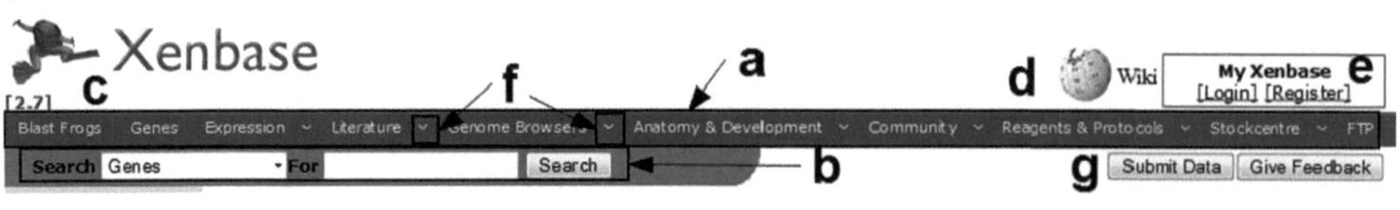

Fig. 1. The header for Xenbase is on every page in Xenbase. It consists of the (**a**) navigation bar, (**b**) search minibar, (**c**) title and version, (**d**) wiki, and (**e**) My Xenbase console. The navigation bar contains links to different topic areas. Clicking the (**f**) chevron to the right of a menu item will open submenu items. The search minibar allows the user to search Xenbase for a variety of different accessions or terms. Clicking the title always takes one to the *Xenopus* homepage and clicking the version will take the user to a description of everything in the current release. The My Xenbase bar allows users to register or login. The wiki icon links to an integrated wiki. There are prominent buttons for (**g**) submitting data and giving feedback.

orthologs and *X. laevis* A and B alloalleles, if known. Many external resources are also linked in. At the time of writing, the gene catalog included 15,865 genes of which 4,711 have manual curation (see Note 3). Users may contribute to Xenbase gene pages by contributing synonyms and entering information about the gene in the Wiki tab.

3. The **expression search** section provides an interface that allows users to search for evidence of gene expression filtered by any combination of gene symbol/clone name, sequence or accession, Nieuwkoop and Faber (NF) development stages (7), tissues (see Note 4), and other variables. Xenbase's gene expression search is described in more detail in Chapter 19.
4. **Anatomy and development** allows users to explore data from the perspective of tissues or development stages. This section includes images of embryos, development stage temperature charts, and cell fate maps. Users can also search for information on each NF development stage or *Xenopus* Anatomical Ontology (XAO) tissue. The page for each tissue includes information on their relationship with other tissues in the XAO (see Note 4), listings of genes and clones expressed in that tissue, and papers containing that evidence. Users can also add information by updating the wiki tab on any anatomy term or development stage page.
5. Xenbase contains a comprehensive **literature** section containing over 40,000 *Xenopus*-related articles that have been imported from Pubmed. The main component of the literature sections allows users to search articles by title, figure legend, abstract, journal, grant funding agency, or Pubmed id. Each article entry contains an abstract entry that has references to genes and XAO terms hyperlinked (see Note 1). Additionally, Xenbase has permission to display gene expression images from a number of journals (see Note 5). In many cases, users can directly view images (often curated for gene expression) from the paper entry in Xenbase. Additionally, Xenbase has implemented the powerful Textpresso text search and mining tool (see Subheading 2.8).
6. The Xenbase **gene browser** is compared with other genome browsers in Subheading 2.2.
7. **Community** provides listings of people, labs, organizations (e.g., companies), and jobs in the *Xenopus* community. Each person's page also includes papers they have published. A person's papers are automatically detected but may be manually modified (8). Anyone can register and create an entry for themselves or their lab in Xenbase.

8. **Reagents** and **Protocols** include a search of *Xenopus* clones, clone libraries, and vector information. Like gene pages, clone pages have an expression tab indicating where they have been expressed and links to suppliers. The expression tab is particularly useful for clones that have not been mapped to a gene. This section also includes pages in the Xenbase wiki for antibodies and protocols. All Xenbase users can add antibodies or protocols.
9. The **Stock Centre** link contains wild type, mutant and transgenic frog lines; clones; and transgenes carried at the European *Xenopus* Resource Centre (EXRC). In the future it will also contain resources from the U.S. *Xenopus* Stock Centre that is currently under development.

2.2. Genome Browsers

Genome browsers are a critical tool for finding and analyzing genomic features and comparing them between organisms. Different genome browsers have different focuses and strengths. Table 1 compares tools and current genome version (4.1 or 4.2) tracks supported by different genome browsers. At the time of writing, version 7.1 of the *Xenopus* genome had not yet officially been released. However, Xenbase GBrowse already supports the 7.1 assembly.

1. The **JGI browser** is operated by the Joint Genome institute (JGI), the organization that sequenced *X. tropicalis* and performed the 4.1, 4.2, and earlier assemblies (9). It has gene models produced by numerous different gene model prediction algorithms; a track called "FilteredModels" containing the best of those models; and a track called "UserModels" that are derived from filtered or other models but have added human annotation. Also, models have GO, KEGG, and KOG annotation. Importantly, as the JGI was not involved in the 7.1 assembly, they will not be supporting it on their genome browser.

   http://genome.jgi-psf.org/cgi-bin/browserLoad/?db=Xentr4
2. **Xenbase GBrowse's** strengths are its integration with Xenbase and focus on development. Xenbase has 4.1 gene model, JGI filtered model, Ensembl, and many preliminary JGI gene model prediction tracks. The Xenbase 4.1 gene models track directly corresponds to Xenbase gene pages which may be derived from JGI or Ensembl models, or Entrez Gene records (see Note 3). Users can click on Xenbase 4.1 gene models track, or either *X. laevis* or *X. tropicalis in situ* clone tracks, and pull up a balloon with thumbnails of gene expression images,

Table 1
Genome browser comparison

	JGI	Xenbase	Ensembl	UCSCD
Assembly versions	3, 4.1	4.1, 7.1	4.1, 4.2	3, 4.1
4.1 Tracks				
Gene model tracks				
JGI filtered	Yes	Yes		
Ensembl	Yes	Yes	Yes	
Xenbase gene models		Yes		Yes
Refseq genes				
Other	Many	Fgenesh, genewise, GeneWisePlus from JGI	Genewise, EST base predictions	Genewise
mRNA Tracks				
ESTs			Yes	Yes
Genbank mRNA	Yes			Yes
X. tropicalis cDNA	Yes	ISH probes	Refseq, EMBl	*Xenopus* Gene Collection
X. laevis cDNA	Yes	ISH probes		
Predicted transcripts	Yes			
Unigene		Yes	Yes	
Gurdon clusters		Yes		
Proteins				
UniProtKB			Yes	
NCBI Refseq	Yes		Yes	
Other tracks				
Start stop codons			Yes	
Transcription start sites			Yes	
tRNA			Yes	
ncRNA			Yes	
CpG islands			Yes	Yes
Restriction enzymes		Yes		Yes
Predicted promoters		Yes		
Affymetrix probes			Yes	
Repeats	Yes		Yes	Yes
BAC end pairs	Yes			
Methylation tracks		Yes		
Other features				
Add track	Yes	Yes (DAS)	Yes	Yes
Alignment tools	BlastN	Blast(Xenbase)	Blast/Blat	Blat
Query tools			Biomart	Table Browser

allowing them to jump immediately from the track to gene expression images. Xenbase is also the only major genome browser to have tracks for the Veenstra laboratory's methylation data (10). While not a browser primarily designed for interspecies comparison, it contains evolutionarily conserved regions from ECRBase (see Subheading 2.3). At the time of writing, Xenbase GBrowse was the leading genome browser to support the pre-release 7.1 assembly. Xenbase already had preliminary gene models, Xenbase gene pages, Gurdon clusters, *X. laevis* and *X. tropicalis* clusters, and Affymetrix probes mapped against the 7.1 assembly and was in the process of mapping additional tracks to the 7.1 assembly.

Assembly 4.1: http://www.xenbase.org/fgb2/gbrowse/Xenopus_tropicalis_4_1/

Assembly 7.1: http://www.xenbase.org/fgb2/gbrowse/Xenopus_tropicalis_7_1/?source=Xenopus_tropicalis_7_1

3. **Ensembl** (11) is a dedicated genome browser, supporting many organisms, with some of the more powerful genome browsing features. Ensembl has a wide choice of tracks including the most genome features mapped to the *X. tropicalis* assembly of any of the genome browsers. Like the UCSC browser, it has strong support for comparative genomics. Table 2 in Note 6 compares comparative genomics tracks between Ensembl and the UCSC browser. However, Ensembl really stands out in comparative genomics with its tool for comparing two separate organisms side by side. Finally, Ensembl incorporates Biomart (12), which allows users to download the Ensembl data tracks.

   www.ensembl.org/Xenopus_tropicalis/

4. The **UCSC genome browser** (13) is a dedicated genome browser, supporting many organisms. It has a wide variety of genomics tracks and, like Ensembl, has strong support for comparative genomics (see Note 6). The UCSC genome browser also has some unique tools. In particular, the table browser tool allows the user to perform powerful queries against tracks on the genome. For example, the user can query intersecting regions of two tracks.

   http://genome.ucsc.edu/cgi-bin/hgGateway?hgsid=206930983&clade=vertebrate&org=X.+tropicalis&db=0

2.3. Other Nucleotide Resources

1. **ECRbase** (14) is a database of evolutionarily conserved regions (ECRs) constructed from whole genome alignments. The database features the ECR Browser, which allows users to browse or search a base genome and look for ECRs. At the time of

Table 2
Genome browser cross-species comparison features

Features	Ensembl	UCSC
Genomic		
Human	Yes	Yes
Chicken	Yes	Yes
Opossum		Yes
Mouse	Yes	Yes
Rat		Yes
Tetraodon	Yes	
Zebrafish	Yes	
Lizard		Yes
Genes		Transmap (29) Refseq Genes
mRNA	European Nucleotide Archive Vertebrates	Transmap mRNA Mappings
Protein	UniProtKB	Human

writing, *X. tropicalis* genome could be compared to cow, *Fugu*, chicken, human (hg18 and hg19), mouse, opossum, chimpanzee, Rhesus macaque, rat, *Tetraodon*, and zebrafish. The ECR Browser also features refseq, Genescan mRNA predictions, and mammalian genome tracks. ECRbase offers downloadable files for promoter, synteny, and ECRs. Xenbase features ECRbase ECR and promoter prediction tracts.

Browser: http://ecrbrowser.dcode.org

Home page: http://ecrbase.dcode.org

2. **Gurdon Clusters** (15) contains clusters of overlapping ESTs. The page for each cluster contains the consensus sequence and a graph showing how it was assembled from component ESTs. This is a useful tool for *X. laevis*, which currently lacks a sequenced genome. A virtual northern blot graph shows the relative abundance of clones from a cluster in a library or libraries drawn from the same tissue or stage.

   (http://genomics.nimr.mrc.ac.uk/cgi-bin/public.exe?dbe=http&dbs=INFO-PUBLIC&uid=guest&species=Xt&current=ESTs&full=Guest+User&src=public&tgt=main&menu=main_intro&project_key=1000000190&version=1)

3. **Rfam** (16) is a database of non-coding cis-regulatory RNA elements. Rfam entries for each family contain descriptions from Wikipedia including clan associations and descriptions of its function and clinical significance. Other sections show species containing the gene family, sequences for each species, visualization of secondary structure, ontology annotations, and literature referencing the ncRNA gene family. Currently, there are approximately 100–200 RNA families directly annotated to *Xenopus*. ncRNA genes can also be viewed as a track on Ensembl.

   http://rfam.sanger.ac.uk/

4. **miRNA** (17) is the central on-line repository for microRNA. It is searchable by name and browsable by organism. At the time of writing, there were 186 *X. tropicalis* entries and 22 *X. laevis* entries. Each entry describes the stem loop sequence and structure and mature sequence. This database also acts as a registry for miRNA naming. *Xenopus* miRNA records are available as a track in the Ensembl gene browser.

   http://www.mirbase.org

2.4. Synteny Tools

1. In addition to the genome browsers and ECRbase, Metazome and Genomicus (18) are valuable tools for viewing synteny. They are described in detail in Chapter 4.

2.5. Protein and Molecular Resources

There are many different protein resources. The following reflects a sample of some of the more popular ones.

1. **UniProtKB** is a central, comprehensive resource for proteins. UniProtKB is composed of the manually curated UniProtKB/Swiss-Prot and computationally annotated UniprotKB/Trembl (see Note 7) entries. Each record indicates whether it comes from the Swiss-Prot or Trembl immediately after the accession. Annotation includes descriptions of protein function, subunit structure, subcellular localisation, expression in terms of tissues and development stages, and Gene Ontology (GO). Additionally, there are sequence annotations for molecule processing, repeat regions, cleavage sites, amino acid modifications, and experimental issues. Each protein entry also includes references to curated literature and many other on-line protein resources. Thus, UniProtKB is an excellent hub for jumping off to other protein resources.

   http://www.uniprot.org/

2. **InterPro** (19) integrates PROSITE, PRINTS, ProDom, Pfam, SMART, TIGRFAMs, PIRSF, SUPERFAMILY, Gene3D, and PANTHER. It contains protein alignments, relationships with

other proteins in InterPro, GO annotation, domain annotation, links to literature, and other resources.

http://www.ebi.ac.uk/interpro/

3. **PFAM** (20) is a database of protein families and superfamilies determined using Hidden Markov Models (HMM). PFAM provides its own summary as well as summaries from Wikipedia and InterPro. It also provides protein domain annotation, HMM model visualizations, interactions, and structure visualizations.

   http://pfam.sanger.ac.uk/

4. **PDB** (21) is a 3D structure database describing experimentally determined protein and other biological macromolecules including nucleic acids and protein-nucleic acid complexes. It contains the atomic coordinates describing the molecule and various tools for visualizing protein structure. At the time of writing, PDB contained approximately 270 *Xenopus* proteins. It also contains a variety of other annotations, literature resources, and links to other on-line protein resources.

   http://www.pdb.org

2.6. Genetic Resources

1. The **Genetic Map of** Xenopus tropicalis (22) is a genetic map of unique simple sequence repeats (SSRs) onto the *X. tropicalis* chromosomes.

   http://tropmap.biology.uh.edu/map.html

2.7. Phenotype Resources

1. **Smith Laboratory Morpholino Screen** provides a site where users can search a screen of morpholinos for 202 genes by gene name, phenotypic class, or specific defects. Each page presents images of the embryo with the morpholino and controls.

   http://www.gurdon.cam.ac.uk/~smithlab/screens/Xenopus-morpholino-pilot/

2.8. Literature

1. **Textpresso** is a powerful literature search and text mining tool that is implemented by, and integrated into, Xenbase. Users can search by keywords and/or ontology terms such as all *Xenopus* genes, XAO terms, gene regulation terms, select GO terms, and many others. Unlike almost all other literature search resources, Textpresso searches the actual paper content and not just the title and abstracts. Textpresso returns sentences from documents that match the user's query. This allows users to do searches like "find at least two genes mentioned in a sentence with a regulatory term" or "find sentences with pax8 and a morpholino-related term" (see Subheading 3.10).

   http://www.xenbase.org/cgi-bin/textpresso/xenopus/home

2. **iHOP** (23) or **information hyperlinked over proteins** is a database of Pubmed abstracts that have been annotated with

gene names. It is useful for manually researching literature for gene interactions. Gene entries in iHOP are organized by pages for different organism's version(s) of each gene. These pages have sections for interactions, definitions, and recent information; with each section containing sentences from article abstracts referencing the subject gene. Sentences are included based on gene references found by matching symbols and synonyms. For each sentence, the user can pull up the full abstract. They can also click on the name of another gene to navigate to the page for that gene. This provides a similar experience to using the literature section in Xenbase, minus the figures and figure legends that are only found on Xenbase. However, in iHOP the user is able to navigate to articles about the same gene in other organisms. Also, for any sentence that contains two gene references, the user is able to add a link to a graph that keeps track of the relationships they are finding between genes.

http://www.iHOP-net.org

2.9. Other Resources

1. **Comparative Toxicology Database** (24) is a database of gene, chemical, and disease interactions in vertebrates, including *Xenopus*, that are curated from literature.

   http://ctd.mdibl.org/

3. Methods

While far too many resources have been described above to provide detailed instructions for using them, this section provides a basic guide on navigating to and making basic use of the above resources. Searching Xenbase for a gene and linking out through the gene page to UniProtKB or one of the genome browsers provides access to the majority of these resources. There are also instructions on accessing unconnected resources: ECRbase, Gene Map of *X. tropicalis*, and iHOP.

3.1. Review a Gene

1. Go to the Xenbase homepage (www.xenbase.org). In the search minibar (Fig. 1), select Genes and enter an accession number (see Note 8). If searching by symbol, using the human gene symbol is most effective (see Note 9). As you type, the search will pull up matching symbols as suggestions. You can choose one of the suggestions or keep typing. Instead of using the minibar, you can also click on the **genes** option in the navigation bar and perform the regular or advanced (see Note 10) search on the genes section landing page.

2. If your search matched only one gene, you will be taken straight to that gene page. Go to step 3. Otherwise, you will have a list of genes to choose from. A camera icon beside a gene indicates that there are images associated with this gene (see Note 11).
3. You will land on the summary tab for the gene page. At the top of the gene page, you will find tabs for different types of information about the gene: Summary, Expression, Gene Literature, Sequence, Interactants, and the Wiki. The top of the page contains information on the gene symbol, name, synonyms (see Note 12), function, and potential interactants and some archetypal expression images for the gene. Below that, the summary tab is divided into sections. The page contains data for *X. tropicalis* and the *X. laevis* gene (*a* and *b* variants if both are known) the data for each gene are organized in columns. There is an accession for the gene page at the top of the page and accessions for each individual gene heading the columns for each gene (see Note 13) .
4. The molecule section has links to Entrez Gene, Ensembl, UniProtKB, and links to Blast or view sequences (see Note 11). Click on the link marked UniProtKB to explore your gene's protein product (see Subheading 3.2).
5. In the genomics section you can see a snapshot of your gene in the genome browser. You can click on JGI, Ensemble, UCSC, or Xenbase GBrowse links to view your gene in any of those browsers. See Subheadings 3.3 and 3.4 for a brief overview of using Xenbase GBrowse and Ensembl, respectively. Chapter 4 contains detailed information on using the UCSC and the JGI genome browser.
6. In the orthology section you can link to human, mouse, and zebrafish orthologs in OMIM, MGI, and ZFIN, respectively. There are direct links to phenotype and morphant data from these organisms when such data exists.
7. If the gene has phenotype data, there will be a phenotype section on the summary page with a link to the Smith laboratory's morpholino screen. (nog2 is an example of such a gene.)
8. To view expression data, click on the expression tab. This page will provide a summary of anatomy terms and the range of development stages for which there is evidence of expression. There are images summarizing expression (the same as the summary images), community submitted images (see Note 14), and images drawn from literature. Click on an image to view more information on it. Summary and literature images are the most likely to be curated. Clicking on the images will open the image, revealing gene expression curation. Users are encouraged to click the thumbs-up or thumbs-down to rate an image's quality. Xenbase uses image quality ratings to better order images as they are presented to the user.

9. The sequences tab has a variety of sequences associated with each gene. The literature tab features articles. Users are encouraged to use the Wiki tab to enter additional useful information about their gene (see Note 15).

3.2. Finding Protein Resources Using Xenbase and UniProtKB

1. Link to the UniProtKB from a Xenbase gene page. (Use the *X. laevis* variant of trove2 for this example.)
2. Check by the accession to see if the UniProtKB resource is from Swiss-Prot (manually curated) or Trembl. On the pages you will notice a wealth of information about the resource.
3. Under the **Cross-references** heading you will find links to the InterPro, Pfam, PDB, and the Comparative Toxicogenomics Databases described in the Resources section of this chapter. You will also find links to many other protein resources.
4. As an example, try going to PDB. Under **Cross-references**, under **3D structure databases,** select **RCSB PDB** in the radiobox and click a link under the column: entry.
5. On the PDB page, look in the right-hand column, in the **Biological Assembly** box to see the 3D structure. Click on the button that says "View in Jmol." You may be asked if you want to run an application. If so, select yes. You will have a 3D image appear in a box (Fig. 2). You can rotate the image around by dragging it with your mouse.

3.3. Using Xenbase GBrowse

1. From a Xenbase gene page (use pax3 for this example), click on the link to GBrowse or click on genome browser snapshot. This will take you to the portion of the genome where the gene in question is located. Alternatively, from the Xenbase home page, select **Genome Browsers -> Xenbase Xentr 4.1** and enter either the gene name or scaffold region in the **Landmark or Region** field.
2. The page is organized from top to bottom with navigation controls, followed by a scaffold and region overviews, detailed view tracks, and track display controls (Fig. 3).
3. To turn tracks on or off, go to the track control section. For this example, turn on the following tracks: Xenbase 4.1 Gene Models, Gurdon XT 7.1 Clusters, Gurdon XL 3 EST Clusters, Fosmids, any ECR tracks, and Promoter predictions. After you open tracks and move them around, GBrowse will open the same tracks the next time you use the browser.
4. You can rearrange tracks by holding the mouse down over a track header and then drag it up or down to where you want it. Icons on the side of the track header allow you to control it.
5. The Xenbase 4.1 Gene Models track corresponds to Xenbase gene pages and may be derived from JGI or Ensembl gene models, or Entrez Gene entries (see Note 3). Click this track to

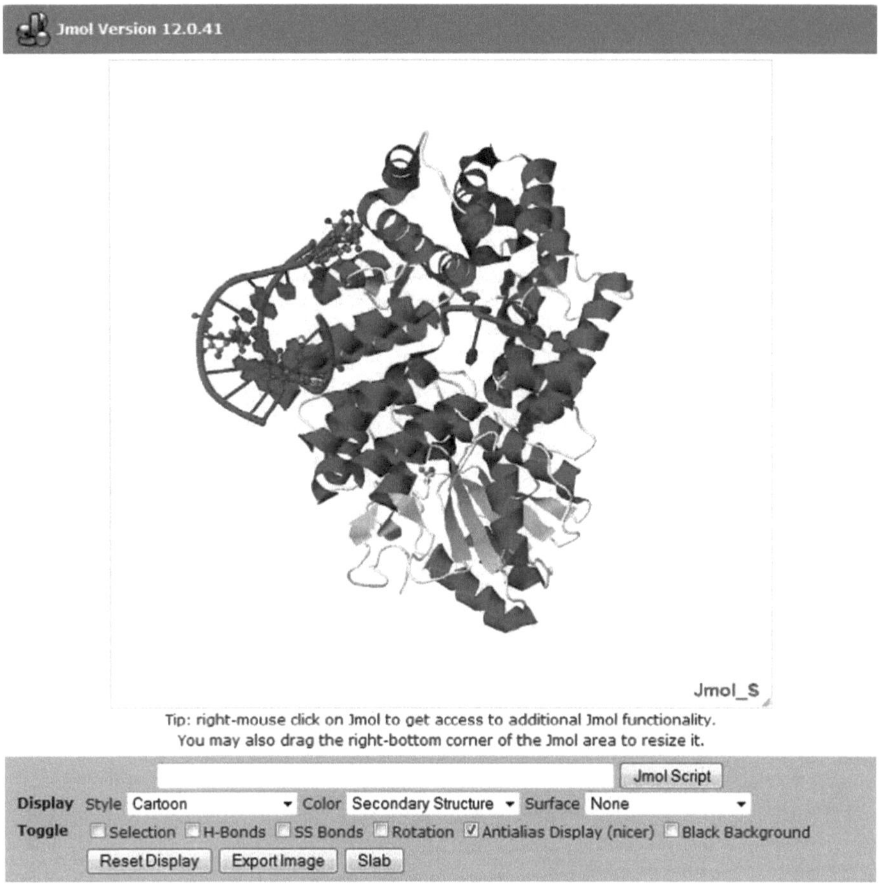

Fig. 2. The JMOL viewer in PDB allows you to manipulate 3D protein assemblies.

open a balloon with links to the gene page, source model, and genome details. If there are gene expression images associated with the gene, then there will be a thumbnail of one of the images in the balloon. Click on the thumbnail to view it or use the arrows to page through additional gene expression images.

6. Click on one of the Gurdon cluster links to jump to the Gurdon EST cluster site.
7. Click on one of the *in situ* clone tracks to view expression images for the clone.
8. Look at the methylation tracks which present the level of methylation in a wiggle format.
9. Click on a fosmid that was used in sequencing the *X. tropicalis* genome. One of the options in the balloon that pops up will be a link to order the fosmid from the EXRC.
10. Examine the positioning of the predicted promoter and ECRs relative to the gene model.

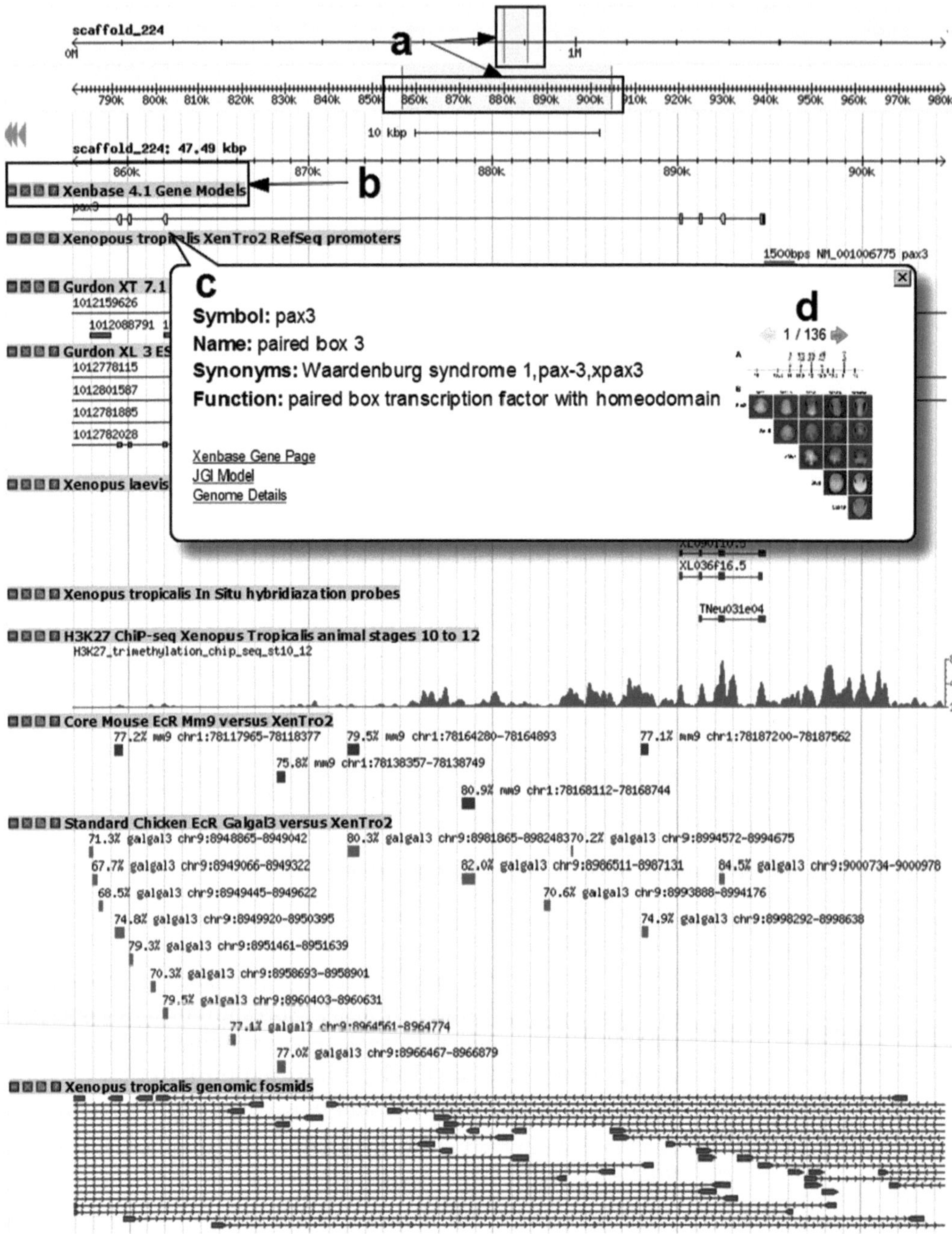

Fig. 3. Xenbase GBrowse genome browser tracks (example pax3 gene): (**a**) *Lines* demarcate the area shown in the detailed view on the scaffold overview and the region tracks, (**b**) a track header that can be dragged to move the track. It also contains controls to *open*/*close*, remove from the display, broadcast with DAS, or access help. (**c**) A balloon with detailed information is opened when clicking on a track (in this case the Xenbase 4.1 gene track). (**d**) You can navigate through in situ images for the gene using *arrow* controls or click on the thumbnail to view a larger image with annotations.

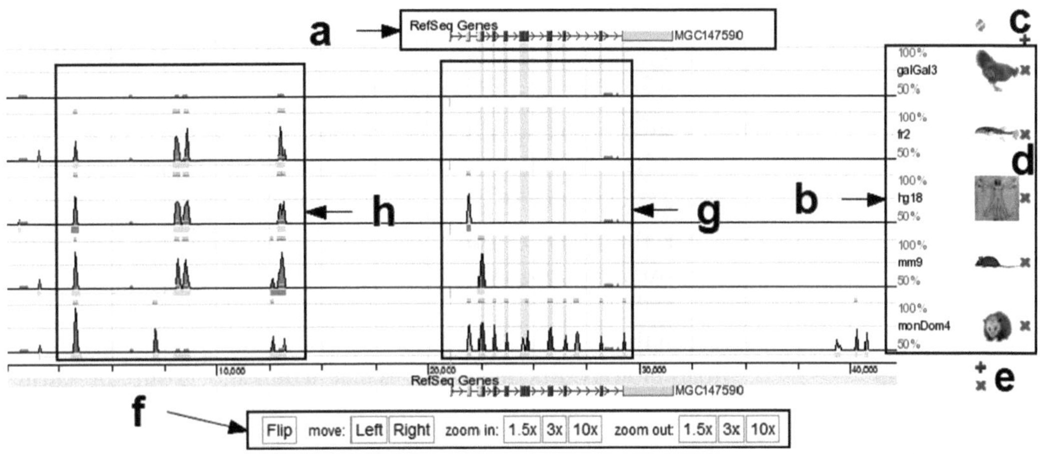

Fig. 4. The evolutionarily conserved region (ECR) Browser (**a**) the *X. tropicalis* model being compared; (**b**) the list of organisms that the *X. tropicalis* model is being compared to; (**c**) the icon for adding additional species to the comparison; (**d**) the icon to remove a species from the comparison; (**e**) icons for adding and removing reference tracks; (**f**) navigation and zoom controls; (**g**) A set of ECRs that line-up with exons in the *X. tropicalis* gene; (**h**) Upstream ECRs that are potentially promoters or enhancers.

3.4. Using Ensembl

1. Click on the Ensembl link on a Xenbase gene page (for this example use pax3 again).
2. To choose what tracks you want, open click the **Configure this page** link in the left hand menu. To access the RFAM and miRNA database described in the resources section of this paper, select **Genes** from the **configure view** and click the ncRNA checkbox to select how you want the track displayed. Then exit the configure browser image by clicking the check-mark in the upper right-hand corner of the window.
3. To compare the *X. tropicalis* genome side by side with another genome, select **Comparative Genomics** -> **Multi-species view** from the **Location-based displays** menu on the left side of the screen.
4. Add species to the view by pressing the **Select species** button on the left. The window that pops up will show selected species on the left and unselected species on the right. Press the plus button beside a species to add it or the minus button to remove a species.

3.5. Using ECR Browser

1. From the ECR Browser home page (http://ecrbrowser.dcode.org/), select “Frog” in the base **genome field** and enter a scaffold position or gene name in the **feature or position** field. Press “submit.”
2. This will take you to a view of that genomic region (Fig. 4). Adjust the focus controls until you can view the full area around the gene or region examined.

3. Add or remove species using the controls on the right side of the screen. Add additional reference tracks (e.g., refseq mRNA) using controls in the bottom right corner.
4. If you are looking at the gene, look for ECRs within the gene boundary as indicators of exons and in the region before the gene as evidence of potential promoter or enhancer regions.
5. Click the **instructions** link on the navigation bar for further information on using ECRbase.

3.6. Using The Genetic Map of Xenopus tropicalis

1. From the Genetic Map of *X. tropicalis* home page (http://tropmap.biology.uh.edu/map), select a linkage group or cluster from the drop boxes and press the "Retrieve_Map" button.
2. This will take you to a page with the map of the chromosome showing the locations of SSRs. Click on an SSR's accession number to view details for it.

3.7. Browsing Anatomy in Xenbase

1. From the Xenbase home page, select **Anatomy & Development** -> Xenopus **Anatomy Ontology** from the navigation bar.
2. Select the **Search Anatomy** tab.
3. On the search anatomy page there are two ways you can search for the anatomy item you want. First, you can use the search field by tissue name, XAO id, or Xenbase Id. If you search by name a suggestion box will provide options that match what you are typing. You can select an option from the suggestion box or you can search for the string you have typed in. If there is more than one match they will be displayed in a list. Alternatively, you can use a tree. For example, to find mesoderm you would select **embryo -> primary germ layer** -> **mesoderm**.
4. The page for the anatomy item will have several tabs at the top: Summary, Anatomy Item Literature, Expression, Attributions, and Wiki. You initially start on the summary page which has the anatomy term's name, synonyms, definition, and relationships with other anatomy term's (see Note 4). You can explore the tabs or link to developmentally related or component tissues.

3.8. Viewing Antibodies in Reagents and Protocols

1. From the Xenbase home page, select the chevron next to the **Reagents and Protocols** and select **Antibodies** from the menu.
2. This will take you to a Xenbase wiki page that has an alphabetical list of curated antibodies. Choose one of the antibodies and you will be taken to a wiki page with additional information.
3. Users are encouraged to add or edit antibodies. You can edit an existing antibody page by selecting wiki edit tab for an antibody. (See Note 15 for more information on editing wikis.)

4. If you wish to add an antibody you would edit the antibody list page and add a link with the name of your antibody. There is a link at the top of the page called **Antibody Reagent Template**. This provides the template that an antibody entry should follow.

3.9. Using the Xenbase to Search for Publications

1. From the Xenbase home page, select **Papers** from the Literature menu on the navigation bar. This will take you to the paper search page.
2. On the paper search page, you can search by many criteria (for this example, search Journal for "Dev Dyn").
3. Select a paper whose link has a camera beside it (indicating it has imported images).
4. On the paper page you will see the abstract with other genes references, anatomy terms, and development stages hyperlinked. You can click on any of these links to explore these items. If authors are registered Xenbase users, their names will also be hyperlinked and clicking on the name will take you to the author's Xenbase person page.
5. Click the Pubmed link to open the Pubmed entry.
6. Go to the **Article images** section and click "show captions." This will allow you to peruse image captions as well as abstracts.
7. As in other sections, clicking on the thumbnail image will open a larger version of the image with annotations of expression.

3.10. Using Textpresso

1. From the Xenbase home page click the chevron to the right of the **literature** menu. Then select **Textpresso.** This will take you to the Textpresso search page.
2. As an example, we will perform a query to find articles with evidence of a regulatory interaction between pax8 and any other gene. In the **Keywords** field type "pax8" (Fig. 5). Ensure the **Exact Match** and **Search Synonyms** are checked. This adds the criterion of "pax8" or synonyms to the search. Next, place your cursor in the first field and press the **List>** button. This will pop-up a menu. Hover over **relationships** and then select the **regulation** option. This adds the criteria that a regulation term must also appear in the sentence. Finally, place your cursor in the next category box. Click the **List >** button again. Select **Biological Concepts** -> **gene**-> **gene** (Xenopus). This requires that the sentence contains an additional gene term. Press the **Search** button.
3. Browse through the results.

3.11. Using iHOP

1. Go to the IHOP home page (www.iHOP-net.org) and enter the name of a gene in the **Search Gene** field. (Use pax8 for this example.)

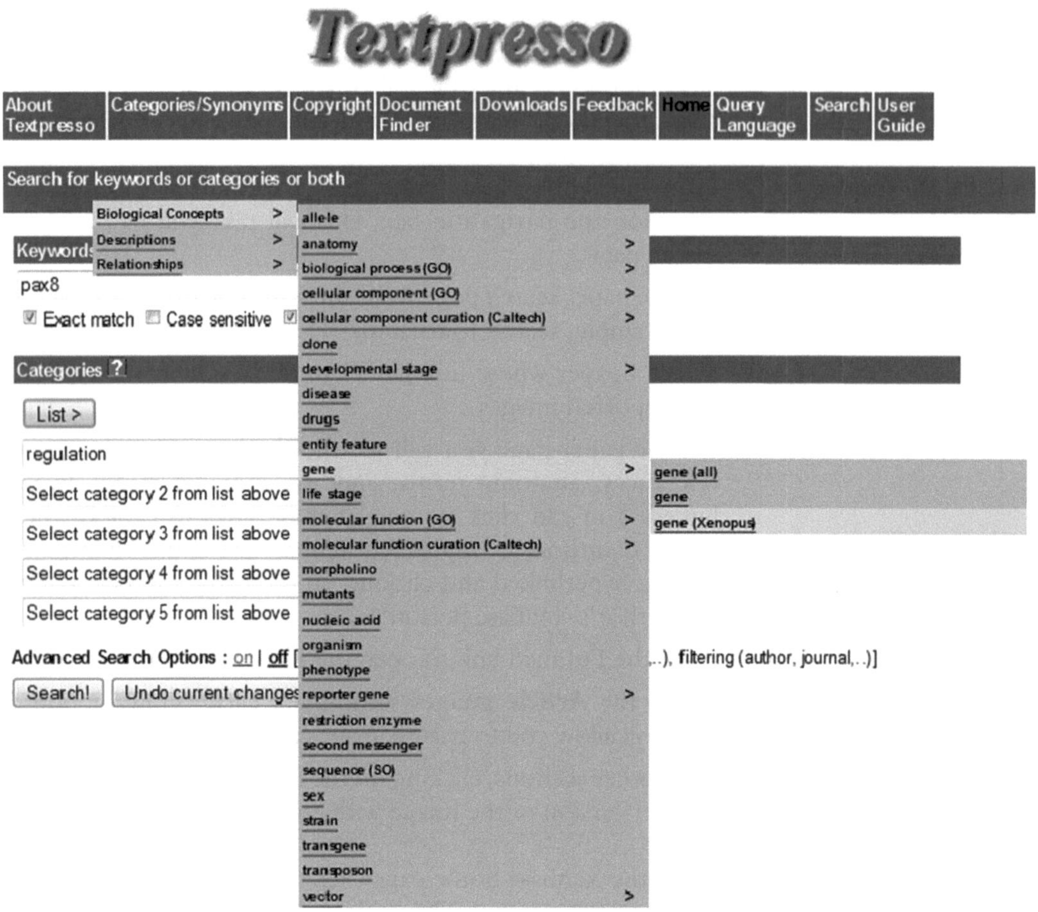

Fig. 5. Textpresso search for pax8 appearing in a sentence with another *Xenopus* gene and a regulation term.

2. This will take you to a page of genes that match your search. These will include *X. laevis*, *X. tropicalis*, and other species' genes. Select the gene that you want.
3. This will take you to the interactions page for your gene. The page will contain sentences that mention your gene and another gene.
4. Press the view abstract icon to view the whole abstract marked up with other gene mentions and MeSH terms.
5. Click the Add to gene model icon beside the sentence and it will add the relationship described in the sentence to a graphical visualization of a gene network.
6. Click a link to another gene to investigate further interactions.

4. Notes

1. Papers are matched to genes by the Xenbase link matching algorithm. The link matching algorithm searches paper titles, abstracts, and captions for gene mentions by finding gene symbols or synonym names using database searches and regular expressions. (Using gene mentions from paper bodies resulted in too many false positives due to mentions in the references section.) Symbols or synonyms likely to produce false positives (e.g., "t" for brachyury) are excluded from the search. Terms that match several genes have disambiguation pages allowing the user to choose the correct gene. XAO terms and development stages are matched in the same way in literature.
2. Potential gene interactants are identified based on co-occurrences of gene mentions in the same paper. The interactants tab indicates how many documents two genes appear in together. Jenssen et al. (25) demonstrated that the co-occurrences of genes references in an abstract are indicative of meaningful biological interactions and the more often genes are mentioned together the more likely this is true.
3. Genes in Xenbase are created from JGI or Ensembl gene models, or *Xenopus* Entrez Gene (26) records. At the time of writing, ~4,000 genes were manually curated and named based on synteny with humans. Manually curated genes are marked as provisional if there is no synteny evidence but only alignment-based putative orthology. Automatically curated genes are marked as "(provisional-machine annotated only)." If a gene has a putative human ortholog, the gene symbol and name will be annotated as "(Predicted: Source is Human Entrez Gene)." Otherwise the naming will come from *Xenopus* Entrez gene records and it will be annotated as "(Predicted: Source is Laevis Entrez Gene)" or "(Predicted: Source is Tropicals Entrez Gene)." Finally, there are a large number of genes whose symbols cannot be determined either via human orthology or *Xenopus* Entrez gene records. They all have the symbol "unnamed" and have a long name drawn from JGI gene model defline records that are in turn based on protein domain annotation and homology derived with a Metazome pipeline.
4. Selecting tissues in the gene selection search is powered by the XAO (27). The XAO stores tissue types, the development stages the tissues generate and terminate at, and the three different types of relationships among them. The anatomy ontology includes **is-a** relationships such as a forelimb is a limb; **part-of** relationships such as the endocardium is part of the heart; and **develops-from** relationships such as prechordal plate develops from the mesoderm. These relationships enable

powerful anatomy-based queries. For example, if a user searches for a gene expressed in the heart, the search would return not only genes annotated as being expressed in the heart but also genes annotated as being expressed in any of the heart's subparts, such as the endocardium. However, a search for genes expressed in endocardium would not return genes annotated as being expressed in the heart because they could be expressed in other parts of the heart excluding the endocardium.

5. At the time of writing, Xenbase had permission to display gene expression images from the journals Cell, Development, Developmental Biology, Developmental Dynamics, Gene Expression Patterns, the Journal of Cell Science, Mechanisms of Development, and PNAS.
6. The Ensembl and USCS genome browsers have the most comprehensive cross-species comparison. Table 2 contains a comparison of features between these two browsers for comparing genomes from different species.
7. UniProtKB was divided into two sections because the number of protein sequences being generated, especially from translation of coding sequences, exceeded UniProtKB's capacity to provide quality human annotation for everything. Thus, Swiss-Prot contains high quality manually curated sequences. Furthermore, all proteins for the same gene are merged into a single Swiss-Prot entry with one canonical sequence. Trembl sequences are curated using a computational pipeline and are not merged into a single entry for each gene.
8. You can search for a gene using a number of criteria such as symbol, name, synonym, ortholog symbols and synonyms, NCBI Accession, Entrez Gene Id, Affymetrix ID, Unigene Id, OMIM ID, GO ID, or Xenbase ID.
9. Xenbase gene naming follows the human genome naming committee (HGNC) conventions. This is consistent with the direction followed by other major model organism databases. The downside of this is that it makes it more difficult to find genes or match genes in literature based on historical names. Therefore, it is especially useful if users help keep synonym lists updated (see Note 12).
10. The Xenbase advanced search allows you to add or subtract criteria. You can add or remove items from search by clicking a plus or minus icon. You can also limit your search to pages that are manually curated or have gene expression data.
11. A number of icons are used throughout Xenbase. A camera icon means an object has images associated with it; a rocket takes the user to a blast search populated with the sequence for that object. Clicking a magnifying glass will bring up a pop-up window with an object's sequence. The user can click on a

thumbs-up or thumbs-down icon to indicate whether the quality of an image is high or low.

12. If you do not see a name that you know is commonly used click on the "add synonym" link to add it. This will aid future searches. You need to be a registered Xenbase user and log in to use this function.
13. To best present the view of a gene, the *X. tropicalis* variant of a gene and the *X. laevis* variant (or both a and b alloalleles) of the gene are all presented on the same gene page. However, each gene variant has its own accession number shown at the top of the gene column. Gene accessions take the form XB-GENE-##### where ##### is a number. There is also an accession for the whole gene page at the top (XB-GENEPAGE-#####). If searching for the gene page by Xenbase accession, you can get to the gene page by linking to any of these accessions.
14. Community images include 2,500 and 18,000 images from Axeldb (28) and the Ueno laboratory's NIBB high throughput *in situ* gene expression screens, respectively. These are described further in Chapter 19.
15. The wiki tab in Xenbase exists on gene, anatomy, and development stage pages. On a wiki page, simply click on the edit tab to add or modify data. You do not need to be logged in. The process used to edit a Xenbase wiki article is the same as is used to edit a Wikipedia article (see http://en.wikipedia.org/wiki/Wikipedia:Cheatsheet).

Acknowledgements

This work is funded by the NICHHD grant 1P41HD064556-01. Thanks to Kevin Snyder for assistance in this work.

References

1. Bowes JB et al (2010) Xenbase: gene expression and improved integration. Nucleic Acids Res 38:D607–D612
2. Donlin, MJ (2009) Using the generic genome browser (GBrowse). Curr Protoc Bioinformatics Chapter 9: Unit 9.9.
3. Muller HM, Kenny EE, Sternberg PW (2004) Textpresso: an ontology-based information retrieval and extraction system for biological literature. PLoS Biol 2:e309
4. Sayers EW et al (2011) Database resources of the National Center for Biotechnology Information. Nucleic Acids Res 39:D38–D51
5. Brooksbank C, Cameron G, Thornton J (2010) The European Bioinformatics Institute's Data Resources. Nucleic Acids Res 38:D17–D25
6. UniProt Consortium (2011) Ongoing and future developments at the Universal Protein Resource. Nucleic Acids Res 39: D214–D219

7. Nieuwkoop PD, Faber J (1994) Normal table of Xenopus laevis (Daudin). Garland, New York
8. Bowes JB et al (2008) Xenbase: a Xenopus biology and genomics resource. Nucleic Acids Res 36:D761–D767
9. Hellsten U et al (2010) The genome of the Western clawed frog Xenopus tropicalis. Science 328:633–636
10. Akkers RC et al (2009) A hierarchy of H3K4me3 and H3K27me3 acquisition in spatial gene regulation in Xenopus embryos. Dev Cell 17:425–434
11. Flicek P et al (2011) Ensembl 2011. Nucleic Acids Res 39:D800
12. Smedley D et al (2009) BioMart–biological queries made easy. BMC Genomics 10:22
13. Fujita PA et al (2011) The UCSC Genome Browser database: update 2011. Nucleic Acids Res 39:D876
14. Loots G, Ovcharenko I (2007) ECRbase: database of evolutionary conserved regions, promoters, and transcription factor binding sites in vertebrate genomes. Bioinformatics 23:122–124
15. Gilchrist MJ et al (2004) Defining a large set of full-length clones from a Xenopus tropicalis EST project. Dev Biol 271:498–516
16. Gardner PP et al (2009) Rfam: updates to the RNA families database. Nucleic Acids Res 37:D136–D140
17. Griffiths-Jones S et al (2008) MiRBase: tools for microRNA genomics. Nucleic Acids Res 36:D154–D158
18. Muffato M et al (2010) Genomicus: a database and a browser to study gene synteny in modern and ancestral genomes. Bioinformatics 26:1119
19. Mulder NJ et al (2003) The InterPro Database, 2003 brings increased coverage and new features. Nucleic Acids Res 31:315
20. Finn RD et al (2010) The Pfam protein families database. Nucleic Acids Res 38:D211–D222
21. Rose PW et al (2011) The RCSB Protein Data Bank: redesigned web site and web services. Nucleic Acids Res 39:D392
22. Wells DE et al (2011) A genetic map of Xenopus tropicalis. Dev Biol 354(1):1–8
23. Hoffmann R, Valencia A (2005) Implementing the iHOP concept for navigation of biomedical literature. Bioinformatics 21(Suppl 2):ii252–ii258
24. Davis AP et al (2011) The comparative Toxicogenomics Database: update 2011. Nucleic Acids Res 39:D1067–D1072
25. Jenssen TK et al (2001) A literature network of human genes for high-throughput analysis of gene expression. Nat Genet 28:21–28
26. Maglott D et al (2011) Entrez gene: gene-centered information at NCBI. Nucleic Acids Res 39:D52
27. Segerdell E et al (2008) An ontology for Xenopus anatomy and development. BMC Dev Biol 8:92
28. Pollet N et al (2000) Axeldb: a Xenopus laevis database focusing on gene expression. Nucleic Acids Res 28:139–140
29. Kent WJ et al (2003) Evolution's cauldron: duplication, deletion, and rearrangement in the mouse and human genomes. Proc Natl Acad Sci U S A 100:11484

Index

Stefan Hoppler and Peter D. Vize (eds.), *Xenopus Protocols: Post-Genomic Approaches, Second Edition*, Methods in Molecular Biology, vol. 917, DOI 10.1007/978-1-61779-992-1, © Springer Science+Business Media, LLC 2012

V

W

X

Z

MIX
Papier aus verantwortungsvollen Quellen
Paper from responsible sources
FSC® C105338

If you have any concerns about our products,
you can contact us on
ProductSafety@springernature.com

In case Publisher is established outside the EU,
the EU authorized representative is:
Springer Nature Customer Service Center GmbH
Europaplatz 3, 69115 Heidelberg, Germany

Printed by Libri Plureos GmbH
in Hamburg, Germany